Biodiversity
An Overview

Biodiversity
An Overview

Editors

Mukesh Kumar

Associate Professor
Department of Botany, Sahu Jain Post-Graduate College
Najibabad, Bijnor, Uttar Pradesh, India
(Affiliated to MJP Rohilkhand University, Bareilly, U.P., India)

Rajan Kumar Gupta

Associate Professor
Department of Botany, Pt. LMS Govt. Post-Graduate Autonomous College,
Rishikesh, Dehradun Uttarakhand, India

Consulting Editor

G.S. Paliwal

Senior Consultant
AFC Ltd., Janakpuri, New Delhi, India

I.K. International Publishing House Pvt. Ltd.

NEW DELHI • BANGALORE

Published by

I.K. International Publishing House Pvt. Ltd.
S-25, Green Park Extension
Uphaar Cinema Market
New Delhi–110 016 (India)
E-mail: info@ikinternational.com
Website: www.ikbooks.com

ISBN 978-93-80578-88-0

10 9 8 7 6 5 4 3 2 1

Published by Krishan Makhijani for I.K. International Publishing House Pvt. Ltd., S-25, Green Park Extension, Uphaar Cinema Market, New Delhi–110 016 and Printed by Rekha Printers Pvt. Ltd., Okhla Industrial Area, Phase II, New Delhi–110 020.

प्रोफेसर धीरेन्द्र पाल सिंह
कुलपति

Prof. D.P. Singh
Vice-Chancellor

काशी हिन्दू विश्वविद्यालय
वाराणसी–221005 (भारत)
BANARAS HINDU UNIVERSITY
(Established by parliament by Notification No. 225 of 1916)
VARANASI-221 005 (INDIA)

FOREWORD

Several small valleys in India offer a wonderful experience to nature lovers and hikers. The vast open hay-fields above the tree line present endless view of the variously coloured Himalayan flowers. The most interesting of them, aesthetically or botanically, are seen in the higher altitudes of 2400m and above. The arrival of spring brings forward an uprising stream of colours, when the 'semal' and the 'palas' with their blazing red flowers put the lower altitude forests on fire. It is also the time for 'burans' to spread its hues at a height of above 2400m, adding colours to the blue and white panorama of snow. The flowers grow not only in the 'Valley of Flowers' in the Garhwal Himalayas, but are also found on different tracks habitually up to great heights as even in the hay-fields, and on rock cracks and moraines.

Thus, biodiversity is essential not only for environmental stability and food security but also for scenic beauty which is depleting at an alarming rate due to demographic, institutional, educational and socio-economic reasons. Realization of adverse impact of environmental degradation and awareness generation efforts has oriented people to come for nature-based tourism.

The paramount importance of assessment and conservation of biodiversity, therefore, can not be overemphasized. There is also a need to set-up a co-coordinated programme for systematic biodiversity management and monitoring, at least in major forest types or eco-regions. It is also desirable to keep regulatory regimes in place and concerted efforts need to be made to attract peoples' participation in a big way in this exercise. Education should be provided and capacity should be built in a manner that people may derive intangible as well as economic benefits from conservation efforts to improve their livelihood.

The rare species call for our special attention as these are the failing species which are becoming genetically weaker and are unable to withstand species competition as well as environmental changes. Recent studies indicate that nearly 35 per cent of vascular plants in India are endemic and about 900 taxa of plants are rare and endangered.

In the present era of scientific development, growing industrialization and drastic surge in transportation facilities along with several other anthropogenic activities like excessive use of polythene and plastic items as well as electronic wastes are leading to environmental pollution hazards. Global warming has attracted the attention of every individual sensitive to environmental conservation and natural resources. This has been witnessed in the form of national and international debates wherein individuals from different walks of life are putting their heads together bringing about multilateral agreements. The Earth Summit of 1992 at Rio de Janeiro, Kyoto Protocol and the United Nations Environment Programme are glaring examples in this regard. India has not only signed the Kyoto Protocol but also has the privilege of hosting in October–November, 2002 the Eighth Conference of the parties to the United Nations Framework Convention on Climate Change (UNFCCC).

The United Nations proclaimed 2010 to be the International Year of Biodiversity, and people all over the world are working to safeguard this irreplaceable natural wealth and reduce biodiversity loss, which is so vital for human well-being. We need to do more. Now is the time to act. The slogan for International Biodiversity Year is:

Biodiversity is life,
Biodiversity is our life.

And its objectives are:

- To raise awareness of the importance of conserving biodiversity for human well-being and promoting the understanding of its economic values.
- To enhance public knowledge of the threats to biodiversity and the means to conserve it.
- To encourage organizations (and through them individuals) to take direct or indirect biodiversity conservation activities.
- To celebrate the achievements of Countdown 2010 partners and other stakeholders.
- To report on the possible failures for not achieving the target
- To prepare the ground for communicating the post-2010 objectives.

It is, therefore, extremely gratifying that the volume entitled “Biodiversity: An Overview” which has been compiled through inviting experts from different parts of India is being brought out by M/s I.K. International Publishing House Pvt. Ltd., New Delhi, in this Year of Biodiversity is bound to broaden the canvas of knowledge of one and all interested in the subject. The Editors, Dr Mukesh Kumar and Dr R.K. Gupta, deserve our warmest compliments for the sincere efforts put in by them. The overall supervision and consistent guidance, furnished by the Consulting Editor, Prof. G.S. Paliwal, a committed Senior Botanist of the country, has certainly enriched the book with the much wanted scientific acumen. These have cumulatively resulted in the collection of such wide ranging and informative articles. I, therefore, commend the book which will surely be welcomed not only in India but also abroad.

(Prof. D.P. Singh)

Prof. S.C. Tiwari
Head, Department of Botany & Microbiology
Dean, Faculty of Science

H.N.B. GARHWAL UNIVERSITY
(A Central University)
Post Box-22
Srinagar-Garhwal-246 174, Uttarakhand
E-mail: profsctiwari@gmail.com

MESSAGE

The Reference Book entitled, "Biodiversity: An Overview" edited by Dr Mukesh Kumar and Dr Rajan Kumar Gupta which is at your hands at this moment consists of the recent concepts and status of biodiversity in the present context.

Biodiversity refers to variety of life forms, i.e., the different plants, animals and micro-organisms. It has been defined as 'the variability among the living organisms from all sources including, *inter alia*, Developed, Agricultural and Natural ecosystems and the ecological complexes in the biosphere of which they are the part. This includes diversity within species, between species and of ecosystems'. During recent years, biodiversity has become a global issue of discussion at all important international fora. The 'Convention on Biological Diversity' a specific document on biodiversity, was discussed and adopted in Rio de Janeiro (Brazil) during the United Nations Conference (June 1992) on Environment and Development, also known as 'Earth Summit'.

The earth summit meeting made the terms 'biological diversity' a political catch phrase. Since then, it has become clear to a wide public which is increasingly being confronted with a dramatic loss of organisms at the level of species. At the same time, we are becoming aware that more than 905 of our cohabitats on this planet currently remain undiscovered. A great number of them would presumably disappear forever without our ever having known them.

Biodiversity provides to the humankind enormous benefits in the form of food, medicines and industrial and other by-products, and has the potential for providing many more yet unknown benefits. It also offers an array of essential services through natural ecosystems. These include maintenance of the gaseous composition of the atmosphere, shaping the climate and producing the soil. Simultaneously, it is essential for the proper ecosystem functioning and stability, the so-called sustainability.

In this book, the erudite editors, having been closely guided by the Consulting Editor, Prof. G.S. Paliwal have collected valuable information on different aspects of biodiversity. This will help the students, researchers and members of the general public to understand the effects of man's activities on various ecosystems. Starting with the salient features of biodiversity and biodiversity hotspots, they have emphasized on the biodiversity of protected groves, wetland flora, mangrove vegetation, orchids, ethnobotany, medicinal plants, pollen diversity, grassland diversity, medico-religious as well as ethno-veterinary plants and the impact of changing water regimes on the depletion of biodiversity.

I hope this treatise will be of immense help to the students, researchers, environmentalists, foresters and all others who owe interest in the biodiversity conservation and utilization of natural wealth. The editors deserve high appreciation to execute this challenging task.

(Prof. S.C. Tiwari)

PREFACE

Biodiversity at global scale is a balance between the rate of speciation and the rate of extinction. At a regional level it is a balance between the rate of immigration/invasion and that of local/regional extinction. These rates determine the persistence of the communities. It can be assessed at genetic, population, species, ecosystem and landscape levels, at three spatial scales, viz., local, regional and global. While for assessment at genetic level, molecular techniques are most profitably used, assessment at other levels depends upon field samplings. Depending upon the objectives, the inventory could focus on single species or multiple taxa. For single species inventories, the focus could be on high impact, indicator, umbrella, flagship, threatened or economically important species.

Biodiversity provides to the humankind enormous benefits in the form of food, medicines and industrial products, and has the potential for offering many more yet unknown benefits. It also provides an array of essential services through natural ecosystems. These include maintenance of the gaseous composition of the atmosphere, shaping the climate and producing the soil. It is essential for the ecosystem function and stability. There are several hypotheses being tested for relationship between biodiversity and ecosystem function.

Biodiversity is not uniformly distributed on the earth. There are broad patterns, such as increasing variations in plants and animals from poles to the equator. It is more concentrated in the tropics because of conditions which permit higher rates of speciation and lower rates of extinction. About 44% of vascular plants and 35% of vertebrates are concentrated as endemic species in 25 hotspots, which account for only 1.4% of the local land area.

The pool of living diversity is dynamic, it increases when new gene variation is produced, a new species is created or a novel ecosystem formed and it decreases when the genetic variation within the species decreases, a species becomes extinct or an ecosystem complex is lost. Today, as ever, human-beings are dependent on their sustenance, health, well-being and enjoyment of life on fundamental biological systems and processes. Humanity derives all of its food, many medicines and industrial products from wild or domesticated components of biological diversity. Biotic reservoirs are source of recreation, and tourism and underpin the ecosystem which provides us with many services.

The Indian region represents an immense variety of climatic and altitudinal zones and is floristically rich. The ecological habitats vary from humid tropical Western Ghats to the hot deserts of Rajasthan and ice-capped mountains of the Himalaya to warm coasts of Peninsular India. These habitats favor a wide variety of plants and animals as well. It is estimated that over 45,000 species of plants are accounted for in India which represent 11% of the known plant species of the world. These are distributed in varied groups, viz., Angiosperms-16657 spp., Gymnosperms-64 spp., Pteridophytes-1022 spp., Bryophytes-2584 spp., Lichens-1600 spp., Fungi-23000 spp., Algae-2500 spp., and Bacteria-850 spp. The flowering plants of India comprise about 6% of the world's known flowering plants. About 315 families (of a total of 400 now defined) and 2250 genera of flowering plants are known to occur in India in different ecosystems.

The loss of biodiversity is a global crisis. There is hardly any reason on the earth that is not facing ecological catastrophes. Biological extinction has been a natural phenomenon in geological history. However, man's intervention has speeded up extinction rates all the more. The destruction of the world's tropical forests, which are disappearing at an alarming rate, is one of today's urgent global environmental issues. The loss of plant/animal diversity has immediate and long-term effects on human survival. The plant genetic resources are being depleted at an alarming pace throughout the world. Current rates of extinction demand immediate concerted efforts for conservation of biodiversity.

We seem to be entering into the phase of mass extinction. Today, we are losing two to five species per hour from tropical forests alone. This amounts to a loss of 16 m populations per year or 1800 populations per hour. If the present trend continues, about 25% of the total 2,50,000 higher plant species will be lost in the next few decades and another 25% by the end of the 21st century. There were around 25,000 species of flowering plants some 125 million years ago, by the late Cretaceous this number had swelled to 100,000 and today there are some 2,00,000 species of plants on Earth. Additionally, while the past extinctions occurred each time over a span of million years or less, the present mass extinction may well occur within a short period of about 200 years, and plants (which make the base of the food chain) will be equally or worse affected.

India's phytodiversity is one of the most significant in the world. India is one of the twelve mega-biodiversity centers in the world and also an important center of origin of agrobiodiversity. It is therefore, very important to study, document and conserve the plant wealth of India and also of the world before its extinction. While it is generally accepted today that the conservation of all biodiversity should be our goal, especially through the preservation and sustainable use of natural habitats, this is an ideal that is unlikely to be achieved and there are convincing scientific, economic and sociological reasons for giving priority to the conservation of the major centers of plant diversity throughout the world.

The strategies to conserve the biodiversity include establishment of protected area network and corridors with emphasis on appropriate levels of management, reduction of anthropogenic pressure on natural populations by cultivating them elsewhere, programmes of augmentation, reintroduction and introduction of target taxa, restoration of degraded habitats, etc. The conservation strategies can be either, *in-situ* conservation of genetic resources within their ecosystem and natural habitat or *ex-situ* conservation of components of genetic material of biological diversity outside their natural habitat. The choice of conservation strategy depends upon the nature of the material to be conserved, i.e., the life cycle and mode of reproduction, size of individual population and ecological status. *Ex-situ* techniques include the establishment of botanical and zoological gardens, banks of pollen, seed tissue culture, DNA, etc. Establishment of forest reserves, national parks, protected areas and on farm conservation of valuable plant varieties is being promoted to facilitate their *in-situ* conservation.

The present book is compilation of 29 chapters in the form of research papers and review articles contributed by eminent scientists and research workers who are acknowledged internationally in their respective fields. The book has been prepared with the intention of providing sufficient depth of the subject to satisfy the academic needs of the reader up to a higher level, which will be comprehensive and interesting. There are articles on the various groups of plants, i.e., angiosperms, sedge vegetation, grassland vegetation, wetland flora, sacred groves, ethnobotany, pharmacognosy of drug-plants, phenology of medicinal plants, changing water regimes, hotspots and biodiversity conservation. In totality, it contains useful information for the postgraduate students, teachers, scientists and researchers dealing with basic and Applied Botany, Ethnobotany, Environmental Biology, Forestry and Agricultural Botany. Moreover, it will prove helpful for Government departments dealing with forestry, social forestry and the cultivation of medicinal plants. We do hope that all those interested in the pursuit of plant diversity and its conservation will find the volume handy and stimulating.

The editors are thankful to all the academicians and scientists whose contributions have enriched this volume. We also express our deep sense of gratitude to our parents whose blessings have always prompted us to pursue academic activities deeply. It is quite possible that in a work of this nature, some mistakes might have crept in inadvertently and for these we owe undiluted responsibility.

We are grateful to all authors for their contribution to present book. The Principals of our institutions deserve special thanks. We are also thankful to Professors R.D. Gaur (former Head), S.C. Tiwari (present Head) and A.B. Bhatt, Department of Botany, HNB Garhwal University, *Srinagar* (Uttarakhand) for their kind co-operation, useful discussions and valuable suggestions for making the volume really informative and comprehensive. Prof. G.S. Paliwal, the Consulting Editor of volume, deserves our special thanks for his manifold help and constant advice throughout the exceution of the Project.

The editors thank most profoundly M/s I.K. International Publishing House Pvt. Ltd., New Delhi, which is known internationally for its reputation in quality publication of scientific books, for kindly accepting our request to publish this volume and bringing out the work to the light of the day. Their prompt action and enthusiastic co-operation during the compilation of this volume is highly appreciated and acknowledged.

The first editor (MK) is thankful to the UGC, New Delhi for providing financial assistance. His wife, Dr (Mrs) Kiran Sharma, PGT Hindi, Kendriya Vidyalaya, BHEL, Ranipur Haridwar, (Uttarakhand) and two sons, Ruchir Sharma and Nishant Sharma deserve special thanks for their affection and multifarious co-operation extended during the accomplishment of this academic and scientific endeavour.

The second editor (RKG) is thankful to UCOST, Dehradun and the UGC, New Delhi for the financial support and wishes to place on record his special thanks to his wife, Mrs Alka and two little daughters, Akriti and Ayushi for their co-operation in all his academic and scientific endeavours. Finally, we will always remain debtor to all our well wishers for their blessings without which this book would not have come to light.

The editors value the generous help offered by Dr (Mrs) Anubha Joshi, Dr K. Avinash Bharti and Manoj Kumar Paliwal in respect to the finalization of the manuscript. We take this opportunity to express our heartfelt gratitude to Professor D.P. Singh, Vice-Chancellor, Banaras Hindu University, *Varanasi*, U.P. and a Botanist himself, for his magnanimity towards writing the 'Foreword' for the volume and Professor S.C. Tiwari, Head, Department of Botany and Microbiology, HNB Garhwal Central University, *Srinagar*, Pauri, Uttarakhand for the generous message.

Mukesh Kumar
Rajan K. Gupta

CONTENTS

CONTRIBUTORS

A.B. Bhatt
Department of Botany and Microbiology, HNB Garhwal Central University, *Srinagar*-246 174, Pauri, Uttarakhand

A.K. Dwivedi
Department of Botany, D.D.U. University, *Gorakhpur*-273 001, Uttar Pradesh

A.K. Paliwal
Department of Botany, Government Post-Graduate College, *Bageshwar*-263 642, Uttarakhand

A.K. Sharma
Additional PCCF, Department of Forests, Van Bhawan, 17 Rana Pratap Road, *Lucknow*-226 001, Uttar Pradesh

A.N. Singh
EIA Division, Directorate of Extension, ICFRE, PO. New Forest, *Dehradun*-248 006, Uttarakhand

A.R. Nautiyal
High Altitude Plant Physiology Research Centre, *Srinagar*-246 174, Pauri, Uttarakhand

Ajay Kumar
EIA Division, Directorate of Extension, ICFRE, PO. New Forest, *Dehradun*-248 006, Uttarakhand

Ajay Shukla
Department of Botany, T.D. College, *Jaunpur*-222 002, Uttar Pradesh

Anamika Tripathi
Department of Botany, Hindu College, *Moradabad*-244 001, Uttar Pradesh

Anoop Badoni
Department of Seed Science and Technology, HNB Garhwal Central University, *Srinagar*-246 174, Pauri, Uttarakhand

Anubha Joshi
Guest Faculty, Department of Botany, Daulat Ram College, University of Delhi, *Delhi*-110 021

B. Choudhary
Department of Botany, Government Post-Graduate College, *Bageshwar*-263 642, Uttarakhand

B.L. Sharma
Department of Botany, NREC College, *Khurja*-203 131, *Bulandshahr*, Uttar Pradesh

Beena Kumari
Department of Botany, Hindu College, *Moradabad*-244 001, Uttar Pradesh

B.D. Tripathi
Department of Botany, D.D.U. University, *Gorakhpur*-273 001, Uttar Pradesh

B.N. Pandey
Department of Botany, Government P.G. College, *Lansdowne*-246 139, Pauri, Uttarakhand

C.J. Prabhakar
Muga Silkworm Seed Organization, Central Silk Board, *Guwahati*-781 001, Assam

C.P. Shukla
Ecology Research Centre, Department of Botany, T.D. Post-Graduate College, *Jaunpur*-222 002, Uttar Pradesh

Chetna Bisht
High Altitude Plant Physiology Research Centre, *Srinagar*- 246174, Pauri, Uttarakhand

D.K. Sharma
NAEB, MoEF, Paryavaran Bhawan, CGO Complex, Lodi Road, *New Delhi*-110 003, India

Deepak Kumar
Department of Botany, Dhampur Degree College, *Dhampur*-246 174, Bijnor, Uttar Pradesh

Dinesh Kumar
Department of Botany, Pt. L.M.S. Government P.G. College, *Rishikesh*-249 201, Dehradun, Uttarakhand

D.P. Paliwal
Research Extension Centre, Central Silk Board, *Bageshwar*-263 462, Uttarakhand

D.P. Semwal
Department of Botany, University of Delhi, *Delhi*-110 007, India

H.N. Singh
Department of Botany, T.D. College, *Jaunpur*-222 002, Uttar Pradesh

G.S. Paliwal
Agricultural Finance Corporation Ltd., NRO, B-1/9, Community Centre, Janakpuri, *New Delhi*-110 058, India

Gayatri Mahar
Agricultural Finance Corporation Ltd., NRO, B-1/9, Community Centre, Janakpuri, *New Delhi*-110 058, India

H. Govindpyari
Department of Botany, University of Delhi, *Delhi*-110 007, India

Harjeet Singh
Agricultural Finance Corporation Ltd., NRO, B-1/9, Community Centre, Janakpuri, *New Delhi*-110 058, India

Hemendra S. Chauhan
Department of Pharmaceutics, Institute of Technology, Banaras Hindu University, *Varanasi*-221 005, Uttar Pradesh

Himanshu Rai
Department of Botany, Pt. L.M.S. Government P.G. College, *Rishikesh*-249 201, Dehradun, Uttarakhand

Iqbal Habib
Department of Botany, Government Degree College, *Budaun*-243 601, Uttar Pradesh

J.K. Pathak
Hydrobiology Laboratory, Department of Zoology, Hindu College, *Moradabad*-244 001, Uttar Pradesh

K.A. Siddiqui
Department of Botany, K.N. Govt. College, *Gyanpur*-225 001, Sant Ravidas Nagar, Uttar Pradesh

K.N. Mishra
Department of Botany, T.D. College, *Jaunpur*-222 002, Uttar Pradesh

Kamlesh Rai
Department of Botany, K.N. Govt. P.G. College, *Gyanpur*-225 001, Sant Ravidas Nagar, Uttar Pradesh

Kshitij Malhotra
Department of Botany, Pt. L.M.S. Government P.G. College, *Rishikesh*-249 201, Dehradun, Uttarakhand

Kumar Avinash Bharti
Department of Botany, NREC College, *Khurja*-203 131, *Bulandshahr*, Uttar Pradesh

Manoj Kumar Paliwal
Department of Botany, I.P. P.G. College, *Bulandshahr*-203 001, Uttar Pradesh

Manjul Dhiman
Department of Botany, Kanahiya Lal DAV Post-Graduate College, *Roorkee*-247 667 (Haridwar), Uttarakhand

Mihir Kumar Das
P.G. Department of Botany, G.M.College (Autonomous), *Sambalpur*-768 004, Orissa

Mohd. Alam
Hydrobiology Laboratory, Department of Zoology, Hindu College, *Moradabad*-244 001, Uttar Pradesh

Mukesh Kumar
Department of Botany, Sahu Jain Post-Graduate College, *Najibabad*-246 763, Bijnor, Uttar Pradesh

N.P. Baduni
Agricultural Finance Corporation Ltd., NRO, B-1/9, Community Centre, Janakpuri, *New Delhi*-110 058, India

Neelam Bhatt
Department of Botany and Microbiology, HNB Garhwal Central University, *Srinagar*-246 174, Pauri, Uttarakhand

P.L. Uniyal
Department of Botany, University of Delhi, *Delhi*-110 007, India

Prabha Sharma
Department of Botany, University of Delhi, *Delhi*-110 007, India

Poonam Paliwal
Department of Botany, I.P. P.G. College, *Bulandshahar*-203 001, Uttar Pradesh

Prakash C. Kuniyal
Department of Botany, Pt. L.M.S. Government P. G. (Autonomous) College, *Rishikesh*-249 201, Dehradun, Uttarakhand

P.S. Negi
Wadia Institute of Himalayan Geology, 33 General Mahadev Singh Road, *Dehradun*-248 001, Uttarakhand, India

R.D. Gaur
Hanumant Puram, Lane-6, Ganga Nagar, *Rishikesh*-249 201, Dehradun, Uttarakhand
Formerly, Professor and Head, Department of Botany, HNB Garhwal Central University, *Srinagar*-246 174, Pauri, Uttarakhand

R.H. Singh
Department of Botany, K.B. Post-Graduate, Collage, *Mirzapur*-231 001, Uttar Pradesh

R.K. Khatri
Zonal Silkworm Seed Organization, Central Silk Board, *Dehradun*-249 201, Uttarakhand

Rajan Kumar Gupta
Department of Botany, Pt. L.M.S. Government P.G. College, *Rishikesh*-249 201, Dehradun, Uttarakhand

S. Bhanumati
Department of Chemistry, Gargi College, University of Delhi, *New Delhi*-110 049, India

S.K. Gupta
DFO Ecotourism, Office of the Chief Conservator of Forests, *Dehradun*-248 001, Uttarakhand

S.K. Katiyar
Department of Botany, Government P. G. College, *Lansdowne*-246 155, Pauri, Uttarakhand

Sanjay Singh
Department of Botany, Pt. L.M.S. Government P.G. College, *Rishikesh*-249 201, Dehradun, Uttarakhand

S.P. Paliwal
Department of Botany, Narain P.G. College, *Shikohabad*-205 135, Firozabad, Uttar Pradesh

Shashi
Department of Botany, D.D.U. University, *Gorakhpur*-273 001, Uttar Pradesh

Shalini Paliwal
Department of Botany, Narain P.G. College, *Shikohabad*-205 135, Firozabad, Uttar Pradesh

Shikha Sharma
Hydrobiology Laboratory, Department of Zoology, Hindu College, *Moradabad*-244 001, Uttar Pradesh

Sushil K. Singh
Department of Pharmaceutics, Institute of Technology, Banaras Hindu University, *Varanasi*-221 005, Uttar Pradesh

Tapendra S. Bisht
Department of Botany and Microbiology, HNB Garhwal Central University, *Srinagar*-246 174, Pauri, Uttarakhand

Tariq Husain
Biodiversity & Angiosperm Taxonomy Division, NBRI, Rana Pratap Marg, *Lucknow*-226 001, Uttar Pradesh

U.K. Chaturvedi
Department of Botany, Bareilly College, *Bareilly*-243 001, Uttar Pradesh

Vandana Sharma
Department of Botany, Kanahiya Lal DAV Post-Graduate College, *Roorkee*-247 667, Haridwar, Uttarakhand

Vidya Dhar Pandey
Department of Botany, Pt. L.M.S. Government P.G. (Autonomous) College, *Rishikesh*-249 201, Dehradun, Uttarakhand

Vinita Katiyar
Civil Engineering Department, Indian Institute of Technology, New Mehrauli Road, *New Delhi*-110 016, India

EDITORS

Dr Mukesh Kumar (b. 1963) obtained his M.Sc. and M. Phil. degrees in Botany, and M.Ed in Education from Meerut University, *Meerut* (now C.C.S. University, Meerut). He took his doctorate degree on Polyhouse Technology from H. N. B. Garhwal Central University, *Srinagar* (Uttarakhand). Presently he is working as an Associate Professor at the Department of Botany, Sahu Jain Post-Graduate College, *Najibabad* (Bijnor) U.P.

Dr Kumar is well-known for his researches in the fields of Polyhouse Technology and Cyanobacteria. He is a member/fellow of several National and International research organizations. Working on various aspects of Cyanobacteria, i.e., distributional pattern, population dynamics and dominance of different genera, he has explored the biodiversity of Cyanophycean Flora of the Sub-Himalayan Belt of Garhwal and Kumaon regions of Uttarakhand State of India. He has accomplished three Research Projects in the fields of his specialization, i.e., Polyhouse Technology and Cyanobacterial Diversity sponsored by the University Grants Commission, New Delhi. Presently he is working on the topic, 'Cyanobacterial Diversity of the Sub-Himalayan Belt of Northern Himalayas' for supplicating to the degree of Doctor of Science of HNB Garhwal Central University, Srinagar (Garhwal), Uttarakhand, India and the States being covered by him include Himachal Pradesh, Jammu & Kashmir and Uttarakhand which have hitherto remained unexplored in respect to the cyanobacterial biodiversity. The UGC, New Delhi has recognized the significance of this work through grant of a major research project to execute it.

He has the credit of presenting his original research findings at the platforms of plant scientists in several national and international conferences/seminars and workshops. He has published more than 37 technical papers in various journals and chapters in several books of repute. He has authored 5 books and supervised several doctoral theses. The great success of his previous books entitled. "Glimpses of Cyanobacteria" "Soil Microflora" has inspired him to work more on the phytodiversity, the most important part of the earth environment.

He is a Fellow of Indian Botanical Society and life-member of several other research organizations. His latest volume "Cyanobacterial Biodiversity of the Sub-Himalayan Belt of Kumaon Region" has been published by M/s Lambert Academic Publishing Gmbh & Co. KG, Federal Republic of Germany as a research monograph. Dr Kumar is a person with broad vision and is keenly engaged in spreading scientific knowledge both in English and Hindi through technical popular writings.

Dr Rajan Kumar Gupta (b 1963) is working as an Associate Professor at the Pt. LMS Govt. P.G. Autonomous College, Rishikesh, Uttarakhand. He obtained his M.Sc. and Ph.D. degrees from Banaras Hindu University and worked on Ecophysiology of Antarctic Cyanobacteria for his Ph.D. degree with Prof. A.K. Kashyap, Centre of Advanced Study in Botany, Banaras Hindu University, Varanasi. For the past twenty years he has been working on various aspects of Antarctic microflora.

Dr Gupta had been deputed by the Govt. of India for his participation as a Biological Scientist in XIth and XIVth Indian Scientific Expeditions to Antarctica during 1991-92 and 1994-95. He has visited several countries like Mauritius, Japan, Nepal, Thailand, South Africa and Belgium for presentation of his works on different aspects of algae. Dr Gupta has worked on different aspects of cyanobacteria, i.e., morphology, ecology and nitrogen fixation, biotechnological applications and published more then 60 technical papers in various national and overseas journals and more than 30 chapters in various books. Dr. Gupta has published three Botany Practical Books and one book on Paryavaran Adhyan (Environmental Studies) and three reference (research) books entitled "Glimpses of Cyanobacteria", "Advances in Applied Phycology" and "Soil Microflora". Three students have been awarded the D.Phil. degree and four are working under his supervision for their D.Phil. degree of HNB Garhwal Central University. He has worked on the use of Cyanobacteria as Biofertilizer in Antarctica as well as in foothills of Garhwal Himalaya and executed a couple of projects on Cyanobacteria of paddy fields of Dehradun District of Uttarakhand. Presently he is working on a project sponsored by the Uttarakhand Council for Science and Technology (UCOST), Dehradun on Diversity of Vasicular Arbuscular Mycorrhiza. The UGC, New Delhi has granted Research Award to Dr. Gupta for carrying out research at any of the Laboratory of the country for a period of two years. Dr. Gupta is member of a number of organizations in India and abroad. He is the Fellow of the Society for Environment & Ecoplanning and International Botanical Society and chaired various sessions in the conferences in India and abroad. He is on the editorial and advisory boards of various journals. Presently Dr. Gupta is teaching Microbiology and Biotechnology at the Department of Botany, Pt. LMS Govt. P.G Autonomous College, *Rishikesh* (Dehradun), Uttarakhand.

Prof. G.S. Paliwal (b 1938) currently a Senior Consultant with the Agricultural Finance Corporation Ltd., New Delhi (formerly Head, Department of Botany, HNB Garhwal Central University, *Srinagar*, Pauri, Uttarakhand, is well-known for his contributions on several facets of Developmental Botany. He has authored 8 books, 05 of which have won State and National level awards. Besides publishing over 200 research papers in the national and international journals, review articles, and monographs, he has supervised over 80 doctoral theses covering a wide panorama of disciplines in Plant Sciences.

He began his research career at the Department of Botany, University of Delhi, *Delhi*, under the able guidance of late Professor P. Meheshwari, FRS, Prof. B.M. Johri and Prof. S.C. Maheshwari, and has the distinction of being the first awardee of the degree of Doctor of Science from the University of Delhi. On account of his flair for learning languages, he has contributed copiously books in Hindi and even authored a dictionary in Hindi-English-German-French. Among other countries, he has paid extensive visits to England, France, Germany, Holland, Canada, Czechoslovakia, Hungary, Poland, Japan, Switzerland and USA. Besides the other Fellowships and Associateships, he is the recipient of the coveted Senior Alexander Von Humboldt Foundation Fellowship of the Federal Republic of Germany. His volume on "Vegetational Wealth of the Himalayas" has won international acclaim on account of its wide coverage of the subject matter, authoritative information and above-all furnishing hitherto unknown details.

Presently he is compiling the following volumes:

1. Tree Wealth of India (in Hindi: भारत की वृक्ष सम्पदा)
2. Medicinal Plants of India: Aspects and Prospects.

1

Biodiversity and Conservation: Salient Features

RD GAUR[1] AND PRAKASH C KUNIYAL[2]

[1]Hanumant Puram Lane-6, Ganganagar, Rishikesh-249 001, Dehra Dun Formerly Professor, Department of Botany, HNB Garhwal Central University, Srinagar-246 174, Pauri, Uttarakhand E-mail: rdgaur2009@gmail.com
[2]Department of Botany, Govt. PG (Autonomous) College, Rishikesh-249 201, Dehra Dun (Uttarakhand) E-mail: kuniyalprakash@gmail.com

ABSTRACT

Biodiversity is one of the most emerging disciplines which is drawing the attention of the people all over the globe since it relates with the well-being of all the living organisms (including human-beings). In the context to accelerating growth and development of human civilization, the contemporary other living farms and physical resources of the globe are under severe stress, causing disharmony with obvious hazards. The living matrix of the world is interconnected with each and every living species, and each species plays an important role; therefore, the survival of all living-beings is essential for the harmonious nature of the earth, besides their role in economic, aesthetic and recreational support. It is no wonder, therefore, that under CBD a large number of nations are associated to deal with the problems of biodiversity loss and conservation, as well as society of all living beings of the world. India is also carrying an important role in the conservation of biological diversity and sustainable development through its own biological diversity act and rules. It has several important sites of protected area network (PAN), and proactive in participation of *ex-situ* conservation. Besides, the protection of biological diversity, India is simultaneously moving ahead with definite measures to promote the improvement of the human societies. Some of the important concerns of biodiversity such as the levels of biodiversity, importance and loss of biodiversity, CBD, biological diversity act and rules, India, biogeographical zones of India, conservation, etc. have been dealt with in the present communication.

Keywords: Biodiversity, biological diversity act and rules, CBD, conservation, *in-situ*, *ex-situ*, international alliances, PAN, sustainable development, threatened taxa.

INTRODUCTION

Nature has enriched the planet earth with millions of living organisms, which are essentially interrelated and interdependent to each other, and finally compose a living web on the earth. Each of the living entities has its own role in maintaining the matrix of life, to grow itself and also to allow others to live in sound harmony. Until the origin of human-beings in the geological history, this interlinked web was spontaneous in function but with the emergence of man there has been a sharp advancement in the complexities of the natural systems.

All the objects of the planet, living or non-living, have been consistently experimented by the human-beings partly due to curiosity and partly for finding wider scope of utilization in the interest of himself. Simultaneously, the continuous process of evolution and civilization of mankind has led to the exploitation of adjacent natural resources. In fact, the natural biosphere has turned to the centre of activities of human intellect, often called as 'noosphere' and subsequently present-day biosphere has more complexities known as 'technosphere' wherein technological development has unamazingly overlapped all spheres of life including simple, natural functions of the global system, either at the individual or the community level. This has led to alteration of environmental conditions of the earth which had prevailed for millions of years. The super power of human activities has resulted in the loss of several organisms and changes into many of the natural phenomena. Thereby throughout the globe an alarming voice arose to recall for the limit of human growth and loss of other physical and natural resources. Biological studies in terms of various organisms (plants, animals, microbes, etc.) have re-validated investigation of a number of living species, the ecosystem analysis with the combination of various species, their interrelationship for perpetuation of new generations and their genetic formations. The whole realm is often termed as biological diversity and in the abstract form as biodiversity (Wilson, 1986). The word '**biodiversity**' is a contraction of biological diversity coined for the first time by an entomologist Wilson (1986). It refers to the range of variations or differentiation among same set of entities. The term is commonly used to describe the numbers, variety and variability of living organisms. Actually it is synonym of **'Life on Earth'.** It is constituted at three distinct levels, i.e., genetic, species and ecosystem, which reflect multidisciplinary approaches of their understanding, keeping pace with time and the prevailing environment.

Biodiversity found on earth today is the result of 3.5 millions of years of evolution. Until the emergence of man, the earth supported more biodiversity than at any period of geological history. Since the advent of human-being, rapid decline began in the number of species, and many of the species have to be categorized as extinct. Extinction is irreversible, thus species loss causes loss of energy, food, ecological linkages, medicinal or other values, as well as degenerate the sound ecosystem functions.

Biodiversity is a multidisciplinary approach of studying various organisms, in holistic attitude it attempts to include the systematics (the rank and position of individual species with proper identity in a particular group), interrelationship (with allied organisms), diversity (species abundance and distribution together with the variations) habitat and ecosystems (most suitable to sustain the different organism), resource utilization, biodiversity assessment, importance of biodiversity in terms of natural services (ecosystem services), and economic importance for the benefit of human-beings (food, fiber, timber, medicine, etc) including aesthetic and recreational values. The other related aspects of importance are species occurrence occasionally unwanted in a particular region commonly known as 'invasive species' and their impact with the introduced communities, loss of biodiversity (due to complexities, many of the species being subjected to several threats for their perpetuation for which proper measures are necessary), and inventorization of species (overall to assess the total number of species in each group of living beings, as till now we know only about one fifth of the living organisms of the world), and finally to develop skills to manage the total biodiversity of the world in different ecosystems in a sustainable manner. The biodiversity has become such an important challenge to the whole world that it has invited attention from various disciplines, people as well as all quarters of the world. In the process there are several conventions, treaties, agreements, rules and laws to deal with these problems, therefore, it has become imperative to involve the knowledge of national, international conventions, agreements, policies and laws for proper understanding and utilization of biodiversity for our lasting future.

LEVELS OF BIODIVERSITY

Genetic Diversity

Heritable variation within and between populations of organisms, ultimately depends on the sequence of four base pairs, as components of nucleic acids constitute the genetic code. New genetic variation arise by gene and chromosome mutations or by recombination. Other variations are related to the amount of DNA per cell, chromosome structure and number. Genetic variation enables both natural evolutionary change and artificial selective breeding (WCMC, 1992; Anonymous, 1999; Stace, 2000). Measurements of genetic diversity carried by allozymes, Restriction Fragment Polymorphisms (RFLRs), Randomly Amplified Polymorphic DNA (RAPD), DNA fingerprints, DNA Sequencing and Polymerase Chain Reaction (PCR), RNA-analysis, Genomic study, Nucleic hybridization, Fluorescent *in-situ* hybridization (FISH), etc.

Species Diversity

Species diversity deals with the variety of living species in different geographical areas, often expressed in terms of species richness or species abundance (Grant, 1981; WCMC, 1992). Species level is generally considered the most natural one. Number of species only provides a partial indication of biological diversity. The ecological importance of a species can have a direct effect on community structure and thus overall biodiversity. Species in a true sense represents the unit of living-beings, as well it is the symbol of communication.

Ecosystem Diversity

Ecosystem diversity deals with the differences between ecosystem types including the habitat diversity and ecological processes in an ecosystem. Patterns of energy flow and biogeochemical cycles (C, N, O, S, P, water, etc.) are distinct in each ecosystem, thereby, sustaining different types of living communities showing variation in their distribution and geographical positions. Ecosystem diversity is evaluated through relative abundance of different species, types of species, and number of species in different size classes at different trophic levels. Thus, it represents quantitative estimation of different species and interaction among various species. Normally, different categories of biotic communities have been expressed by different authors from different zones, and at the world level sometimes biomes, biogeographical regions, life zones, eco-regions etc. are referred. Majority of authors differentiate regions of world with high endemism and species richness termed 'hot spots' or centre of origins (Myers, 1990; Holdridge, 1967; Bailey, 1989a, b; Udvardy, 1975, 1984). Ecosystem diversity is measured through:

(A) **Alpha Diversity:** Species diversity within a community or habitat, comprises two components, i.e., species richness and evenness.

(B) **Beta Diversity:** It is the intercommunity diversity expressed by the rate of species turnover or per unit change.

(C) **Gamma Diversity:** Refers to the overall diversity at the landscape level

ECOSYSTEM SERVICES

For a sound environmental system a balanced composition of its components, i.e., soil, air, water and living-beings is most essential, which is sustained by the organisms at different levels, i.e. various consumers, producers, and others including scavengers and decomposers, etc. All the pollinators, natural cycle (carbon, nitrogen, water and oxygen) alongwith ecological food-chain etc. are essential to keep the harmony for sound environment. All these ecological cycles are controlled at different levels by variable sets of environment and organisms, some of the common ecosystem services maintained by the biodiversity are formation and maintenance of soil, of air quality through definite sets of gaseous composition, maintenance of water quality through different types of microbes and oxygen.

Various organisms are responsible for detoxification and decomposition of the natural or man-made waste, thus cleaning our soil and atmosphere. In nature, there is an automatic cycle to control various pests and diseases, likewise our plants are benefitted by pollinators at proper time to succeed in producing fruits and seeds (constitute basic food for all beings). It is rightly claimed that rich biodiversity is a secured bond for the food of growing populace of the world (Swaminathan, 1992). Other ecosystem services have a control to stabilize the weather and climate, in this reference pollution hazard and disturbance in atmospheric gaseous combination, acid rains (including depletion of ozone in stratosphere) lead to global warming, an yet another major threat for the future prospectives. A good interactive combination of living organisms needs a sound ecosystem. However, disturbances due to biodiversity loss reduce the capabilities to mitigate natural hazards such as earthquakes, floods, cyclones, volcanic eruptions, droughts, cloud burst, climate change, etc. In a true sense, the biological web of different organisms preserves the integrity of earth, suitable for all living-beings (WCMC, 1998; Gaur, 2005, 2007b).

BIODIVERSITY LOSS

Biodiversity loss is a natural evolutionary process; having been accelerated on account of anthropogenic stresses, habitat loss, invasion of aliens, pests and pathogens, etc. All these events are more or less associated with the human interference to the nature. Developmental processes as urbanization, industrialization, road network, hydroelectric power, exploitation of economic plants, deforestation, associated with pollution, global warming, climatic change tend to modify the original habitats, resulting into great loss of biological resources (Norse and Mc Manus,

1980; Wilson, 1986; Turner, 1996; Gaur, 2007a; Gaur and Kuniyal, 2009). An estimated 75 per cent of biodiversity loss is accounted for due to habitat modification and 20% of the loss is due to over-exploitation, whereas the remaining 5% shared by pests and pathogens and invasion of unwanted weeds (Anonymous, 1997; IUCN, 1999, 2000; Gaur, 2008). Following are some of the major impacts of biodiversity loss (Anonymous, 2004):

- Reduction of productivity of ecosystem, thereby declining nature's reservoir of goods and services essential to all living-beings
- Destabilization of ecosystems and weakens their ability to deal with natural disasters, i.e., floods, droughts, cloud burst, cyclones and population explosion and climate change
- Threatening resource systems of energy, food, medicine, etc.
- Reduction of opportunities of recreation and tourism industries
- The decline of biodiversity leaves us little number of choices for resolving future problems

CONVENTION ON BIOLOGICAL DIVERSITY (CBD)

International community concerns about the unprecedented loss of biodiversity emerged at UN conference 1972 on human environment, Stockholm. This conference resolved to establish United Nations Environment Programme (UNEP). The world commission on environment and development (The Brundland Commission) called for a new era of environmentally sound economic development and sustainable development became the theme of UN Conference on Environment and Development (UNCED) held at Rio de Janerio (June, 1992). November, 1990, UNEP began negotiating sessions on the objectives of International Treaty of CBD. CBD was presented at Earth Summit in Rio de Janerio (June, 1992) where it was signed by 153 nations and European community (Anonymous, 1999, 2004; Singh and Singh, 2004; Gaur, 2009). CBD has following objectives:

(i) Conservation of biological diversity.
(ii) Sustainable use of biological resources.
(iii) Sharing the benefits arising from the bioresources, in a fair and equitable manner.
(iv) Sovereign rights on bioresources.
(v) Facility to access genetic resources for environmentally sound uses based on mutually agreed terms and conditions.

Importance of CBD

(i) It recognizes for the first time that conservation of biodiversity is the common concern of the globe and is an integral part of developmental processes.
(ii) The agreement covers all ecosystems, species and genetic resources.
(iii) It links traditional conservation efforts by using bioresource sustainability.
(iv) It also covers recent fields of biotechnology, transfer of skill, biosafety, etc.
(v) Conservation will bring us significant environmental, social and economic benefits, and ultimately to develop mutual harmony in the world.

Contentious Issues of CBD

(i) Granting of access to genetic resources and conditions pertaining to their use.
(ii) Equitable distribution of benefits arising out of the use of genetic resources between the country of origin and exploiting country.

(iii) Provision of biotechnology safety relating to GMOs.
(iv) Fair and favourable conditions to access and to transfer the technology through negotiations.
(v) Global list of species and CITES (to restrict commercial use of certain species of endangered nature).

BIOLOGICAL DIVERSITY ACT, INDIA (2002)

India is rich in biological diversity, associated traditional and contemporary knowledge system related thereto (Biological Diversity Act, 2002). India is a party to the United Nations Convention on Biological Diversity, signed at Rio de Janerio on the 9th day of June, 1992. The said convention came into force on 29th December, 1993 (Biological Diversity Act, 2002). This act includes 12 chapters and 65 sections therein, covering preliminary aspects, regulation of access to biological diversity, National Biodiversity Authority (NBA), State Biological Board (SBB), Biodiversity Management Committee (BMC), finance-accounts, and duties of various concerns, the Central and State Government, etc. In 1994 the Central Government initiated preliminary work under the Chairmanship of Dr M.S. Swaminathan and in 1997 a discussion at national level was organized. Following in the year 1998, a draft was circulated to all the States, NGOs and others, and in the same year a discussion was held under the Ministry of Environment and Forests. Ultimately, in February 2002 the cabinet approved it and the bill was passed by Lok Sabha as well as the Rajy Sabha, and finally approved by His Excellency, The President of India (Anonymous, 2004, The Biological Diversity Act, 2002 and Biological Diversity Rules, 2004).

The Biological Diversity Rules, 2004

Ministry of Environment and Forests notification dated April, 15 2004, enforced Biological Diversity Rules. The rules include various functions or definitions of the act, authority, biodiversity management committee, chairperson, member secretary, etc. It also embodied the allowances, terms and conditions for various services. In great details it dealt with procedure of access to biological resources and associated traditional knowledge, including restriction on activities related to access to biological resources. Procedure for seeking approval for transferring results of research from India, procedure of seeking prior approval before applying intellectual property protection, criteria for equitable benefit, have been properly addressed (Anonymous, 2004).

Importance of Biological Diversity Act

It ascertains sovereign rights of India on biological resources and suggests proper measures to check biopiracy. It also enlists measures for the protection of bioresources or biodiversity outside the Protected Areas. The prime regulating authority for the proper use of resources is National Biodiversity Authority (NBA), India to safeguard the interest of the nation and the produce. There are different measures for the protection of biodiversity through SBB and BMCs. Biological Diversity Act, India (2002) recommended to record all the biological resources through Biodiversity Register, from Village Panchayat to the national level, for which local participation of individuals to SBB, BMC, NBA are to be involved. Sustainable utilization and sharing of benefits as well as Provision of Intellectual Property Right (IPR) have been laid down in the biodiversity rules. There are scopes for financial assistance through National Biodiversity Fund (NBF), State Biodiversity Fund (SBF), Local Biodiversity Fund (LBF) and Global Environmental Facility (GEF) to deal with the approaches on various problems of biodiversity.

OTHER BIODIVERSITY RELEVANT ACTS OF CENTRAL GOVERNMENT

From time to time the Central Government of India has promulgated reservations; laws and rules with respect to various aspects, more or less connected with the safety of biodiversity. Some of these are:

- Fisheries Act, 1987
- Destructive Insects and Pest Act, 1927
- Agriculture Produce (Grading and Marketing) Act, 1937
- Import and Export Control Act, 1947
- Prevention of Cruelity to Animals Act, 1960
- Seed Act, 1966 Amend. 2004
- Wildlife Protection Act, 1972; Amend. 1991
- Water (Prevention and Control) Pollution Act, 1974
- Forest Conservation Act, 1980
- Air (Prevention and Control) Pollution Act, 1981
- Agriculture and Process Food Products Export Development Authority Act, 1985-86
- Environment Protection Act, 1986
- New Seed Development Policy, 1988
- Indian Patent Act, 1970 (2005) Plant Variety Protection Act, etc.

GLOBALIZATION: INTERNATIONAL TREATIES

World communities are no more isolated, CBD and earlier efforts of United Nations and other organizations, attracted worldwide attention in the form of treaties and facilities, connected with the CBD. International communities came closer to each other, added with the information techniques, thus the whole world became a closet core. The basic aims of these treaties include education, social upliftment, better quality of human life, biodiversity conservation with sustainable use, etc.

In the ongoing globalization process India is involved with several international conventions, agreements and treaties. Some of the important ones include BCIS (Biodiversity Conservation Information System); CBD (Convention on Biological Diversity); CEL (Convention of International Law, IUCN), CITES (Convention of International Trade in Endangered Species Wild Fauna and Flora); CSD (Commission on Sustainable Development); FAO (Food and Agriculture Organization); FCCC (Framework Convention on Climate Change); GATT (General Agreement on Tariffs and Trade); GISP (Global Invasive Species Progammes); GEF (Global Environmental Facility); IBP (International Biological Programme); GWP (Global Water Partnership); ICIMOD (International Centre for Integrated Mountain Development); IBPGR (International Bureau of Plant Genetic Resources); MAB (Man and Biosphere); SCOPE (Scientific Committee on the Problems of Environment); SPP (Social Policy Programme); TRAFFIC (Trade Record Analysis of Flora and Fauna in Commerce); WTO (World Trade Organization); WWC (World Water Council), etc.

Majority of these units are funded and facilitated through UNDP (United Nations Development Pogramme); UNEP (United Nations Environment Programme); UNESCO (United Nations Environmental Scientific and Cultural Organization); IUCN (International Union of Conservation of Nature and Resources); IMF (International Monetary Fund); WB (World Bank); WWF (World Wildlife Fund), etc. (Anonymous, 1997, 1999, 2000; Gaur, 2007b).

BIODIVERSITY: SALIENT FEATURES OF INDIA

The sovereign mountain of Himalaya, Bay of Bengal, Indian Ocean and Arabian Sea are the peculiar boundary lines of India, eventually the core land mass is heterogeneous in lithographic, physiographic, edaphic, climatic and geological combinations. The drainage system is widely variable from snow-capped Himalaya, to perennial rivers to annual or ephemeral drainages, with the addition of different forms of precipitation. All these factors are responsible to host a variety of biota (BSI, 1983, 2005; Forest Survey of India, 1997; Khoshoo, 1988, 1994, 1995; Chatterjee, 1939; Champion and Seth, 1968; Gaur, 2008). It is to be underlined that India represents about 7.8% of the total biodiversity coupled with the fact that the species inventory can reach still higher per cent of representation. Following are the selective features:

India is one of the 12 megabiodiversity centres of the world and its 2.5% land area of the world accounts for 7.8% of the global species. The whole country is divisible into 10 biogeographical zones. Two 'hotspots' with species richness and endemic climates are Eastern Himalaya and Western Ghats. The total number of species recorded is more than 1,27,000, and about 31% are endemic and about 4,00,000 species are expected yet to be identified. About 28% Bryophytes, 15% Lichens, 21% Fungi, 16% Algae, 13% Aves, 12% Pisces of the world (BSI, 1983; Mittermeier and Werner, 1990; Mac Neeley *et al.*, 1990; Myer, 1990; ZSI), are represented in India (Table 1).

Table 1: Comparison of number of species identified in the world and their representation in India

Taxa	World	India	Percentage (%)
Protista	31290	2577	8.24
Mollusca	70000	5050	7.21
Arthropoda	1065000	60383	5.67
Other invertebrates	87121	8329	9.56
Protochordata	2173	116	5.34
Pisces	21723	2546	11.72
Amphibia	5145	206	4
Reptilia	5680	485	8.54
Aves	9673	1228	12.69
Mammalia	4629	372	8.03
Angiosperms	250000	17500	7
Gymnosperms	650	48	7.38
Pteridophytes	10000	1200	12
Bryophytes (Mosses)	7000	2000	28.57
Liverworts	7500	850	11.33
Lichens	13500	2021	14.97
Fungi	70000	14500	20.71
Algae	40000	6500	16.25
Bacteria	8050	850	10.55
Viruses	4000	-	-
Total	**1713134**	**126761**	**209.76**

Agro-biodiversity

Agro-biodiversity has an essential role in maintaining the biodiversity, as several plants, their varieties, wild relatives, animals, pisces, insects, birds, microbes are important associates of agrarian system. Furthermore, plant genetic diversity is the prime subject of modern times, either to introduce desirable genes in a particular crop or to modify the species and at the same time it provides food security for the growing human populace (Swaminathan, 1992; Anonymous, 1999). The global plan of action of CBD stresses the scope of systematic survey, inventorization, collection, evaluation, documentation and safe conservation of crop plants and their relatives. In this context, India represents a rich vegetational wealth, with 356 domesticated species of economic importance and 326 species of their wild relatives (Anomymous, 1999).

The NBPGRI, through its headquarters and network of 11 stations and 30 other institutes, has been maintaining germplasm collection of 1,13,360 accessions of agricultural crop species, medicinal or aromatic plants, etc. The national Genebank (NBPGRI) has the conservation facility of over one million samples. Agro-biodiversity conservation and sustainable use are the prioritized issues of Indian agricultural (Anonymous, 1999). Salient features of agro-biodiversity of India include two of the eight Vavilovian centres of origin and diversity of crop plants, and about 20 crop plants originated from here. It is regarded as a rich centre of origin of 32-50 thousand varieties of rice, pigeon pea, mango, turmeric, ginger, sugarcane, cardamom, colocasia, etc. It has 350 cultivated major and minor crops, and 326 wild relatives of these crops. It is second only to China in regard to providing herbal medicines.

Biogeographical Zones of India

India is one of the best examples to represent a series of tropical warm vegetation, to subtropical and temperate like (montane) and high altitude alpine zones. The island vegetation is of special interest that composed of typical mangroves of shallow water, some tidal vegetation and special submerged plants. Likewise the Himalayan mountains sustain rich vegetation especially adapted for harsh climatic and edaphic conditions, drought, dry desert like conditions due to extremely low temperature. No doubt, the Western Ghats and NE zone are very special and significant in representing species richness as well as endemism. The desert, semi-arid zones and Gangetic plains have wide spectrum of edaphic, climatic and topographic features, thereby host a large biota. Similarly, the Deccan Peninsula represents its peculiar vegetation in the form of scrubs or dry-deciduous forests, semi-evergreen, including the Shola vegetation. Orchids of Himalaya and Western Ghats are quite peculiar whereas forest communities with primitive type of trees prevail in Eastern Himalaya and NE zone. Diversity of *Rhododendrons, Bamboos, Laurels,* and *gymnosperms* are peculiar to the region. Associated with the vegetation, these biogeographical zones have specific wildlife habitation, including microorganisms, insects, mycorrihzal elements, etc. (Rodgers and Panwar, 1988; Rao, 1994; Khushoo, 1988, 1995; Gaur, 2005; Champion and Seth, 1968; Chatterjee, 1939; BSI, 1997; Myer, 1990; Myer *et al.*, 2000).

Table 2 and two adjoining maps of India represent localities of different biogeographical zones together with salient features of vegetation.

Table 2: Biogeographical zones of India

S. No.	Biogeographical Zones	Vegetation
1	Trans-Himalaya	Cushion and bushes
2	The Himalaya	Forest and alpine vegetation
3	The Deserts	Scanty vegetation, sclerophyllous type.
4	Semi-arid Zone	Deciduous forest, open scrubs, savana
5	Gangetic Plain	Monsoon deciduous forests
6	Western Ghats	Evergreen, semi-evergreen forest, shola vegetation
7	Deccan Peninsula	Dry deciduous forest and scrubs
8	Coastal Zone	Submerged and mangrove vegetation, swamp forests, palms
9	North-East Zone (Transition zone between India, Indo-China and Indonesia)	Tropical, temperate, sub-alpine forests rich in orchids, Bamboo and *Rhododendron*, etc
10	Islands	Mangrove vegetation, tidal forests, submerged or stand vegetation

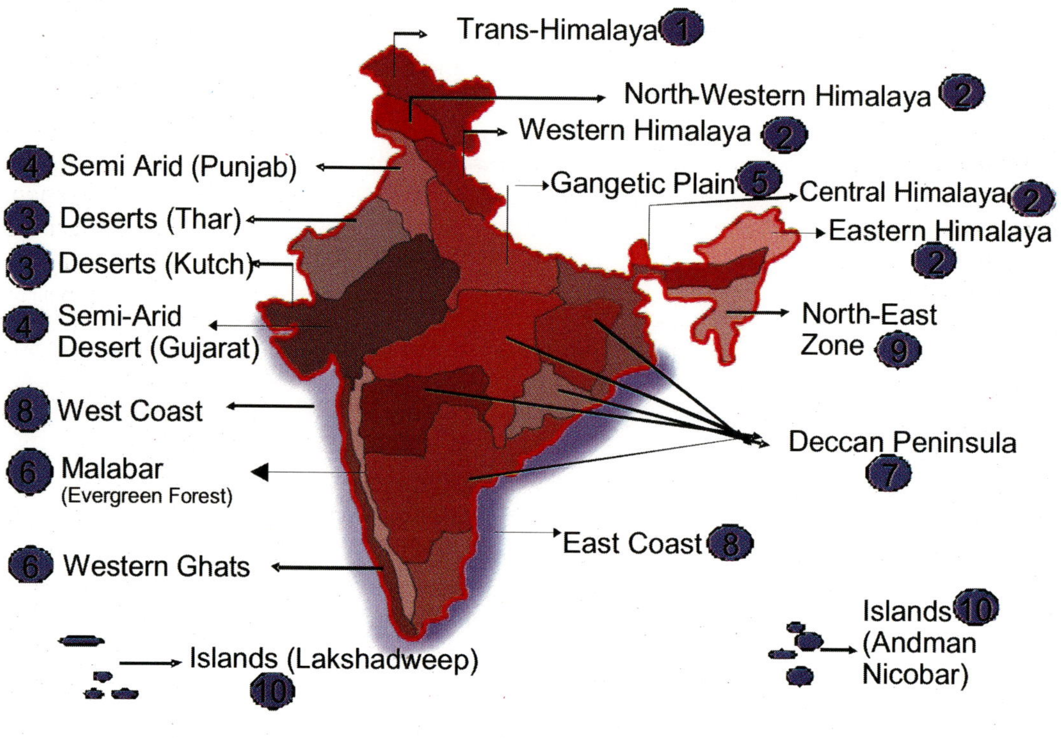

Map: Biogeographical zones of India.

Ethnobiological Features

India, with a rich repository of vegetation wealth has also been blocked with intrinsically deep rooted traditional knowledge on plant utilization and environmental management. Our old treatises include several plants of economic, medicinal, aesthetic or recreational importance. Most of the primitive Indian societies were forest dwellers, and acquired their culture, customs, ethos from the interactions with forests or vegetation of their surroundings. About 550 tribal communities, several folk and some expertise indigenous medicinal practitioners are the important sources of traditional ethnobotanical knowledge, only part of which is tapped and still there is a vast gap to explore more knowledge before it is likely to get vanished. With the modern development, ethnobotanical lores pave ways to several new innovations and discoveries (Gaur, 1999). The main features of ethnobiology of India are:

- India is the reservoir of traditional indigenous knowledge as it represents 4000 endogamous communities, 550 tribal types, and the population of about 1.2 billion.
- So far 9,500 plant species of ethnobotanical importance of which 7,500 medicinal plants and 3,000 edible plants have been included besides a large number of animals used for ethnozoological importance.
- Of the total land, 15% of geographical area is occupied by 550 tribal communities. India has unique ecosystems of forests, wetlands, grasslands, deserts, high altitude meadows, coastal and marine communities, etc. which include distinct forms of plants, animals and microbes of ethnobiological importance.

CONSERVATION OF BIODIVERSITY

Conservation is a noble attitude towards the well-being of all life-forms of the globe. It needs proper assessment to settle the priority of organisms requiring greater attention for conservation. The concept of threatened taxa and conservation are only a few decades old and in fact for earlier to five decades, we really do not know what type of species became extinct. In this context, our knowledge is just as an appearance iceberg. IUCN categorized different types of threats based on the size of the population and distribution of an individual species (Ehrlich and Wilson, 1991; Lucas and Synge, 1978; IUCN, 1999). These are as follows:

Extinct: Species not definitely located in the wild during the past 50 years.

Critically Endangered: The taxon whose number has reached a critical level or whose habitat has been drastically reduced to move for extinction.

Endangered: Taxa in danger of extinction and whose survival is unlikely of the causal factors if continue to operate.

Vulnerable: Taxa believed likely to move into endangered category in the future. It also includes the taxa whose populations are decreasing because of over-exploitation extensible destruction of habitat and other environmental disturbances.

Rare: Taxa with small world population that are not at present endangered or vulnerable but are at risk.

Intermediate: Taxa known to be endangered, vulnerable or rare but where still enough information is not available to assign it to the appropriate category.

Insufficiently Known: Taxa which are suspected but not definitely known to belong to any of the above categories due to paucity of information.

The Botanical Survey of India undertook detailed survey and published more than 600 plant species under threatened categories (Nayar and Sastry, 1987, 1988, 1990). Still there are several plant and animal species to be notified in the Red Data Books Theatened Category.

Conservation requires basic aims and objectives for future use, since, we still do not know the actual biological material that may prove of great value in the future. As *Pencillium, Aspergillus,*

Streptomyces, etc. were once upon a time the neglected organisms but their great potential led to the distinct antibiotic remedial measures. Similarly, several of algae have proved to be of great value in medicine, nutrition, food and industries. In recent years several of the microorganisms have been shown to possess a proactive role in medicine, hygiene, food, reducing the pollution, sewage treatment, gene therapy and several new biological avenues. Therefore, conservation requires distinct perception for ethical, aesthetic or materialistic reasons. In all, conservation of all beings add to the pleasure and quality of life on the earth (Anonymous, 1999b, 2000; Norse and Mc Manus, 1980; BSI, 1997).

In situ Conservation

Throughout the world, one of the most important methods of *in-situ* conservation is the concept of protected areas. According to IUCN, the protected area is an area of 1000 ha or more, managed by competent authority. ICUN Commission on National Parks and Protected Areas (CNPPA) has developed a classification of different types of PAs. These are based on the methods of their keeping, aims and scopes of conserving entities, etc. Some of the important PAs are Scientific Reseve, National Park, Natural Monument, Wildlife Sanctuary, Protected Landscapes, National Biotic Area, Biosphere Reserve, World Heritage Site, Resource Reserve, etc (WCMC, 1992).

Protected Areas Network (PAN) of India

About 5% of the geographical area is under PAN, of which there are 15 Biosphere Reserves, 92 National Parks, About 540 Wildlife Sanctuaries, 28 Conservation Reserves, 28 Tiger Reserves; 28 World Heritage Sites and 25 Ramsar sites (wetland ecosystems; *please see* Tables 3, 4 and 5)

The following table lists the Biosphere Reserves of India alongwith the information of the years of their establishment, location, status, categories, and the areas covered:

Table 3: Important biosphere reserves of India

S.No.	Years of establishment	Names	Locations	States	Categories	Area (Km^2)
1.	2008	Kuhch Biosphere Reserve	Part of Kutch, Rajkot and Surendranagar Districts	Gujarat	Semi-Arid	12454
2.	1989	Gulf of Mannar	Indian part of Gulf of Mannar between India and Sri Lanka	Tamil Nadu	Coasts	10500
3.	1989	Sunderbans	Part of delta of Ganges and Barahamaputra river system	West Bengal	Gangetic Delta	9630
4.	1988	Nanda Devi	Part of Chamoli, Pithoragarh and Almora Districts	Uttarakhand	West Himalaya	5860
5.	1986	Nilgiri Biosphere Reserve	Part of Wynad, Nagarhole, Bandipur and Mudumalai, Nilambur, Silent Valley and Siruvani Hills	Tamil Nadu, Kerala and Karnataka	Western Ghats	5520
6.	1998	Dehang Debang	Part of Siang and Debang valley	Arunachal Pradesh	East Himalaya	5112
7.	1999	Pachmarhi	Part of Betul, Hoshangabad and Chhindwara districts	Madhya Pradesh	Semi-Arid	4926

Table 3: Contd.

S.No.	Years of establishment	Names	Locations	States	Categories	Area (Km2)
8.	1994	Similipal	Part of Mayurbhanj district	Orissa	Deccan Peninsula	4374
9.	2005	Achanakamar – Amarkantak	Part of Anuppur, Dindori and Bilaspur districts	Chhattisgarh Madhya Pradesh	Semi-Arid	3835
10.	1989	Manas	Part of Kokrajhar, Bongaigaon, Barpeta, Nalbari, Kamrup and Darrang Districts	Assam	East Himalaya	2837
11.	2000	Kanchanjunga	Parts of Kanchanjunga Hills	Sikkim	East Himalaya	2620
12.	2001	Agasthyamalai	Neyyar, Peppara and Shenduruny Wildlife Sanctuary and their adjoining areas	Kerala	Western Ghats	1701
13.	1989	Great Nicobar	Southernmost islands of Andaman and Nicobar group of Islands	Andaman and Nicobar Islands	Islands	885
14.	1988	Nokrek	Parts of Garo Hills	Meghalaya	East Himalaya	820
15.	1997	Dibru-Saikhowa	Parts of Dibrugarh and Tinsukia Districts	Assam	East Himalaya	765

Table 4: List of wetlands of international importance under Ramsar convention.

S.No.	Name of Wetlands	Dates of Declaration	States
1.	Ashtamudi Wetland	Aug. 2002	Kerala
2.	Bhitarkanika Mangroves	Aug. 2002	Orissa
3.	Bhoj Wetland	Aug. 2002	Madhya Pradesh
4.	Chandertal Wetland	Nov. 2005	Himachal Pradesh
5.	Chilika Lake	Oct. 1981	Orissa
6.	Deepor Beel	Aug. 2002	Assam
7.	East Calcutta Wetlands	Aug. 2002	West Bengal
8.	Harike Lake	Mar. 1990	Punjab
9.	Hokera Wetland	Nov. 2005	Jammu and Kashmir
10.	Kanjli	Jan. 2002	Punjab
11.	Keoladeo National Park MR	Oct. 1981	Rajasthan
12.	Kolleru Lake	Oct. 2002	Andhra Pradesh
13.	Loktak Lake MR	Mar. 1990	Manipur
14.	Point Calimere	Aug. 2002	Tamil Nadu
15.	Pong Dam Lake	Aug. 2002	Himachal Pradesh
16.	Renuka Wetland	Nov. 2005	Himachal Pradesh
17.	Ropar	Jan. 2002	Punjab
18.	Rudrasagar Lake	Nov. 2005	Tripura
19.	Sambhar Lake	Mar. 1990	Rajasthan
20.	Sasthamkotta Lake	Aug. 2002	Kerala
21.	Surinsar-Mansar Lakes	Nov. 2005	Jammu and Kashmir
22.	Tsomoriri	Aug. 2002	Jammu and Kashmir
23.	Vembanad-Kol Wetland	Aug. 2002	Kerala
24.	Wular Lake	Mar. 1990	Jammu and Kashmir
25.	Upper Ganga River (Brijghat to Narora Stretch)	Nov. 2005	Uttar Pradesh

MR-sites under Montreaux Record.

Source: The Ramsar Convention on Wetlands (www.ramsar.org)

Table 5: World heritage sites in India

S. No.	Names	Regions	Description
1	Kaziranga Wild Life Sanctuary	Assam, India	It is located in the flood plains of the Brahmaputra River's south bank, was declared a World Heritage Site by UNESCO in 1985 for its unique natural environment. Rhinoceros is the peculiar mammal of the heritage.
2	Manas Wild Life Sanctuary	Assam, India	It is located in the plains of the Manas River's in the foot hills of the Himalayas on the border tith Bhutan (contiguous with the Manas Wild Life Sanctuary in Bhutan It was declared as a World Heritage Site by UNESCO in 1985 for its unique natural environment. The sanctuary is the habitat of several species of p!ants and about 21 most threatened species of mammals.
3	Mahabodhi Temple Complex at Bodh Gaya	Bihar, India	Mahabodhi Temple Complex at Bodh Gaya (Buddha Gaya) was inscribed in UNESCO World Heritage List as a unique property of cultural and archaeological importance.
4	Humayun's Tomb	Delhi, India	Humayun's Tomb, Delhi, the first tomb built with several innovations, set at the centre of luxurious gardens with water channels, was the precursor monument to the Taj Mahal (built a century later). It was built in 1570 and was inscribed as a UNESCO World Heritage Monument in 1993 for its cultural importance.
5	Qutb Minar and its Monuments	Delhi, India	Qutb Minar and its Monuments, Delhi, located to the south of Delhi, is a complex with the Qutb Minar of a red sandstone tower of 72.5 metres (238 ft) height with a base of 14.32 metres (47.0 ft) reducing to 2.75 metres (9.0 ft) diameter at the top. It was inscribed under the UNESCO World Heritage List in 1993 for its unique representation of the Islamic architectural and artistic excellence.
6	Red Fort Complex	Delhi, India	It represents the glory of the Mughal rule and is considered the Highpoint of Mughal architectural, artistic aesthetic creativity. It was included in 2007 as UNESCO World Heritage site.
7	Churches and Convents of Goa	Goa, India	Churches and Convents of Goa are monuments inscribed by UNESCO under the World Heritage List in 1986 as cultural property which were built by the Portuguese colonial rulers of Goa.
8	Champaner-Pavagadh Archaeological Park	Gujarat, India	It was inscribed as aUNESCO World Heritage Site in 2004 as a natural site. The Kalikamata Temple on top of the Pavagadh Hill is considered to be an important shrine, attracting large numbers of pilgrims throughout the year.
9	Group of Monuments at Hampi	Bellary District, Karnataka, India	Hampi subsumes the ruins of Vijayanagara, which was the former capital of the Vijayanagara Empire. Dravidian temples and palaces abound in Hampi which is an important Hindu religious centre, has the Virupaksha Temple and several other monuments, inscribed under category in the UNESCO World Heritage List in 1986.

Table 5: Contd.

S. No.	Names	Regions	Description
10	Group of Monuments at Pattadakal	Bijapur, Karnataka, India	The Group of monuments in Pattadakal designated under UNESCO World Heritage List, in 1987, cover a remarkable series of nine Hindu temples, as well as a Jain sanctuary.
11	Buddhist Monuments at Sanchi	Madhya Pradesh India	Buddhist Monuments at Sanchi, located 45 kilometres (28 mi) from Bhopal . It was inscribed as a World Heritage Site byUNESCO on January 1989 for its unique cultural importance.
12	Rock Shelters of Bhimbetka	Madhya Pradesh, India	The rock shelters, discovered only in 1957, comprise a group of "five clusters of rock shelters" with paintings that are inferred to date from the "Mesolithic period right through to the Historical period". It was inscribed as a World Heritage Site by UNESCO in 2003 as a unique cultural property.
13	Khajuraho Group of Monuments	Madhya Pradesh, India	Khajuraho Group of Monuments attributed to the Chandela dynasty of Rajput descent, it was inscribed by UNESCO as a World Heritage Site, a cultural property on October 1982 for its unique original artistic creation.
14	Ajanta Caves	Maharashtra, India	Ajanta Caves listed under UNESCO World Heritage List 1983 as a cultural heritage site, are Buddhist caves that depict richly decorated paintings.
15	Ellora Caves	Maharashtra, India	Ellora Caves also known as Ellora Complex are a cultural mix of religious arts of Buddhism, Hinduism and Jainism. This cultural property has been inscribed under the UNESCO World Heritage List 1983.
16	Elephanta Caves	Maharashtra, India	The Elephanta Caves are a network of sculpted caves located on Elephanta Island, or Gharapuri (literally "the city of caves") inMumbai Harbour. Renovated in the 1970s, the caves were designated an UNESCO World Heritage Site in 1987 to preserve the artwork.
17	Chhatrapati Shivaji Terminus (formerly Victoria Terminus)	Maharashtra, India	Chhatrapati Shivaji Terminus is a historic railway station in Mumbai, which serves as the headquarters of the Central Railways. On July 2004, the station was nominated as a World Heritage Site by UNESCO.
18	Sun Temple, Konârak	Puri, Orissa, India	Konark Sun Temple is built in the form of the chariot of Surya (Arka), the sun god with 24 wheels, and is heavily decorated with symbolic stone carvings and lead by a team of six horses. It is a World Heritage Site inscribed in 1984 as cultural property.
19	Keoladeo National Park	Rajasthan, India	Keoladeo National Park in Rajasthan is located within theIndus-Ganges It was declared a national park in 1982. It was inscribed in the UNESCO World Heritage List in 1985.
20	Jantar Mantar	Jaipur, Rajasthan, India	The Jantar Mantar in Jaipur is a collection of architectural astronomical instruments, built by Maharaja (King) Jai Singh. It has been inscribed as cultural property on the UNESCO World Heritage List as "an expression of the astronomical skills and cosmological concepts of the court of a scholarly prince at the end of the Mughal period.

Table 5: Contd.

S. No.	Names	Regions	Description
21	Great Living Chola Temples	Tamil Nadu, India	The Great Living Chola Temples, built by kings of the Chola Empire stretched over all of Tamil Nadu. The site was inscribed under UNESCO World Heritage List in 2004 as Cultural heritage.
22	Group of Monuments at Mahabalipuram	Mahabalipuram, Tamil Nadu, India	The Group of Monuments at Mahabalipuram, in Tamilnadu, about 58 km from Chennai, were built by the Pallava kings. It was inscribed under the UNESCO World Heritage list in 1984 as a cultural heritage.
23	Agra Fort	Uttar Pradesh, India	Agra Fort, also known as the Red Fort of Agra, which representedMughal opulence and power as the centre piece of their empire was inscribed in the UNESCO World Heritage List in 1982.
24	Taj Mahal	Uttar Pradesh, India	Taj Mahal, one of the Seven Wonders of the World is a mausoleum – a funerary mosque. It was built by Emperor Shahjahan in memory of his third wife Begum Mumtaz Mahal who had died in 1631. It was inscribed in the UNESCO World Heritage List in 1983.
25	Mountain Railways of India	(i) Himalayan Railway (1999), Darjeeling, West Bengal (ii) Nilgiri Mountain Railway (2005) Ooty, Tamilnadu 3. Kalka-Shimla Himachal Pradesh (2008)	The Mountain Railways of India represents a collective listing of the Darjeeling Himalayan Railway (1999), the Nilgiri Mountain Railway (2005), and the Kalka-Shimla Railway (2008), under the UNESCO World Heritage Site. The World Heritage UNESCO recognition "outstanding examples of bold, ingenious engineering solutions for the problem of establishing an effective rail link through a rugged, mountainous terrain.
26	Nanda Devi and Valley of Flowers National Parks	Chamoli District, Uttarakhand, India	Nanda Devi and Valley of Flowers National Parks nestled high in West Himalaya, is renowned for its meadows of endemic alpine flowers and outstanding natural beauty. It was initially. It was inscribed under the UNESCO World Heritage List in 1988 with extension in 2005
27	Sundarbans National Park	West Bengal, India	The Sundarbans National Park, the largest estuarine mangrove forest in the world is a National Park, Tiger Reserve, UNESCO World Heritage Site and a Biosphere Reserve located in theSundarbans Ganges River delta bordering the Bay of Bengal, inWest Bengal. It was inscribed on the UNESCO World Heritage list in 1987 as a natural property.

Source: wikipedia (en.wikipedia.org)

Ex-situ Conservation

Ex-situ conservation involves field gene bank, on farm conservation, botanical/herbal gardens, arboreta, orchidium, fern house, net or polyhouse, tissue culture, etc. where complete plant or parts of plants are maintained as germplasm.

Gene bank, seed bank, pollen bank, tissue culture repository, DNA library, cryo-bank are some of the ex-situ strategies for plant or plant parts conservation. IBPGRI, ICRISAT, ICAR,

CAZRI and several universities and insititutes are employed with the conservation of important biota (Anonymous, 1999; Gaur, 2008).

Anthropogenic Pressure

Anthropogenic activities are responsible to a great extent for the loss of biodiversity. There is a great need to evaluate the process of development, human livelihood, ambitions, and settlement for the secured future of all beings. A medium to strict code subject to the apprehension of bioresource loss, be developed with the wishes of the people to conserve biodiversity. Over-exploitation of natural resources and habit modification by humans are the greatest threats to the biodiversity, need and strict consciousness.

Restoration or Rehabilitation of Threatened Species

IUCN defines re-introduction on the international appearance of an organism into parts of its native range from which it has disappeared or became extirpated as a result of human activities or catastrophes. The intention is establishing of self maintaining, viable population existing under the pressure of natural selection (Mac Neely *et al.,* 1990; Anonymous, 1997, 2000).

COMMON BIODIVERSITY ISSUES

Effective management and restoration of ecosystem require:

(i) Restoration of traditional knowledge and modern practices following some examples, the process involved with the developments of policies, integrated water, land and management.

(ii) Development of case studies on forest restoration as well as imparting training and awareness material to the ecosystem approaches and dissemination of the best possible practical guidelines on various issues of biodiversity.

(iii) For sound ecosystem of a region, control on the species invasion through proper management is warranted.

(iv) For sustainable utilization assessment of bioresources of national and international industrial output is highly devisable.

(v) Considering different types of holistic and other biotic or abiotic components there is need to develop methodology of rural development through pertinent ecological principles and resources.

Information Network

In common name with the present-day innovations, to develop and implement proper information network on biodiversity; is the most essential requirement. Its various components such as the Ecosystem information service, Species information service, Protected areas resources service, Environmental laws information service, Economics of biodiversity, Conservation through digital libraries, etc. need to be included in the information for dissemination of knowledge, awareness and motivation to all the concerned people. Improved mechanism of capacity building (particularly taxonomist to conduct survey and record of various species) at various levels of learning, teaching and practice is highly advisable.

CONCLUSIONS

Biodiversity is a multidisciplinary, holistic approach of study of life on the planet earth. The only life supporting planet earth sustains the life of all living-beings through a balanced sharing of various components. On the one hand, biodiversity is aimed to identify and search the individual species, on the other, it has advanced towards the concept of analysis of genome, biotypes, proteomics, and GMOs. Nevertheless, the future of biodiversity is dependent on the classical approach of taxonomy as well as modern knowledge of genetic engineering (isolation, multiplication of genes, and gene transfer, etc.). Undoubtedly, our primary approach of naming of a species, its role in the whole ecosystem and interrelationship with the other organisms are pre-requires to process further as well as value adding for bioresource utilization and bioprospecting. It is not only the biologists but all the individuals who have to play a key role in the promotion of knowledgeable approach of biodiversity and sustainable development. In essence we need to be responsive and take proper measures to check biodiversity loss to conserve this bounty of nature.

REFERENCES

Anonymous, (1997). IUCN guidelines for the prevention of biodiversity loss due to biological invasion. Species **31, 32**: 28-42

Anonymous, (1999). Jai Vigyan National Science and Technology Mission on conservation of Ago-biodiversity. National Bureau of Plant Genetic Resources, New Delhi

Anonymous, (2000). World Conservation Congress: Agenda and Documentation. IUCN. Amman, Jordan

Anonymous, (2004). The Biological Diversity Act, 2002 and Biological Diversity Rules, 2004 National Biodiversity Authority, India. Frontline Offset Printers, Chennai, India

Bailey, RG (1989a). Ecoregions of the continents, US Department of Agricultures Forest Service. Washington, DC

Bailey, RG (1989b). Explanatory Support to ecoregions of the continents. Environ Conserve. **16:** 307-309

Botanical Survey of India (BSI), (2005). Plant Diversity in India (Status and Conservation). Leaflet, Kolkata, India

Botanical Survey of India (1983). Flora and Vegetation of India: an outline. BSI, Howrah, India

Champion, HG & Seth, SK (1996). Forest Types of India. The Manager Publications, Delhi

Chatterjee, D (1939). Studies on the endemic flora of India and Burma. *J. Asiat. Soc.* Bengal. **11-5:** 19-67

Ehrlich, PR & Wilson, EO (1991). Biodiversity studies: Science and Policy. *Sci.* **253:** 768-762

Forest Survey of India (1997). State of Forest. Ministry of Environment and Forests. Dehra Dun

Gaur, RD (1982). Dynamics of vegetation of Garhwal Himalaya. pp. 347-413. *In:* Paliwal, G.S. (Ed.), *Vegetational Wealth of Himalayas.* Puja Publishers, New Delhi

Gaur, RD (1999). *Flora of the District Garhwal, North West Himalaya* (with Ethnobotanical Notes). Trans Media, Srinagar, Uttarakhand

Gaur, RD (2005). Biodiversity and Conservation. Special Address to National Seminar on Biodiversity. March, 2005. G.G.D. University, Bilaspur, Chhattishgarh

Gaur, RD (2007a). An integrated vision of science and society in Uttarakhand. Samay Vigyan Shodh Patrika Special volume 207-213

Gaur, RD (2007b). Biodiversity and river valley projects in Uttarakhand. *Proc. Natl. Acad. Sci.* **77B:** 253-262

Gaur, RD (2008). Biodiversity: concepts and status. Keynote address in national seminar on biodiversity conservation and its application October 10 October 13, 2008), Govt. PG College, Gopeshwar, Uttarakhand, pp.1-11

Gaur, RD & Kuniyal, P (2009). Preliminary studies on the impact of climate change on the biodiversity of Uttarakhand. pp. 163-188. *In: "Climate Change at the Third Pole",* Vandana Shiva & Bhat Vinod K (Eds.) (Navdanya/ Research Foundation for Science Technology and Ecology), New Delhi

Grant, V (1981). Plant Speciation 2nd Edition. Columbia University, Press, New York

Holdridge, LK (1967). Life Zone Ecology. Tropical Science Centre, San Jose, USA

IUCN, (1999). Draft guidelines for the application of IUCN red list criteria at national and regional level. Species **31, 32:** 58-70

IUCN, (2000). World Conservation Congress; Agenda and Documenttion, Jordan, Aman

Khoshoo, TN (1988). Environmental Concerns and Strategies. Ashish Publ. New Delhi

Khoshoo, TN (1994). India's biodiversity: tasks ahead. *Curr. Sci.* **67:** 578-580

Khushoo, TN (1995). Census of Indian biodiversity: tasks ahead. *Curr. Sci.* **69:** 14-17

Lucas, GL & Synge, H (1978). The IUCN Plant Red Data Book, Norges, Switzerland

Mac Neely, Miller, KR, Reid, WV, Mittermeir, RA & Werener, TB (1990). Conserving the World Biological Diversity, IUCN. Gland, switzerland

Mittermeier, RA & Werner, TB (1990). Wealth of plants and animals united megadiversity countries. Tropicus **4:** 4-5

Myers, N (1990). The Biodiversity Challenges: expanded hot spots analysis. The Environmentalists **10:** 243-256

Myers, N, Mittermeier, CG, Da Fonseca, GAB, & Kent, J (2000). Biodiversity Hot Spot for conservation priorities. *Nature* **403:** 853-858

Myers, N (1990). The biodiversity challenge:expanded hot spot analysis. *The Environmentalist* **10:** 243-256

Nayar, MP and Sastry, ARK, (1987, 1988, 1990). Red Data Book of Indian Plants. BSI Howrah, India

Norse, A and Mc Manus (1980). Forests and Biodiversity: Biodiversity and Conservation **3:** 3-20

Rao, RR (1994). Biodiversity in India: Floristic Aspects. Bishen Singh-Mahendra Pal Singh, Dehradun

Rodgers, WA & Panwar, SH (1988). Biogeographical Classification of India. FRI, New Forests, Dehradun

Singh, MP & Singh, DB (2004). Biodiversity why? A lot of hue and cry at present. Agrobios Newsletter **2:** 14-17

Stace, CA (2000). Cytology and cytogenetics as a fundamental taxonomic resource for 20th & 21st centuries. Taxon **50:** 451-477

Swaminathan, MS (1992). Biodiversity: Implication for global food security. Macmillan, Madras

Turner, IM (1996). Species loss in fragnment of tropical rain forests: a review of the evidence. *J. Appl. Ecol.* **33:** 200-209

Udvardy, MDF (1975). A classification of the biogeographical provinces of the world IUCN, Occasional paper 18, Gland, Switzerland

Udvardy, MDF (1984). A classification of the biogeographical provinces of the world. IUCN, Gland, Switzerland

Wilson, EO (1986). Biodiversity. National Academy Washington, USA

World Conservation and Monitoring Centre (WCMC), (1992). Global Biodiversity: Status of Earth's Living Resources. *In:* Groombridge, B. (Ed.) Chapman & Hall, London

2

Biodiversity Hotspots: A Tool for Biodiversity Conservation

HIMANSHU RAI AND RAJAN KUMAR GUPTA

Department of Botany, Pt. L.M.S Govt. Post-Graduate College, Rishikesh–249 201 (Dehradun) Uttarakhand, India

ABSTRACT

Unsustainable human impact has put nearly all the life-form's survival on stake. In order to protect biodiversity, the concept of hotspot gives us a tool to tackle conservation in a very effective way The concept of biodiversity hotspots was originally suggested by Myers (1988 & 1990). Recently the concept has been revised in the light of modern habitat conditions by Mittermeier in the book "Hotspots revisited" (2004). A biodiversity hotspot must contain at least 1,500 species of vascular plants as endemics and must have lost at least 70% of its original habitat.

The "Hotspots revisited" (2004), recognizes 34 global "biodiversity hotspots". Three of these regions fall in India: 1. Himalaya 2. Indo-Burma 3. Western Ghats and Sri Lanka. The Indian hotspots are the most vulnerable regions due to population pressures, overexploitation of natural resources, global warming and climate change and need immediate attention. The concept of hotspots is shaping global conservation efforts in a very fundamental way by helping in analyzing, planning and implementation of these endeavours by government, researchers, private firms and conservation Initiatives.

Keywords: Biodiversity hotspots, climate change, global warming, natural resources, species conservation.

INTRODUCTION

Unsustainable consumption of natural resources by humans is driving nearly all the floral and faunal forms to extinction. Extinction, which is a natural process, has got aggravated in the form of increased rate (possibly several thousand times the natural rate) solely, due to human impacts. Mass extinctions of this magnitude have only occurred five times in the history of our planet; the last at the end of Mesozoic era (about 65 million years ago) that brought the end of the dinosaur age. Wilson (2002) estimates that at the current rates of human destruction of the biosphere, one-half of all species of life will be extinct in 100 years.

In this grim situation it must be our first priority, to identify areas of crucial biodiversity degradation. Here the concept of **"Biodiversity hotspots"** may be of great help, which recognises areas of our planet having critically threatened biodiversity and need instant attention. With very few resources spare for conservation, the hotspots make the policy-makers easy to utilize their economic resources with maximum conservation impact.

DEVELOPMENT OF HOTSPOT CONCEPT AND PRESENT SCENARIO

The concept of biodiversity hotspots was originally suggested by **Myers** in two articles in "The Environmentalist" (1988 and 1990), revised after thorough analysis by Myers in "Hotspots: Earth's Biologically Richest and Most Endangered Terrestrial Ecoregions" (1999), and again revised by **Mittermeier** in the book "Hotspots revisited" (2004). The book has been brought out by Washington, D.C.-based **Conservation International** a nonprofitable initiative, detailing a four-year analysis of its global hotspot strategy for biodiversity conservation.

To qualify as a biodiversity hotspot, a region must meet two strict criteria:

- It must contain at least 1,500 species of vascular plants as endemics.
- It has to have lost at least 70% of its original habitat.

In the 1999 analysis, published in the book *Hotspots: Earth's Biologically Richest and Most Endangered Terrestrial Ecoregions*, and a year later in the scientific journal *Nature* (Myers *et al.*, 2000), 25 biodiversity hotspots were identified. These areas held endemics which were 44% of the world's plants and 35% of terrestrial vertebrates in an area that formerly covered only 11.8% of the planet's land surface. In "Hotspots revisited" (2004), 34 global "biodiversity hotspots" (Table 1; Fig. 4) are recognized, each have lost at least 70% of their original habitat and each contain more than 1500 endemic vascular plant species. These sites support nearly 60 per cent of the world's plant, bird, mammal, reptile, and amphibian species, with a very high share of endemic species. The habitat extent of this land area has reduced by 87.8% of its original extent, thus these hotspots cover only 1.4 %of Earth's land surface.Over 50% of the world's plant species and 42% of all terrestrial vertebrate species are endemic to the 34 biodiversity hotspots (Figs. 1, 2 and 3).

Table 1: List of redistributed biodiversity hotspots (Mittermeier *et al.*, 2004)

1. Atlantic Forest (South America)	18. Madrean Pine - Oak Woodlands (North America)
2. California Floristic Province (North America)	19. Maputaland - Pondoland – Albany (Africa)
3. Cape Floristic Region (South Africa)	20. Mediterranean Basin (North Atlantic Ocean-Europe)
4. Caribbean Islands (Cuba, West Indies Dominican Republic)	21. Mesoamerica (North America)
5. Caucasus (West Europe)	22. Mountains of Central Asia (Asia)
6. Cerrado (South America)	23. Mountains of Southwest China (Asia)
7. Chilean Winter Rainfall - Valdivian Forests (South America)	24. New Caledonia (Coral Sea-South Pacific Ocean)
8. Coastal Forests of Eastern Africa (Africa)	25. New Zealand

Table 1: Contd.

9. East Melanesian Islands (North Pacific Ocean)	26. Philippines (Asia)
10. Eastern Afromontane (Africa)	27. Polynesia – Micronesia (Polynesia)
11. Guinean Forests of West Africa (Africa)	28. Southwest Australia (Australia)
12. Himalaya (Asia)	29. Succulent Karoo (Africa)
13. Horn of Africa (Africa)	30. Sundaland (Asia)
14. Indo-Burma (Asia)	31. Tropical Andes (South America)
15. Irano-Anatolian (Asia)	32. Tumbes – Choc (South America)
16. Japan (Asia)	33. Wallacea (Asia)
17. Madagascar and the Indian Ocean Islands (Africa)	34. Western Ghats and Sri Lanka (Asia)

IMPACT OF HOTSPOT CONCEPT

The concept of hotspots has helped global conservation efforts in a very fundamental way:

- It has boost up scientific analysis of biodiversity depletion and factors affecting it as this concept recognises area of unique and fragile habitats and give researchers a point to concentrate. After the introduction of hotspot concept by Myers in 1988 there has been several citation of this concept which are still on the rise.
- More importantly, the impact of the hotspots concept in terms of investment in conservation has been dramatic. Policy makers (Global organizations-UN, Governments, etc.) and many conservation initiatives have adopted hotspots as their primary global investment strategy.
- The hotspots concept has also entered the mainstream as a tool for private sector businesses. For e.g., preference is given to pulp and paper vendors that protect natural forests in the biodiversity hotspots and high-biodiversity wilderness areas.

Thus, nowadays hotspots have become the centre stage of nearly all the conservation efforts may be it Governments, UN agencies, Private Concerns, Conservation initiatives and Scientific Community.

BIODIVERSITY HOTSPOTS: THE INDIAN CONTEXT

With the publication of "Hotspots revisited" (2004) the number of biodiversity hotspots in India has been increased to **three** regions from previous **two** (Myers *et al.*, 1999), which are:

1. Himalaya
2. Indo-Burma
3. Western Ghats and Sri Lanka

The hotspot of north-eastern states is now divided into two regions. The Himalaya (covering states of West Bengal, Sikkim, Assam, and Arunachal Pradesh), and Indo-Burma (covering the states of Manipur, Mizoram, Tripura, Meghalaya, Nagaland and south Assam) which alongwith Western Ghats form the three hotspot zones in India.

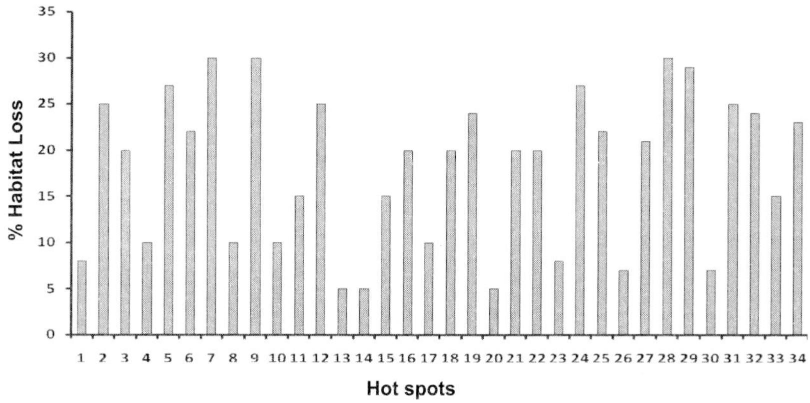

Fig. 1. Percentage habitat loss in each of the 34 redistributed biodiversity hotspots, (Based on the data from Mittermeier *et al.*, 2004). Numbers on the X axis correspond to the serial number of hotspots in Table No.1.

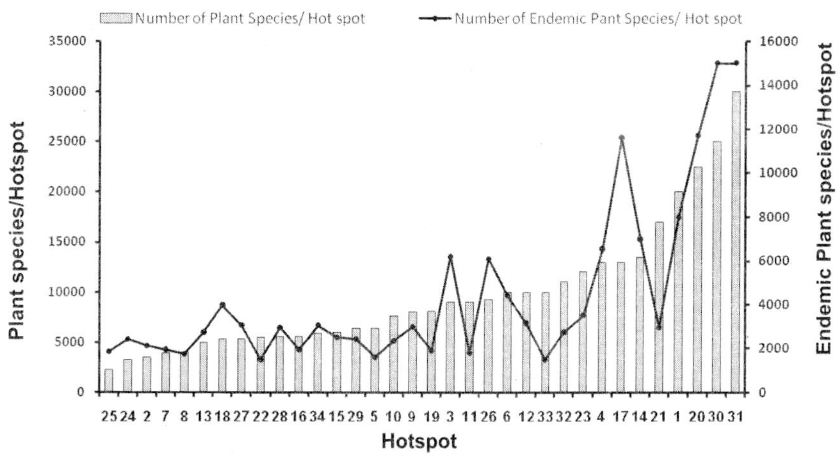

Fig. 2. Plant distribution dynamics of each of the 34 redistributed biodiversity hotspots, showing clear concentration of maximum number of Plants and Plant endemics in Asia (Based on the data from Mittermeier *et al.*, 2004). Numbers on the X axis correspond to the serial number of hotspots in Table No.1.

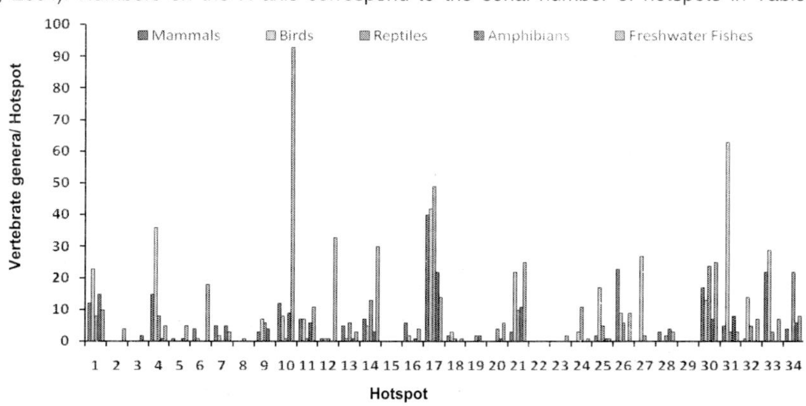

Fig. 3. Distribution dynamics of endemic Vertebrate genera in each of the 34 redistributed biodiversity hotspots, (Based on the data from Mittermeier *et al.*, 2004). Numbers on the X axis corresponds to the serial number of hotspots in Table No.1.

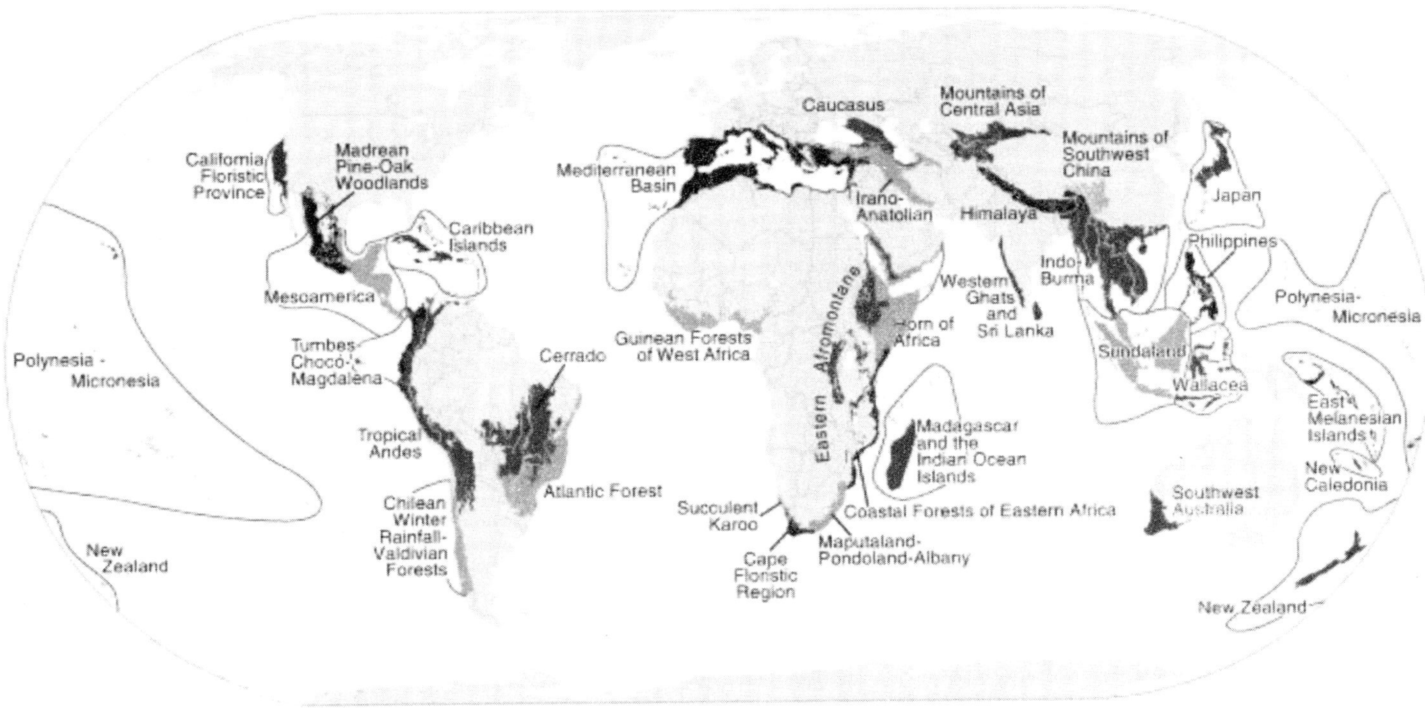

Fig. 4. Thirty-four biodiversity hotspots, of which three Himalayas, Indo-Burma and Western Ghats lie in India and Sri Lanka (R. Mittermeier *et al.*, hotspots Revisited 2004.)

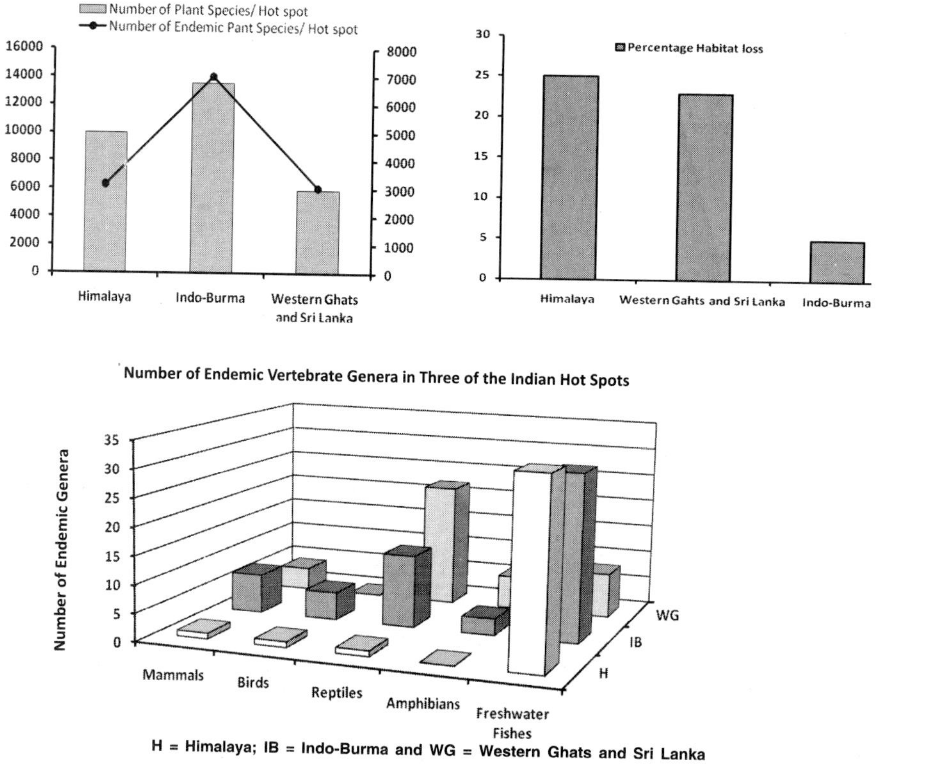

Fig. 5. The vital statistics of each of the three redistributed biodiversity hotspots of the Indian subcontinent (Based on the data from Mittermeier *et al.*, 2004).

The Himalaya

Fig. 6. Himalyan hotspot is one of the most diverse ecological ecosystems having, grasslands, forests, forests, and meadows. (R. A Mittermeier *et al.*, Hotspots Revisited 2004.)

Covering about 741,706 km^2 area this hotspot stretches in arc shape from northern Pakistan, Nepal, Bhutan and the north-western and north-eastern states of India. Home to world's highest peak Mt. Everest or Sagarmatha, this immense mountain range has been divided into two regions: the Eastern Himalaya, which covers parts of Nepal, Bhutan, the north-east Indian states of West Bengal, Sikkim, Assam, and Arunachal Pradesh, south-east Tibet (China), and northern Myanmar; and the Western Himalaya, covering the Kumaon-Garhwal, north-west Kashmir, and northern Pakistan.

Due to the abrupt rise of the Himalayan Mountains from less than 500 meters to more than 8,000 meters, diversity of ecosystems that range, in only a couple of hundred kilometers, from alluvial grasslands (among the tallest in the world) and subtropical broadleaf forests along the foothills to temperate broadleaf forests in the mid hills, mixed conifer and conifer forests in the higher hills, and alpine meadows above the treeline, is seen.

The Himalayan hotspot is home to 10,000 species of plants, of which about 3,160 are endemic, 980 bird species of which 15 are endemic and 300 mammal species of which a dozen are endemic.

The Indo-Burma Hotspot Region

The Indo-Burma hotspot encompasses 2,373,000 km^2 of tropical Asia and is situated in the east of the Ganges-Brahmaputra lowlands. Formerly the Himalaya chain and the associated foothills in Nepal, Bhutan and India, were considered in the Indo-Burma hotspot, but now it has been excluded from it and forms the Indo-Chinese subregion of Himalayan hotspot.

Western tragopan (*Tragopan melanocephalus*)

(Courtesy: John Corder) A male Snow leopard (*Uncia uncial)* in Ladakh, (Courtesy: Ami Vitale)

Sikkim Orchid (*Dendrobium nobile)*

Fig. 7. Some of the endemic species of Himalaya Hotspot, which are also among the most threatened and need immediate efforts for conservation.

The Indian states of Manipur, Mizoram, Tripura, Meghalaya, Nagaland and south Assam constitute this region. This area has some of the most densely forested regions of the country. The pattern of biological diversity in Indo-Burma is the result of the interaction of topography, climate changes, soil characteristics, and the patterns of seasonal rainfall. The hotspot contains many centers of endemism, which are due to montane barriers, areas of lowland wet evergreen forest that were isolated at some stage, and river basins.

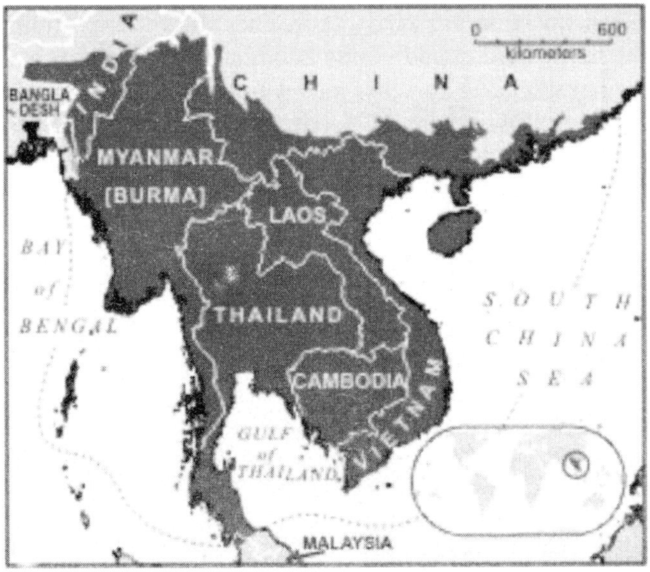

Fig. 8. Indo-Burma region's endemism is due to topographical isolation and unique rainfall pattern. (R. A Mittermeier *et al.*, Hotspots Revisited 2004.)

This hotspot is home of 13,500 species of plants, of which about 7,000 (52%) are endemic, 1,260 birds species of which more than 60 of are endemic and 430 mammal species of which more than 70 species and seven genera are endemic.

Teak (*Tectona grandis*).

Golden-crested Myna, *Ampeliceps coronatus* (Hideo Tani)

Fig. 9. Indo-Burma region's biodiversity is threatened by over-exploitation.

The Western Ghats and Sri Lanka Region

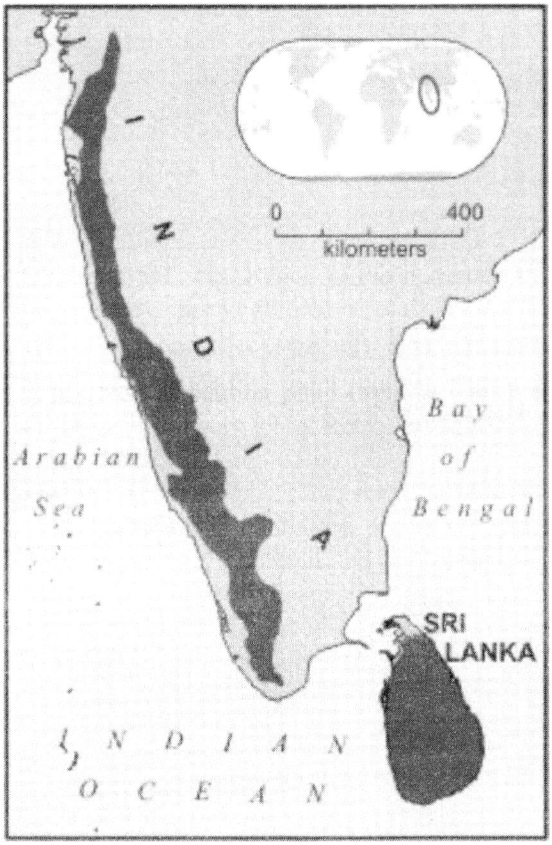

Fig. 10. Geographical extent of the hotspot of Western Ghats and Sri Lanka in Indian Peninsula. (R. A Mittermeier *et al.*, Hotspots Revisited 2004.)

The Western Ghats (locally known as Sahyadri Hills) of south-western India and the highlands of south-western Sri Lanka, are strikingly similar in their geology, climate and evolutionary history. They cover an area of about 160,000 km^2 and stretch for 1,600 kilometers in the states of Gujarat, Maharashtra, Karnataka and Kerala. The wide varied rainfall patterns in the Western Ghats, alongwith the complex geography, produce a great variety of vegetation types which supports a large variety of life-forms.

The Western Ghats harbour approximately 5,000 species of vascular plants belonging to nearly 2,200 genera; about 1,700 species (34 per cent) are endemic, more than 450 bird species, of which about 35 are endemic; about 140 mammal species, of which less than 20 are endemic; 260 reptile species, of which about 175 (66 per cent) are endemic and of 175 amphibian species, of which roughly 130 are endemic. The Ghats are known for their amphibian diversity, some monotypic genera (e.g *Nasikabatrachus sahyadrensis*) which occur in this region.

Among mammal species, the most prominent are the lion-tailed macaque (*Macaca silenus*), Nilgiri tahr (*Hemitragus hylocrius*), which lives in the montane grasslands of the Western Ghats.

Nilgiri Thar (*Hemitragus hylocrius*)
– (Courtesy: V.S. Suresh)

Nasikabatrachus sahyadrensis
– (Courtesy: S.D. Biju)

Fig. 11. Western Ghats show highest amphibian biodiversity

The Indian hotspots are the most vulnerable habitats due to:

- High population pressures.
- Indiscriminate selective logging of trees.
- Destruction of biodiversity due to monoculture agricultural practices.
- Damming of rivers in these habitats resulting in habitat destruction.
- Overexploitation of natural resources in these regions have reached critical levels.
- Global warming and climate change which is altering the rain pattern and snow line in higher altitude, resulting in permanent change in the habitat.

Biodiversity hotspots act as the indicators of the health of our natural habitats and any change or fragility induced will ultimately trigger a chain reaction which ultimately effect us, the most advanced life-form on the planet.

CONCLUSION

Biodiversity hotspots by virtue of their conceptualization on the basis of total floral and faunal biodiversity, total endemic species and loss of habitats serve as one of the most acceptable classification of the major biodiversity centers on our planet. This categorization can be useful in identifying area of our planet which needs immediate attention. The majority of the biodiversity is concentrated in the tropical and subtropical regions of the planet. India which now harbours three rich biodiversity hotspots is facing various threats and pressure influencing the very existence of these hotspots. The Indian hotspots like other such hotspots in Southeast Asia are facing the sustainability problem as this area also holds a very densely populated human population which draws its livelihood from them. Further, the urbanization and utilization of major river systems for energy production is negatively affecting these areas. These regions must be protected and some comfortable and sustainable mechanism must be derived to release the pressure on them.

REFERENCES

Alnutt, T, Wikramanayake, E, Dinerstein, E, Loucks, C, Jackson, R, Hunter, D & Carpenter, C (2002). Composition of the alpine Himalayan protected areas network and its contribution to biodiversity conservation. In: E.D. Wikramanayake, E. Dinerstein, C. Loucks, D. Olson, J

Biju, SD & Bossuyt, F (2003). New frog family from India reveals an ancient biogeographical link with the Seychelles. *Nature* **425:** 711-714

Birdlife International. (2003). *Saving Asia's Threatened Birds: A Guide for Government and Civil Society.* Cambridge: Birdlife International

BirdLife International (2004). *Threatened Birds of the World 2004.* CD Rom. Cambridge, U.K.: BirdLife International.

E.O. Wilson, Harvard University, *The Future of Life* (2002)

Ganeshaiah, KN (2003). *Sasya Sahyadri: Distribution, Taxonomy and Diversity of Plants of Western Ghats.* Bangalore, India: University of Agricultural Sciences

Ghosh, AK (1996). Faunal Diversity. In Gujral, G.S. and Sharma, V. (Eds.), *Changing Perspectives of Biodiversity Status in the Himalaya.* New Delhi: The British Council

Gimaret-Carpentier, C, Dray, S and Pascal, JP (2003). Broad-scale biodiversity pattern of the endemic tree flora of the Western Ghats (India) using canonical correlation analysis of herbarium records. *Ecography* **26:** 429-444

Leakey, Richard and Roger Lewin, (1996). The Sixth Extinction: Patterns of Life and the Future of Humankind, Anchor, ISBN 0-385-46809-1

Mittermeier, RA, Myers, N, Thomsen, JB, da Fonseca, GAB & Olivieri, S (1998). Biodiversity hotspots and major tropical wilderness areas: approaches to setting conservation priorities. *Conservation Biology* **12:** 516-520

Mittermeier, RA, Robles Gil, P, Hoffmann, M, Pilgrim, J, Brooks, T, Mittermeier, CG, Lamoreux, J & da Fonseca, GAB (2004). *Hotspots Revisited.* Mexico: CEMEX

Myers, N & Mittermeier, RA (2003). Impact and acceptance of the hotspots strategy: Response to Ovadia and to Brummitt and Lughadha. *Conservation Biology* **17:** 1449-1450

Myers, N (1988). Threatened biotas: 'Hotspots' in tropical forests. *The Environmentalist* **8:** 1-20

Myers, N (1990). The biodiversity challenge: Expanded hot-spots analysis. *The Environmentalist* **10:** 243-256

Myers, N (2003). Biodiversity hotspots revisited. *BioScience* **53:** 916-917

Myers, N, Mittermeier, RA, Mittermeier, CG, da Fonseca, GAB & Kent, J (2000). Biodiversity hotspots for conservation priorities. *Nature* **403:** 853-858

Rodgers, WA, Panwar, HS & Mathur, VB (2002). *Wildlife Protected Area Network in India:* A review. Dehradun: Wildlife Institute of India

San Diego Zoo. (2007). *Animal Bytes: Snow Leopard.* Retrieved 5 May, 2007

Sherpa, M, Wangchuk, S & Wikramanayake, ED (2003). *Linking Protected Areas for Ecosystem Conservation: A Case Study from Bhutan.* A Paper presented at the World Parks Congress, Durban, South Africa

Snow Leopard Trust. (2006). *Habitat and Range.* Retrieved 27 November, 2006

Snow Leopard Trust. (2007). *Physical Features.* Retrieved 5 May, 2007

Snow Leopard Trust. (2007). *Prey.* Retrieved 5 May, 2007

Solheim, WG, II. (1972). An earlier agricultural revolution. *Scientific American.* **266(4):** 34-41

UNDP. 1998. *Ecoregional Co-Operation for Biodiversity Conservation in the Himalayas*

UNESCO World Heritage Centre Nanda Devi and Valley of Flowers National Parks. Brief Description. Retrieved 27 November, 2006

Wikramanayake, ED, Dinerstein, E, Loucks, C, Olson, D, Morrison, J, Lamoreux, J, McKnight, M, & Hedao, P (Eds.). 2002. *Terrestrial Ecoregions of the Indo-Pacific: A Conservation Assessment.* Washington, D.C.: Island Press

WWF & ICIMOD. 2001. *Ecoregion-based Conservation in the Eastern Himalaya. Identifying Important Areas for Biodiversity Conservation.* Katmandu: WWF Nepal Program

3

Perceptions of Villagers on Climate Change and its Adverse Effect on Biodiversity Depletion and Agro-Ecosystem of Selected Villages of District Rewari, Haryana

NP Baduni[1]; DK Sharma[2]; Gayatri Mahar[1] and GS Paliwal[1]

[1]*Agricultural Finance Corporation Ltd., NRO, B-1/9, Janakpuri,* New Delhi-*110 058*
[2]*NAEB, MoEF, Paryavaran Bhavan, CGO Complex, Lodhi Road,* New Delhi-*110 003*

ABSTRACT

It has been repeatedly emphasized that the mountain and hill ecosystems support the livelihood of 10% of world's population since these are direct resources of water, forest, minerals and even agricultural products. The climate change is causing serious damage to these mountains and accordingly not only the hill people but even the inhabitants of the adjoining plains are experiencing the profound effect of this disturbance of ecosystem. It is extremely desirable, therefore, that the adverse effects are examined and the adaptation of the local people to the changed scenario is assessed. In the present paper, we have analyzed the situation in Rewari district of Haryana and covered seven villages which form part of the Aravalli Hills, in North-Western India. The objectives envisaged to determine: (i) the change in composition of floral and faunal species (ii) the climatic variations and its impact on agricultural practices (iii) documentation of the knowledge and experiences of the local people about the pattern of climate change and its impact on forest, agriculture and water regime. The authors have followed a participatory approach to reach the various conclusions wherein change in composition of forest vegetation, alteration in behaviour in respect to flowering and fruiting due to change in weather pattern; change in the common plant/animal species and size variations of leaves, flower colour, phenology of budding and even quantities of trees, herbs and shrubs has been recorded. We have also noticed changes in the cropping pattern and their productivity due to reduced availability of groundwater, lower annual rainfall, and higher average annual temperature. The local community of the studied region possesses knowledge on the climatic variation and its adverse impact on their natural resources.

Keywords: Aravalli mountain range, change in forest and agricultural sector, cropping pattern, groundwater, phenology, participatory approach.

INTRODUCTION

The crucial role of mountain ecosystems towards supporting livelihood of about 10% of the world's population is well recognized. According to United Nations Organization (2001), the world's population depends directly on mountain resources such as water, forests, agricultural products and minerals. Intergovernmental Panel on Climate Change (IPCC) pointed out that climate change is causing serious damage to natural resources, socio-economic conditions and livelihood of the people. It is feared that the impact of global climate change in mountain will have profound effect not only on hill people but also those inhabiting the adjoining plains. Thus, these not only directly affect the human-beings but also disturb the entire ecosystem of the forests.

Like other mountain ecosystems of the world, the Aravalli Hills which stretch about 615 km from South West Gujarat through Rajasthan, Haryana and Delhi have also not been left untouched by the effect of climate change. The range has been playing an important role in protecting the fertile plains on their eastern side against invading sand drifts from the Thar desert in Rajasthan. In Haryana, the Aravallis cover an area of about 1,46,000 hectares, of which the major portion falls under the category of common lands under the authority of elected village institutions, such as the Panchayat and all social groups have an access to these lands. The increasing pressure of human and livestock population has taken a heavy toll of the vegetation, leaving the Aravallis highly degraded with concomitant adverse environmental impact. This situation has caused misery to the poor section of the community particularly the women-folk, who have to travel long distances to collect fodder, fuel and non-wood products for their subsistence (Srivastava and Kaul, 1994). In this region, the impact of climate change has been witnessed in all the sectors with profound impact on forest and agricultural resources. The decreasing level of rainfall and rainy days is closely related to deforestation. The increasing rate of soil erosion and depleting groundwater level are well-established phenomenon of deforestation. All these factors contribute towards a major shift of lifestyle of the local communities. The Aravalli forest ecosystem is very fragile and climate change will not only alter the composition of forest ecosystem itself but it will also change the basic physiological changes in the species and between the two species also. The climate change has already started affecting timing of reproduction in plants as well as animals, shifting of plant species, migration of animals, flowering period of the species, pollination, etc. (Singh, 2007). These all processes will have far reaching consequences on the ecosystem.

The agricultural sector is particularly vulnerable to present-day climate variability, including multiple years of low and erratic rainfall (Karen, 2004). The agricultural sector in India is infuenced by more than changing climatic conditions. Widespread promotion of Green Revolution technologies during the 1960s increased agricultural yields in India for some crops and farmers by introducing high-yielding varieties that depend on inputs, including irrigation, chemical fertilizers, and pesticides (Goldman and Smith, 1995; Freebairn, 1995). In the State of Haryana also the yield of agricultural products has increased considerably due to the introduction of modern techniques. Simultaneously, however, the traditional cropping pattern has undergone changes over the last few decades.

Realizing the magnitude of the problem and urgent need to revive and restore the natural resources and its adverse impact of deforestation on the inhabitants of Aravalli hills and nearby areas in Haryana, a number of Government initiatives have been taken from time to time. The "Rehabilitation of Common Lands in the Aravalli Hills" and Haryana Community Forestry Project has been amongst the major programmes being run by the Government. These programmes have resulted in great success due to the active role of local communities. Besides this, the traditional communities possess good knowledge on changing pattern of climate change and have accordingly adopted different strategies to cope up with these effects.

In view of the changing scenario, it is necessary to assess the impact of climate change on forest ecosystem *vis-a-vis* the livelihood pattern of the inhabitants. Also, the strategies adopted by the local inhabitants to cope up with these effects need to be documented properly. The present paper attempts to highlight the adverse effect of climate change on forest ecosystem and adaptability of villagers residing in the Aravalli hill area. Thus, the chief objectives of the present study are: (i) to observe the change in composition of floral and faunal species the biodiversity (ii) to record the climatic variations and its impact on agricultural practices (iii) to document the knowledge and experiences of the local people about the pattern of climate change and its impact on forest, agriculture and water regime.

METHODOLOGY

Study area

The study has been conducted in the Rewari district of Haryana which lies between 27° 57' 17" to 28° 27' 47" N latitude and 76°17' to 76° 51' E longitude. The district has an area of 1,559 sq km and is divided into two subdivisions, i.e.. Rewari and Kosli which are further divided into 3 revenue tehsils, i.e., Rewari, Bawal and Kosli with five Development Blocks, namely, Rewari, Jatusana, Khol, Bawal and Nahar. Rewari is predominantly a rural district with, 82.2% of its population living in rural areas (Census, 2001). The literacy rate of the district is 75.25 with the male literacy of 88.45%. The district has a total of 63,744 land-holdings with an area coverage of 1,24,707 ha. Besides, there are as many as 71,816 landless families in the district. The total cultivable area in the district is about 1.43 lakh ha of which 1.25 lakh ha (87.4%) is already under use. The major source of the district income originates from the agriculture sector (73.25%) which is followed by animal husbandry (16.5%) and horticulture (9.1%). Mustard and wheat are the main crops in rabi season and bajra belongs to the kharif season. The main source of irrigation in the district is tubewell, which provides water to the major part of the irrigated area. However, their desperate use has depleted the groundwater to the level of over-exploited category.

Two types of forests have been witnessed in the State, i.e., tropical dry deciduous forests in northern districts in the Shiwalik ranges and tropical thorn forests of Aravalli in the southern districts. The tropical thorn forests are predominant in the district covering 3,970 hectares constituting merely 2.33% of the total geographical area. The forests confer manifold ecological benefits to the economy alongwith a great bearing on groundwater occurrence, soil erosion, floods and environment which are a major source of raw materials and shelter for fauna.

As per the 2001 Census, there are in all 399 revenue villages and 347 panchayats in the district. Block Rewari is the largest in terms of geographical area, whereas block Nahar is the smallest. Out of total 399 villages in the district, 41 villages across 3 blocks are located adjoining Aravalli Mountain Range. Among them a total of seven villages, i.e., Harzipur, Khori, Khaleta, Manethi, Tankari, Kanuka, Gudiyaani from three forest ranges, i.e., Rewari, Nahar and Bawal have been selected for the present study. The selection criteria are: (i) representation of a wider and diverse geographical area within the district, (ii) size and forest cover available, (iii) dependence of forest fringe communities on the forest itself.

Data Collection

The data generated here are based on primary as well as secondary information. The former has been collected by conducting Focal Group Discussions (FGD) which concentrated on documenting the knowledge and experiences of the local people and the officials of Forest Department about the pattern of change in climatic conditions and its effect on forest ecosystem as well as other livelihood resources in the selected villages. Efforts were made to cover the respondents of all the age groups including women. In each village, the respondents were categorized into: (a) age group 20-40 years (b) age group year 41-50 and above, (iii) women. Apart from these, other set of interactions were held with a group of 3-5 representatives of the Forest Department including Forest Guards, Foresters and Rangers. Transect walks with village elders and forest officers were also undertaken to assess the composition of forest and assessing the natural regeneration prospectus in the adjacent forest areas of the sample villages. Resource mapping of each selected village has been achieved through structured questionnaire and participatory approach, consisting of a mix of secondary and primary information. The objective has been to understand the changes in status of natural resources over the past few decades by means of a natural resource map. The village profile covered a comprehensive list of physical resources, natural resources, demographics, facilities, services, infrastructure, etc.

Besides the primary information, secondary information has also been collected from various sources which include the State and District Statistical Handbook, Forest Department, Agriculture Department, Department of Revenue/Land Records, Haryana Agricultural University, Regional Research Station, Bawal.

OBSERVATIONS

Change in Forest Ecosystem

Change in Composition of Plant Species and Phenology

Forest degradation has taken different shapes, particularly in the open forest communities, based on outcome of the human activities such as overgrazing by cattle, over-exploitation (for firewood or timber), repeated fires, or natural causes such as attacks by insects, diseases, plant parasites and the declining annual rain. In most instances, degradation not only shows a decrease in the area of forest vegetation but also results in a gradual reduction of the total biomass, change in species composition and soil degradation. Unsustainable logging practices have also contributed to this phenomenon since removal of the mature trees has not been accompanied by their regeneration and replanting and continued drought like situation persists. The local community has also observed changes in composition of plant species which has emerged from Focal Group

Discussion (FGD) and micro-plan of the Forest Department. Further, the plantation drives conducted during the implementation of various afforestation projects brought about change in the plant species composition in the district. According to the villagers, it has replaced traditional Kikar (*Acacia nilotica*) with Vilayati kikar (*Prosopis chilensis*). According to Labrada (2006) the canopy of *Prosopis* spp. takes over the ground cover, making it difficult for other grasses to grow. Pods of this leguminous plant have high sugar content, which causes teeth decay in animals ingesting them. *Prosopis juliflora* foliage has allelopathic effects on seed germination and seedling growth of Bermuda grass (*Cynodon dactylon*) cultivars of *Zea mays,* four cultivars of *Triticum aestivum* and *Albizia lebbek* (Prasad, 2006). Further, specimens of several timber-yielding species have died out in the area; a few which have been left out are either diseased or infected by termites and other insects exhibiting copies deforestation and stunted growth. The observation of sample villagers collected through FGD on changes in composition of plant species during the past six decades is compiled and presented in Table 1 below:

Table 1: Changes in composition of plant species during the past six decades

S. No.	Local Names	Scientific Names	Habit	Uses	1950[1]	1991[2]	Current Status[1]
1.	Ronj	*Acacia leucophloea* Willd.	Tree	Fuelwood, medicine	Common	Rare	Very rare
2.	Khairi	*Acacia senegal* Willd	Tree	Edible (Gum), fodder	Common	Rare	Non-existent
3.	Kikar, Babool	*Acacia nilotica* (Linn.) Delile	Tree	Fodder, fuelwood	Common	Common	Rare
4.	Ullu Neem	*Ailanthus excelsa* Roxb.	Tree	Fodder, fuelwood, medicine	Common	Common	Rare
5.	Siris	*Albizzia lebbek* Benth.	Tree	Fodder, fuelwood, medicine	Common	Rare	Non-existent
6.	Dhok	*Anogeissus pendula* Edgew	Tree	Fuelwood and edible (Gum)	Common	Rare	Very rare
7.	Neem	*Azadirachta indica* A. Juss.	Tree	Fodder, fuelwood, medicine	Common	Common	Rare
8.	Hingot	*Balanites aegyptiaca* (Linn.) Delile	Small tree	Fodder	Rare	Very Rare	Non-existent
9.	Dhak	*Butea monosperma* (Linn.) Kuntze	Tree	Fuelwood, fodder, medicine, Dye	Common	Rare	Very rare
10.	Amaltas	*Cassia fistula* Linn.	Tree	Medicine	Common	Rare	Very rare
11.	Gugal	*Commiphora wightii* Bhandari	Small Tree	Medicine	Common	Very rare	Non-existent
12.	Lasura	*Cordia dichotoma* Forst.	Small Tree	Edible vegetable and Pickle)	Common	Very rare	Non-existent

[1]*Source* – Focus Group Discussion,
[2]*Source* – Aravalli Micro-Plan

Table 1: Contd.

S. N.	Local Names	Scientific Names	Habit	Uses	1950[1]	1991[2]	Current Status[1]
13.	Shisham	*Dalbergia sissoo* Roxb.	Tree	Timber, fuelwood	Common	Common	Rare
14.	Papri/Kanju	*Holoptelea integrifolia* Planch.	Tree	Medicine, fuelwood, fodder	Common	Rare	Very rare
15.	Karanji/Kanji	*Pongamia pinnata* Pierre	Tree	Uses for making shades	Common	Common	Very rare
16.	Jand/Saingri	*Prosopis cineraria* Druce	Tree	Medicine	Common	Common	Non-existent
17.	Jaal	*Salvadora oleoides* Decne	Tree	Veterinary medicine for camels	Common	Very rare	Non-existent
18.	Babul	*Acacia jacquemontii* Benth.	Tree	Medicine	Rare	Very rare	Non-existent
19.	Dhouli Mundi, Bui	*Aerva javanica* Juss. ex Schult	Herb	Medicine	Rare	Very rare	Non-existent
20.	Bansa	*Adhatoda vasica* Nees	Shrub	Medicine	Common	Rare	Non-existent
21.	Phogalo	*Calligonum polygonoides* Linn.	Shrub	Edible	Rare	Very rare	Very rare
22.	Karir, Kair	*Capparis decidua* Edgew.	Shrub	Edible (vegetables and pickles)	Common	Common	Rare
23.	Ber	*Ziziphus mauritiana* Lam.	Shrub/ small tree	Fruits	Common	Rare	Very rare
24.	Jhar Ber, Mallah, Beri	*Ziziphus nummularia* Burn. Wight and Arn.	Shrub	Fruits	Common	Rare	Very rare
25.	Kheef	*Leptaedenia pyrotechnica* (Forsk.) Decne	Shrub	Fodder	Common	Very rare	Non-existent
26.	Kateli	*Solanum surattense* Burm.f.	Shrub	Veterinary Medicine	Common	Very rare	Non-existent
27.	Gokhru	*Tribulus terrestris* Linn.	Herb	Medicine	Common	Rare	Very rare
28.	Ollakata/ Shatavar	*Asparagus racemosus* Willd.	Herb	Medicine	Common	Very rare	Non-existent
29.	Ranpyaz	*Urgenia indica* Kunth.	Herb	Medicine, veterinary medicine	Common	Very rare	Very rare
30.	Vilayati Kikar	*Prosopis chilensis* Stuntz	Tree	Fuelwood	Non-existent	Very rare	Common
31.	Ailanthus	*Ailanthus altissima* (Mill.) Swingle	Tree	Fuelwood, fodder	Non-existent	Very rare	Common

The study of plant phenology provides knowledge about the pattern of plant growth and development as well as the influence of environment and seasonal variation on flowering and fruiting behaviour (Zhang *et al.*, 2006). Thus, besides observing the change in species composition, phenotypic variation has also been noticed by the local communities especially at the bud initiation stage, foliage formation, flowering, fruiting, colouring, leaf-fall and variation of leaves size, etc. (Table 2). Similar type of phenological changes have also been reported in some of the species, viz. *Ziziphus nummularia* and *Cordia* spp. by Yadav and Gupta (2009) while studying Sariska Tiger Reserve in north-eastern Rajasthan where the leaf flushing is one month shorter per episode.

Table 2: Change is phenology in some of the species

Species	Phenological variations observed
Jhand *Prosopis cineraria* (L.) Druce	• The tree currently bears fruits in August and September as against its fruiting in May and June a few decades back. • There are more thorns on the trees now as against their more leafy nature earlier with a dense canopy.
Lasura *Cordia dichotoma* Forst.f.	• Fruiting in August and September as against April-May earlier.
Kair *Capparis decidua* Edgew.	• Nearly 40% of the plant population now fruits twice a year in April-May and October-December. Earlier, *Kair* fruits were seen only during April-May.
Jhaad Beri *Ziziphus nummularia* (Burm.f.) Wight & Arn.	• Leaf has become perceptibly smaller in size.
Neem *Azadirachta indica* A. Juss.	• The fruits of the tree (*Nimbori)* present an increasingly bitter taste rendering it inedible.
Vilayati Kikar *Prosopis chilensis* Stuntz.	• Leaf has become perceptibly smaller in size.

Changes in Wildlife Composition

The change in plant species composition has also changed the wildlife composition of the region. Animals and birds like hyena, deer, wolf, vulture, leopard, fox and black buck are almost becoming extinct from the area. As per the villagers' observation, during 1950s to 1970s the area was frequented by tigers, panthers, leopards, black bucks, chinkaras, rabbits, eagles, fox and deer. These animals were sometimes even spotted in the vicinity of the villages. The villages were also inhabited by a number of birds. The then dense forest cover played an important role in providing sufficient food, water and shelter to the animals and birds. Increased human interference for mining, industrialization, farming and exploitation for fodder and changes in climatic conditions has affected the wildlife population in the region to a great extent. The forest density has been reduced significantly over the past few decades which has adversely affected the natural food chain. Furthermore, the shrinking water bodies have resulted in inadequate drinking water for animals. All in all, a major decline has been recorded in the population of wildlife in the area. As per the villagers observation the diminishing rainfall during last 30-40 years has resulted in major water bodies in the region getting dried up. The shrinking source of drinking water for animals has forced them to permanently migrate from the area. A summary of changes in status of major animal and bird species in the area over the past few decades has been presented in Table 3.

Table 3: Change in wildlife composition in the area

S. No.	Names of Species	Local Names	Binominal Names/ Families	Status		
				1950	1991	Current Status
1.	Hyena	Geedar	*Hyaenidae* canis aureus	Common	Very rare	Non-existent
2.	Deer	Hiran	*Cervidae muntiacus muntjak*	Rare	Very rare	Non-existent
3.	Wild Cat	Bidhav	*Felis silvestris*	Common	Rare	Very rare
4.	Wolf	Bhedia	*Wild Canid Canis lupus*	Rare	Extinct	Non-existent
5.	Blue Bull	Nilgai	*Boselaphus tragocamelus*	Common	Common	Rare
6.	Rabbit	Jangli Khargosh	*Leporidae lepus nigricollis*	Common	Common	Very rare
7.	Porcupine	Sheh	*Hystrix indica*	Common	Very rare	Very rare
8.	Snakes	Saanp	*Snakes*	Common	Rare	Very rare
9.	Eagle	Cheel	*Aquila rapax*	Common	Common	Very rare
10.	Vulture	Giddh	*Accipitridae (Aegypiinae) Cathartidae*	Common	Common	Non-existent
11.	Wild Dog	Jarak	*Cuon alpinus*	Common	Rare	Very rare
12.	Leopard	Cheeta	*Panthera pardus*	Very Rare	Extinct	Non-existent
13.	Wild Boar	Jangli Suar	*Sus scrofa*	Common	Rare	Very rare
14.	Fox	Lomri	*Canidae vulpes bengalensis*	Common	Very rare	Non-existent
15.	Black Buck	Kala hiran	*Antilope cervicapra*	Rare	Extinct	Non-existent
16.	Crow	Kaowa	*Corvus corvus*	Common	Common	Very rare

Source: Data available from Aravalli Micro-Plan prepared in 1991 by the Forest Department, However, the data of 1950 and current status is based on FGD

CLIMATE VARIATION

Change in Rainfall Pattern

The district experiences extreme arid to semi-arid climate, characterized by hot-dry and windy summers, cold winters and humid-warm monsoon months. In addition to low quantum, the erratic rainfall aggravates the problem to a much higher degree. About 80-85% of total annual rainfall is brought about by the south-western monsoon, which is spread over between July and September. Weather remains almost dry throughout October and March, except a few occasional light showers. The rainfall pattern is most vulnerable to climate change. It has plunged from 596 mm in the 1990s to 421 mm during the last 7 years, a decrease of approximately 29%. However, average rainfall in the past 30 years was around 498 mm, which is much lower than the average rainfall of 1990s (Figure 1). In 1990s, average rainfall was comparatively high being over 600mm in the years of 1995-97 and 1999. The local inhabitants viewed that the ill effects caused by erratic and decreasing rainfall are: (i) declining regeneration of endemic species in forest, (ii) decreased forest density, (iii) depleting groundwater table resulting in scarcity of water for irrigation and drinking, (iv) non-sowing of some major pulse crops like *chana, moong, moth* which require adequate rainfall, (v) depleting traditional water bodies, like ponds, *johads*, etc., resulting in scarcity of drinking water for the cattle, (vi) depleting water bodies and forest density has also resulted in decreased wildlife.

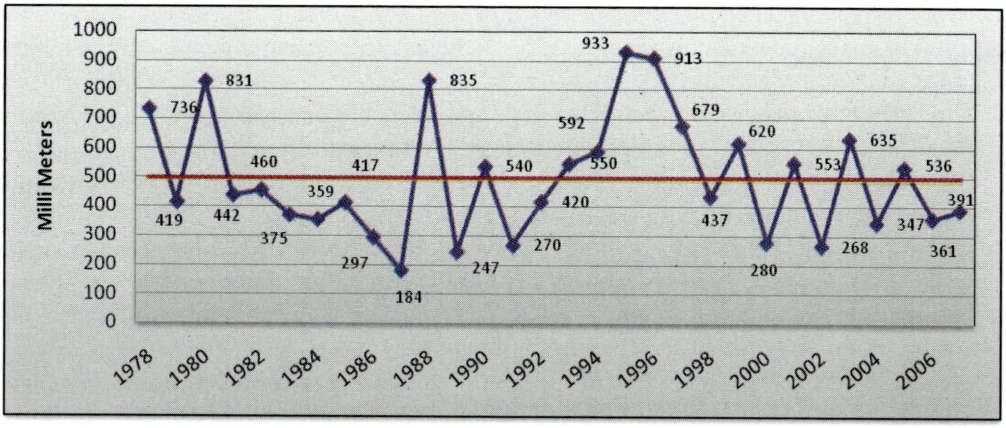

Fig. 1. Average rainfall in the district.
(*Source:* Aravalli Micro-plan and District Statistical Handbook)

Change in Groundwater Level

Around 45% of the geographical area of the district falls within the 10-20 m depth of watertable range. In the past three decades (June 1974 - June 2004), the water table is receding at an alarming rate of 30cm per annuum in the district due to overdrafting of groundwater causing failure of shallow tubewells. Looking at 2007 figures, the depth of water table varies from 13.52 m in the Rewari Block to 39.32 m in Khol Block (Figure 2). The decline in water level

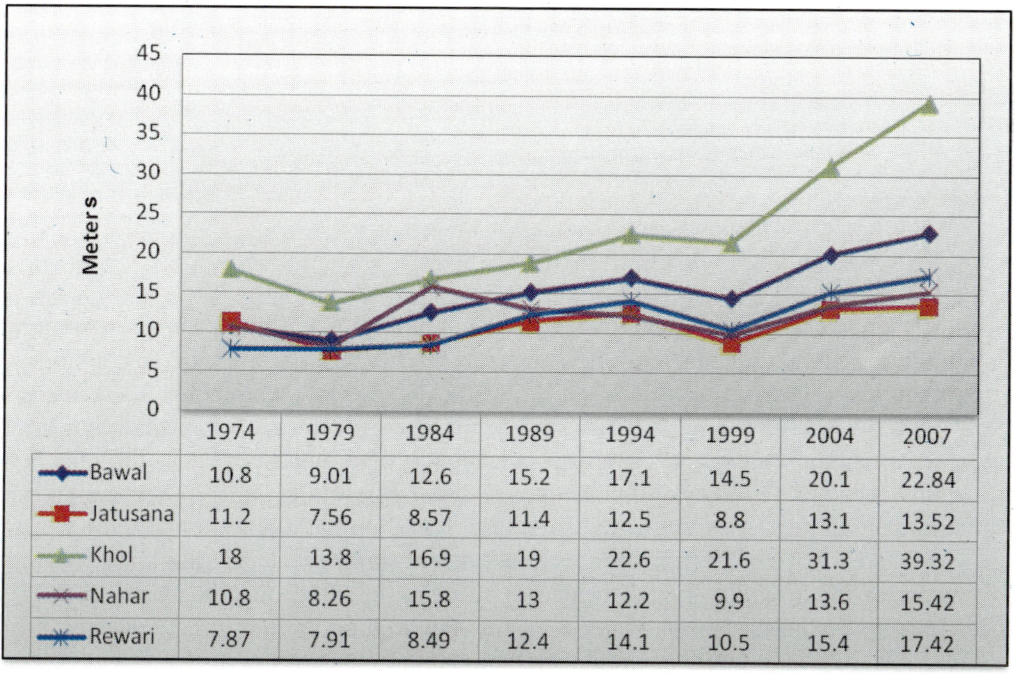

	1974	1979	1984	1989	1994	1999	2004	2007
Bawal	10.8	9.01	12.6	15.2	17.1	14.5	20.1	22.84
Jatusana	11.2	7.56	8.57	11.4	12.5	8.8	13.1	13.52
Khol	18	13.8	16.9	19	22.6	21.6	31.3	39.32
Nahar	10.8	8.26	15.8	13	12.2	9.9	13.6	15.42
Rewari	7.87	7.91	8.49	12.4	14.1	10.5	15.4	17.42

Fig. 2. Block-wise level of groundwater table.
(*Source:* Comprehensive District Agriculture Plan, Rewari)

is much drastic (62 cm/annuum) in Khol Block of the district due to over-exploitation of water, being good in quality and lesser recharge due to low quantum of rains in the block, and minimum in Jatusana (6.0 cm per annum) due to recharge by seepage of Jawahar Lal Nehru canal in the area. The water table decline between 1994-2007 is less than 45 cm per annum in the district as a whole and 1.29 m/annuum in the Khol Block of the district.

Inadequate and erratic occurrence of rainfall has placed excessive pressure on groundwater for irrigation. The increased use of advanced irrigation equipment has further compounded the situation by fostering a water crisis. In the 33 year period (1974-2007), the average depth of groundwater (pre-monsoon) has receded to 22m from 12m below the ground level (Figure 3). Thus, the average groundwater depth has plunged almost 10m in the last two decades. There was a respite from the constant decline during 1995-1999, when sufficient quantum of rainfall was received to recharge groundwater in the area. The water table declined from 1994-2007 to less than 45 cm per annum in the district. The situation in some parts of Jatusana Block is comparatively better; due to recharge by seepage from JLN canal.

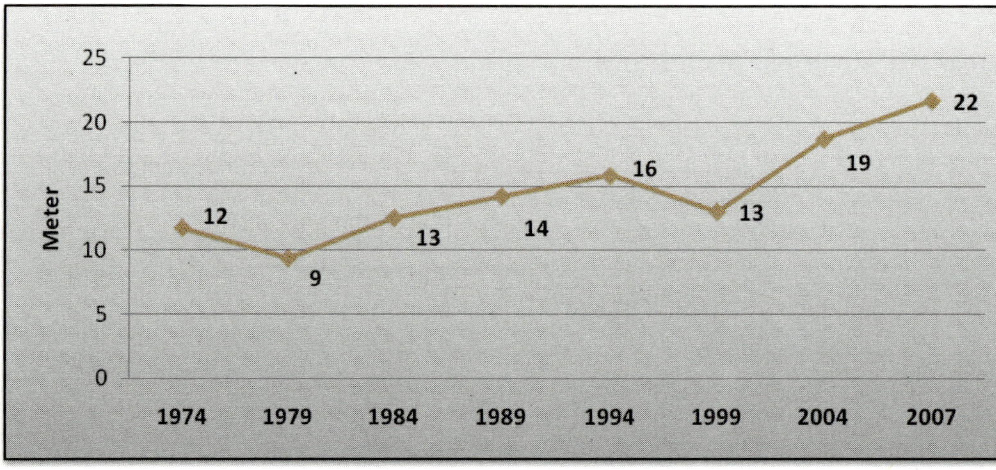

Fig. 3. Average depth of groundwater In the district.
(*Source:* Comprehensive District Agriculture Plan, Rewari)

The irrigation water forms a very scarce input and has become a major hurdle in the growth of agriculture since the major source of irrigation is tubewell which has overexploited groundwater during recent years. The normal rainfall in the district is also low, not enough to recharge the depleted groundwater. Therefore, it is essential to manage groundwater efficiently with the use of advanced irrigation innovations like drip irrigation system and ensure that the rain falling over a region is tapped as fully as possible through water harvesting, either by recharging it into the groundwater aquifers or storing it for direct use. Based on the discussion with the villagers, it has been shared that until a few decades back, a few surface water bodies were capable of supplying water for irrigation, which helped in reducing pressure on groundwater. Today, no surface water bodies hold adequate water even for consumption by the cattle. However, though the respondents were fully aware about the alarming rate of groundwater depletion and excessive pressure on groundwater for irrigation, they did not highlight any concern to take constructive measures to recharge groundwater and use advanced irrigation techniques which would utilize less water. This lack of sensitivity seems reinforced by successful changeover from traditional rainfed crops like gram and pulses to irrigated crops like wheat, mustard and vegetables.

Thus, it becomes imperative for the farmers not only to adopt techniques for efficient water use but also pay attention towards rainwater harvesting and its recharging, preferably to be taken up on a community level. Out of the several methods of water harvesting, open well method is more appropriate under the present conditions. This proposed recharge measure will not only check decline in water level, but will result in rise of water level. It will also bring additional land under irrigation.

Agriculture Sector

Shift in cropping pattern

The region has witnessed a major shift in cropping pattern, productivity and area under different crops over the past few decades. Mustard and wheat are the major crops of Rabi, and bajra and guar are the main crops of Kharif. Cultivation of crops like gram, *moong* and *moth* has minimized. Out of the net sown area in the State, 82.5% area was sown more than once in 2005-06 as compared to 34.4% in the year 1966-1967. As per the official figures available with the Department of Agriculture, average cropping intensity in the district is around 145%. However, there is a block-wise deviation in average, dependent on availability of irrigation resources and quality of groundwater in the block. Nahar and Jatusana demonstrate an above average cropping intensity, reason being Jawahar Lal Nehru Canal and higher water table compared to other blocks (Table 4). On the other hand, Khol demonstrates lowest intensity as a result of low groundwater table.

Table 4: Cropping intensity in the district

Names of the Blocks	Geographical Area (ha)	Net Sown Area (ha)	Gross Cropped Area (ha)	Cropping Intensity (%)
Rewari	32,213	27,080	37,580	138.7
Bawal	31,329	23,158	31,489	135.9
Khol	28,927	23,167	29,937	129.2
Nahar	26,782	20,114	33,372	165.9
Jatusana	31,679	26,342	42,054	159.6
Total	**1,50,930**	**119,869**	**174,432**	**145.5**

Source: Comprehensive District Agriculture Plan, Rewari

In the sample villages almost hundred per cent cultivable land is put to use, leaving no land for open grazing. Factors fostering cropping intensity are improved irrigation facilities, high-yielding-variety seeds, fertilizers, growing number of nuclear families, inadequate alternate source of income and decreasing land-holding. A family with small land-holding has no other option than making maximum use of available piece of land. Increased land use is also creating an imbalance in nature, fostering climate change.

Changes in Area, Productivity and Yield of Mustard and Wheat

As far as cropping intensity of major crops is concerned, the area and productivity of wheat has increased significantly. It occupied 47,000ha with an average productivity of 42.2 q/ha in the year 2006-07 whereas it was only 39.4 thousand ha in 1990-91. However, the average yield per hectare has augmented by 12.5% since 1991 (Figure 4). Similarly, the area under mustard cultivation has increased by 31.5% (71,000 ha) since 1991. As per the available data since past 17 years, the average yield of mustard is around 14.63 quintals/hectare. However, the figure demonstrates a below average yield in 1991-92, 92-93 95-96, 2002-03, 2005-06 mainly due to variation in climatic conditions. In the year 1996-97, the yield was 18 quintals/hectare which is much higher than the 17-year average 14.63 (1991-2007). The factors behind the increase of area and productivity of wheat are availability of high-yielding variety of seeds. However, limited options are available in terms of commercial crops, effective price support and public stocking policies, as well as utilization of residue as fodder. As against this, in case of mustard factors like high commercial value, requirement of less irrigation, intercropping of mustard with potato and multiple cropping with *Carissa carandus* and easy availability of improved seeds are responsible for a higher yield.

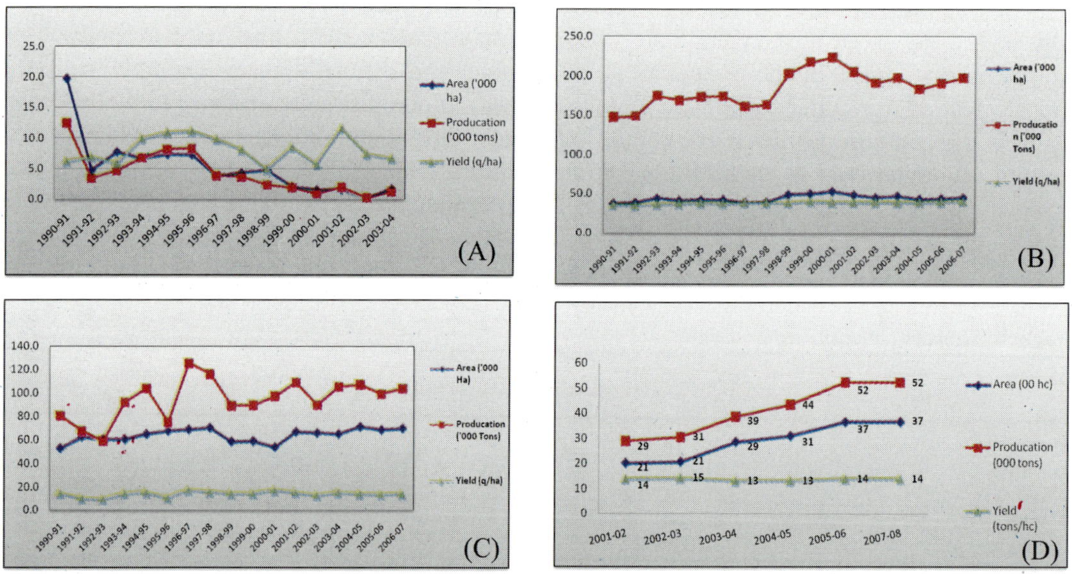

Fig. 4: Changes in area, production and yield of different crop in the district, **A:** Gram, **B:** Wheat; **C:** Mustard, **D:** Vegetables.

Change in Area, Productivity and Yield of Gram

During the years 1990-1991 to 1991-92 gram cultivation declined by 75 per cent (Department of Agriculture). Only five thousand ha land was brought under gram cultivation in the district, which was 20 thousand/ha in the preceding year. It has witnessed a regular downturn through 1998-99. In the year 2002-2003 the area under this was only 500 ha. Thus, although one of the major Rabi crops in the area, a few decades back, gram is now on the verge of becoming an extinct crop in the district. The area under gram cultivation was nearly non-existent during the year 2004-2005. Till a couple of decades ago, barley, gram, *moong* and *moth* used to be the chief crops grown during Rabi season in the region. Low rainfall with its erratic distribution, scarce water resources for irrigation, poor groundwater quality coupled with arid and semi-arid conditions resulted in relatively low production during Rabi season. This has resulted in a major shift in the cropping pattern. After 1996, Rewari has witnessed a major downfall in gram cultivation.

Changes in Area, Productivity and Yield of Vegetables

Figure 4 (D) brings out the growth of vegetable crops in terms of area under cultivation and production during the past seven years. However, the per hectare yield has remained almost static, around 14 tons per hectare. A growth trend is visible in area under vegetable cultivation since this is economically viable as compared to the traditional crops in relatively small land-holdings.

The current study also shows that the fertilizer consumption has grown over 150% in the past few years in the district (Figure 5). Besides irrigation, fertilizer is the second most important input for cultivating high-yielding varieties. During the past two decades fertilizer consumption in the district has augmented from 12.07 tons to 31.39 tons with the growth rate of 1.14 tons per year. Depleting rainfall and limited source of irrigation have triggered high fertilizer consumption in the district. The FGD held with the villagers revealed that they are not concerned enough about balanced use of fertilizers, although they are aware that non-judicious use of inorganic fertilizers is increasing dependability of soil on them. Unplanned fertilizer application has also resulted in problems related to soil health and nutrient imbalance. This also indicates that no effective attempts are being made for soil testing and nutrient restoration. The situation is further compounded by the incidence of double-cropping that does not allow cultivable land to regain its nutrient balance naturally.

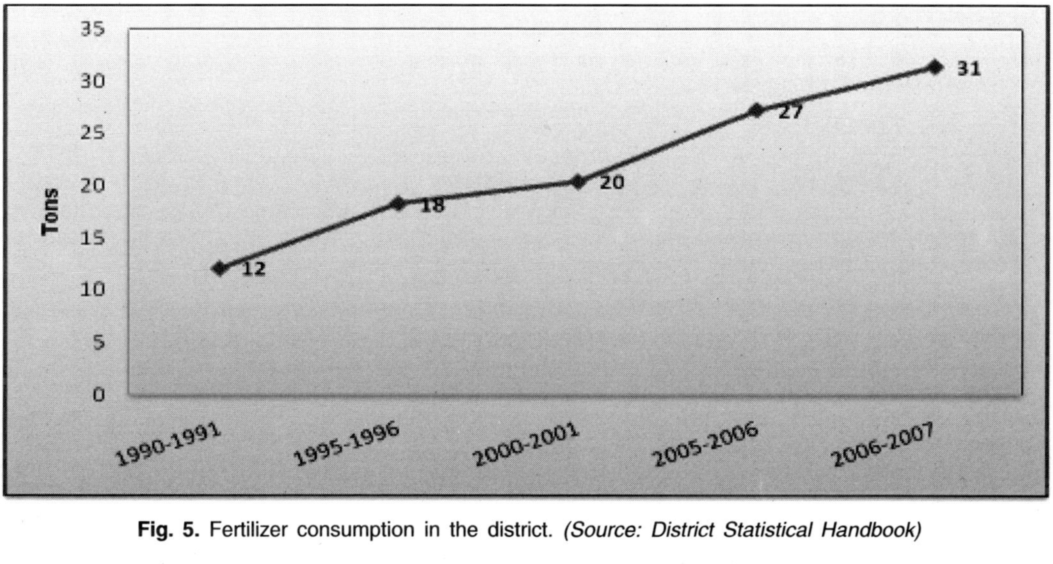

Fig. 5. Fertilizer consumption in the district. *(Source: District Statistical Handbook)*

CONCLUSION

The study reveals that the covered region has also witnessed the impact of climate change in terms of changing the composition of floral and faunal species. Some of the native species now have become extinct or have been replaced due to the invasion of alien species. Besides, phenological changes have also been observed by the inhabitants in some of the common species found in the region. With the change in floristic composition and decrease of forest density due to climatic variation have adversely affected the faunal population. The rainfall pattern has significantly changed over the last few decades which has resulted in adverse impact on natural regeneration, crop diversity and intensity. However, the modern inputs like improved irrigation facilities, high-yielding variety seeds, and fertilizers have increased the productivity of crops. The use of tubewells in irrigation practices has increased leading to over-exploitation of groundwater during recent years. These patterns have brought about significant changes in socio-economic status of the people and have led to the loss of traditional knowledge. Unplanned fertilizer application has resulted in problems related to soil health and nutrient imbalance. The changing soil and water quality are also a cause of major illness among the community.

Thus to mitigate the adverse impact of climate change on the ecosystem, the use of advance technologies for recharging of underground water sources and adoption of large-scale irrigation programmes, viz. sprinklers, drip irrigation and ridge and furrow irrigation can have a wider impact towards minimum use of groundwater. The use of sustainable agricultural practices and limited number of fertilizers should be promoted among the community. Also encouragement of plantation of native species under afforestation programme in the region will have a direct effect on conserving the floral and faunal diversity of the region. Besides, awareness about the climate change and its impact on the ecosystem need to be created among the local inhabitants to ensure their participation as the mitigation strategy. At the outset, there is a great need to study entire process of climate change in totality based on long-term data which would ensure that climate change has started affecting our biodiversity.

REFERENCES

Freebairn, D (1995). Did the Green revolution concentrate incomes? A quantitative study of research reports. *World Development* **23:** 265-279

Goldman, A & Smith, J (1995). Agricultural transformations in India and Northern Nigeria: exploring the nature of Green revolutions. *World Development* **23:** 243-263

Karen O'Briena, Robin Leichenkob, Ulka Kelkarc, Henry Venemad, Guro Aandahla, Heather Tompkinsa, Akram Javedc, Suruchi Bhadwalc, Stephan Bargd, & Lynn Nygaarda, & Jennifer Westa. (2004). Mapping vulnerability to multiple stressors: climate change and globalization in India. *Global Environmental Change* **14:** 303-313

Labrada Ricardo (2006). *Prosopis juliflora* and *Parthenium hysterophorus*, major invasive plants in tropical areas. Abstract. International Symposium on Biology, Ecology and Management of World's Worst Plant Invasive Species. 10-14 December, p. 36

Prasad, MNV (2006). Bioprospecting *Prosopis juliflora* (Sw) DC.–an invasive phreatophyte. Abstract. International Symposium on Biology, Ecology and Management of World's Worst Plant Invasive Species. 10-14, December, p. 39

Singh, US (2007). Various threats to Biodiversity and impact of climate change. International Biodiversity Day-2007: Souvenir, pp. 43-48

Srivastava, JPL & Kaul, RN (1994). Greening of the common lands in the Aravallis, Aravalli Project, Forest Department, Government of Haryana, Gurgaon

Yadav, DS & Gupta, SK (2009). Observation on the phenology of woody species of Sariska Tiger Reserve in North-eastern Rajasthan *Indian For* **135:** 1707-1714

Zhang, G, Song, Q & Yang, D (2006). Phenology of *Ficus racemosa* in Xishuangbanna, Southwest China. *Biotropica* **38:** 334-341

4

Protected Groves: A Traditional Way of Conserving Biodiversity in Garhwal Himalaya, Uttarakhand

TAPENDRA S BISHT, NEELAM BHATT AND AB BHATT

Department of Botany and Microbiology, HNB Garhwal Central University, Srinagar–246 174, Pauri, Uttarakhand
E-mail: bhattab2011@gmail.com

ABSTRACT

The phenomenon of beliefs in sacred and protected groves is ancient. The tracts of virgin forest harbouring rich biodiversity, protected by the local people based on the indigenous culture, and religious beliefs and taboos are called sacred groves. They are the repositories of rare and endemic species and can be regarded as the remnants of the primary forest left untouched by the local inhabitants and protected by them due to the belief that the deities reside in these forests. These groves harbour rich medicinal plant resources and play a significant role in the conservation of biodiversity. India has a long tradition of prudent use and wise conservation of all resources that are useful to the people.

The existence of sacred groves in India most likely dates back to an ancient pre-agrarian hunter-gathering era, and their presence has been documented since the early 1800s. Believing trees to be the abode of gods and ancestral spirits, many communities set aside sanctified areas of forest and established rules and customs to ensure their protection. It is also believed that the presiding deities administered punishment, often death, to individuals who violated the rules, and sometimes to the entire community in the form of disease or crop failure. As a result of this protection, preserved over countless years, sacred groves are now important reservoirs of food, fodder and medicinal plant wealth and also contribute towards regional as well as global biodiversity. These are the storehouses of medicinal plants valuable to village communities as well as modern pharmacopoea, and they contain wild relatives of crop species that can help to improve cultivated varieties. These groves provide for the water needs of nearby communities.

It is intended to build a comprehensive database documenting biodiversity status of sacred groves. Information such as culture and traditions, conservation history and stories associated with them would also be recorded. Development of such factual databases would support the establishment of strategies for conservation and protection of these unique heritage ecosystems. Such models would strengthen and support the national and regional biodiversity conservation programs elsewhere in developing nations where similar types of ecosystems exist.

Keywords: Biodiversity conservation, folklores, protected groves, sacred groves, taboos.

INTRODUCTION

In India as elsewhere in several parts of the world a number of communities practice different form of nature worship. One significant tradition of nature worship is that of providing protection to patches of forests dedicated to deities or ancestral spirits. These patches of forests are known as sacred or protected groves. The institution of sacred groves is very ancient and once was widespread in most parts of the world with over 50,000 sacred groves so far reported from different parts of India. These are rich heritage of India, and play an important role in the religious and sociocultural life of the local people. Being a variety of ecosystems by themselves, these perform all the ecological functions and threatened, endangered, and rare species find safe refuge in them. The groves are repositories of biological wealth of the nation with rare and endemic species as well. India is one of the world's top 12 megadiversity countries with rich variety of biological community types that includes coral reefs and alpine meadows, rainforests and desert scrubs. As many as 16 different major forest groups, which are further subdivided into 20 subgroups and 160 subdivisions (Champion and Seth, 1968a) have been identified here. The vegetation varies from tropical, evergreen forests on the west coast and northeast to alpine forests in the Himalaya in the north. Between these, the country has semi-evergreen, deciduous, sub-tropical broad-leaved, subtropical pine and subtropical montane temperate forests. Bamboos and Rattans also constitute important species occurring widely in the Indian forests.

Himalaya has been a perennial source of attraction, curiosity and challenge to human intellect throughout the ages. Uttarakhand possesses luxuriant and varied vegetation within the Himalayan region. Almost every plant has economic value from either a nutritional, aesthetic, or medicinal viewpoint. In fact, large percentage of crude drugs in the Indian market comes from this Himalayan area (Badoni, 1990). According to available evidences, the Garhwal Himalaya alone has more than 3,500 species of flowering plants, most of which are in forest and alpine meadows (Rawat *et al.*, 2001). In India out of an estimated 15,000-16,000 flowering plant species, about 1,500 (10%) have already come under the various categories of threatened plants (Rao, 1994). The same number of species is used in Indian traditional system of medicine, i.e., 600 species with more than 8,000 herbal remedies in Ayurvedic, 500 species in Unani, and 550 species in phyto-pharmaceutical industries. Nearly thirty species from the Garhwal Himalaya have been listed in various categories under threat in the Indian Red Data Book (Nayar and Sastry, 1990) of which 24 are from high altitude alpine region.

Garhwal Himalaya in India commonly referred to as Dev Bhumi (land of the Gods) houses a number of important religious shrines like Badrinath, Kedarnath, Yamunotri and Gangotri, etc. besides the sacred confluence of five tributaries of holy Ganga. It is interesting to note that many a time, an entire landscape represented by a variety of species and ecosystems had been considered sacred or protected and conserved as such in pristine condition by forbidding the use of any resource from it. This strategy seems to be quite analogous to the present day's concept of species conservation through sanctuaries, national parks, and biosphere reserves. The phenomenon of beliefs in protected or sacred groves is ancient. The tract of virgin forest harbouring rich biodiversity, protected by the local people based on grounds of indigenous cultural and religious beliefs is called protected or sacred grove. Being repositories of rare and endemic species and can be regarded as the remnants of the primary forest left untouched by the denizens due to the belief that the deities reside in these forest. The examples of forest community conservation are the well known *Chipko, Raksha Sutra,* and *Maity Movements*. These are among the pioneer social movements towards the conservation and regional environment initiated in the Central Himalaya. All these movements have one common objective, i.e., to conserve the serenity of the environment by maintaining the natural forest wealth. The *Chipko* and *Raksha Sutra* are aimed to prevent the forests by deforestation whereas the *Maity Movement* aimed at raising new forest through plantation programs. Many traditional societies all over the world revered and worshipped nature and considered certain plants and animals as sacred. Some communities also followed the practice of setting aside certain patches of land or forest as "sacred groves" dedicated to a deity or village God, protected and worshipped. In India sacred groves are found all over the country and abundantly along the Western Ghats, West coast and in several parts of Kerala, Karnataka, Tamil Nadu, Maharashtra, Madhya Pradesh, Rajasthan, Orissa and Himachal Pradesh.

Protected groves in different states are locally known by different names. Various sectors and tribes in India made efforts at their own to preserve the biodiversity at local and regional levels. Many plants and animals are considered sacred from historical times for e.g. peepal (*Ficus religiosa*), banyan (*Ficus bengalensis*) and khejadi (*Prosopis cineraria)* are traditionally revered and, therefore, rarely cut. More than a hundred such species of plants are considered sacred viz., sandalwood (*Santalum album)*, betel nut palm (*Areca catechu)*, aam (*Mangifera indica),* coconut tree (*Cocos nucifera)*, juniper (*Juniperus* sp.), champaka (*Michelia champaca),* lotus (*Nelumbium speciosum),* tulsi (*Ocimum tenuifolium)*, dub (*Cynodon dactylon*), and silangi (*Osmanthus fragrans*), etc. Such traditional and cultural attitudes, though based on religious faith, have made significant contributions to the protection and propagation of various species of trees and plants in India and especially in Uttarakhand.

INVENTORY OF SACRED AND PROTECTED GROVES

The existence of sacred or protected groves in India dates back to an ancient pre-agrarian hunter-gathering era, and their presence has been documented since the early 18th century. Believing trees to be the abode of Gods and ancestral spirits, numerous communities set aside sanctified areas of forest and established rules and customs to ensure their protection. These rules varied from grove to grove but often prohibited the felling of trees, the collection of any material from the forest floor, and the killing of animals. As a result of these protective restrictions, preserved over countless years, sacred groves are now important reservoirs of biodiversity, being often the last refuge for endemic and endangered plant and animal species. They are storehouses of medicinal plants valuable to village communities as well as modern pharmacopoeia, and contain wild relatives of crop species that can help to improve cultivated varieties. These also provide for the water needs of nearby communities besides improving soil stability, preventing topsoil erosion and providing irrigation for agriculture in drier climates.

A vast diversity is witnessed amongst India's protected groves. Some contain only a few trees, whereas the others are hundreds of acres in size. Sometimes groves overlap with larger forested areas, in contrast to others which exist as islands in open plains or desert. Even their names vary from region to region. In several groves, villagers perform annual rituals and ceremonies to appease the presiding deity and ensure the well-being of the community. There are also large, "pan-Indian" groves that involve people from many parts of the country. In some groves, all forms of resource extraction are strictly prohibited, whereas in others people may collect material such as fallen branches and dried leaves from the forest floor, or bark, seeds and fruits from the trees.

The role of sacred groves in the conservation of biodiversity has long been recognized (Kosambi, 1962; Gadgil and Vartak, 1975; Haridardan and Rao, 1985; Khan *et al.* 1997). All forms of vegetation in the sacred groves are supposed to be under the protection of the reigning deity of that grove, and the removal of even a small twig is considered to be a taboo (Vartak and Gadgil, 1973). It is believed that sacred virgin forests date back to several thousands of years when human society was in the primitive state. Gadgil and Vartak (1976) have traced this historical link of the sacred groves to the pre-agricultural, hunting and gathering societies. Hughes and Chandran (1997) have presented an overview on the distribution of sacred groves around the earth in Asia, Africa, Australia, Europe, and America.

In India, the earliest documented work on sacred groves is that of Brandis (1897). Fifty years later Gadgil and Vartak (1973) conducted floristic and ethnobotanical studies on the sacred groves of Maharashtra. Burman (1992) has reported the existence of sacred groves all along the Himalaya from the northwest to northeast, western Himalaya of Kumaun and Garhwal, Darjeeling and Meghalaya. Ramakrishnan (1996) also reported sacred groves from different parts of India, known by different names given to them in ethnic terms. Several scholars have been working on conservation of sacred groves through socio-cultural practices in different parts of India (Gadgil and Vartak (1975, 1976), Boojh and Ramakrishnan (1983), Khiewtam and Ramakrishnan (1989), Rodgers (1994), King (1997), Tiwari *et al.* (1998).

However, little information is available on the conservation of the biodiversity through protected and sacred groves in Garhwal Himalaya, Uttarakhand. People of Garhwal follow ancestral worship and animism in the form of deity worship, with the central focus of worship on forest patches which signify sacred (protected) groves. The identified protected groves range from a few to several hectares in area at different natural ecosystems. The protected groves help in conservation of biodiversity. But unfortunately, due to population explosion and various encroachment activities, protected groves have also become degraded although the extent of degradation in the groves is less as compared to other forests. Degradation of groves not only signifies loss of rich and relict vegetation but also the loss of rich cultural diversity. Therefore, there is an urgent need to make an extensive inventory of the groves, their biodiversity and ethnobotanical importance in Uttarakhand Sector of Himalayas and analyse the role of associated cultural and religious beliefs, and their conservation.

Affection towards nature was a zoolatry (worshipping of animals), totem (considering plants and animals sacred), etc, which in turn led to a sort of prudent conservation. Religious beliefs, traditions and customs of Indians bear an allegiance in restricting the exhaustive use of natural resources.

CLASSIFICATION OF PROTECTED GROVES

The protected groves in India can basically be classified as *traditional protected groves* and *temple groves.* In a traditional protected grove the village deity is said to reside and is represented by an elementary symbol. In the later category a grove is created around a temple and conserved. Sometimes the groves are also found around the burial or cremation grounds as the relatives of deceased usually plant a tree in the sweet memory of departed soul. Besides, several protected groves also came in existence as a result of awakening towards environment by various communities. Several protected groves of Rajasthan to mention, are good example of environmental awareness.

ECOLOGICAL SIGNIFICANCE

The protected groves are important from the ecological viewpoint. These help in conserving the biodiversity. They are often the last refuge of endemic species in the geographical region. Recharge of aquifers, viz., ponds, springs and streams is also associated with groves well conserved. Similarly, a grove ameliorates soil fertility, helps in soil conservation and prevent its erosion. The large protected groves regulate and support the elimate of the vicinal locality. A rich faunal wealth is always associated with rich plant diversity since protected groves and dense forests provide the suitable habitat for wild life as well.

DISTRIBUTION OF PROTECTED GROVES IN INDIA

Since there has been no comprehensive survey of protected groves in India, their exact number and area are unknown. At least 13,720 protected groves have been reported in several parts of the country, but experts estimate that the actual number may exceed far 50,000. The area of these groves range from a few square meters to several hectares situated in different altitudinal gradients and natural ecosystems which helps in conservation of biodiversity. Unfortunately, due to population explosion, various anthropogenic pressures, and development activities, sacred groves have also become the victim of encroachment and exploitation, though the extent of degradation in the sacred groves is less when compared with the other forests. Degradation of groves signifies loss of rich vegetation as well as the loss of rich cultural diversity.

In India, the sacred groves are found all over the country and are in abundance along the Western Ghats in the states of Kerala and Karnataka. The following table gives the total number of sacred groves documented so far in India (Table 1).

PROTECTED GROVES OF UTTARAKHAND (GARHWAL HIMALAYA)

Uttarakhand is an area of great natural beauty with forests, lush valleys and the snow-covered Himalayan peaks that include the world famous Nanda Devi Biosphere Reserve. It is a favourite tourist destination, attracting national and international visitors. Apart from tourism, the economy of Uttarakhand is mainly based on agriculture. The industrial sector is very limited. Some measures are being taken to develop the region's hydroelectric potential. Along with major rivers and their tributaries about 45 large or small scale dam projects are either completed or are under the process of development. viz., Maneri Bhali II, Lakhwar Dam, Vishnu Ganga Project, Kotli Bhel Hydroelectric Project and many more. These projects on the one hand are reflection of overall

Table 1: Statewise documentation of protected (sacred) groves in India

States	Local terms used for Sacred Groves	No. of documented sacred groves
Andhra Pradesh	-	750
Arunachal Pradesh	*Gumpa* Forests (Sacred Groves attached to Buddhist monestries)	65
Assam	*Than, Madaico*	40
Chhattisgarh	Sarna, Devlas, Mandar, Budhadev	600
Gujarat	-	29
Haryana	-	248
Himachal Pradesh	*Devbhumi*	5,000
Jharkhand	*Sarana*	21
Karnataka	*Devara Kadu*	1,424
Kerala	*Kavus*	2000
Madhya Pradesh	*Devkot, Matikot, Devsthali, Budhadev*	21
Maharashtra	*Devrais*	1,600
Manipur	*Gamkhap, Mauhak* (sacred bamboo reserves)	365
Meghalaya	*Law Lyngdhoh*	83
Orissa	*Jahera, Thakuramma*	322
Pondicherry	*Kovil Kadu*	108
Rajasthan	*Orans, Kenkris, Jogmaya*	9
Sikkim	*Gumpa* Forests	56
Tamilnadu	*Kovil Kadu*	503
Uttarakhand	*Deothan Bugyal* (sacred alpine meadows)	25
West Bengal	*Garamthan, Harithan, Jahera, Sabitrithan.*	670

* Courtesy Malhotra *et al.* (1991).

development of the region but excessive disruption of water flow in the river system tends to cause biodiversity loss due to several hydrological aspects and fragile geomorphic features of the Himalaya (Gaur, 2007).

There has been no systematic effort to document the protected groves of Uttarakhand. At the outset an exhaustive survey of six districts of Garhwal was undertaken to identify the protected groves. According to a preliminary survey conducted in district Rudraprayag the smallest district of Garhwal Commissionary, there are more than 20 protected groves identified so far. Of these a few protected groves viz., Banjyani, Hariyali Devi, Jakh devta, Kodima van Satoli, Mathinakhal and Umra Narayan groves are well documented (Plates 1, 2). The prominent sacred groves of other districts include Binsar, Kot, Pabau, Nandasain and Tarkeshwar in district Pauri; Bhojwasa, Chirwasa, Dakhtya and Dev van in Uttarkashi; Ansuya devi, Dewal, Khalla van in Chamoli and, Dobakoti, Ghoghas, Jald van, Kandar, Karuna van, Jurana van, Kaintholi and, Kukuru Shah protected groves in district Tehri (Plate 3). An extensive survey is required to identify all the protected groves of not only these districts but at the state level and the number would be very high. The information given in the present paper is based on some of these sacred groves visited frequently during 2006-09.

(a)

(b)

Fig. 1a and 1b. Umra Narayan Temple (above) with Adjoining Protected Grove (below) in District Rudraprayag.

(a)

(b)

Fig. 2a, 2b. Mathinakhal Protected Grove (above), Hariyali Devi Protected Grove (below) in District Rudraprayag.

(a)

(b)

Fig. 3a, 3b. Panoramic view of Karas (above) and Ghoghas Protected Grove below in District Tehri.

FLORISTIC AND FAUNAL ELEMENTS IN THE GROVES OF UTTARAKHAND HIMALAYA

For the study, six protected groves identified during our survey of Garhwal Himalaya were selected of which three fall in district Rudraprayag and remaining three in district Tehri. The protected groves of district Rudraprayag in general represent wet temperate vegetation whereas the groves falling in district Tehri represent subtropical to dry temperate. Since the climatic and topographic features of these protected groves differ moderately, their impact on the flora of a particular grove is obvious. To give an overall picture of the groves of Garhwal Himalayan Sector, the floristic information has been compiled together.

The dominant flora of these groves is characterised by the prominent vegetation comprising tree species of *Albizia procera, Buchanania lanzan* (katkafal), *Celtis australis, Dalbergia sissoo, Ficus carica, F. clavata, F. cunia, F. rumphii, Lyonia ovalifolia, Morus alba, Pinus roxburghii, Pyrus pashia and Toona ciliata* (tun) at lower altitudes *and Quercus semicarpifolia, Rhododendron arboreum* at higher elevations. The shrub species, viz., *Berberis asiatica, Bergera koenigii, Carissa spinarum, Cotinus coggygria, Desmodium cephalotes, Eupatorium adenophorum, Euphorbia royleana, Inula cappa, Lantana camara, M. paniculata, Nyctanthus arbor-tristis, Reinwardtia indica* (pheonli*), Rhus parviflora, Rubus ellipticus* (hissar), *Rumex nepalensis,* etc. have been collected from sub-tropical to temperate protected groves. *E. adenophorum* and *L. camara* species are the exotic ones that have invaded the area during the past 2-3 decades and are the chief biotic factors for the decline of floral diversity of the protected groves. The herbaceous layer underneath mainly comprises *Geranium wallichianum, Hedychium spicatum, Reinwardtia indica, Strobilanthus dalhausianus, Swertia chirayita, Thalictrum foliolosum,* in temperate areas. etc. Some species considered sacred by the hill people and reported from various protected groves of Garhwal are presented in Table 2.

The faunal wealth inhabiting the protected areas include the avian fauna viz., *Aguila heliacea* (Garrur), *Columba intermedia* (pigeon), *Corvus splendens* (crow), *Francholinus francholinus* (titar), *Glaucidium radiatum* (owl), *Ictinaetus malayensis* (eagle), *Lophophorus impejanus* (monal), *Picus canus* (woodpecker), *Passer domesticus* (sparrow), *Psittacula eupatria* (parrot), *Streptopelia decaocto* (Ghugati) and blackbird, bluetit, chaffinch, koel, magpie, pheasant, patridge, quail, robin, skylark, swallow, thrush, etc. Reptiles are represented by *Calotes versicolor* (lizard), *Naja naja* (cobra), *Varanus monitor* (Gaulu), etc. The protected forests have a rich mammalian fauna with leopard, wild boar, Himalayan bear, viz., *Canis aureus* (wolf), *Felis chaus* (jungle cat), *Melursus ursinus* (bhalu), *Panthera pardus* (bagh), etc. Similarly the prominent mammals frequently seen in the protected and neighbouring areas are: *Hystrix indica*, *Herpestis auropunctatus* (neola), *Lepus nigricollis* (hare), *Muntiacus muntjak* (barking deer), *Rhinolopus contrus* (bat), *Selenaractos tibetanus* (bear) and *Vulpes bengalensis* (Indian fox), etc.

RITUALS, TABOOS AND FOLKLORES

There are rituals, taboos and folklores associated with the protected groves and the deity. The interesting features are:

- Women are strictly prohibited from entering in some of the sacred forests.
- The people are not allowed to go beyond interactive zone.

Table 2: Sacred species identified in various protected groves of the Garhwal Himalaya

Plant species	Local names	Materials in use	Religious cultural value	Ecological features
Ocimum tenuifolium	Tulsi	Leaves	Worshipped as goddess	Insect repellant
Pinus roxburghii	Kulain	Fuel, timber and manure	Worshipped on the occasion of marriage ceremony	Tolerant fire adapted
Cedrus deodara	Deodar	Timber, oil of medicinal value	Planted in the holy premises	Susceptible to fire
Aegle marmelos	Bel	Edible fruit	Leaves offered to Lord Shiva	Grows on dry bare slopes
Ficus benghalensis	Barh	Fodder for goats	Worshipped on new moon day	Grows open areas
Ficus religiosa	Peepal	Fodder	Temple premises	Open non-forest area
Opuntia elatior	Nagphani	Edible fruit	Protect from evil spirit	Degraded area
Euphorbia royleana	Sulla	Fencing	-do-	-do-
Zanthoxylum alatum	Timru	Teeth cleaner and spices	-do-	Wide ecological amplitude
Dendrocalamus hamiltonii	Bans	Building material handicrafts	To carry dead bodies in hills	Degraded open area
Ziziphus numularia	Ber	Edible fruits, and fencing	Offered to Lord Shiva	Degraded forest
Datura stramonium	Datura	Narcotic	Flower fruit offered to Lord Shiva	Scrub
Cannabis sativa	Bhang	Edible seed and fibre	-do-	Grows up to 1500masl
Sesamum indicum	Til	Oil and medicinal value	Offered to sacred pyre	Kharif season crop
Amaranthus caudatus	Marchha	Edible grain	Offered to lord Ganesh	Kharif season crop
Fagopyrum tataricum	Phafar	Flour	Chappatis offered to Goddess	Kharif season crop
Cynodon dactylon	Doob	Fodder and medicine	Used to offer water to God, libation	Open areas
Erianthus munja	Babla	Fodder, thatching material	Rope and brooms used	Exposed calcarieous rocks
Prunus cerasoides	Paiyan	Fuel	Leaves and wood, used in sacred pyre	Degraded forest in villages
Cedrella toona	Tun	Timber	Planted in temple premises	Associated sp. on sandy soil
Bauhinia vahlii	Malu	Fodder and edible seed	Planted in temple premises	Shaded area at lower altitudes
Quercus incana	Banj	Fodder, fuelwood and timber	Planted in temple premises	Climax species at higher elevation
Plumeria acutifolia	Temple tree	Medicine	Planted in temple precinct	Up to 1000 m elevation
Osmanthus fragrans	Silingi	Medicinal, highly aromatic	Flowers offered to God	1400-1800 m
Jasminum grandiflorum	Chameli	Used in perfumery, medicine	Flowers offered to deities	Wild in open scrub patches of the groves
Saccharum nurida	Babla	Fodder, thatching material	Rope and brooms used	Exposed calcareous rock

- Fetching fodder or fuelwood from sacred forest is strictly prohibited. The myth prevails that use of agricultural tools in any form (knife, sickle, etc.) on the plants or animals will be a step to hurt the sentiments of deity. The forest angels in turn are unhappy and their wrath may be exhibited not only in the form of turning the person speechless, mad or deformed but it can also lead to disaster in the family of the offender or even to the entire village.
- If someone making a pilgrimage comes across a snake in midway, he will not abort the journey but restart only after worshipping the '*heet*' God.
- One week before the worship of the deity the villagers stop taking onion, garlic, egg, meat, etc. The same holds true if anyone wants to make a trip to the sacred forest.

From Himalayan states of north India to down south in Tamil Nadu and Kerala, from Gujarat to North-East the entire landscape of India is full of forests and innumerable sacred groves protected and preserved for centuries by the local communities. It is only in the recent time that an impending crisis is looming large over their very existence. The wide scale urbanization coupled with deforestation has created havoc to the intricate ecological balance and is uprooting local cultures and traditions at a fast pace.

Despite the dreaded destruction and devastation, there is still a lot which can be protected and conserved. There have been several efforts to revive the degraded protected groves in some pats of India for instance; the Braj Foundation in Western U.P. has taken up the task of protecting, reviving and rejuvenating 48 sacred groves. The volunteers of Braj Rakshak Dal – the community mobilization wing of the Braj Foundation, keep a vigil to ensure that no further damage is done to these groves. The Eco-Task force and a few NGOs working on environmental aspects in Uttarakhand have taken up the task of rejuvenating the degraded groves also.

IMPORTANCE OF PROTECTED GROVES

The traditional Indian society recognizes individual species as objects of worship, based on accumulated empirical knowledge and their identified value for one reason or the other. The protected groves are multifaceted social institutions and symbolize the dynamic social forces linked with access and control over resources. These possess a great heritage of diverse gene pool of many forest species having socio-religious attachment and possessing medicinal values. Sacred groves are ecologically and genetically very important. They are the abodes of rare, endemic and endangered species of flora and fauna. The socially worshipped multipurpose tree, *Quercus* spp. (Oak) is an important fodder and fuelwood species. It is also considered to serve a variety of functions and is an important component of the mountain forest ecosystem. It helps in improving the soil fertility through efficient nutrient cycling, conserving soil moisture through humus build up in the soil and partly through a deeply placed root system which has root biomass uniformly distributed throughout the soil profile. The value of sacred groves is immense as they are good sources of a variety of non-wood products, fatty oils, species like pepper, cinnamon and nutmeg, medicinal plants, *etc*. The faunal wealth of sacred groves is also worth mentioning. Besides these, the amenity value, role of sacred groves in water conservation and their effect on microclimate of the region, *etc.* are also important and deserve special attention in future studies. The ethnobotanical importance of a few species present in these protected groves has been summarised in Table 3.

Table 3: Ethnobotanical importance of some significant species identified in various protected groves and neighbouring areas in Garhwal Himalaya

Botanical names	Vernacular names	Diseases/aliments	Parts used/modes of application
Aconitum atrox	Meetha Bish	Rheumatism, Paralysis, Dyspepsia, Phthisis and fever	Paste of rhizomes fried in ghee, powder of the purified rhizomes externally used.
Actaea acuminata	Mamira	Bronchial	Decoction of root
Anemone rivularis	Angeli	Food poisoning	Paste of leaves
Angelica glauca	Choru	Flatulence, colic	Root-stocks
Arisaema wallichianum	Meen erysipelas	Erysipelas and Scabies	Root-paste (externally)
Arnebia benthamii	Laljari balchari	Cuts and wounds	Juice of fresh roots
Asparagus filicinus	Jhirni	Sexual disability, urinogenital disorders	Powder of dried tuberous roots
Bergenia ciliata	Shilphori	Kidney stones, sores, jaundice	Decoction of roots/rhizomes
Caltha palustris	Kushnya	Abscess	Leaf juice
Dactylorrhiza hatagirea	Salampanja	Cuts and wounds	Powder of the root
Delphinium denudatum	Nirbishi	Contusions	Root paste
Dioscorea bulbifera	Genthi	Bronchial coughs	Tubers
Dioscorea deltoidea	Tairu	spermatonorrhoea	Rhizomes
Euphorbia hirta	Dudhibari	Piles	Entire plant with curd
Fumaria indica	Pitapapra	Fever	Juice of entire plant
Gentiana capitata	Bumlya	Urinary infection	Root decoction
Geranium wallichianum	Neenai	Dysentery and diarrhoea	Root decoction
Hippophae rhamnoides	Dhooplakkar	Cardiac trouble	Fruit juice
Megacarpea polyandra	Barmoola	Fever, stomachache	Root
Morina longifolia	Bishkandara	Snake-bite	Root-decoction
Origanum vulgare	Bantulsi	Whooping cough	Extract of leave
Paris polyphylla	Satwa	Diarrhoea	Root powder
Picrorrhiza kurroa	Kutki	Severe coughing, fever	Root powder

PRESENT STATUS OF HIMALAYAN PROTECTED GROVES

The protected groves are indicators of the phenomenon of ethno-environmental management. Our ancestors were fully aware that the natural resources that sustained them must be conserved for the sustenance of future generations. But, at present, fast growth of infra-structural facilities and on-farm activities is the prime cause of deteriorating quality status of the groves. As in the forest, many of the tree species are valuable timber species, they have been largely extracted for timber during the past few decades and thus, subsequently replacing the climax forests (Oak) to early successional pine forests. This has done considerable ecological damage in the region, making the soil more acidic and adversely affecting nutrient cycling and soil fertility. Growth of tourism industry is also deteriorating the faith towards deity and groves. Sacred groves are the victims of this grim tragedy. The groves located near the settlements are disappearing at a fast rate and only a few sacred groves are in their pristine condition. These are Umra Narayan, Hariyali, Kukurushah, Dobakoti, Tapovan, Binsar and Tarkeshwar in Garhwal Himalaya. Other groves are disappearing, as the vicinal forests are being cleared and utilized for construction and repairing of deity houses. Most of the temple groves are seen disappearing due to inevitable factors like animal grazing and human interference.

Sacred groves are a social institution, which permits management of biotic resources through people's participation. A scientific understanding of the sacred groves would be significantly important for designing strategies for rehabilitation of degraded landscapes, involving local people's participation, and training for promotion of traditional and social norms. There is a need for preservation, restoration and proper management of existing groves. Various traditional approaches to conservation of nature require a belief system, which includes a number of prescriptions and proscriptions for restrained resource use. These forestlands need proper conservation and protection by formulating consistent conservation strategies in order to save them from the verge of further degradation. Proper legislative support and specific policies should be provided. Mushrooming infrastructure facilities in the area are deteriorating the proper functioning of social institutions, which reflect that sacred groves are no longer getting the privilege they had in the past. Human interference should be regulated by encoding various indigenous practices along with scientific implications rather than only old religious prescriptions and proscriptions.

THREATS TO THE PROTECTED GROVES

The threats vary from one region to the other and even from one grove to the other. However, the overall common threats identified so far are:

i. The traditional belief systems which were fundamental to the concept of sacred and protected groves are gradually disappearing. These systems and their rituals are now considered mere superstitions.

ii. Protected groves in many parts of our country have been destroyed due to rapid urbanization and developmental interventions such as roads, railway tracks and dams including commercial forestry. Encroachment has led to the shrinkage of some of the largest groves in the country.

iii. Many groves are suffering due to '*Sanskritisation*' or the transformation of the primitive forms of nature worship into formal temple worship.

iv. Invasion by exotic weeds such as *Eupatorium adenophorum, Eichhornia crassipes, Lantana camara, Parthenium hysterophorus and Prosopis juliflora* is a serious threat to many groves of lowland areas.

v. Pressures due to increasing livestock and fuelwood collection have also added to the misery of these green oases.

MEASURES FOR CONSERVATION

The exercise of listing the protected and sacred groves has been highly rewarding. Some of the treasures of biodiversity preserved due to the correct attitude of people towards them have been highlighted. It is quite possible that the exercise has left out some groves and a measure of success of this effort will lie in that more people should come forward to reveal more such treasures. It is also apparent that more taxonomic work is required particularly on fauna.

It is clear from a number of the reports received that many groves are facing threats of destruction and even extinction. Most of the protected groves are already in a degraded state. The force of modernization and increase in demand for fuel, fodder, etc. are not only passing by these groves but also making inroads into them. The local people get sucked into this incursion, often unwittingly. The sanctity of the groves no longer impresses some of them who feel deprived of economic resources and mistakenly make unsustainable drawls on these natural resources. We are at a point of time when we have to restore to the groves their dignity and utility while ensuring alternatives provided to the local population. We have to marry science and tradition to conserve the groves having rare and endangered flora and fauna and these should be protected to the highest extent possible. Where the threats are great, some of the genetic material amenable to cultivation should also be grown elsewhere under care. For such *ex-situ* cultivation, a careful linkage between industry and local villages should be established such that the benefits of both traditional knowledge and current efforts in the country to conserve biodiversity and foster the rights of tribal, traditional knowledge holders (like the vaidyas and women) and farmers.

Some of the groves already have a measure of legal protection when they are inside reserve forest or notified sanctuaries and national parks. Regretfully most of them are outside, either on temple or on private land or government land designated for other uses. It is necessary to look at the Forest Act, Wildlife Protection Act, Revenue codes, Endowments Act, acts relating to wakf properties, etc., and see if amendment can be brought about to recognize and notify these groves and ensure that interference in groves attracts legal redress and penalties. If this is done through different enactments and different authorities, then there should be overall coordination at the district and state levels to ensure that similar approaches are adopted.

It is to be underlined that protection cannot be ensured through laws and rules alone as the groves are scattered in nature and clandestine activities cannot be checked except at an excessive cost to the exchequer. It is, therefore, necessary to look at ways of strengthening public participation in both protection and management of the groves. The involvement of NGOs where existing and taking constructive interest in this will be equally helpful. The structures that exist under schemes like Joint Forest Management and Eco-development could be explored for the purpose. At village level there should be specific allocation of fuel, fodder, etc., for people who would otherwise raid the groves. Very strong involvement of women in these schemes and proper extension education to villagers on scientific value of groves must be ensured. It is also necessary to ascertain that in environmental impact assessment (EIA) exercises due attention should be given to these sacred and protected groves wherever projects are likely to impinge on them.

Protected groves are the oases of Himalayan biodiversity and an integrated and collective effort of the folk, naturalists, environmentalists, NGOs and Government should be made to conserve their floral and faunal wealth as well as sanctity.

ACKNOWLEDGEMENTS

The authors are thankful to the Head, Department of Botany and Microbiology for laboratory facilities and Professor G.S. Paliwal, Senior Consultant, AFC Ltd., Janakpuri, *New Delhi*-110058 for going through the manuscript critically. For the article the information has been gathered from several sources and we gratefully acknowledge all of them.

REFERENCES

Badoni, AK (1990). Remarks on the high altitude medicinal plants of Garhwal Himalaya. *J. Himalayan Studies and Regional Dev.* **13-14:** 37-45

Bisht Tapendra Singh (2005). Environmental conservation strategies in GarhwalHimalaya-An ecological consideration of few protected groves. Ph.D. Thesis submitted to HNB Garhwal University, Srinagar

Boojh, R & Ramakrishan, S (1983). Sacred Groves and Their Role in Environmental Conservation. pp 6-8. *In: Strategies for Environmental Management.* Souvernir Vol. DST, U.P., Lucknow

Brandis, D (1897). *Indian Forestry,* Working. Oriental Institute, Working

Burman, RJJ (1992). The institute of sacred grove. *J. Anthropological Society* **27:** 219-238

Champion, HG & Seth, SK (1968). *A Revised Survey of the Forest Types of India.* New Delhi

Gadgil, M & Vartak, VD (1975). Sacred groves of India- a plea of the continuous conservation. *J. Bombay Nat. Hist. Soc.* **72:** 313-320

Gadgil, M & Vartak, VD (1976). Sacred groves of Western Ghats of India. *Econo. Bot.* **30:** 152-160

Gaur, RD (2007). Biodiversity and river valley projects in Uttarakhand. *Proc. Nat. Acad. Sci.* India. **77(B):** 253-262

Haridasdan, K & Rao, RR (1985). *Forest Flora of Meghalaya,* Vol. 1. Bishen Singh-Mahendra Pal Singh, Dehra Dun

Hughes, DJ & Chandran, SMD (1997). *Sacred Groves around the Earth*: An overview. Paper presented in the workshop on the role of sacred groves in conservation and management of Biological Resources, KFRI, Trichi

Khan, ML, Menon, S & Bawa, KS (1997). Effectiveness of the protected area network in biodiversity conservation: a case study of Meghalaya state. *Biodiversity and Conserva.* **6**: 853-868

Khiewtam, RS & Ramakrishnan, PS (1989) Socio-cultural studies of the sacred groves at Cherrapunji and adjoining area in North-Eastern India. *Man in India* **69:** 64-71

King, OI (1997). Botanical and Socio-culture studies on Sacred Groves in Cuddalore District, Tamilnadu, India.

Kosambi, DD (1962). *Myth and Reality.* Popular Press, Bombay, India

Malhotra, Kailash C, Gokhale, Y, Chatterjee, S & Srivastava, S (1991). Cultural and Ecological Dimensions of Sacred Groves in India. INSA, New Delhi

Nayar, MP & Sastry, ARK (1990). *Red Data Book of Indian Plants.*Vol. III. Botanical Survey of India, Calcutta

Ramakrishan, PS (1996). Conserving the sacred: from species to landscape. *Nature and Resource* **32**: 11-19

Rao, RR (1994). *Biodiversity in India (Floristic aspects).* Bishen Singh-Mahendra Pal Singh, Dehradun

Rawat, DS, Bhandari, BS & Gaur, RD (2001). Vegetational wealth, pp. 70-92. In: Kandari, OP & Gusain, OP (eds). *Garhwal Himalayas Nature, Culture and Society.* Transmedia. Srinagar (Garhwal)

Rodgers, WA (1994). The sacred groves of Meghalaya. *Man in India* **74**: 339-348

Shah, NC (1971) Medicobotany of Dronagiri—The mythic hill in Kumaon (U.K.). *J. Res. Indian Med.* **8:** 47-59

Singh, BT (2009). Environmental conservation strategies in Garhwal Himalaya- An ecological consideration of few protected groves. Ph.D. Thesis submitted to HNB Garhwal University, Srinagar

Tiwari, BK, Barik, SK & Tripathi, RS (1998) Biodiversity value, status and strategies for conservation of sacred groves of Meghalaya. India. *Ecosystem Health* **4:** 7-10

Vartak, VD & Gadgil, M (1973). Dev Rahati: An ethnobotanical study of forests preserved on grounds of religious beliefs. Abstract. *Proc. Indian Science Congress* **60:** 341

5

Phytodiversity: A Medicinal Aspect

SUSHIL K SINGH AND HEMENDRA S CHOUHAN

Department of Pharmaceutics, Institute of Technology, Banaras Hindu University, Varanasi–221 005

E-mail: sksingh.phe@itbhu.ac.in

ABSTRACT

Biodiversity signifies variation at molecular, genetic, species and habitat level of life. Plants constitute 22% of global biodiversity. Plants have immense role in living healthier life and form the basis of many traditional medicine systems and therapeutic armamentarium. The uses of plants for primary health care have progressively increased globally in recent times. Most pharmaceutical industries focused their research for obtaining potential chemotherapeutic agents/leads from the plants. It has been estimated that about 877 new chemical entities (NCEs) were discovered during the period of 1981-2002 and 69.0% of these NCEs were obtained from the natural products or are their synthetic/semi-synthetic analogues. The role of plants in medicine and discovery of anticancer drugs has been presented. Economic aspects of medicinal plants, bringing about over-exploitation, has resulted in uncontrolled harvesting of medicinal plants, threats for the phytodiversity and perspectives for the protection of plants are briefly discussed.

Keywords: Anticancer agents, drug discovery, natural product, phytodiversity.

INTRODUCTION

Biodiversity means variety of life, occurring in close association of the habitat or ecosystem. Plants constitute an important part of this ecosystem, keep balance between environmental changes and living conditions; hence, provide the conditions and processes that sustain human life (Lister, 1998). Plant biodiversity is a relatively broad term used for the variation at all levels (i.e. molecular, genetic, species and habitat) of life and includes patterns in nature up to the landscape level (Gugerli *et al.*, 2008). It is roughly estimated that plant kingdom represents 22% of global biodiversity that accounts for the prevalence of approximately 300,000 known plant species (Basso *et al.*, 2005). However, these species are not evenly distributed on the earth and about 70% of them are present in twelve countries, viz. Australia, Brazil, China, Columbia, Ecuador, Indonesia, Madagascar, Mexico, Peru and India (Shanahan and Masood, 2009). India, one of the biggest hotspot regions of biodiversity includes more than 8 per cent of world's known plant species, despite the fact that its total land area is only 2.4% of the geographical area of the world. India has over 49,000 plant spread species distributed throughout, of which 4,900 are endemic and distributed in 10 biogeographic regions of the country. These species are predominant source of food, spices, flavours and fragrances, vegetable oils, soaps, natural rubber, gums, resins, insecticides, industrial and agricultural raw materials, livelihood and medicines (See below Figure 1; Ramakrishnappa, 2002).

Fig. 1. Applications of phytodiversity in pharmaceutical industry.

History of Role of Plants in Medicine

Natural drug products have played a predominant role in the pharmaceutical healthcare system. Plants have shaped the foundation of many traditional systems, which are in existence since many centuries (Houghton, 1995). The fossil carbon dating showed that the use of plants by the human-being is about 6,000 years old (Solecki and Shanidar, 1975). However, the cuneiform document found from Mesopotamia entails the probable use of plants as medicine commenced in 2600 BC (Wilson and Reynolds, 1990). The upshot of uses of plants as medicine and the practices of herbal therapy in present era are due to shift of medicinal knowledge in the generations of ethnic community. These led to formation of several therapeutic armamentariums. For example, in India, several thousand years old sacred text-book of Hindus, 'Rig-Veda' enlists the description of many plants as medicines. In 1,000 BC, the Charak Samhita' and 'Sushruta Samhita' documented the use of plants and polyherbal formulations for healthcare needs. The traditional medicinal knowledge of India is preserved in the form of two advanced systems of medicine, i.e. Ayurveda and Siddha (Samy *et al.*, 2008; Zysk, 1992). Ancient Egyptian drug lore was documented in the form of Ebers Papyru (1550 BC), which comprehensively enlists 876 prescriptions and 500 substances obtained from mineral, animal, and plant sources (Borchardt, 1999). Huangdi Nei Jing is the first document on traditional Chinese medicines. This was written between 300 BC and 100 BC, and enlists the diagnosis and treatment for many disorders and diseases (Unschuld, 2004). Hippocrates, father of modern medicine, used several plants to treat various ailments in Greece (Hart, 1965). Eminent Greek physician Dioscorides had compiled his empirical knowledge on medicinal values of more than 600 plant species in the form of text 'De Materia Medica', and is still serving as reference. In 16th century, Martin de la Cruz and Juannes Badianus wrote a famous 'Badianus Manuscript' which contains the description of over 184 plant species of medicinal value in Aztec language. Pharmaceutical preparations based on Unani and Homeopathy systems of medicine also utilize plants for the treatment of a variety of diseases (Janick, 2003).

Plants are still an indispensible source of medicines. A study conducted during 1959-1980 in United States, revealed that about 25% of total prescriptions contain plant extracts or active principles derived from higher plants. The value of medicinal plants is evident by the WHO Report (2003), which estimates that plant constitute 11.1% of about 252 essential chemicals necessary for human health. It has been proved that plants play a safe and important role in the health care (Balandrin *et al.*, 1985). It is estimated that 70-80% of people worldwide rely chiefly on herbal medicines to meet their primary healthcare needs and for India this is 65% of the rural population (WHO, 2003).

Plant as a Source of Drug

Nature biodiversity is a great source of wide range of chemical compounds, which are formed in response to protecting themselves from the attack of predators and herbivorous. These defense chemicals are the plant secondary metabolites which include a wide range of high molecular weight compounds such as defense proteins, polysaccharides, and polyterpenes and low molecular weight secondary metabolites such as steroids, triterpenes, lignins, phenolics, flavonoids, alkaloids, etc. When a plant is attacked or serves as feed for any form of life, ranging from bacteria to virus, insects to craniates and sometimes by bying plant species. A good amount of some of these secondary metabolites reach into rival species, and it reacts with one or more molecular and/or cellular target of the organism, and produces inhibitory action, which may modify the biological process of cell, tissue, organ or even the complete body. Thus, bringing physiological changes or toxicity or sometimes even killing the invading species. Therefore, the secondary metabolites are presumed to be bioactive compounds. The biological activity of these compounds is contingent on orientation of functional group, stereochemistry and extent of interaction that occurs with the receptor/target site. The comprehension of chemistry, physiological role and use of correct amounts of these bioactive compounds can be made for some important pharmacological activities, which benefits the human health (Paré and Tumlinson, 1999; Bohlmann *et al.*, 1988; Jander and Howe, 2008).

Only 20-30% of plant species on the planet have been deployed for the extraction and isolation of their bioactive constituents. More than 1,00,000 low molecular weight secondary metabolites have been identified till date. Among these structurally known secondary metabolites there are more than 21,000 alkaloids, 25,000 terpenoids, 7,000 phenolics and polyphenols, 1,500 polyacetylenes, fatty acids, waxes, and 200 carbohydrates The structure of these compounds varies in size and shape, within the class of molecule due to alteration in the functional groups present therein (Wink, 2008). This chemodiversity present in the millions of plant species is greater than the synthetic library ever formed. Pharmaceutical industries speculate this chemo-diversity as an important resource for enormous bioactive compounds. Currently, pharmaceutical research interest rests in the search of potential chemotherapeutic agents from the plants and exploits natural compounds as templates for lead optimization programmes intended to make safe and effective drugs. Study of Newman *et al.* (2003) revealed that about 877 new chemical entities (NCEs) were discovered during the period of 1981-2002 and 69.0% of these NCEs were obtained from the natural products or were their synthetic/semi-synthetic analogues. More than 50% of small-molecules discovered as drug in the last twenty years are of natural origin (some examples have been listed in Table 1) and over 20 natural products/natural product-derived drugs have been launched into the United States, Europe and Japan market during the last six years (Newman *et al.*, 2003; Newman and Cragg, 2007). Nowadays, most of the pharmaceutical industries are working on natural product discovery; considering the treatment of not only common infectious diseases but also other molecular targets and diseases.

Table 1: List of some molecules/leads derived from plants used as drug and offered in the form of various formulations by pharmaceutical industries

Drugs	Sources	Indication/Uses
Atropine	*Atropa belladonna*	In organophosphate poisoning, for mydriasis.
Caffeine	*Camellia sinensis, Coffea canephora,*	stimulant
Colchicine	*Colchicum autumnale*	Gout
Curare	*Strychnos toxifera*	muscle relaxant
Digoxin	*Digitalis purpurea*	cardiac arrhythmias
Diosgenin	*Dioscorea deltoidea*	Antifertility
Ephedrine	*Ephedra sinica*	In hypotension, appetite suppressant, decongestant,
Gossypol	*Gossypium species*	Male contraceptive
Hesperidin	*Citrus species*	Capillary fragility
Hydrastine	*Hydrastis canadensis*	Hemostatic, astringent
Hyoscyamine	*Hyoscyamus niger*	Anticholinergic
Morphine	*Papaver somniferum*	Analgesic
Nicotine	*Nicotiana tabacum*	Action on the Central Nervous System
Pilocarpine	*Pilocarpus microphyllus*	Glaucoma, xerostomia
Quinine	*Cinchona officinalis*	Anti malarial
Reserpine	*Rauwolfia serpentine*	Antipsychotic and antihypertensive
Salicylic acid	*Salicilin*	Analgesic
Scopolamine	*Datura species*	Sedative
Sennosides A, B	*Cassia species*	Laxative
Silymarin	*Silybum marianum*	Hepatoprotective agent
Silymarin	*Silybum marianum*	Antihepatotoxic
Sparteine	*Cytisus scoparius*	Anti-arrhythmias
Theophylline	*Theobroma cacao*	Diuretic, bronchodilator
Yohimbine	*Pausinystalia yohimbe*	Aphrodisiac

Plants in Drug Discovery: An Example of Anticancer Drug Research

The surge for the discovery of bioactive constituents from natural sources (plant extracts, marine and microbial) has led to many new drugs benefitting human health. Strategies employed in natural product research are concerned with sample selection/collection, isolation, structure elucidation, biological evaluation, semisynthesis, dereplication, biosynthesis, as well as optimization of downstream processing. The generation of good quality lead compounds from natural sources is a costly, tedious and time consuming process due to the structural complexity of the natural compounds (Gunatilaka and Kingston, 1988). Therefore, the big pharmaceutical industries down regulated the plant-based drug discovery programme. But, reappearance of diseases due to drug resistance is the main driving force to continue such programmes. Significance of plants as life-saving drugs has been evaluated and accounted by many scientists (Baker *et al.*, 2007). Secondary metabolites from plants continue to provide many bioactive compounds, and some of them are used as drug. Cancer is the major cause of death globally and a cause of deep concern for the society. Here, we present some examples of drug/lead derived from plants as the anticancer agents. Anticancer drug vincristine and vinblastine from plant were serendipitously discovered at Eli Lilly. Vinca alkaloid with indole skeleton structure binds with the β-tubulin, arrests mitosis by interfering in assembling and disassembling process of microtubules and hence causes the death of metaphase cell population. Vincristine inhibits the incorporation of certain amino acids into the protein of tumour cells and also the synthesis of soluble RNA (sRNA), and hence inhibits the protein synthesis in tumor cells. Further, it increases the level of oxidized glutathione and cAMP, and brings changes in the metabolism of lipid. *Catharanthus* alkaloid also inhibits the calcium-calmodulin-regulated cAMP phosphodiesterase. These alkaloids are used in combination therapy for treatment of leukemias, lymphomas, breast and other type of cancers, and have severe toxicity. It has been observed that daily high dose of vincristine produces dose-related sensory, motor, and autonomic neuropathy and central nervous system dysfunctions. In some patients with advanced stage of cancer, it may cause the leukopenia and thrombocytopenia. Many semi-synthetic derivatives of vinca alkaloid have been synthesized for the optimization of lead. Some vinblastin analogues, viz. Vinorelbine, Vinflunine, Anhydrovinblastine are in different phases of clinical trials, whereas, Vindesine has been approved for its clinical applications (Kingston, 2009).

Combretastatin A-4, isolated from the root bark of the *Combretum caffrum* tree also inhibits the polymerization of the tubulin during metaphase of cell cycle. However, this compound suffers with a drawback of severe toxicity. A series of compounds have been synthesized using template structure of Combretastatin A-4, and among them Combretastatin A4 phosphate, Combretastatin A-1, OXi4503, and AVE8062A are in the clinical trials. Well-known antitussive drug Noscapine (isolated from opium plant) was also found to exhibit action on confirmation, assembly and dynamics of tubulin protein. Finding of this new indication for old drug suggests its use in the cancer. However, the drug is in phase I of clinical trial (Ortiz *et al.*, 2007).

Taxanes are a new class of structurally complex diterpene alkaloid. First compound discovered in this class is taxol, also known by its generic name paclitaxel, which was isolated from the bark of *Taxus brevifolia* in 1960s. Paclitaxel was approved in 1992 for the treatment of metastatic ovarian cancer and later on for breast cancer. This compound stabilizes microtubules and reduces their dynamicity, promoting mitotic arrest and cell death. Upon assembly of the α/β-tubulin heterodimer, GTP bound to β-tubulin is hydrolyzed to GDP reaching steady-state equilibrium between free tubulin dimers and microtubules. The binding of taxol to β-tubulin in the polymer results in cold-stable microtubules at the expense of tubulin dimers, even in the absence of exogenous GTP (Kingston 2009; Xiao *et al.*, 2006). Taxol is the most potent among all anti-cancer drugs obtained from the plants. However, low solubility, toxicity, emergence of resistance, and relatively low yield of paclitaxel from its natural resources urge the development of taxanes. Hundreds of taxanes have been either isolated or synthesized so far. The isolation of 10-deacetylbaccatin III in high concentration from the leaves of *Taxus* plant, has brought a major breakthrough in the development of taxanes. Docetaxel, another FDA approved anticancer drug has been synthesized by using 10-deacetylbaccatin III as an intermediate. Docetaxel has been successfully used to treat breast cancer, prostate cancer and non-small cell lung cancer. In the past decade, pharmaceutical companies have developed some more semi-synthetic taxanes with better bioavailability and activity against taxaol resist cancer. However, these compounds are underway of clinical development (Hartwell, 1982). BMS-184476 and BMS-188797 have been developed by Bristol-Myers Squibb as semisynthetic derivatives of paclitaxel. These compounds exhibit potent antiproliferative activity both in Taxol-resistant human tumor cell lines and Taxol-resistant tumor models (Kim *et al.*, 2001). A new paclitaxel analogue, IDN5109 was found to exhibit more potent activity than the compounds in multidrug resistance-positive human cancer cell lines, MCF-7 ADR and CEM VBL. This compound is in the phase II of clinical trial (Nicoletti *et al.*, 2000).

Clinical efficiency of Chinese medicinal plant *Indigofera tinctoria* L. against chronic myelocytic leukemia was discovered at Institute of Hematology, Chinese Academy of Medical Sciences. Further studies on the plant revealed that the activity was due to its minor constituent 'indirubin', which inhibits the cell cycle regulating cyclin-dependant kinase and glycogen synthase kinases-3. This compound has been found to display some side effects like abdominal pain and diarrhoea, decrease of platelet count in some patients during clinical trial. However, N-methyl isoindigotin which was synthesized using template of this molecule, have fewer side effects as compared to indirubin (Newman *et al.*, 2002).

Some plant derived molecules have been found to exhibit anti-cancer activity by their inhibitory action on DNA topoisomerase I and II. DNA topoisomerases are the enzymes which maintain the topological state of DNA and regulate DNA replication process (Rothenberg, 2007). The inhibition of these enzymes results in cell cycle arrest. The first topoisomerase I inhibitor was isolated from the stem bark of Chinese plant *Campthotheca acuminata* by Wall *et al.* (1966), and identified as camptothecin and has shown significant anti-leukemic activity in mice. Clinical failure of this molecule was accompanied with unpredictable toxicity and solubility problems. Success of later studies in making its water-soluble analogues and finding the mechanism of action of camptothecin, have led to resurgence of lead optimization programme. This ensued in the development of topotecan and irinotecan, which are approved by the USFDA for the treatment of ovarian and metastatic colorectal cancer, respectively (Rahier *et al.*, 2005). Beta-lapachone, isolated from the heartwood of the lapacho tree in 1882 by Paterno, also inhibits topoisomerase I. In contrast to camptothecin, beta-lapachone does not involve instabilization of the cleavable complex, yet possibly has direct interaction with topoisomerase I (Li *et al.*, 1993).

Currently, topoisomerase II inhibitors have similar interest in Cancer Research. Development of ellipticine and etoposide have made a significant breakthrough in this area. Ellipticine is an alkaloid found in several species of the family Apocynaceae and exhibits antineoplastic activity by DNA intercalation and topoisomerase II inhibition. This compound has failed in clinical trial due to severe cardiovascular toxicity and haemolysis. However, Elliptinium (an ellipticine analogue) has remarkable diminished toxicity than the parent drug and is being marketed in France (Stiborová *et al.*, 2004; Itokawa *et al.* 2005). Etoposide is a semisynthetic derivative of podophyllotoxin which is obtained from *Podophyllum*. Etoposide has been successfully employed for the treatment of small cell lung cancer, testicular cancer, leukemias, lymphomas, and other cancers. Teniposide, another thiophene derivative analogue has also similar therapeutic profile (Lee *et al.*, 2005). Emergence of drug resistance and low bioavailability are the major problems associated with these drugs, which draw the attention of researchers for further development of this novel pharmacophore. Till date, several plant-derived anticancer compounds and their synthetic/semi-synthetic analogues are either in clinical trials or approved as a drug. However, it is always desirable to search new, safe and effective drugs, which can also be helpful to combat drug resistance.

Interaction of Plants with Environment: Medicinal Aspect

Cultivation pattern of medicinal plant species in nature is scattered and limited by ecological boundries and other dependent factors like soil type, water content, altitude and climatic conditions. It has been observed that light, shortage of organic food content, water, change in humidity, attack of herbivorous or competitive species, and pollution alter the chemical composition of secondary metabolites in plants. (Figure 3) Such effects of environmental conditions on the production of secondary metabolites have been reported and well reviewed.

Climate change has either direct impact on the plant-herbivorous relationship or the physiology of the plant itself (Figure 4 Interface of plant and environment is shown). For example, intensity and wavelength of light have predominant role in the height, thickness, area of the leaf and plant. Ultraviolet light causes damage to epidermal cells of plants by acting on the DNA of the cell or by producing reactive oxygen species. Therefore, ultraviolet radiation (UV-B) on plant triggers the production of phenolics particularly flavanoids, which have potential antioxidant and free radical scavenging activities. UV-B radiation catalyzes the action of phenylalanine lyase and chalcone synthatase enzymes, which are involved in the phenyl propanoid pathway. Overproduction of flavanoid and lignin in leaves and stem against the UV-B stimuli helps the plants to filter the radiation and protect them from the harm of radiation. Similarly, increased synthesis of lignin, and tannins is also observed. Lignins are known to provide rigidity, flexibility to plant and protection from microbial degradation, while tannins make the taste unpleasant for the herbivores. Soil is the basic source for the nutrients like nitrogen, phosphorus, potassium, etc. supply for the plants. Variation in nutrient content of soil influences plant metabolism, eventually is resulting in change in the production of plant secondary metabolites. For example, Salidroside used for its antidepressant, anxiolytic activities, and is an active constituent of *Rhodiola sachalinensis* which varies according to organic matter, pH and nitrogen content of soil. High yield is obtained in the rich organic matter, low pH and high levels of exchangeable nitrogen and total nitrogen content of soil (Yan *et al.*, 2004). Similarly, increased salinity of soil has been found to produce high content of vinca alkaloid. Composition of atmospheric gases has an indirect effect on interplay between plants and surrounding species. Stiling and Cornelissen (2007) reported that rise in atmospheric CO_2 causes disturbance in the C: N balance in the

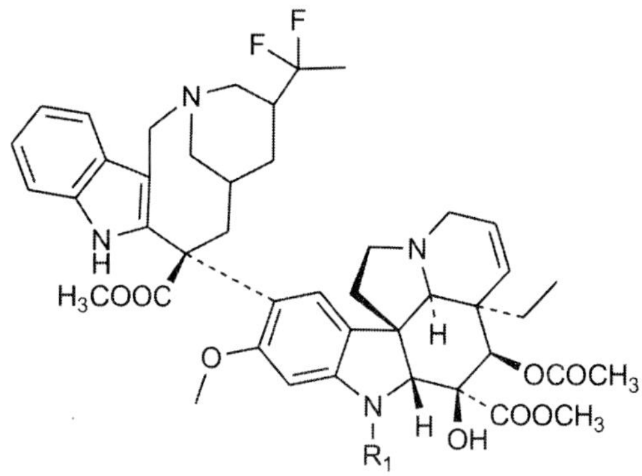

Vinflunine

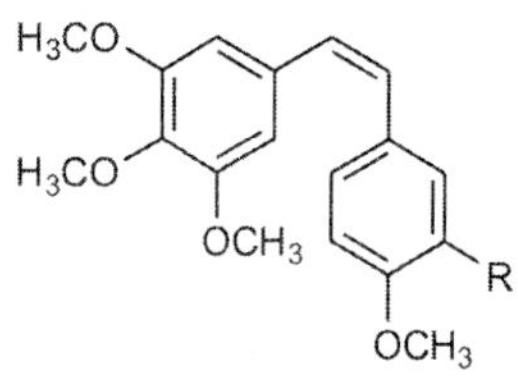

Combretastatin A-4 R = OH
Combretastatin A4 PO_4 R = OPO_3Na_2
AVE8062A R = $NHCOCH(NH_2)CH_2OH$

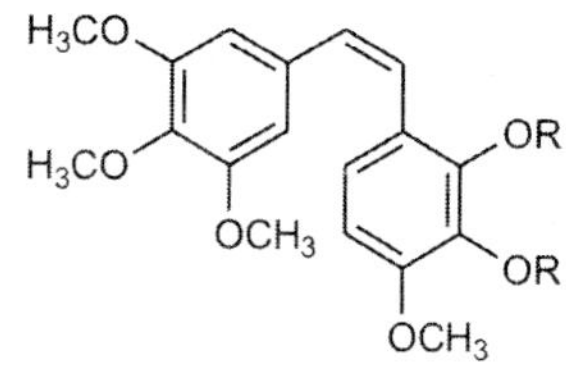

Combretastatin A-1 R = H
OXi4503 R = OPO3Na2

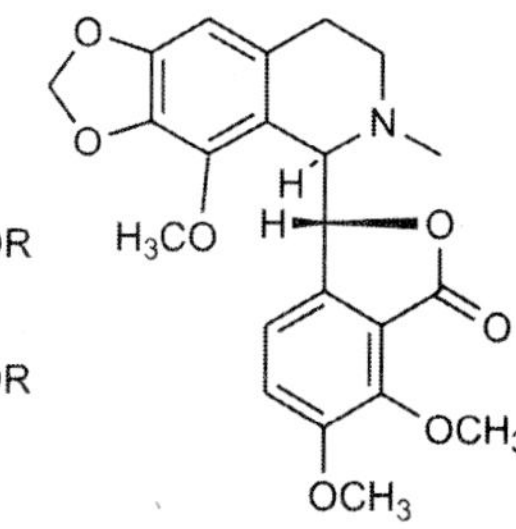

Noscapine

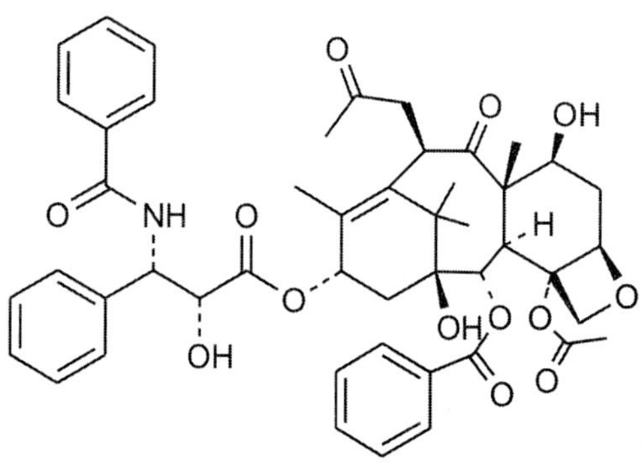

Taxol

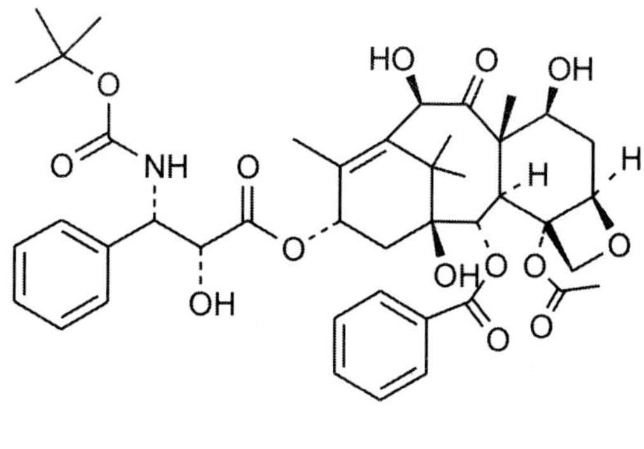

Docetaxel

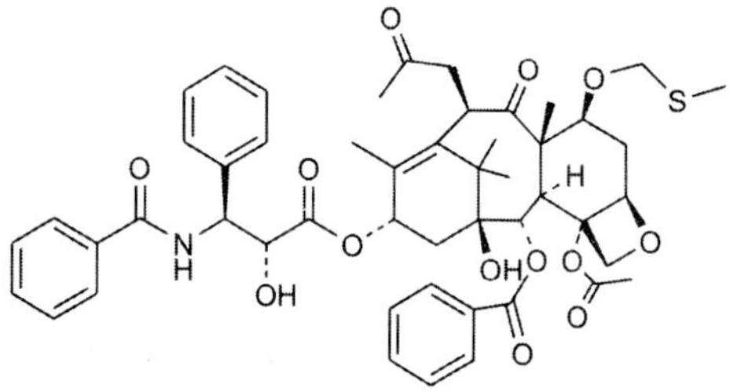

BMS-184476

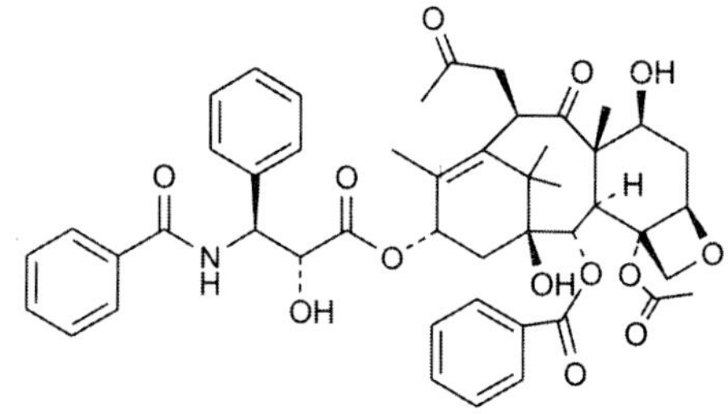

BMS-188797

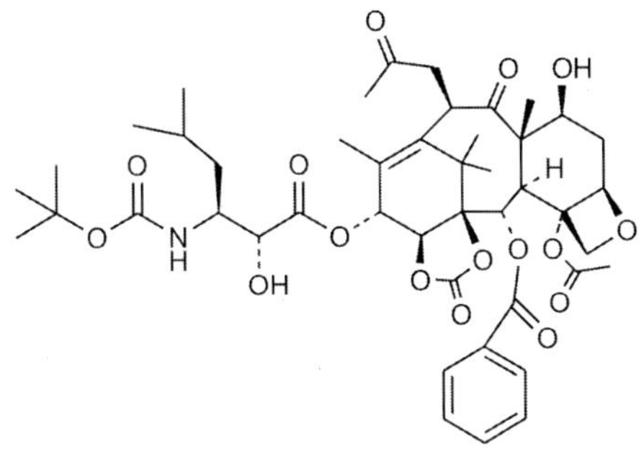

IDN5109

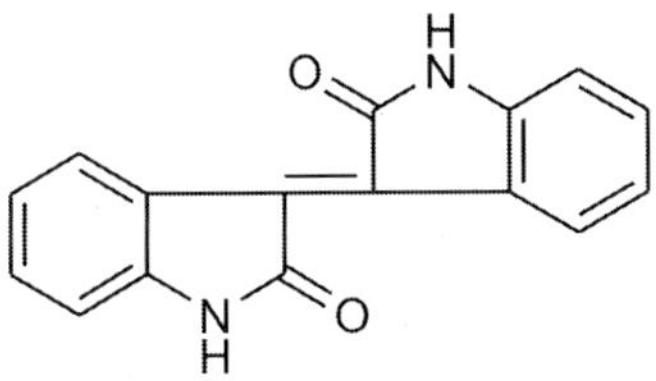

Indirubin

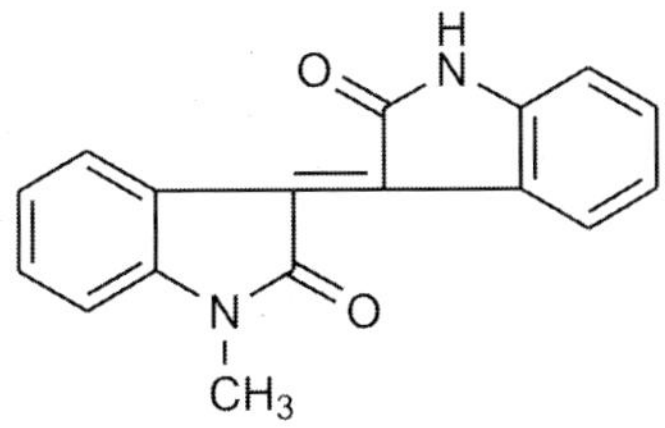

N-methyl indirubin

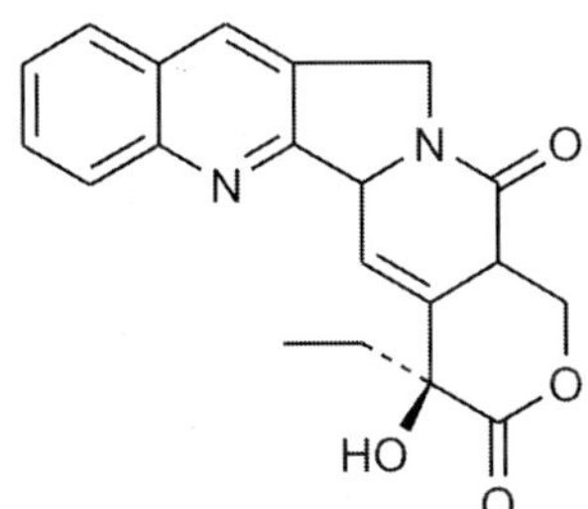

Camptothecin

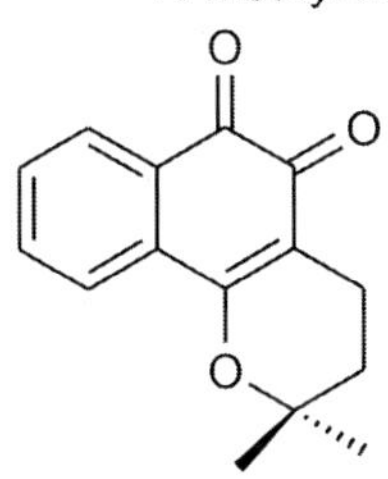

beta-lapachone

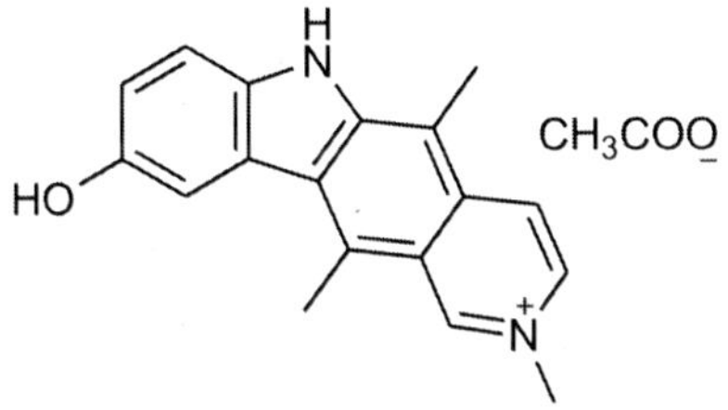

Elliptinium

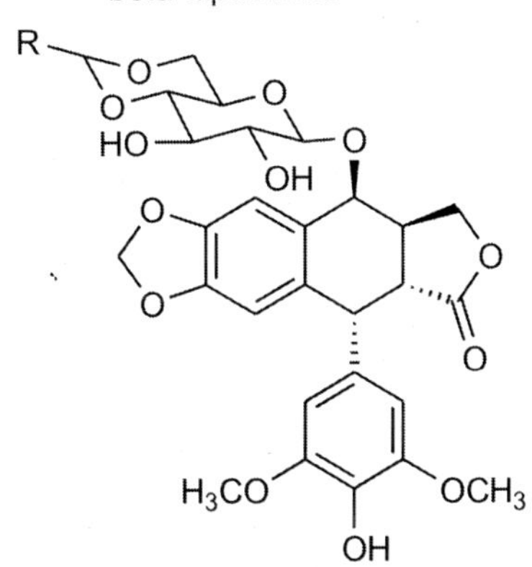

Etoposide R = CH3
Teniposide R = -2 thipophene

Fig. 2. Chemical structure of some plant-derived anticancer agents.

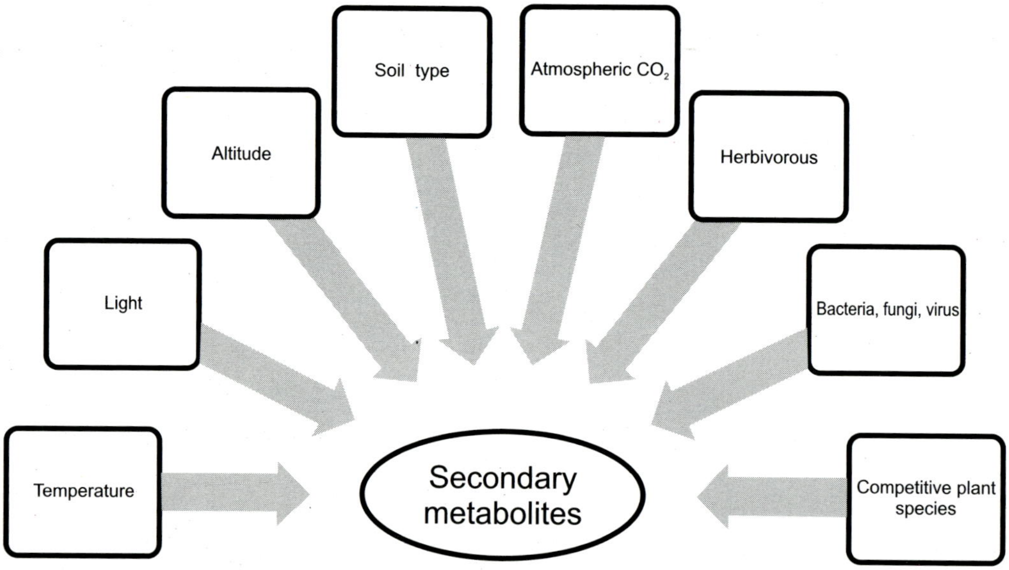

Fig. 3. Environmental factors affecting secondary metabolite production in plants.

plants and affects the secondary metabolite production. Impact of altitudinal variation on secondary metabolite contents of plants is well recognized. Spitalert *et al.* (2008) reported the significant variation present in the ratio of ortho-dihydroxy flavonoids: other flavonoids content in *Arnica montana* L. cv. ARBO (homeopathic medicine) plant grown at 600 and 1400 m height, respectively. Relationship of plants and insects depends upon mutual benefits. During pollination, plants enhance the productivity of volatile oil and terpenes to attract the insect, which helps in carrying the pollen grains from one plant to another plant. The effect of volatile oil on pollination has been comprehensively studied (Dudareva and Pichersky, 2000, Kephart *et al.*, 2006). The effect of the regional, seasonal and altitudinal variation on taxol content of *Taxus brevifolia* have been shown in Table 2. This reveals that relatively high yield of taxol can be obtained from plants grown in warm and dry regions (Wheeler *et al.*, 1992).

Fig. 4. Interface of phytodiversity and environment.

Table 2: Taxol content (% dry wt) of *Taxus brevifolia* individual tree samples (printed with kind permission from American Chemical Society)

Population/Accession	Lat	Long.	Region[a]	Taxol Content Shoots[b] April'	Bark[b] Sept.	April'	Sept.
WASHINGTON							
Cowlitz	47"35'	121'40'	CM				
3334				0.010	0.007	0.005	0.001
3332				-	0.016	0.008	0.002
3335				-	0.033	0.003	0.001
3336				-	-	0.005	0.001
3337				0.008	0.007	0.004	-
Wind River	44'15'	12 1'50'	CM				
3328				0.010	0.004	0.010	0.001
3329				-	0.017	0.006	0.004
3325				-	0.001	0.005	0.001
3326				0.006	0.007	0.002	0.001
3327				-	-	0.013	-
McKensieBridge .	44'15'	12 1'50'	CM				
3351				0.010	0.003	0.008	0.001
3352				0.009	0.008	0.009	0.006
3353				-	0.023	0.001	0.001
3354				-	0.013	0.001	0.001
3355				0.010	-	0.004	-
Butte Falls	42'30'	122'35'	WD				
3360				0.008	0.004	0.010	0.004
3339				-	0.002	0.005	0.002
3362				-	0.001	0.005	0.002
3338				0.008	0.008	0.003	0.001
3361				-	-	0.004	-

[a] = CM = cool, moist; WD = warm, dry.
[b] = Shoots and bark were collected in April and September of l989 and assayed during winter of 1989-1990. April collections from a fifth population in Idaho were not analyzed.

Medicinal Plants: Economic Aspect

There is an increasing trend in consumption of medicinal plants worldwide, due to common belief that the herbs are safe, have no/minimal side effects, easy accessibility and inexpensive means of treatment (Figures 5, 6). Many countries like China and United Arab Emirates are almost dependant on medicinal plants for their healthcare need. In the secretariat report on Convention on Biological Diversity, global sale of herbal product was about US$60,000 million in 2002. Since the approximation in trade of medicinal plants is very difficult because domestic trade is not documented, hence reliability of such data is always questioned.

An estimate of import value of medicinal plants of developed and developing countries was more than 1,200 million USD during 1997-2002. Noteworthy expenditure (about 600 million USD) on import use was done by developed countries. The rise in global trade of medicinal plant also affects Indian herbal industries. Currently in India, about 9,500 registered herbal industries and numerous unregistered small units exist trying to meet the demand of medicinal

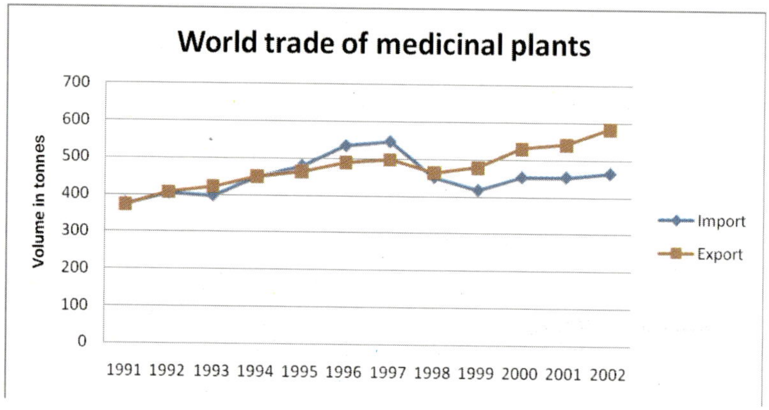

Fig. 5. Trade of medicinal plants during 1991-2002.

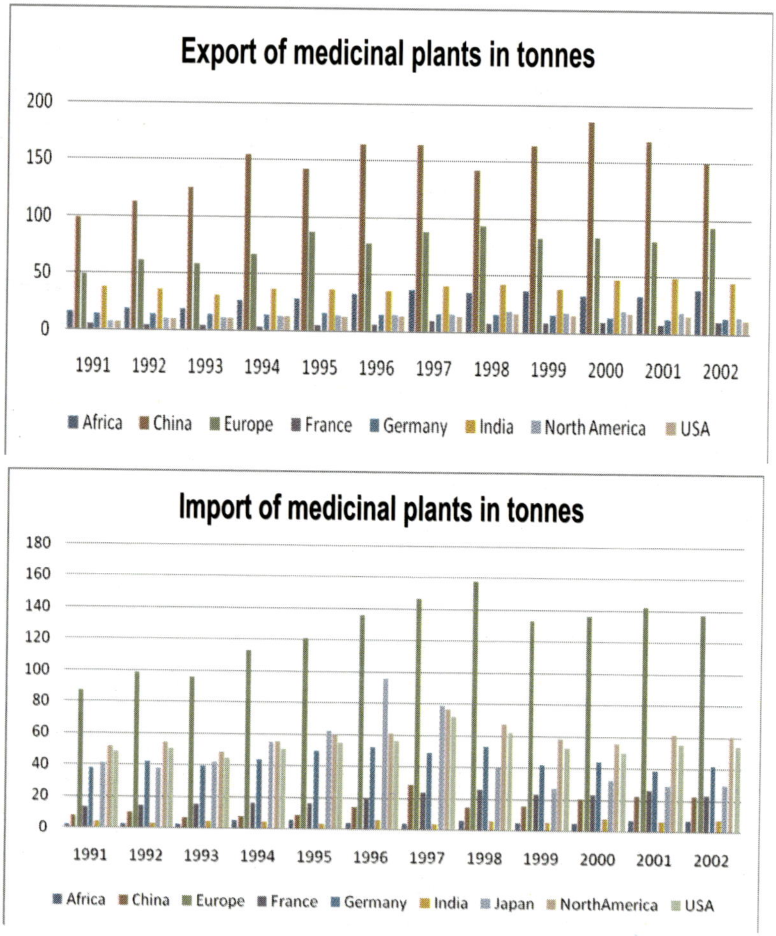

Fig. 6. Trend in trade of medicinal plants in major buyer/seller countries during 1991-2002.

plant products. However, medicinal plant export-import is relatively smaller than China, Europe (Figure 6). About 960 medicinal plant species are traded by India. The supply of medicinal plants is done by either harvesting from agriculture fields or by wild plant collection by the local communities (Anoyn; Titz, 2004; WHO, 2003b).

Biodiversity Loss

The value of biodiversity as a source of pharmaceutically active substances has been the subject of a number of studies. The growing dependence on the use of medicinal plants, high demand in world market has resulted in their overexploitation. These situations create the danger of extinction of some medicinally important species, for example uncontrolled, indiscriminate harvesting for the lucrative market prices for *Pilocarpus jaborandi* shrubs (its bioactive constituent pilocarpine is used to reduce intraocular pressure associated with glaucoma) led to its depletion from the forests of eastern Brazilian Amazonia. Moreover, the persistence of high demand of *T. brevifolia* and *T. baccata* for their very low-yielding phytoconstituent taxol (anti-tumor activity) may endanger the species. There are many more examples of either endangered or extinct plant species, may be because of deforestation, loss of habitats, over-exploitation pollution, invasion of exotic species and climate change. The conditions become worrisome, when extinction of some species might result in the loss of other plants, animals or microbes in a habitat. Such loss of biodiversity has now been well recognized.

Biodiversity loss is a deep concern, raising many arguments for conserving natural habitats in general and tropical forests in particular and urges the need for sustaining the close nexus of 'Biodiversity–Environment–Human health'. A complex relationship of factors shape these choices at all political, social, and economic levels. Poor government policies, undefined lands, over-harvesting of trees, and economic perspective are adversely affecting the ecosystems (Sheldon *et al.*, 1998; Cassis, 1998).

Policy Framework for Sustainable Use of Medicinal Plants

There is an urgent need for conservation of our ecosystem and this can be achieved only with the help of strong political and social will power. Following are some steps that should be kept in mind to facilitate and strengthen a process of improved national decision-making about biodiversity, including the protection and preservation of threatened natural resources (Rao, 2000).

- Effective monitoring studies and/or research on conservation techniques for the sustainable use of medicinal plants.
- Developing effective databases for the information on medicinal plants biodiversity.
- Developing public and private sector relationship, and creating more funds for the protection of scarce species.
- Improving policies for the export-import of medicinal plants, to prevent over-exploitation of plants.
- Defining comprehensive, adequate and representative protected areas for cultivation. Promoting the cultivation of medicinal plants and investigating new non-destructive harvesting techniques for the collection of medicinal plants.

REFERENCES

Anonymous, trade in medicinal plants, http://www.fao.org/docrep/008/AF285E/AF285e00.htm

Baker, DD, Chu, M, Uma Oza, U & Rajgarhia, V (2007). The value of natural products to future pharmaceutical discovery, *Nat. Prod. Rep.* **24:** 1225-1244

Balandrin, MF, Klocke, JA, Wurtele, ES & Bollinger, WH (1985). Natural plant chemicals: sources of industrial & medicinal materials. *Science* **228:** 1154-1160

Basso, LA, da Silva, LHP, Fett-Neto, AG, Junior, WF de, Al de S. Moreira, Palma, MS, Calixto, JB, Filho, SA, dos Santos, RR, Soares, MBP & Santos, DS (2005). The use of biodiversity as source of new chemical entities against defined molecular targets for treatment of malaria, tuberculosis, & T-cell mediated diseases: a review. *Mem. Inst. Oswaldo Cruz.* **100:** 575-606

Bohlmann, H, Clausen, S, Behnke, S, Giese, H, Hiller, C, Philipp, UR, Schrader, G & Barkholt V Apel, (1988). K, Leaf-specific thionins of barley—a novel class of cell wall proteins toxic to plant-pathogenic fungi & possibly involved in the defence mechanism of plants. *EMBO J.J.* **7:** 1559-1565

Borchardt, JK (1999). The Beginnings of Drug Therapy: Ancient Egyptian Medicine, *Drug News Perspect* **12:** 123

Cassis, G (1998). Biodiversity Loss: a human health issue. *MJA* **169:** 568-569

Dudareva, N & Pichersky, E (2000). Biochemical & molecular genetic aspects of floral scents. *Plant Physiol.* **122:** 627-633

Gugerli, F, Englisch, T, Niklfeld, H, Tribsch, A, Mire, Z, Ronikier, M, Zimmermann, NE, Holderegger, R & Taberlet, P (2008). Perspectives in plant ecology, *Evolution and Systematics* **10:** 259-281

Gunatilaka, AAL & Kingston, DGI (1998). DNA damaging Natural product with potential anticancer activity, in Studies in Natural Product Chemistry : Structure and Chemistry (Part C), Elsevier Publication, pp. 463

Hart, GD (1965). Asclepius: God of Medicine, *Canad. Med. Ass.* **92:** 232

Hartwell, JL (1982). Plants Used Against Cancer. Quarterman, Lawrence, MA

Houghton, PJ (1995). The Role of Plants in Traditional Medicine and Current Therapy, *J. Altern. Complement Med.* **1(2):** 131-143

Itokawa, H, Wang, X & Lee, K-H, (2005). Homoharringtonine and related compounds. In: Cragg, GM, Kingston, DGI and Newman, DJ (Eds.). Anticancer Agents from Natural Products. Brunner-Routledge Psychology Press, Taylor & Francis Group, Boca Raton, FL, pp. 47-70

Jander, G & Howe, G (2008). Plant Interactions with Arthropod Herbivores: State of the Field. *Plant Physiology* **146:** 801-803

Janick, J (2003). Herbals: the connection between Horticulture and medicine, *Hor. Technolgy* **13:** 29

Kephart, S, Reynolds, RJ, Rutter, MT, Fenster, CB & Dudash, M.R. (2006). Pollination and seed predation by moths on *Silene* and allied Caryophyllaceae: evaluating a model system to study the evolution of mutualisms. *New Phytol.* **169:** 667-680

Kim, JS, Amorino, GP, Pyo, H, Cao, Q, Price, JO & Choy, H (2001). The novel taxane analogs, BMS-184476 and BMS-188797, potentiate the effects of radiation therapy *in-vitro* and *in vivo* against human lung cancer cells, *Int. J. Radiat. Oncol. Biol. Phys.* **51:** 525-534

Kingston, DGI (2009). Tubulin-Interactive Natural Products as Anticancer Agents, *J. Nat. Prod.* **72:** 507-515.

Lee, K-H & Xiao, Z (2005). Podophyllotoxins and analogs. In: (Cragg, GM, Kingston, DGI & Newman, DJ) (Eds.). Anticancer Agents from Natural Products. Brunner-Routledge Psychology Press, Taylor & Francis Group, Boca Raton, FL, pp. 71-88

Li, CJ, Averboukh, L & Pardee, AB (1993). beta-Lapachone, a novel DNA topoisomerase I inhibitor with a mode of action different from camptothecin, *J. Biol. Chem.* **268:** 22463-22468

Newman, DJ & Cragg, GM (2007). Natural products as sources of new drugs over the last 25 years, *J. Nat. Prod.* **70:** 461-477

Newman, DJ, Cragg, GM & Snader, KM (2003). Natural products as sources of new drugs over the period 198 1-2002. *J. Nat. Prod.* **66:** 1022-1037

Newman, DJ, Cragg, GM, Holbeck, S & Sausville, EA (2002). Natural products as leads to cell cycle pathway targets in cancer chemotherapy. *Curr. Cancer Drug Targets* **2:** 279-308

Nicoletti, MI, Colombo, T, Rossi, C, Monardo, C, Stura, S, Zucchetti, M, Riva, A, Morazzoni, P, Donati, MB, Bombardelli, E, D'Incalci, D & Giavazzi, R (2000). IDN5109, a Taxane with Oral Bioavailability and Potent Antitumor Activity, *Cancer Research* **60:** 842-846

Ortiz, J, Bukowski, RM & Remick, SC (2007). Novel vascular targeting/disrupting agents: Combretastatin A4 phosphate and related compounds, *Curr. Oncol. Rep.* **7:** 1523-3790

Paré, PW & Tumlinson, JH (1999). Plant Volatiles as a Defense against Insect Herbivores, *Plant Physiol. October,* **121:** 325-332

Rahier, NJ, Thomas, CJ & Hecht, SM (2005). Camptothecin and its analogs. In: (Cragg, GM, Kingston, DGI & Newman, DJ) (Eds.). Anticancer Agents from Natural Products. Brunner-Routledge Psychology Press, Taylor & Francis Group, Boca Raton, FL, pp. 5-22

Ramakrishnappa, K (2002). Biodiversity and the Ecosystem Approach in Agriculture, Forestry and Fisheries

Rao, PK (2000). Sustainable development: economics and policy, Wiley-Blackwell publisher, 1

Rothenberg, ML (2009). Topoisomerase I inhibitors: Review and update, *Annal. Oncol.* **8:** 837-855

Samy, RP, Pushparaj, PN & Gopalakrishnakone, P (2008). A compilation of bioactive compounds from Ayurveda, *Bioinformation* **3(3):** 100-110

Sheldon, JW *et al.*, (1998). Diversity and Distributions, **4:** 93-94

Solecki, R & Shanidar, IV (1975). A Neanderthal flower burial in northern Iraq. *Science* **190:** 880-881

Spitaler, R, Winkler, A, Lins, I, Yanar, S, Stuppner, H & Zidorn, C (2008). Altitudinal Variation of Phenolic Contents in Flowering Heads of *Arnica montana* cv. ARBO: a 3-Year Comparison, *J. Chem. Eco.* **34:** 369-375

Stiborová, M, Sejbal, J, Dohalská, LB, Aimová, D, Poljaková, J, Forsterová, K, Rupertová, M, Wiesner, M, Hudeèek, J, Wiessler, M & Frei, E (2004). The Anticancer Drug Ellipticine Forms Covalent DNA Adducts, Mediated by Human Cytochromes P450, through Metabolism to 13-Hydroxyellipticine and Ellipticine N^2-Oxide, *Cancer Research* **64:** 8374-8380

Titz, A (2004). *Medicinal Herbs and Plants – Scope for Diversified and Sustainable Extraction.* 22-26 July 2004, Bangalore

Unschuld, PU (2004). Huang Di Nei Jing Su Wen: Nature, Knowledge, Imagery in an Ancient Chinese Medical Text, *J. Altern. Complement Med.* **10:** 191-196

Wheeler, N, Jech, K & Masters, S (1992). Effects of genetic, epigenetic, and environmental, factors on taxol content in *Taxus brevifolia* and related species, *J. Nal. Pro.* **55:** 432-440

WHO (2003b). *WHO guidelines on good agricultural and collection practices (GACP) for medicinal plants* , p. 1. Geneva

Wilson, JV & Reynolds, EH (1990). Texts and documents. Translation and analysis of a cuneiform text forming part of a Babylonian treatise on epilepsy. *Med. Hist.* **34:** 185-198

Wink, M (2008). Ecological Roles of Alkaloids, In Modern Alkaloids Structure, Isolation, Synthesis and Biology, Wiley-VCH Publication, pp. 2-15

World Health Organization (WHO), (2003a). Doc. no. A56/18s

Yan, X, Wu, S, Wang, Y, Shang, X & Dai, S (2004). Soil nutrient factors related to salidroside production of *Rhodiola sachalinensis* distributed in Chang Bai Mountain, *Enviro. Exp. Bot.* **52:** 267-276

Zysk, KG (1992). Religious medicine: the history and evolution of Indian medicine, Transaction Publishers, pp. 11

6

Biodiversity Conservation and Ecotourism–Opportunities and Issues in Uttarakhand

SK GUPTA

DFO, Ecotourism Office of Chief Conservator of Forest, Dehra Dun-248 001, Uttarakhand
E-mail: santoshforester86@gmail.com

ABSTRACT

Biodiversity is essential for environmental stability and food security as well as scenic beauty which is depleting at an alarming rate due to demographic, institutional, educational and socio-economic reasons. Realization of adverse impacts of environmental degradation and awareness generation efforts have oriented people to come for nature-based tourism. The Uttarakhand State, also known as 'Dev Bhoomi' owing to sacred locales, is endowed with bounty of nature in the form of pristine forests and areas of natural beauty and invites a good number of tourists every year. The present study attempts to elaborate the need for biodiversity conservation alongwith certain environmental, administrative and socio-cultural issues in generation of employment opportunities for local people through ecotourism.

Keyword: Biodiversity conservation, ecotourism, environment pollution, responsible tract, community-based tourism.

INTRODUCTION

The forests provide an abode to a vast array of animal kingdom because of their sheltering and livelihood services through multifarious uses. Local communities depend on forests nearby their habitations for fuel, fodder, timber, food, fibres and several other kinds of minor forest produces. India is an agrarian economy where forests play a very important role not only in providing livelihood opportunities but bringing in monsoon to foster the rainfed agriculture. These regulate local and global climate and influence earth's energy budget. In fact, forest security is as important as food security for the survival of life on the planet. Forests have also been considered a center of mental peace and wisdom and secured a very high position in our spiritual life since times immemorial. But the overemphasis on revenue-earning policies in forest management, based on species-specific sustained yield of timber without paying adequate heed to the requirements of local people has led to alienation of local communities from forest conservation activities which has hampered the ever-existing symbiotic relationship between forests and the communities (Gupta, 2002). In the present era of scientific development, growing industrialization and drastic surge in transportation facilities alongwith several other anthropogenic activities like excessive use of polythene and plastic items as well as electronic wastes are leading to environmental pollution hazards. Global warming has attracted the attention of every individual sensitive to environment conservation and natural resources. This has been witnessed in the form of national and international debates putting their heads together by individuals from different walks of life and multinational agreements. The Earth Summit of 1992 at Rio de Janeiro, Kyoto Protocol and the United Nations Environment Programme are such examples in this regard. India has not only signed the Kyoto Protocol but also has the privilege of hosting the eighth conference of the parties to the United Nations Framework Convention on Climate Change (UNFCCC) in October–November, 2002 (Rana, 2003). Owing to the grave concerns expressed by environmentalists, social scientists and perceptive foresters, the situation has started to turn the side from timber production to biodiversity conservation, policing practice in forest management to involvement and participation of local communities, and stress on more revenue earnings to increasing employment and income opportunities for local people. Realizing the adverse impacts of environmental degradation and due to awareness generation efforts made by various agencies people have started to understand the sensitivity of the problem and come towards the natural areas for nature-based tourism. The United Nations Organization declared 2002 as the International Year of Ecotourism and has adopted "Tourism and Biodiversity" as the theme of the World Tourism Day 2010 as a measure to create awareness amongst the masses especially tourists about depletion of natural resources and the impacts thereof.

NEED FOR BIODIVERSITY CONSERVATION

The western Himalayan region in which Uttarakhand is situated also adds to the richness of country's biodiversity, which is indispensable to agricultural production and food security since it helps in stability of farming systems. About 5,000 plant species have been used as food and another 75,000 are expected to be edible (Trivedi, 2003). It is a valuable ingredient of environment and provides safeguard against anticipating futuristic changes in environmental conditions. It renders certain environmental services for survival of life such as availability of breathing gas, regulation of environmental temperature, soil and water conservation, aesthetic value, etc. alongwith source of employment for a large number of people. It also feeds several industries viz. pharmaceuticals, colours and paints, petroleum, all industries based on wood and non-wood forest products since the biological resources constitute raw materials for them. Though sociocultural and spiritual environment in our country is conducive for biodiversity conservation as people worship several species, viz. Aam (*Mangifera indica* L.), Bargad (*Ficus benghalensis* L.), Pipal (*Ficus religiosa* L.) Kela (*Musa paradisia* L.), Malijhan (*Bauhinia vahlii* L.), Tulsi (*Ocimum tenuifolium* L.), etc., yet it has been realized with serious concerns that biodiversity is being lost from natural habitats at an alarming rate. The basic causes of loss of biological resources are demographic, institutional, educational and socio-economic. There is a growing demand of biological resources due to increasing population and economic development. People fail to recognize long-term consequences of their actions often due to lack of knowledge. Alienation of local communities from forest protection deeds alongwith loss in community feeling and soaring selfishness further aggravate the situation. Educational and sensitization programmes to create awareness among masses about judicious use of biological resources for maintenance of ecological balance are required to counter the threats to environmental stability. Such programmes need to aim at employment and income generation to local people and their involvement in management and execution so that it may awaken their sense of belonging about natural resources and strengthen the weakening symbiotic relationship between them.

Man has always been inquisitive to know more and more about different parts of the planet. Tourism used to be considered from times immemorial as an effective means for extension and association of different civilizations, cultures, education, technology, business, etc. Therefore, people use to go to other areas for coming out of their own peripheries. It is considered one of largest sector to provide employment and income opportunities directly or indirectly, locally as well as distantly since it generates employment opportunities in the form of different jobs, viz. transportation, hotels and home stays, cooking, housekeeping, tour operation and guides, service providers, cultural activities, business, etc. Development of tourism destinations, infrastructure, and basic amenities further add to the employment opportunities and mobilization of local economy. Ecotourism, also used as synonymous to nature tourism, village or community-based tourism, herbal tourism, scientific or education tourism, etc., is recognized conducive to biodiversity conservation and fits very well within the larger gamut of sustainable tourism.

ECOTOURISM

Ecotourism is relatively a novel concept, firstly used in 1982-83 by Hector Ceballos Lascurain, which incorporates conservation of natural resources and employment and income generation for local people. The term has been defined differently by various organizations and authors (*see also* Bhattacharya, 2005; Lindberg *et al.*, 1999). The most acceptable definition of ecotourism has been given by the International Ecotourism Society – "A responsible travel to natural areas that conserves the environment and sustains well-being of local people". The Commonwealth Department of Tourism has defined it as "nature based tourism that involves education and interpretation of the natural environment and is managed to be ecologically sustainable". The World Tourism Organization has given the definition of ecotourism as "Tourism that involves travelling to relatively undisturbed areas with the specified object of studying, admiring and enjoying the scenery and its wild plants and animals". Another prevalent and comprehensive definition of ecotourism given by the IUCN states that "it is environmentally responsible travel and visitation to relatively undisturbed natural areas, in order to enjoy and appreciate nature (and any accompanying cultural features–both past and present) that promotes conservation, has low visitor negative impact and provides for beneficially active socio-economic involvement of local populations". Several other authors have also commented on ecotourism definition but, all in all, ecotourism encompasses: (i) minimum negative impacts (ii) education (iii) building environmental and cultural awareness and respect (iv) gaining positive experience both for visitors and hosts (v) providing economic benefits and empowerment of local people. In essence, it must be environmentally, socially, culturally and economically sustainable. Thus, every tourist coming to forest areas cannot be regarded as ecotourist. The prime objectives of ecotourism comprise: (i) to increase awareness among the masses about importance of environmental conservation and judicious use of natural resources (ii) to empower local communities to manage ecotourism and generate additional livelihood options for them through conservation of efforts to strengthen symbiotic relationship between natural resources and local communities (iii) to provide visitors with a personal experience of nature and culture in a way that leads to greater understanding and appreciation (iv) to stimulate private efforts to make positive contribution to conservation of nature, natural and cultural resources and promote local development.

OPPORTUNITIES

The Uttarakhand State came into existence from the erstwhile Uttar Pradesh State on November 8, 2000 and covers about 2/3rd of its total area under forest. The total geographical area of the State is approximately 53483 sq. km. out of which 344650.444 sq. km. is under forests of different categories. Of the total geographical area, 46.06% falls under reserved forest, protected forest area is 18.48%, private and other forests 0.25%, and non-forest area 35.21%. Also, out of the total forest area, 70.05% is under the control of Forest Department Van Panchayats 15.73%, Revenue Department 13.76%, and private and others 0.46% (Anonymous, 2009). Obviously, a big chunk of population of the State depends on forests nearby their habitations. The State is recognized as a megadiversity region on the world map. There are six national parks and six wildlife sanctuaries alongwith two conservation reserves. These include the world famous Corbett National Park, Valley of Flowers National Park, Nanda Devi National Park, Binsar Wildlife Sanctuary, Assan Conservation Reserve, etc. The Nanda Devi Biosphere Reserve consisting of the Valley of Flowers National Park and the Nanda Devi National Park has been declared as 'World Heritage Site' by the International Union for Conservation of Nature (IUCN). These places fetch a big number of wildlife and nature tourists every year. The State is endowed with bounty of nature in the form of pristine forests and areas of natural beauty, viz. high Himalayan snow-covered peaks, glaciers and alpine meadows at higher elevations, temperate and sub-temperate forests of conifers and associated broad-leaved species in the middle, subtropical to tarai and bhabhar forests in the lower ranges of the Himalaya and the Shiwalik alongwith upper Gangetic plains (Arha and Singh, 2008). The third highest peak in the world, Nanda Devi lies in the State. The climate varies from tropical at the foothills and plain areas to temperate and permanent ice and snow at higher elevations. Uttarakhand is also termed as 'Dev Bhoomi' because of the highly sacred places, viz. Kedarnath, Badrinath, Gangotri, Yamunotri, Poornagiri temples as well as Haridwar and Rishikesh situated in its different parts alongwith the route to Kailash Mansarovar and attracts a huge number of religious tourists every year from different parts of the country and abroad. Mussoorie, Nainital and Ranikhet are highly favourable hill stations of a large number of tourists. The state possesses ample scope for mountain trekking and adventure tourism. Auli, known as the Switzerland of Uttarakhand, is the only place in the world where skiing can be enjoyed at more than 3,000m height having scenic beauty of more than 6,000m high peaks in front. Rafting and beach camping on the banks of the Ganga, Tons and Ghaghara rivers have become very popular among the adventure tourists and offer ample scope for extension of this activity in other areas of the State, as well.

ISSUES

Ecotourism is steadily gaining pace particularly in the 21st century. According to the World Tourism Organization ecotourism is considered the fastest growing market in the tourism industry with an annual growth rate of 5% worldwide, representing 6% of the world's gross domestic product and 11% of all consumers' spending. However, in the absence of a well-defined ecotourism policy, unplanned and uncontrolled ecotourism may lead to adverse impacts which may be ecologically as well as socioculturally very significant (Mieczkowski, 1995). In an effort to adopt community management mechanism for natural resource conservation, it is imperative to strengthen socio-cultural integrity which is the foundation for the sustainability of indigenous management practices (Anonymous, 1999). Certain issues before executors and policy makers in the Uttarakhand State for proper management of ecotourism may be described are:

1. Environmentally, quantum of tourism to eco-fragile areas is a very important issue. The unplanned and uncontrolled ecotourism may lead to mass tourism which is bound to create environmental problems such as change in local climate, lateral shift in species composition, augmentation in biotic pressure on natural resources, disturbance to wildlife habitats, etc. The loss of biodiversity and environment pollution are the most critical issues related to environment and development as biological resources and clean environment are very essential not only for life but for the ecotourism too. The tour companies and operators advocate for big infrastructure development for high class boarding and lodging facilities, easy and comfortable transportation, sanitation works, etc. but their contribution towards environment conservation is certainly a debatable issue on the basis of experiences gained as yet. Several truck loads of polythene and non-degradable garbage left over by tourists as their footprints were lifted from Hemkund Saheb, Gomukh and Gangotri, Dhanolti areas of the State which were creating environmental pollution hazards. Resort owners and managers on the periphery of the Corbett National Park provide high class staying facilities aiming at more money from tourists residing in their resorts where per capita energy consumption is manifold to the average energy consumption being done by the local people. Alike situation is also witnessed at other ecotourism places in different parts of the country where unplanned tourism development and execution is leading to leisure tourism in lieu of the ecotourism which is not a sojourn. An ecotourist, instead of high class boarding and lodging facilities, wants to enjoy the nature and experience wilderness and is interested to gather more information about the local environment and natural resources, culture, traditions, food habits, historical descriptions, heritages, etc. and be provided with literature related to these, experiences of local food and cultural shows, facilities of eco-games for ladies and children during their stay. There is an obvious need to sensitize tour operators, resort managers alongwith tourists about the importance of environment conservation and proper use of natural resources.

2. Administratively, the prevailing attitude of forest personnel who are accustomed to a specific style of functioning based on regulatory mechanism, policing and high ego creates a big bottleneck in proper execution of ecotourism activities. Moreover, they are glutted with different types of activities besides development and execution of ecotourism activities etc. at the same time, viz. protection of forest and wildlife resources, conservation of natural resources, huge burden of paper work and meetings, hospitality of higher officials and VIPs, legal proceedings, association of local communities in forest management alongwith their social obligations. Ecotourism is a very comprehensive activity consisting of transportation, hospitality, boarding and lodging, infrastructure development related to ecotourism, marketing, etc. and each one is a specialized subject in itself. Being a relatively new concept about which the forest personnel are not adequately skilled to fulfil the high expectations of tourists. It is a general complaint that ecotourism is not on priority list of forest officials and they are reluctant to book accommodation in forest rest houses for tourists which may be true sometimes but occurs due to imminent emergency demand by higher authorities and VIPs. Moreover, the authority to book accommodation in forest rest houses lies with divisional offices whereas the tour operators want instant booking in a forest rest house by the Forest Rest House Attendant. This floppy situation is certain to be misutilized at times. It is proper to get accommodation booked from the divisional authorities in advance but with the receiving attitude by forest authorities too. To provide patronage to local communities during execution of ecotourism activities particularly in community-based ecotourism is very much required. It has been experienced that communities are short of tourists for sustenance in the heritage 'Corbett Village, Chhoti Haldwani' where communities have established home stays with very neat and clean boarding and lodging facilities alongwith opportunities to acquire valuable information about the renowned hunter cum conservationist, Jim Corbett at a reasonable price. The place is not far from the Corbett National Park. The flux of tourists in the Corbett National Park is so big that a sizable number of tourists are unable to get booking inside the park and they reside in resorts. The park authorities can develop some mechanism to incorporate the accommodation capacity of the home stays of that village in their management and can provide such tourists to the villagers who can be accommodated there. Training institutions and colleges may also be directed to plan excursions to such villages which are engaged in ecotourism activities so that trainees may have an opportunity to get experience of community-based ecotourism and communities can be benefitted.
3. Socioculturally also, unplanned and uncontrolled tourism is certain to produce detrimental impacts as it not only augments biotic pressure on natural resources but creates several stresses on local resources related to availability of food, water and energy, sanitation, environmental pollution, staying facilities as well as law and order. The presentation of local cultures and providing lifestyle experiences are important parts of ecotourism. The Uttarakhand State has a variety of rich cultural traditions unique to the respective areas and habitant communities are very sensitive about them. Tourists need to observe, study and enjoy these but behave properly and with respect to them so that community sentiments are not hurt. Local people have also to be vigilant against pollution of their innate culture by exotic influences. Here, the role of tour operators and resort managers is very critical since the tourists get information about the area which they visit at first from them. They

provide facilities alike to exotic lifestyle and culture in their resorts and opportunities to consume more energy and resources whereas they have to focus on creating itineraries with a view to explore new areas providing opportunities to local business and give best possible experiences to tourists, minimizing the adverse impacts on natural and cultural resources (Horizontes, 1996). The situation is leading to more commercialization rather than conservation of natural resources. The stress on merely monetary advantages is not sufficient to promote willingness among the village communities towards conservation of natural resources. Development of infrastructure, amenities and pathways leading to local temples, cultural fair places and heritage sites also need to be undertaken alike being done for the tourist destinations. It must not be forgotten that all works related to any help in conservation of environment and natural resources such as forest fire control, check on illicit felling and poaching will be performed by local people only.

Training and capacity building programmes for community members as well as frontline forest personnel alongwith awareness generation programmes for masses including tourists, tour operators and resort managers about the importance of environmental stability and proper use of natural resources are very much required. Guidelines to be followed by tourists during their visit be provided to them in the form of pamphlets as well as displayed on signboards particularly at camping places. These should include:

(i) Please enjoy the nature without disturbing the biodiversity of the area and be an ambassador back home.
(ii) Respect local traditions and culture and behave decently.
(iii) Be cautious about your safety and remain on trek while trekking, do not go beyond the earmarked area and try to escape the time slot for visitation.
(iv) Pay attention about sanitation, pick your garbage and polythene with you instead of throwing it and use dustbins.
(v) Be fair and frugal while using water and energy.
(vi) Do not make noise, feed and disturb wildlife.
(vii) Never chase wild animals for ultimate photograph and come between a parent animal and its kids.
(viii) Please cooperate with forest personnel for environment conservation and inform them immediately about any fire or unwanted incident.
(ix) Use of certain items such as firearms, searchlights and inflammables, etc. be restricted inside the forest area, and likewise.

There is an urgent need to promulgate ecotourism policy at the national as well as state level for effective planning and management of ecotourism activities. Management of ecotourism through public-private partnership is a very delicate matter as it frightens to produce certain complexities related to protection of forest and wildlife resources, infrastructure and environment conservation. Ecotourism is certainly an eco-friendly activity but with adequate precautions. Forest personnel also need to change their attitude to facilitate ecotourists in the best possible manner with welcoming attitude. It has to be seriously regarded that monetary advantages must not overcome the concerns relating to the environmental stability and sociocultural integrity.

REFERENCES

Anonymous (1999). Biodiversity Conservation Network – Evaluating LINKAGES Between BUSINESS, the ENVIRONMENT, and Local COMMUNITIES – Final Stories From the Field. World Wildlife Fund. Inc., Washington DC

Anonymous (2003). Sustainable Development of Ecotourism – A Compilation of Good Practices in SMEs. World Tourism Organization

Anonymous (2009). Uttarakhand Forest Statistics, 2008-09. Uttarakhand Forest Department, Dehra Dun

Arha RS & Singh Latika (2008). Glimpses of Uttarakhand. A B D Publishers, Jaipur

Bhattacharya AK (2005). Introduction **In:** Ecotourism and Livelihoods - Capacity Building for Local Authorities. *In:* Bhattacharya AK (Ed.) .Concept Publishing Company, New Delhi

Gupta SK (2002). Aspects and Sustainability of Joint Forest Management System. Bishen Singh. Mahendra Pal Singh, 23-A, Connaught Place, Dehra Dun

Horizontes (1996). A Brief History of Horizontes. San Jose, Costa Rica, USA

Lindberg, K, Wood, M & Engeldrum, D (1999). Ecotourism – A Guide for Planners and Managers. Vol. 2. Natraj Publishers, Dehra Dun

Rana, AK (2003). Editorial, special issue on climate change and forestry- part I. Indian For. Vol. No. 6

Mieczkowski, Z (1995). Environmental Issues of Tourism and Recreation. University Press of America, Inc., New York

Singh, HS (2003). Plant biodiversity conservation: challenges for the new millennium, pp. 1-4. *In:* Trivedi, PC & Sharma KC. (eds) Biodiversity Conservation. Aavishkar Publishers and Distributors, Jaipur

Trivedi PC (2003). Biodiversity and Human Welfare, pp. 260-276. *In:* Trivedi, PC & Sharma, KC (eds). Aavishkar Publishers and Distributors, Jaipur

7

Distribution, Phylogeny and Phytochemistry of *Ephedra*: A Traditional Medicinal Plant

Vandana Sharma and Manjul Dhiman

Department of Botany, Kanahiya Lal DAV PG College, Roorkee–247 667, Haridwar, Uttarakhand
E-mail: manjul.dhiman@rediffmail.com

ABSTRACT

Ephedra is considered as Somlata in Indian ancient Ayurvedic literature and has also been known as an important traditional Chinese medicine for several thousand years. There are about 50 species of *Ephedra* occurring worldwide. However, its phylogenetic position is a matter of concern to plant biologists because of its important position in evolutionary history of gymnosperms and angiosperms. There are differences in its systematic position based upon morphological features and molecular studies. A review is proposed to assess the distribution, phylogeny and phytochemistry of this important medicinal plant which is going to be exhausted due to its excessive use.

Keywords: Biodiversity, *Ephedra*, ephedrine, molecular markers, phylogeny, phytochemistry.

INTRODUCTION

Ephedra (Joint fir) is an interesting genus of the seed-bearing non-flowering plants belonging to the highly evolved order of gymnosperms, Ephedrales (Bhatnagar and Moitra 1996). Events such as double fertilization, advanced anatomical features and entemophyllous pollination in a few species, further add to the antiquity of this plant. *Ephedra* may be treated as closest relative of the angiosperms, however; molecular studies suggest that this is not the ancestor of angiosperms (Dowd *et al.*, 1998, Martin and Dowd 1986; Chase *et al.*, 1993; Doyle *et al.*, 1994; Nickrent and Soltis 1995).

Ancient Indian Ayurvedic literature reveals *Ephedra* as the **Soma** plant and in China, it is being used as TCM (Traditional Chinese Medicine) for more than 5000 years and is popularly known as **Ma Huang**. The plant shows extreme xerophytic characters and is mostly shrubby 'switch plant' less than 2m in height (Fig. 1A and B). A few species are lianas; others spread by underground rhizomes; *E.compacta* is a compact, profusely branching type usually up to 30cm in height (rarely reaching up to 50cm), whereas, one species (*E. triandra*) is a small tree. The stem of *E. triandra* reaches a diameter of 30cm and a height of several meters. *E. campylopoda*, which has pendulous branches, has been found growing naturally between the stones.

Ephedra resembles to that of *Equisetum* in morphology. The internodes are longitudinally ridged and the ridges on the successive internodes alternate. The leaves are in whorls of three (sometimes four) at each node. There are determinate branches, which are borne in the axil of some of the leaves of indeterminate branches show opposite and decussate pairs of leaves. The leaves are small, scaly and thin and are fused into a basal sheath. They are green when young but turn brown at maturity and are shed. Accessory buds arise below and at the base of axillary shoots. Since the leaves are scaly and reduced, the photosynthesis is performed through the green stem.

Ephedra is typically dioecious. The male and female strobili are borne in the axils of leaves of determinate shoots and also in whorls on the nodes of older branches. The male plant (Fig. 1C) bears male strobilus consisting of a central axis and a few opposite and decussately arranged bracts. Each bract bears a microsporangiate flower in its axil-except the lowest pair which is usually sterile. Each flower consists of an unbranched sporangioiphore, bearing at its tip 2-6 bilobed, sessile, microsporangia. The sporangiophore is enclosed in a pair of perianth leaves and it grows out of the latter during maturation of the sporangia. The female strobili (Fig. 1D) consist of four to seven pairs of opposite and decussate green bracts, fused at the base forming a cup-like structure. In the axil of each of the uppermost pair of bracts, there is an ovule. Each ovule consists of two envelopes. The inner envelope, or the integument, is thin and fused with the nucellus except in the upper region, which protrudes to form a long micropylar tube. The outer envelope is thicker and completely free from the integument. The two subtending bracts of megasporangiate strobilus form the outer fleshy covering of the mature seed. This may be red as in *E. gerardiana* (Fig. 1E), ivory white as in *E. foliata* (Fig. 1F), or orange as in few other species.

Systematics of the genus *Ephedra*

Ephedra belongs to class Gnetopsida having 3 different genera; *Ephedra, Gnetum* and *Welwitschia.* These are grouped in monotypic orders, viz. Ephedrales, Gnetales and Welwitschiales, respectively. The genus *Ephedra* is divisible into three sections based on the nature of bracts in ovulate cones. These sections are *Alatae* Stapf, *Asarca* Stapf, and *Pseudobaccatae* Stapf (Stapf, 1889). Section *Alatae* was defined as having the cone bracts laterally winged and dry at maturity, whereas in section *Asarca* the bracts were described as dry but without prominent membranous wings, and in section *Pseudobaccatae* (now known as section Ephedra since it includes the type species), the bracts are unwinged and become fleshy at maturity. The sections are further subdivided into seven "tribes," whose correct rank under the International Code of Botanical Nomenclature is

Fig. 1. A and B: Plants of *E. gerardiana* growing in natural habitat. **1C.** Male plants of *E. gerardiana* bearing male strobili. **1D.** Female plant of *E. gerardiana* having several female cones. **1E.** Mature seeds of *E. gerardiana.* **1F.** Mature seeds of *E. foliata.*

either subsection or series (Stapf 1889). However, this grouping is not supported by pollen and wood characters (Price, 1996; Ickert Bond and Wojciechowski 2004). Recently, Price recognized as much as 50 species based upon geographic distribution. A list of currently recognized taxa of *Ephedra* and their geographic distribution is given in Table 1 (Price 1996). There is still a major need for broad-based systematic studies making use of both morphological and molecular characters to critically assess the delimitation of species and sectional or subsectional groups.

DISTRIBUTION

There are about 50 species of *Ephedra* (Stevenson 1993, Price 1996, Caveney *et al.*, 2001) spreading worldwide, in Europe, temperate Asia, South America and Afghanistan to Bhutan (2400-5000m) adapted to semiarid and desert environment. These are widely distributed in both Eastern as well as Western Hemisphere (Fig. 2).

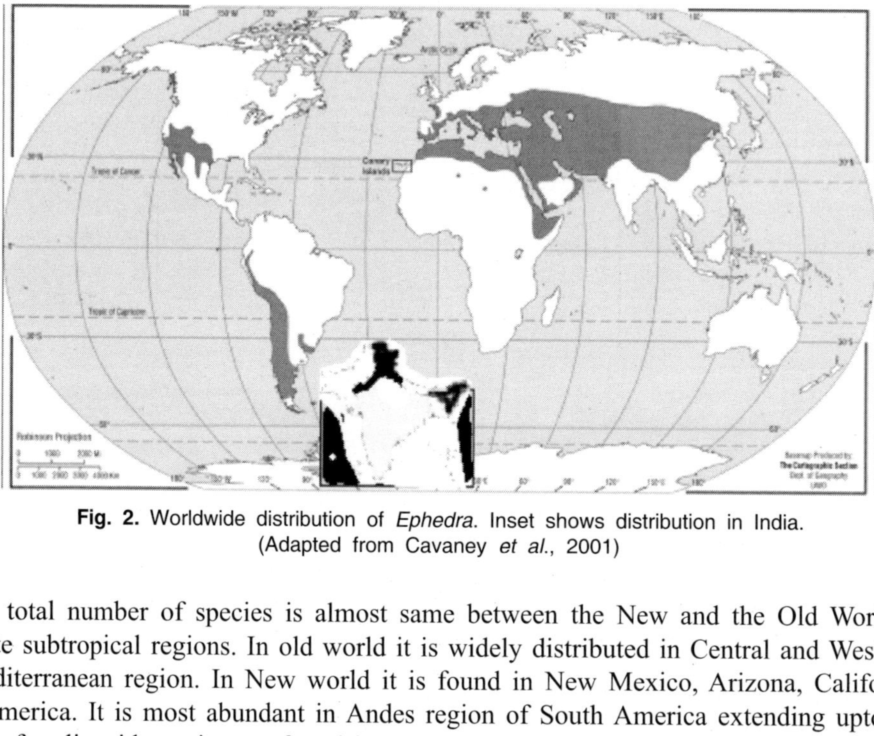

Fig. 2. Worldwide distribution of *Ephedra*. Inset shows distribution in India. (Adapted from Cavaney *et al.*, 2001)

The total number of species is almost same between the New and the Old World in the temperate subtropical regions. In old world it is widely distributed in Central and Western Asia and Mediterranean region. In New world it is found in New Mexico, Arizona, California and North America. It is most abundant in Andes region of South America extending upto atlantic island. A few lianoid species are found in temperate regions along the Mediterranean coastline and humid montane sites in North Africa and Southwest Asia (Freitag and Maier-Stolte, 1989). About 25 species of *Ephedra* are found in drier regions of the Old World extending westwards from Central Asia across Southwest Asia and into Mediterranean Europe and North Africa (Freitag and Maier-Stolte, 1994). The status and distribution of some of the Old World species of *Ephedra*, particularly those of the eastern Mediterranean and Southwest Asia and adjacent arid regions, have been revised and updated by Freitag and Maier-Stolte (1989, 1994). In the New World, twelve species of *Ephedra* are found ranging from the Southwestern United States to the

Table 1: Taxa of *Ephedra* based upon geographical distribution (After Price 1996)

A. Eurasian and North African species

Section *Alatae*

1. *E. alata* Decaisn- N Africa to SW Asia
2. *E. holoptera* Riedl- Iran
3. *E. lomatolepis* Schrenk- Central Asia
4. *E. prezewalskii* Stapf- NW China, Mongolia
5. *E. strobilacea* Bunge- Central Asia to Afghanistan

Section *Ephedra*

6. *E. aphylla* Forssk.- N Africa, Near East
7. *E. altissima* Desf.- N. Africa
8. *E .distachya* L.-Widespread in Europe, Central Asia

   subsp. *distachya* [including *E. monostachya* L.;

   subsp. *mon-ostachya* (L.) Riedl]- S Europe to Central Asia, W Siberia

   subsp. *helvetica* (C.A. Meyer) Ascherson & Graebner [*E. helvetica* C.A. Meyer; *E. vulgaris* Rich.]- Central Europe (Alps)
9. *E. equisetina* Bunge- Russia, Central Asia, N China
10. *E. fedtschenkoae* Paulsen [including *E. regeliana* Florin; *E. monosperma* C.A. Meyer var. *disperma* Regel]- Russia, Central Asia to NW China
11. *E. foeminea* Forssk. [including *E. fragilis* var. *ampylopoda* (C.A. Meyer) Ascherson & Graebner; *E. campylopoda* C.A. Meyer]- Mediterranean Europe, N Africa
12. *E. foliata* Boiss, ex C.A. Meyer [including *E. alte* C.A. Meyer; *E. ciliate* Fisch. and C.A. Meyer ex C.A. Meyer; *E. kokanica* Regel; *E. peduncularis* Boissier]- Egypt to India, Central Asia
13. *E. fragilis* Desf.- Mediterranean Europe, N Africa
14. *E. gerardiana* Wallich ex C.A. Meyer- Himalayan region to SW China
15. *E. intermedia Schrenk and C.A. Meyer*

   var. *intermedia*- SW, Central Asia to Siberia, NW China

   var. *persica* Stapf- Afghanistan to Iran

   var. *tibetica* Stapf- Afghanistan to W China
16. *E. lepidosperma* Cheng- N China (Ningxia)
17. *E. likiangensis* Florin- SW China
18. *E. major* Host [including E. *nebrodensis* Tineo; *E. scoparia* Lange]

   Subsp. *major*- Mediterranean Europe

   Subsp. *Procera* (Fisch. and C.A. Meyer) Bornm. [*E. procera* Fisch. and C.A. Meyer) - Greece, Asia Minor, Caucasus, E to the Himalayan region
19. *E. minuta* Florin- W China
20. *E. monosperma* C.A. Meyer [including *E. minima* Hao]- Siberia to NW China
21. *E. oxyphylla* Riedl- Afghanistan
22. *E. pachyclada* Boissier- Iran, Afghanistan
23. *E. sarcocarpa* Aitch. & Hemsl.- Afghanistan, Pakistan
24. *E. saxatilis* (Stapf) Royle ex Florin- W China
25. *E. sinica* Stapf- N China, Mongolia

B. North American species
Section *Alatae* Stapf
26. *E. torreyana* S. Watson- SW U.S., N Mexico
27. *E. trifurca* Torrey ex S. Waston- N Mexico, SSW U.S.
Section *Asarca* Stapf
28. *E. aspera* Engelmann ex. S. Watson (incl. *E. peninsularis* I.M. Johnston)- SSW U.S., N Mexico
29. *E. californica* S. Watson- California, Baja California
30. *E. cutleri* Peebles [including *E. coryi* var. *viscida* Cutler; *E. viridis* var. *viscida* (Cutler) Benson]- SW U.S.
31. *E. fasciculata* A. Nelson [including *E. clokeyi* Cutler]- SW U.S.
32. *E. funerea* Coville and Morton- Death Valley area, California/Nevada
33. *E. viridis* Coville- SW U.S.
Section *Ephedra*
34. *E. antisyphilitica* Berl. ex C.A. Meyer [including *E. occidentalis* Torrey ex Parl; *E. texana* Reed]- SW U.S. (Texas-Okala.) N. Mexico
35. *E. compacta* Rose- Central Mexico
C. Eurasian and North African species
36. *E. coryi* Reed- SW.U.S.
37. *E. Nevadensis* S. Watson- W. U.S.
38. *E. pedunculata* Engelmann ex S. Watson- Texas, N Mexico
D. South American species
39. *E. boelckei* Roeg- Argentina
40. *E. multiflora* Phil. ex Stapf- N Chile, Argentina
Section Ephedra
41. *E. americana* Humb. and Bonpl. ex Willd.- Andean South American, Argentina
42. *E. breana* Phil. [including *E. wraithiana* I.M. Johnston]- Argentina, Chile
43. *E.chilensis* K. Presl [including *E. andina* Poepp. ex C.A. Meyer; *E. americana* var. andina (Poepp. ex C.A. Meyer) Stapf; *E. bracteata* Miers; *E. chilensis* Miers: *E.* dumosa Miers]- Andean South America
44. *E. frustillata* Miers [including *E. nana* Dusen; *E. patagonica* Phil. ex Stapf]- S Argentina, Chile
45. *E. gracilis* Phil. ex Stapf- Andean Chile
46. *E. ochreata* Miers- Argentina
47. *E. rupestris* Benth- Andean South America
48. *E. triandra* Tul- Argentina to S Brazil
49. *E. trifurcate* Zoellner- Chile
50. *E. tweediana* Fisch. and C.A. Meyer- Uruguay, N Argentina

central plateau of Mexico, and twelve more species occur in South America in an area from Ecuador to Patagonia (Hunziker, 1949; Price, 1996).

Due to a heavy demand and consumption, Japanese market began to search the crude drug from other sources of *Ephedra* from other countries (Kakiuchi *et al.*, 2007) and therefore surveys have been conducted to explore *Ephedra* resources and their genetic diversity from China (Mikage *et al.*, 2003; Long *et al.*, 2004, 2005), Mongolia (Takahashi and Mikage 2003),

Nepal (Mikage and Kondo 1996, 1998) and Russian far east (Kakiuchi *et al.* 2007). Yang and Fu (2003) described a new species *E. rituensis* of the section *Pseudobaccatae* from China.

In India, *Ephedra* is represented by nine species namely *Ephedra foliata, Ephedra gerardiana, Ephedra intermedia, Ephedra nebrodensis, Ephedra regeliana, Ephedra saxatilis, E. pachyclada, E. przewalskii* and *E. sumlingensis* occurring from North West Himalayas to Sikkim (See Fig. 2 Inset) as follows:

1. *E. foliata* Boiss. It is quite common in deserts of Punjab, Sindh and Rajasthan.
2. *E. intermedia* Schrenk. var. *tibetica* Stapf. It is found in N.W. Himalayas- Kashmir, Punjab, Himachal Pradesh and Jaunsar.
3. *E. gerardiana* Wall. It is found in drier regions of temperate and alpine Himalaya from Kashmir to Sikkim, Chamba, Lahaul, Spiti and Ladakh.
4. *E. nebrodensis* Tin var. *procera* Stapf. It is more common in Kashmir.
5. *E. regeliana* Florin (*E. major*). It grows well at Ladakh and Lahaul.
6. *E. saxatilis* Royale var. *sikkimensis* (Stapf) Florin. It occurs in Sikkim and adjacent countries.
7. *E. pachyclada*
8. *E. przewalskii*
9. *E. sumlingensis* (Sharma and Uniyal). It is reported from Sumling, Spiti district of Himachal Pradesh (Sharma and Uniyal 2008)

Porwal *et al.* in 2003 have attempted to map and stratify *E. gerardiana* using remote sensing and GIS to gather geographical information of the species in complex terrain conditions of Lahaul and Spiti region of India. This has resulted into bioprospecting the species in specified areas. Due to excessive usage and collection, the *Ephedra gerardiana* (high alkaloid contents) is listed as an endangered medicinal plant, nationally and internationally (Gupta and Sethi 1984, Badola and Pal 2002).

PHYLOGENY

The significance of the Gnetales to the solution of problems of angiosperm origin has been emphasized by Markgraf (1930), who held the angiosperm to be a taxon derived from gymnosperms and one that branched off from ancestors of modern gymnosperms so long ago as to be of no particular relationship to present-day conifers. Fagerlind (1947) reinterpreted the morphology of *Gnetum* flower, demonstrated it to be homologous with those of *Ephedra* and *Welwitschia*, and considered the three taxa to have had a common ancestor which was derived from the same stock as the single ancestor postulate to have given rise to the proangiosperms from which all modern angiosperms evolved polyphyletically (Lawrence, 1951). However, molecular studies are of different opinion (Dowd *et al.*, 1998; Martin and Dowd 1986; Chase *et al.*, 1993; Doyle *et al.*, 1994; Nickrent and Soltis 1995).

Use of molecular marker for determining the phylogeny of gymnosperms was initially attempted by Hasebe *et al.* in 1992. They sequenced *rbcL* gene of several seed plants and obtained phylogenetic tree based upon amino-acid sequences using the neighbour joining and maximum likelihood methods. They contradicted the results of earlier morphological data and stated monophyly of gymnosperms containing *Gnetum* group. Same view was also supported by Boivin *et al.* (1996) based upon *chl B* gene comparison of various gymnosperms, bryophytes and algae. They described monophyly of the conifers-*Ginkgo*-cycads clade, where conifers were sister to *Ginkgo* and cycads. Further they suggested an early divergence of seed plant lineage leading to *Ephedra*.

Price (1996) carried out *rbcL* sequence in several species of *Gnetum*, *Ephedra* and various other extant seed plant groups. He strongly supported monophyly of Gnetales regardless of whether only one or both of reported sequences for these species.

Evolutionary significance of Gnetales was also described by Bowe *et al.* in 2000 based upon sequenced data sets of *cox1* and *atpA* mitochondrial gene. They strongly supported a gnetophyte-conifer grouping. The gene sequence also suggested Gnetales as derived conifers, sister to Pinaceae.

Huang and Price (2003) assessed the origin period and age of *Ephedra* using chloroplast *rbcL* gene sequences. The *rbcL* marker gene was compared with members of Pinaceae and *Ginkgo*. Based upon the time of a landmark event used to estimate the rate of nucleotide substitution of the lineage (molecular clock hypothesis), the age of extant *Ephedra* was assumed to be nearly 8-32 million years ago.

Ickert Bond and Wojciechowski in 2004 described the phylogenetic relationship alongwith character evolution and historical biogeography in *Ephedra* based upon sequence data analysis of nr DNA *ITS* 1 and plastid *rps* 4 gene. On the basis of sequence analysis, they showed that in *Ephedra*, the length of nr DNA *ITS* 1 varied from 1,081 to 1,143 bp which is drastically much shorter in outgroups (*Gnetum, Welwitschia and Pinus*). By using parsimony and maximum likelihood (ML) methods, they supported monophyly of *Ephedra* and not its subdivision into traditional sections such as Ephedra, Asarca and Alatae. Phylogenetic results also indicated a derivation of New world clade including two main subclades of North and South American species of *Ephedra*.

Evolutionary history of *Ephedra* was described by Rydin *et al.* in 2004 based upon sequence data of two chloroplast gene (*rbcL* 7and *rps4*) and three nuclear ribosomal gene (*26S, ITS, 18S*). They sampled 27 representatives of *Ephedra* from all major continents. A comparison of fossil study and molecular data indicated that there is a very little variation in key reproductive structures of fossils and extant species of *Ephedra*.

Huang *et al.* in 2005 sequenced the nr DNA *ITS-1*, *matK* and *rbcL* gene from species of *Ephedra*. They indicated that phylogenetic relationships in *Ephedra* can be better predicted by geographic region of origin rather than by ovulate cone character. On the basis of sequence data, they suggested a possible early divergence of a New world clade of *Ephedra* from among the Old world groups.

Phylogenetic position of *E. rhytidosperma* was described by Wang *et al.* in 2005 based upon sequence data of *mat k* and *rbcL*, ribosomal gene *18S* and *ITS* region of nuclear ribosomal DNA. They analysed DNA sequence from 16 taxa representing 15 species of genus *Ephedra* by using maximum parsimony (MP), neighbour-joining (NJ), minimum evolution (ME) and maximum likelihood (ML) methods. The result showed that *E. rhytidosperma* is closely related to *E. equisetina* forming a monophyletic group based upon all the phylogenetic trees (Fig. 3).

Ghafoor *et al.* in 2007 assessed the genetic variability using molecular markers occurring in different biotypes of *Ephedra* growing in Pakistan. They collected six genotypes from diverse geographic zones of Pakistan and amplified DNA of all genotypes using nine *RAPD* primers and studied genetic variability at molecular level. On the basis of various *RAPD* markers including functional genes and ribosomal spacer region, they estimated a moderate to high level of genetic polymorphism (6 to 49%) among the various genotypes. The study suggests estimation of genetic distances and diversity in different biotypes of *Ephedra* using molecular markers.

Kakiuchi *et al.* in 2007 analysed the DNA sequences of the *ITS* 1 and 2 of nuclear ribosomal DNA and a non coding sequence of chloroplast DNA (*trn L/F*) to assess the genetic diversity and species confirmation of *Ephedra* resources growing in northern area of Pakistan. Based upon the molecular markers and external morphological features, they confirmed that *Ephedra intermedia* was the major species whereas *Ephedra gerardiana* and *Ephedra prezewalskii* were present sporadically in the area. Further the study also suggested more heterogeneity of *E. intermedia* in DNA sequences of *ITS* 1 region in compare to species growing in neighbouring area.

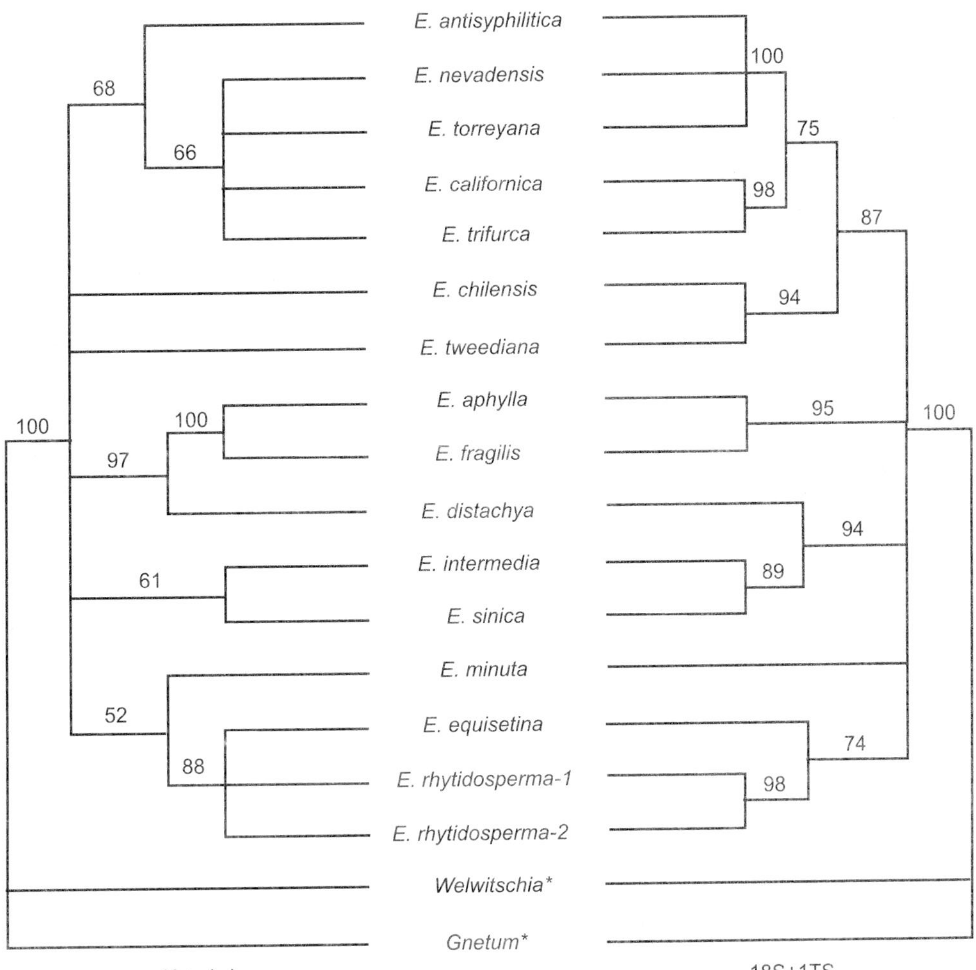

Fig. 3. The phylogenetic tree of *Ephedra* based upon *matK* + *rbc*L chloroplastic genes and *18S* + *ITS* region of nr DNA. (Reprinted from Wang *et al.*, 2005)

The results of phylogenetic analyses show that the evidences from the molecular data do not support the traditional taxonomy of *Ephedra* based on morphological characters. However, molecular data can be utilized to assess various geographical races and subspecies.

PHYTOCHEMISTRY

The major active ingredients of *Ephedra* are alkaloids that constitute 0.5 to 2.5 % of total mass, and are referred to as ephedrine type alkaloids (Blumenthal and King, 1995). The six optically active alkaloids that have been isolated from *Ephedra* species are (-)-ephedrine, (+)-pseudoephedrine, (-)-*N*-methylephedrine, (+)-*N*-methylpseudoephedrine, (-)-norephedrine, (+)-norpseudoephedrine (Fig. 4).

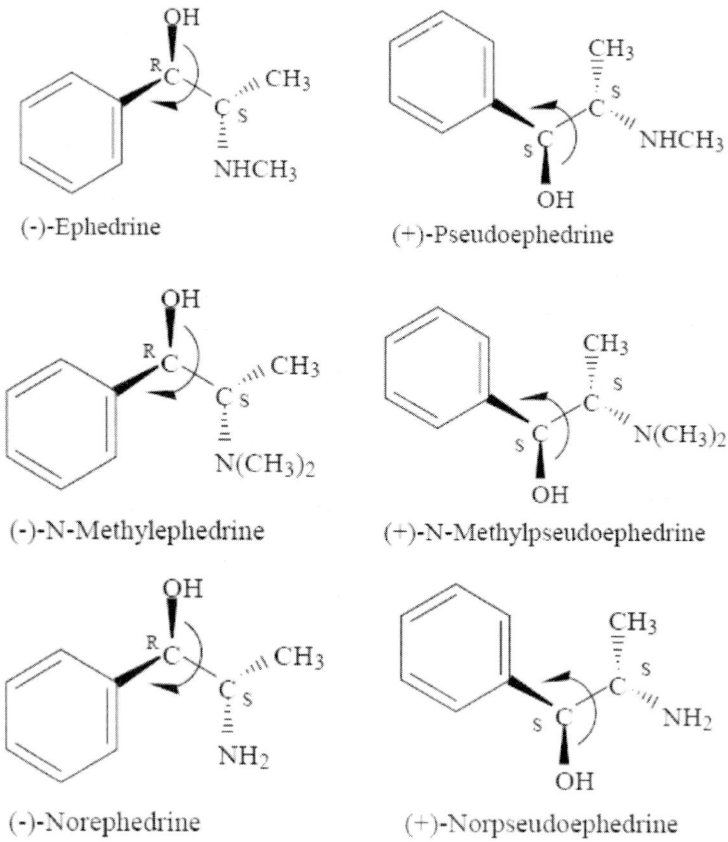

Fig. 4. Molecular structures of the ephedrine analogue. (adopted from Finar 1997)

Physiologically, the action of *Ephedra* is similar to that of adrenaline and it can be taken orally. (-)-Ephedrine is the major isomer that comprises 30 to 90% of total alkaloid fraction accompanied by (+)-pseudoephedrine, with trace amount of other ephedrine type alkaloids (Blumenthal and King, 1995). There is variation in the alkaloid contents of different species of *Ephedra*. The total contents of ephedrine-type alkaloids depend on the species of *Ephedra*, time of year of harvest, weather conditions and altitude where the plants grow (Blumenthal and King, 1995), and can exceed by two per cent (Bruneton, 1995). Evans (1989) suggested that there is variation in chemical races within the species *Ephedra distachya*. Therefore, environmental and genetic factors influence the alkaloid yield of the ephedrine. Besides the above, distribution of alkaloids in various parts of the plant also varies (O'Dowd *et al.* 1991). This variation depending upon environmental conditions explains why some *Ephedra* containing dietary supplements of the same brand often show alkaloid content markedly differing from label claims and also variation among lots when analyzed chemically (Gurley *et al.*, 2000).

Preparation of ephedrine-type alkaloids from the crude plant material involves an acid/base extraction procedure (Reti, 1953). In addition to the extraction from plants, ephedrine – type alkaloids can also be chemically synthesized, and most of the ephedrine and pseudoephedrine used in western medicine has been manufactured synthetically. However, these synthetic ephedrine-type alkaloids differ from the natural forms in that they are racemic, i.e., optically inactive, because they are made up of two enantiomorphic isomers (Abourashed *et al.*, 2003).

The molecular formula of ephedrine is $C_{10}H_{15}NO$ and since on oxidation ephedrine forms benzoic acid, the structure contains a benzene ring with only one side-chain. When treated with nitrous acid, ephedrine forms a dibenzoyl derivative suggesting that one hydroxyl group must be present (one benzoyl group is accounted for amino group). Finally, when heated with hydrochloric acid, ephedrines form methylamine and propiophenone (Finar, 1997).

The starting point for ephedrine is phenylalanine (Fig. 5) and this is synthesized by the shikimic acid route. Methionine has been shown to be the source of both *O*- and *N*-methyl groups of many alkaloids. Formate can also act as a source for *N*–methyl, and is the source for *C*-methyl groups.

Pharmacologically, *Ephedra* has been the main botanical source of the active alkaloids l-ephedrine (E) and d-pseudoephedrine (PE) for thousands of years, with records of its medicinal

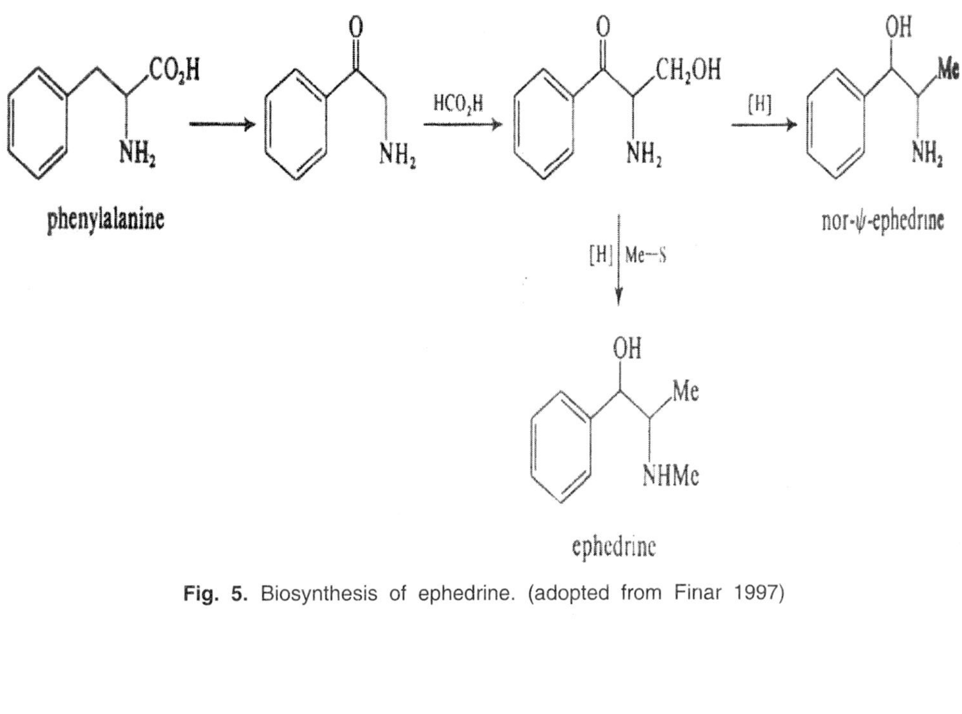

Fig. 5. Biosynthesis of ephedrine. (adopted from Finar 1997)

use dating to 5000 years B.C. (Evans 1989). The alkaloids E and PE remain important drugs today- the current world consumption of d-pseudoephedrine salts (PE-sulphate and PE-hydrochloride) stands at 1000-2000 tonnes per annum with a value of approximately $100-200 millioin (O'Dowd *et al.* 1998). Synthesized ephedrine alkaloids are found in hundreds of prescriptions, and over the counter pharmaceutical products as antisthmatic, bronchodilators, antihistamines, decongestants, appetite suppressants, and weight-loss aids. Although it is currently possible to manufacture ephedrine from benzaldehyde via a yeast fermentation and catalytic reduction, several species of *Ephedra* continue to be exploited for these high-demand pharmaceutically active compounds. Therefore, the continued importance of these well-known *Ephedra* alkaloids appears assured for some time to come, and there is the prospect that other pharmacologically active agents may be found in *Ephedra.*

APPLICATIONS

Ephedra has been used for more than 5,000 years in China and India to treat conditions such as cold, fever, flu, headache, asthma, wheezing, and nasal congestion. It has also been an ingredient in many dietary supplements and used for weight loss, increased energy, and enhanced atheletic performance.

Ephedra and its alkaloids are effective bronchodilators in the treatment of mild to moderate asthma and hay fever. Ephedrine is a potent bronchodilator that, in appropriate doses, can be administered safely alongwith therapeutic doses of theophylline without the fear of progressive tolerance or toxicity.

Ephedra preparations are also useful as a weight loss aid. Although ephedrine suppresses appetite, its main mechanism for promoting weight loss appears to be by increasing the metabolic rate of adipose tissue. Its weight reducing effects are greatest in those individuals with a low basal metabolic rate and/or decreased diet-induced thermogenesis.

As a therapeutic agent for the treatment of obesity (Zgourides *et al.*, 1989), l-ephedrine has been linked with toxic psychosis (Whitehouse and Duncan, 1987) with the result that the demand is increasing for the stereoisomer, d-pseudoephedrine (PE), which has fewer side effects than l-ephedrine and a weaker, longer-lasting effect upon the central nervous and cardiac system. A recent development is the introduction by Schering-Plough of a new combination product containing pseudoephedrine sulphate and Loratadine, a non-sedating antihistamine (Sanchez Garcia and Trejo Tapia, 1993).

Unlike other herbs, the active ingredients are well characterized and consist of ephedrine and related alkaloids (mostly ephedrine, pseudoephedrine, norephedrine, and norpseudoephedrine). Ergogenic effects of *Ephedra* are similar to ephedrine, and include increased energy, time to exhaustion, power output, running speed, and weight loss (Bell *et al.*, 2002; Jacobs *et al.*, 2003; Shekelle, 2003; Pittler and Ernst, 2004). Ephedrine is a sympathomimetic substance that stimulates both α- and β-adrenergic receptors (Abourashed *et al.*, 2003). Stimulation of α_1-adrenergic receptors produces contraction of vascular smooth muscle, increased contractile force of the heart and arrhythmias, glycogenolysis, gluconeogenesis, hyperpolarization, and relaxation of intestinal smooth muscle. Stimulation of α_2-adrenergic receptors decreases insulin secretion, platelet aggregation and the release of norepinephrine from the nerve terminals, and cause contraction of vascular smooth muscle. Stimulation of β_2-adrenergic receptors increases rate of contraction of heart, velocity of conduction through the atrioventricular node, and rennin secretion. When used in therapeutic doses, stimulation of β_2-adrenergic receptors causes relaxation of smooth muscle of blood vessel and bronchi (Mack, 1997). Ephedrine also has central nervous system (CNS) stimulant effects similar to those of amphetamines, but less pronounced (McEvoy, 2000).

In Japan, *Ephedra* plants are consumed at a rate of some 300 tons a year due to their inclusion as a crude drug in several Kampo (sino-japanese medicine) prescriptions, and almost all of these are imported from overseas, mainly from China. However, the stability of the *Ephedra* supply from China is unpredictable, since the wild habitats of *Ephedra* plants in China have been damaged in recent years by the increase in desertification caused by climate change and the industrialization of the country (Kakiuchi *et al.,* 2007). The species *Ephedra gerardiana* has immense medicinal potential in Indian and Chinese medicines. Its excess use as medicine has intensified the pressure of landscapes bearing this species. This has necessitated bioprospection and active planning to ensure safe conservation of the existing gene pool and sustainable utilization of this land resource (Porwal *et al.,* 2003).

EPILOGUE

The species of *Ephedra* differ strikingly from the other gymnosperms both in habit and habitat. They are not valuable as timber, but are important for their aesthetic and medicinal value. Another dimension to this aspect is added by the realisation that the natural habitats of gymnosperms inhabiting tropical and warm temperate areas are undergoing changes at a rapid pace. This has led to a steady decline in their natural populations and very few plants can be located in the wild.

In India, high drug-yielding species, namely, *E. nebrodensis* and *E. gerardiana* grow at higher altitudes. Due to over-exploitation of the plant from wild populations, *E. gerardiana* is now listed as an endangered species. As already mentioned, *Ephedra* is one of the source of the alkaloids mainly l-ephedrine and pseudoephedrine. Biotechnological tools such as callus culture can be possibly used for ephedrine biosynthesis in reactors. It may also be feasible to screen *Ephedra* populations and then clonally propagate the elite plants (Dhiman *et al*., 1998). All this can help to meet the demand of this important traditional medicinal plant.

REFERENCES

Abourashed, EA, El-Alfy, AT, Khan, IA & Walker, L. (2003). *Ephedra* in perspective-a current review. *Phytother Res.* **17:** 703-712

Badola, HK & Pal, M (2002). Endangered and medicinal plant species in Himachal Pradesh. *Curr. Sci.* **83:** 797-798

Bell, DG, McLellan, TM & Sabiston, CM (2002). Effect of ingesting caffeine and ephedrine on 10 km runs performance. *Medicine & Science in Sports & Exercise* **34:** 344-349

Bhatnagar, SP & Moitra, A (1996). Gymnosperms. New Age International Ltd. New Delhi

Blumenthal, M & King, P (1995). Ma Huang: Ancient herb, modern medicine, regulatory dilemma; A review of the Botany, Chemistry, medicinal uses, safety concerns, and legal status of *Ephedra* and its alkaloids. Herbal Gram, 22

Boivin, R & Richard, M (1996). Phylogenetic inferences from chloroplast *chl B* gene sequences of *Nephrolepis exaltata* (Filicopsida), *Ephedra altissima* (Gnetopsida), and diverse land plants. *Mol. Phylogenetic & Evol.* **6:** 19-29

Bowe, L Michelle, Coat, Gwenaele, & DePamphilis, Claude W. (2000). Phylogeny of seed plants based on all three genomic compartments: Extant gymnosperms are monophyletic and Gnetales closest relatives are conifers. *Proceedings of National Academic Sciences* **97:** 4092-4097

Bruenton, J (1995). Pharmacognosy, phytochemistry, medicinal plants (Paris, Lavosier).

Caveney, S, Charlet, DA, Frietag, H, Maier-Stolte, M & Starratt, N (2001). New observation on the secondary chemistry of world *Ephedra* (Ephedraceae). *Am. J. Bot.* **88:** 1199-1208

Chase, MW, Soltis, DE, Olmstead, RG, *et al.,* Morgan, D, Les, DH, Mishler, BD, & Duvall, MR *et al.* (1993). Phylogenetics of seeds plants: an analysis of nucleotide sequences from the plastid gene *rbcL. Ann. Mo Bot Gard.* **80:** 528-580

Dhiman, M, Moitra, S, Singh, MN & Bhatnagar, SP (1998). *In-vitro* studies on some non-coniferous gymnosperms. *Curr. Sci.* **75:** 1113-1115

Doyle, JA, Donoghue, MJ & Zimmer, EA (1994). Integration of morphological and ribosomal RNA data on the origin of angiosperms. *Ann. Mo. Bot. Gard.* **81:** 419-450

Evans, WC (1989). Alkaloids. In: Trease and Evan's Pharmacognosy. Balliere Tindall, London. P. 99

Fagerlind, F (1947). Strobilus and Blute von *Gnetum* und die Moglichkeit, aus ihrer Struktur den Blutenbau der Angiospermen zu deuten. *Arkiv Bot.* **33:** 1-57

Freitag, H & Maier-Stolte, M (1989). The *Ephedra* species of P. Forsskal: identity and typification. *Taxon* **38:** 545-556

Freitag, H & Maier-Stolte, M (1994). Ephedraceae. In Browicz, K. (ed.), Chorology of trees and shrubs in south-west Asia and adjacent regions. Bogucki publishers, Poznan, Poland

Ghafoor, S, Shah, MM, Ahmed, H, Swati, ZA, Shah, SH, Pervez, A & Farooq, U (2007). Molecular characterization of *Ephedra* species found in Pakistan. *Gen. and Mol. Res.* **6:** 1123-1130

Gupta, R & Sethi, KL (1983). Conservation of medicinal plant resources in the Himalayan region. In: Jain, SK and Mehra, KL (Eds.). Conservation of Tropical Plant Resources. Botanical Survey of India, Howrah pp. 101-109

Gurley, BJ, Gardner, SF & Hubbard, MA (2000). Content versus label claims in *Ephedra*-containing dietry supplements. *Am. J. Health Sys. Pharm.* **57:** 963-969

Hasebe, M, Kofuji, R, Ito M, Kato, M, Iwatsuki, K & Ueda, K (1992). Phylogeny of gymnosperms inferred from *rbcL* gene sequences. *J. Plant Research* **105:** 673-679

Huang, J & Price, RA (2003). Estimation of the age of extant *Ephedra* using Chloroplast *rbcL* sequence data. *Mol. Biol. Evol.* **20:** 435-440

Huang, J, Giannasi, DE & Price, RA (2005). Phylogenetic relationships in *Ephedra* (Ephedraceae) inferred from Chloroplast and nuclear DNA sequences. *Mol. Phylogenetics and Evol.* **35:** 48-59

Hunziker, JH (1949). Sinopsis de las species argentines del genero *Ephedra. Lilloa* **17:** 147-174

Finar, IL (1997). Stereochemistry and the chemistry of natural products. Addison Wesley Longman limited England

Ickert-Bond, SM, & Wojciechowski, ME (2004). Phylogenetic relationships in *Ephedra* (Gnetales): Evidences from nuclear and chloroplast DNA sequence data. *Systematics Botany.* **29:** 834-849

Jacobs, I, Pasternak, H & Bell, DG (2003). Effects of ephedrine, caffeine, and their combination on muscular endurance. *Medicine & Science in Sports & Exercise* **35:** 987-994

Kakiuchi, N, Inoue, K, Kurita, Y, Ohkubo, K, Tsuda, Y & Mikage, M (2007). Survey of *Ephedra* resources in the northern areas of Pakistan and their genetic diversity. *J. Nat. Med.* **61:** 357-365

Lawrence, MH George, (1951). Taxonomy of vascular plants, Oxford & IBH Publishing Co. Pvt. Ltd, New Delhi

Long, CF, Kakiuchi, N, Takahashi, A, Komatsu, K, Cai, SQ & Mikage, M (2004). Phylogenetic analysis of the DNA sequence of the non coding region of nuclear ribosomal DNA and chloroplast of *Ephedra* plants in China. *Planta Med.* **70:** 1080-1085

Mack, RB (1997). "All but death, can be adjusted". Ma Huang (ephedrine) adversities. *N.C. Med. J.* **58:** 68-70

Markgraf, F (1930). Monographie der Gattung *Gnetum. Bull. Jard. Bot. Buitenzorg, ser 3,* **10:** 405-511

Martin, PG & Dowd, JM (1986). A phylogenetic tree for some monocotyledons and gymnosperms derived from protein sequences. *Taxon* **35:** 469-475

McEvoy, GK, ed. (2000). AHFS drug information (Bethesda, MD, authority of the Board of Directors of the American Society of Hospital Pharmacists)

Mikage, M & Kondo, N (1996). Micro-botanical studies of *Ephedra* plants from the Himalayan region Part I. Anatomical studies on the herbal stems and botanical origin of Tibetan crude drugs " Tshe" and "Balu". *J. Jpn. Bot.* **71:** 323-332

Mikage, M & Kondo, N (1998). Micro-botanical studies of *Ephedra* plants from the Himalayan region Part II. Anatomical characters of herbal stems, and the estimation of the origin locality of Tibetan crude drugs "Tshe" and "Balu". *J. Jpn. Bot.* **73:** 155-164

Mikage, M, Takahashi, A, Chen, HB & Li, QS (2003). Studies of *Ephedra* plants in Asia. Part 1. On the resources of *Ephedra* plants in China. *Nat. Med.* **57:** 202-208

Nickrent, DL & Soltis, DE (1995). A comparision of angiosperm phylogenies from nuclear *18S* r DNA and *rbcL* sequences. *Ann. Mol. Bot. Gard.* **82:** 208-234

O'Dowd, N (1991). *In-vitro* culture, micropropagation and secondary metabolite production of *Ephedra*. Thesis, Trinity College, University of Dublin, Ireland

O'Dowd, NA, McCauley, PG, Wilson, G, Parnell, JAN, Kavanaugh, TAK & McConnel, DJ (1998). *Ephedra* species: *in-vitro* culture, micropropagation and the production of ephedrine and other alkaloids. In: Bajaj, YPS (ed.), Biotechnology in agriculture and forestry **41**. Medicinal and aromatic plants X. Springer-Verlag, Berlin, Germany

Pittler, MH & Ernst, E (2004). Dietary supplements for body-weight reduction: A Systematic review. *Am. J. Clin. Nutrition* **79:** 529-536

Porwal, MC, Sharma, L & Roy, PS (2003). Stratification and mapping of *Ephedra gerardiana* Wall. in Poh (Lahul and Spiti) using Remote Sensing and GIS. *Curr. Sci.* **84:** 208-212

Price, RA (1996). Systematics of the Gnetales: a review of morphological and molecular evidence. *Int. J. Pl. Sci.* **157:** S40-S49

Reti, L (1953). *Ephedra* Bases. In The Alkaloids, Manske, RHF & Holmes, HL (eds.). (New York, NY, Academic Press)

Rydin, C, Pedersen, KR & Friis, EM (2004). On the evolutionary history of *Ephedra* cretaceous fossil and extant molecules. *Proceedings of National academy of Sciences* **101:** 16571-16576

Sanchez, Garcia, M, Trejo, & Tapia. D (1993). Efficacy and Safety of ioratidine plus pseudoephedrine, pediatric solution in children with allergic rhinitis. *Invest. Med. Int.* **20:** 92-98

Schaneberg, BT, Crockett, S, Bedir, E & Khan, IA (2003). The role of chemical fingerprinting: application to *Ephedra*. *Phytochemistry* **62:** 911-918

Sharma P & Uniyal PL (2008). *Ephedra sumlingensis* (*E gerardiana*) Sharma and Uniyal—A new species from Himachal Pradesh, India. Bull. Bot. Surv. Indi **50:** 1-6

Shekelle, PG & Hardy, ML (2003). Efficacy and safety of *Ephedra* and ephedrine for weight loss and athletic performance: a meta-analysis. *JAMA*. **289:** 1537-1545

Stapf, O (1889). Die Arten der Gattung *Ephedra*. Denkschr Kaiseri Akad Wiss Wien Math Nat Wiss Ki. **56:** 1-112

Stevenson, DW (1993). Ephedraceae. In flora of North America Editorial Committee (eds.), Flora of North America, vol. **2**: 428-434. Oxford University Press, New York, New York, USA

Takahashi, A & Mikage, M (2003). A field research for *Ephedra* plants in Inner-Mongolia, China. *Newsl. Himal. Bot.* **31:** 1-7

Whitehouse, AM & Duncan, JM (1987). *Ephedra* psychosis rediscovered. *British J. Psychiatry* **150:** 218-281

Yang, Yang, & Fu, Dezhi. (2003). A new species of *Ephedra* (Ephedraceae) from China. *Novon*. **13:** 153-155

Zgourdies, GD, Warren, R & Englert, M (1989). Ephedrine induced thermogenesis as an adjunct to cognitive restructuring and covert conditioning: a proposal for treatment of obese individuals. *Prepetual motor skills* **69:** 563-572

8

Quantitative Assessment and Ethnobotanical Uses of Plants of Koti–Kanasar Forest Range (Jaunsar), Uttarakhand

DP Semwal, Prabha Sharma, H Govindapyari and PL Uniyal

Department of Botany, University of Delhi, Delhi-110 007
E-mail: uniyalpl@rediffmail.com

ABSTRACT

The present paper highlights the structure, composition and local uses of plant species in Koti-Kanasar area, situated in Dehradun district of Uttarakhand State in north-west Himalaya. These forests constitute immense plant wealth in terms of biodiversity and in providing necessary ecosystem services like preserving wildlife and forest soil, sustaining human life, stabilizing climate, optimizing water-yield and purifying air and water. *Quercus* is the most abundant taxon among angiosperms with its three species *Q. glauca* Thunb., *Q. leucotricophora* A. Camus and *Q. semecarpifolia* J.E. Sm. being equally distributed at altitude ranges 1500–2700 m in this region. *Pinus roxburghii* Sargent and *Daphne papyracea* Wall. ex. Steud. were found to be other dominant species. In site I a total of eight species are recorded, out of this *Pinus roxburghii* Sargent has high frequency (60%) and high density (2.2 trees/100 m^2). In site II, a total of six species is present, out of this *Cedrus deodara* (Royle ex D. Don) has high frequency (50%) and high density (2.5 trees/100 m^2).

Conifers form the uppermost storey of the forest that includes species like *Abies pindrow* Royle, *Cedrus deodara* (Royle ex D. Don), *Cupressus torulosa* D. Don, *Picea smithiana* (Wall.) and *Pinus roxburghii* Sargent. Second storey is formed by Oak. Various herbaceous plants are observed along the altitudinal gradient. The most abundant being *Chenopodium album* L., *Desmodium elegans* DC., *Impatiens glandulifera* (Royle), *Oxalis corniculata* L., *Stellaria himalayensis* Maj. and *Viola canescens* Wall. Other herbs are *Anemone rivularis* Buch.-Ham.ex DC., *Berberis aristata* DC., *Commelina paludosa* Bl., *Galium acutum* Edgew., *Polygonum convolvulus* L., *Prinsepia utilis* Royle, *Ranunculus hirtellus* Royle ex D. Don, and *Sarcococca saligna* (D. Don.).

Keywords: Anthropogenic activities, bioresources, diversity, evolution.

INTRODUCTION

Biodiversity is the product of spontaneous evolution during millions of years. The concept of diversity relates to the species richness of the community. Biodiversity depends upon the intensity, predictability and scale of these interactions and essential biodiversity services to humans (Obute, 2010; Suleiman *et al*, 2010).

India is one of the leading countries in Asia in terms of the wealth of traditional knowledge systems related to the use of plant species. Mountains are globally important as biological rich landscape with complex climate and weather regions. These provide key resources such as minerals, forest products, recreation and life support to the biota. A common characteristic of mountain system is the sociocultural and ethnolinguistic variability of human population.

Forests are the most indispensable aspect of environment and have always been a source of material and food products. This requirement has never ceased and pressure on natural resources has become greater with the growth of the population (Govindapyari *et al.* 2010). Consequently, forests have been indiscriminately cleared for agriculture, grazingland and other commercial purposes. Successful forest management requires base line data on the dynamics of forest communities and methods to build up interdisciplinary approaches to formulate any conservation strategy (Rawat, *et al.*, 2008; Semwal *et al.* 2010). Further, we still lack basic information on the composition and structure of community and the extent to which a particular forest is threatened due to unscientific management. For long-term conservation and sustainable utilization of resources, quantitative studies in such areas are essential. Such ecological studies would help environmentalists, planners and managers in looking towards forest management in a more pragmatic and positive way.

Objectives

The main objective of the study is to assess the structure and composition of forest communities present in and around Kanasar and conducted under the following heads:

1. Species composition.
2. Rate of change of species diversity according to the altitudinal gradient.
3. Association among various species.
4. Resilience in the community.
5. Distribution of various species.
6. Mapping damage of forest due to anthropogenic activities.
7. Species regeneration rate.

Study Area

Kanasar area is situated in Dehradun district of Uttarakhand State in north-west Himalaya. The locality harbours the catchment areas of Tamsa (Tons) and Yamuna. Its altitude ranges between 2200-2800 m and latitude 30° N to 78° E. The area includes long and narrow strips of low-lying plains of 10-25 km width. It is bound by the snowy range on the north and the Shiwalik runs parallel to the hills of Kanasar. Two different sites, Site I and Site II were selected for the quantitative analysis of vegetation.

Site I

The site I was selected at a higher elevation at the hilltop and north-west facing gentle slope. Terrain is mostly grassy along with forest tree species. At the top of the hill large areas of forest have been cleared and these are being used for agricultural purposes. Litter is comparatively limited and composed mostly of needles of *Cedrus deodara, Picea smithiana* and *Pinus roxburghii.* The forest community is two-storied, with the top storey composed of *Cedrus deodara* and the under storey composed by *Sarcococca saligna* (D. Don.) and *Rhododendron arboreum* Sm. Other recorded herbaceous taxa are *Ainsliaea aptera* DC., *Artemisia japonica* Thunb., *Berberis aristata* DC., *Daphnae payracea* Wall ex. Steud. and *Geranium wallichianum* D. Don ex Sw. *Cedrus deodara* Royle ex D. Don is the dominant tree species.

Site II

The site II was selected at hill base as starting point (uphill), with a valley and the north-west facing slope. The terrain is rocky and soil very loose. Rocks are obliquely oriented and sedimentary in nature (Valdiya, 1980). The area is interrupted by roads and hence 'minor gaps' have been created by the passage construction. These gaps have severely affected the species composition by habitat fragmentation and disrupted the continuity of the vegetation. In the regions where conifers are dominant, landslides are very common and hence explain the presence of dry and loose soil.

The soil is generally very moist under the broad-leaved forest. The litter is very thick being about 5 cm, deep. It is composed of coniferous needles as well as *Quercus* leaves. The forest is two-storied, upper storey being formed by species of *Cedrus* and *Quercus* and under storey by *Rhododendron arboreum* Sm. and *Pinus* saplings. At this site *Cedrus* and *Quercus* are co-dominant. The dominant herbs are *Bergenia ligulata* (Wall.), *Desmodium elegans* DC., *Polystichum aculeatum* (L.) Roth and *Vicia sativa* L.

Climate

The climate of area differs from the adjacent Gangetic range in respect to (i) shorter or less severe summer, (ii) more or less higher precipitation and (iii) colder and more prolonged winter. The climate factors, i.e., precipitation, temperature, relative humidity slope, aspects, drainage and vegetation, etc. and wind in association with elevation proximity to Greater Himalayas, create variations in climate at micro levels. Lower peaks (1500 m) show subtropical climate whereas higher peaks (2400 m) show temperate climatic conditions. Precipitation during south west monsoon is nearly 200 mm and snowfall occurs from December to February.

Vegetation

Western Himalaya is cradle of about 5000 species of flowering plants and home of nearly 800 endemic species. These forests constitute immense plant wealth in terms of biodiversity and in providing necessary ecosystem services like preserving wild life and soil, sustaining human life, stabilizing climate, optimizing water-yield and purifying air and water. Vegetation is represented by broad-leaved forests at lower altitudes and mixed and pure conifer forest at higher altitudes (Dhar *et al*., 1997; Gaur 1999; Singh, 2002).

Quercus is the most abundant, its three species *Q. glauca, Q. leucotricophora* and *Quercus semecarpifolia* (Fig. 1A) being equally distributed at a heights of 1500 and 2400 m in Western Himalaya. Conifers form the uppermost storey of the forest that includes *Abies pindrow, Cedrus deodara, Cupressus torulosa, Picea smithiana* and *Pinus roxburghii* (Semwal, *et al*., 2007). Various herbaceous plants are observed along the gradient of altitude and topography. The most abundant being *Chenopodium, Desmodium, Fragaria nubicola, Impatiens, Oxalis, Stellaria* and *Viola.* Other herbs were *Anemone, Berberis, Commelina, Galium, Polygonum, Prinsepia utilis, Ranunculus* and *Sarcococca.*

The forest flora develop in span of millions of years passing through the change in climate. The breakdown of floristic diversity in short span is of great concern. In addition to the impact of biotic and climatic pressures, the composition, distribution and habitat fragmentation is most profound aspect to study the Western Himalayan forest (Ram, *et al*., 2004). In general, maximum soil surface is covered by grasses, associated with almost all types of plant community. They are ecologically very useful in soil conservation. Similarly, the weeds and herbaceous species that occupy large area are useful in terms of biological production by trapping the sun energy. Several types of fungi and microorganisms as well as some angiosperms are adapted to parasitic mode of life and in process instead of depending on one host; they are evolving towards multiple choices of hosts.

METHODOLOGY

Phytosociology

Phytosociological study aimed to provide detailed information about the composition, structure, species diversity, growth and pattern of distribution. The data on the frequency, density, abundance and relativity of the species were taken into account while characterizing a community. The Transect Method has been used to study the above said parameters. Quadrat was taken as a sampling unit being a representative of the whole area (Misra, 1968). Population of each quadrat was recorded for synthetic analysis of vegetation.

Belt transect method was used to achieve the objectives. Transect is divided into segments of suitable length. This method is useful to analyse vegetation structure, composition, diversity through ecological gradient (Simpson, 1949; Singh, 2006; Shannon and Weaver, 1963; Semwal and Mehta, 1996; Semwal *et al*., 2006). Each segment was taken as quadrat (10 × 10 m) (Magurran, 1988). Other aspects such as basal area, age and height of tree, canopy cover and physical and chemical features of soil have also been estimated. The pH of soil samples have been estimated by a rapid field test. Potassium (K^+) and Phosphate PO_4-P has estimated by Purdue Method or Cobalti Nitrite test. Organic matter (%) was estimated by using Walkely and Black's rapid titration method (Singh and Singh, 1992).

Observations

Species Composition

The area is richly represented by the bryophytes, pteridophytes and herbs which make the ground flora. Open areas are mostly inhabited by grasses and ferns. Mosses were found on

Fig. 1. A - Oak mixed forests B – Deforested pine C – *Anaphalis sp.* D – Cattle grazing.

the soil, rocks and tree bark, where the filtered light is available. Some of the dominant taxa found in the area are as follows:

Bryophytes - *Cryptoleptodon flexuosus* (Harv.) Ren. and Card., *Gymnostomum calcareum* Nees and Hornsch., *Hydrogonium leucodontoides* (Gangulee) Chopra., *Hymenostylium recurvirostre* (Ehrh.) Lindb., *Hypohila involuta* (Hook.) Jaeg., *Leucodon sciuroides* (Hedw.) Schwaegr.

Pteridophytes- *Adiantum capillus-veneris* L., *Asplenium adiantoides* (L.) C. Chr., *Dryopteris gamblei* (C. Hope) C. Chr., *Onychium japonicum* (Thunb.) Kunze, *Polystichum aculeatum* (L.) Roth, *Pteris critica* L., *Selaginella adunca* A. Braun ex Hieron.,

Gymnosperms- *Abies pindrow* Royle, *Cedrus deodara* (Royle ex D. Don), *Cupressus torulosa* D. Don., *Picea smithiana* (Wall.), *Pinus roxburghii* Sargent and *P. wallichiana* Jacks.

Angiosperms- *Abelia triflora* R. Brown ex Wall., *Agrimonia pilosa* Ledeb., *Anaphalis contorta* (D. Don.), *Artemisia japonica* Thunb., *Aster peduncularis* Wall., *Berberis aristata* DC., *Brunella vulgaris* L., *Cardamine scutata* Thunb., *Desmodium cephalotes* (Roxb.) Wall., *Elsholtzia pilosa* (Benth.), *Ficus sarmentosa Buch.-Ham.*, *Galium aparine* L., *Hedera nepalensis* Koch.., *Oplismenus burmanii* (Retz.), *Polygonum nepalensis* Meissn., *Prinsepia utilis* Royle, *Quercus leucotrichophora* A. Camus, *Rubus niveus* Thunb., *Rumex nepalensis* Spreng., *Rubia manjith* Roxb.ex Flem., *Salvia nubicola* Wall.ex Sw., *Thymus linearis* Benth.

Bioresources

There is enormous wealth of plants which have multifarious usage and have been used as sources of various products employed in medicinal, cosmetic industries and for aesthetic purposes since times immortal. Some of these are listed below:

1) *Abelia triflora* R. Brown ex Wall. (Caprifoliaceae): Wood is used for making walking sticks. Also serves as a useful nurse plant for deodar seedlings.
2) *Agrimonia pilosa* Ledeb. (Rosaceae): Methanolic extract of underground parts possesses antibiotic activity. Also used as an astringent and tonic. Bark is rich in tannin.
3) *Ainsliaea aptera* DC. (Asteraceae): Root is bitter and is used as medicine to relieve stomachache.
4) *Armillaria mellea* Vahl. Fr. (Physalacriaceae): Edible fungus.
5) *Artemisia japonica* Thunb. (Asteraceae): Leaves and flowers yield an essential oil having Thujone like flavour.
6) *Anaphalis contorta* (D. Don) (Asteraceae): Yields essential oil, i.e., antibacterial, flower heads and hairs are employed for stopping bleeding (Fig. 1C).
7) *Asplenium adiantoides* (L.) C. Chr. (Aspleniaceae): Used for enlarged spleen, jaundice, malaria, rhizomes of this species are anthelmintic.
8) *Aster peduncularis* Wall. (Asteraceae): Roots used for cough and pulmonary infections, also in malarial fever and haemorrahage.
9) *Berberis aristata* DC. (Berberidaceae): Dried stem used as bitter tonic during intermittent fevers. Dried berries are edible. Root bark contains alkaloid *Berberin.* Roots and stem also yield a yellow dye.
10) *Bergenia ligulata* (Wall.) (Saxifragaceae): Rhizomes astringent, diuretic, laxative, used in diarrhoea, spleen enlargement. Rhizomes yield tannin.

11) *Brunella vulgaris* L. (Lamiaceae): Expectorant and antispasmodic.
12) *Cardamine scutata* Thunb. (Brassicaceae): Leaves and flowers used in salad. Contains an essential oil.
13) *Chenopodium album* L. (Chenopodiaceae): Used as laxative and has anthelmentic property. Yields an essential oil. Plant may serve as a field indicator for (Mg^{2+}) since it greatly stimulates growth of plant.
14) *Clematis buchananiana* DC. (Rannunculaceae): Leaf paste externally applied for skin ailments.
15) *Cupressus torulosa* D. Don. (Cupressaceae): Wood is suitable for hulls and decks of boats and general construction work.
16) *Desmodium cephalotes* (Roxb.) Wall. (Fabaceae): Suitable for fodder and paper making.
17) *Desmodium elegans* DC. (Fabaceae): This is an important Nitrogen fixing plant in coniferous forests.
18) *Elsholtzia blanda* Benth. (Lamiaceae): Used for choleric diarrhoea. Contains an essential oil.
19) *Ficus heterophylla* L. (Moraceae): Roots are edible, root juice given in colic, root bark used for cough.
20) *Fragaria indica* Andr. (Rosaceae): Fruits are made into jams, jellies, etc. Leaves yield essential oil.
21) *Galium aparine* L. (Rubiaceae): Fruits used as a substitute for coffee, infusion of herb used as diuretic, fruit yield a purple dye.
22) *Hedera nepalensis* K. Koch (Araliaceae): Leaves and berries are stimulant and cathartic. Berries and seed have glycoside, i.e., haemolytic and acts as irritant to the elementary canal, vaso constriction, lower blood pressure.
23) *Ilex dipyrena* Wallich (Aquifoliaceae): Leaves occasionally used as fodder. Yields cheap timber for household items.
24) *Impatiens thomsonii* Hook. (Barsamiacea) Seeds and seed oil edible.
25) *Kalanchoe pinnata* (Lam.) Pers. (Crassulaceae): Leaf juice acts as purgative and insecticide plant poisonous to cattle.
26) *Oplismenus burmanii* (Retz.) P. Beauv. (Poaceae): Much relished by cattle (Fig. 1D), particularly when the grass is young and green, also yield good hay.
27) *Phoebe lanceolata* (Wall. ex Nees) (Lauraceae): Wood used for furniture, Leaves as fodder Berries used to cure wounds.
28) *Polygonum strigosum* R. Br. (Polygoniaceae): Eaten either raw or cooked, yields a yellow dye.
29) *Polystichum aculeatum* (L.) Roth (Polypodiaceae): Fronds used in preparation of curries.
30) *Princepia utilis* L. (Rosaceae): Seeds yield a semi-drying fatty oil, which is used for cooking purposes and as illuminant. Wood used for walking sticks and as fuel.
31) *Pteridium aquilinum* L. (Polypodiaceae): Rhizomes are boiled and eaten at times of scarcity. Starch in flour is bitter, but can be removed by washing. Contains bitter *sapoxin* toxin.
32) *Pteris cretica* L. (Pteridaceae): Young fronds eaten as a flavouring agent. Decoction of fresh fronds given for dysentery. Juice of Rhizome applied to glandular swelling of neck.
33) *Pteris ovalifolia* D. Don. (Pteridaceae): Reported to be poisonous, contains andromedotoxin.

34) *Quercus leucotrichophora* A. Camus (Oak) (Fagaceae): Wood is used for building and agricultural implements. Leaves used as cattle fodder, used in diuretic in gonorrhea and as astringent.
35) *Rannunculus arvensis* L. (Rannunculaceae): Used for the treatment of asthma, skin ailments.
36) *Rhododendron arboreum* Sm. (Ericaceae): Wood used as timber, flowers eaten raw and made into sauce, jams and jellies, Bark used for respiratory and digestive disorders.
37) *Rhus cotinus* L. (Anacardiaceae): Fruits are used as a substitute for Rennet in the preparation of curd.
38) *Rubia cordifolia* L. (Indian Madder) (Rubiaceae): Source of Indian Madder. Long being employed for dying coarse cotton fabrics, blankets. Extracts forms a constituent of dry *Septilin.*
39) *Rubus niveus* Thunb. (Rosaceae): Fruits eaten as dessert, also used for making jam.
40) *Rumex nepalensis* Spreng. (Polygonaceae): Infusion of leaves given in colic and applied to syphilitic ulcers.
41) *Salvia aegyptiaca* L. (Lamiaceae): Seeds used for diarrhoea and hemorrhoids, suspected of causing abortion in cattle.
42) *Sarcococca saligna* (D. Don.). (Buxaceae): Leaves used in rheumatism.
43) *Sanicula* Buch.-Ham. ex D. Don. (Apiaceae): Astringent, used in diarrhoea and dysentery.
44) *Sedum multicaule* Wall. ex Lindl. (Crassulaceae): Leaves has insecticidal properties.
45) *Smilax aspera* L. (Liliaceae): Roots rich in tannin, tubers used in venereal diseases.
46) *Swertia angustifolia* Buch. -Ham. ex D. Don. (Gentianaceae): It has tonic properties.
47) *Tagetes minuta* L. (Asteraceae): Yield a strong aromatic essential oil used as antiseptic.
48) *Thymus linearis* Benth. (Lamiaceae): Shoots employed for flavouring, yields oil, i.e., *Thyme,* which is used in soap perfumes. Thymol is an alkaloid, which is a powerful germicide.
49) *Urtica ardeus* (Urticaceae): Used for ropes, mats, seed yield edible oil.
50) *Urtica dioica* L. (Urticaceae): Memostatic, used in uterine haemorrhage, bleeding from nose, and vomiting of blood. Infusion of leaves and roots used as hair stimulant. Stem yields fibre.
51) *Velleriana jatamansi* Jones (Valerianaceae): Root has medicinal properties. They smell like doop.
52) *Verbascum Thapsus* L. (Scrophulariaceae): Leaves and fruits used in diarrhoea and pulmonary diseases of cattle.
53) *Viola biflora* L. (Violaceae): Roots emetic, used as substitute and adulterant. Flowers used in infection of the chest as a pectoral.
54) *Vitis himalayana* (Royle) Brandis. (Vitaceae): Fruits are edible.

Quantitative Analysis

In the present research work randomly selected Quadrats were studied to find out the quantitative assessment of tree species. It is found that *Pinus roxburghii* and *Daphnae papyracaea* present only in site I while other species found in both the sites (Table 1). In site I a total of eight species is present out of this *Pinus roxburghii* has high frequency (60%) and high density (2.2 trees/100 m^2). In site II a total of six species is present out of this *Cedrus deodara* has high frequency (50%) and high density (2.5 trees/100 m^2).

Table 1: Quantitative analysis of prominant tree and shrub species in two sites of Kanasar

Plant species	Site I		Site II	
	Frequency (%)	Density 100 m^2	Frequency (%)	Density 100 m^2
(1) *Cedrus deodara* (Royle ex D. Don)	40	1.5	40	2.5
(2) *Quercus leucotrichophora* A. Camus	20	1.6	30	1.2
(3) *Picea smithiana* (Wall.)	30	1.2	30	2.1
(4) *Pinus roxburghii* Sargent	60	2.2	–	–
(5) *Pinus wallichiana* Jacks	50	1.3	20	1.4
(6) *Sarcococca saligna* (D. Don)	40	2.1	30	0.8
(7) *Daphnae papyracaea* Wall ex. Steud.	20	1.8	–	–
(8) *Berberis aristata* DC.	30	1.9	20	0.9

Successional Trends

Communities are dynamic entities that keep on changing more or less regularly over time and space. These are never found to be in complete balance with their physical environment. Such process of change in successive community developing one after another, over the same area in definite sequence of communities over period of time is known as Ecological Succession. There are major categories of ecosystem change which can be identified:

1. Long-term change in physical environment-This type of change normally occurs very slowly and generally we cannot observe the consequences of biotic communities changing during our lifetime.
2. Changes in genetic constitution of organism- As the result of natural selection, genetic change occurs rapidly in resource to enter physical or biotic selection pressures, but occurs on a longer time scale, in response to slow but directional changes in climate, soil conditions, etc. Natural selection is constantly altering the genetic constitution of population in a manner that increases their genetic fitness.
3. Changes in type, number and grouping of organisms occupying an area and concomitant changes in certain features of the physical environment. This type of change occurs in both newly exposed previously uncolonized physical environments and colonize previously occupied.

Present study when compared with the earlier studies presume that there is drastic change in the topography, type of vegetation specific to each area, which might have arisen due to above already described factors and most importantly being interference by man. The major portion being occupied by conifers and the dominant species are *Cedrus deodara* and *Pinus wallichiana.* Quercus is found to be dominant among angiosperms. Climax stage could not be concluded, but it might be that above plants are the parts of sub-climax stage and *Pinus*, *Cedrus* and *Quercus* are found to be dominant species. It was also observed that forest showed disturbed community due to gaps that might have been created by several environmental and human factors. Barren area thus created act as land for new successional changes which further give rise to totally new communities.

Factors which create gaps includes landslides, lumbering, felling of trees, excessive grazing, etc. The gaps led to drastic change in not only in the intensity of sunlight reaching the forest floor but also soil pH, climate, and change in soil texture. All this leads to death of young plants that are adapted to earlier environment. Also young seedlings growing under the shade of trees die on exposure to naked sunlight. This barren land becomes a site of development and establishment of pioneer species populations. Together with these mixed forests exclusive conifers trees were also seen. Most of the trees seen were; *Pinus wallichiana,* and *P. roxburghii.* Some species of *Pinus* have been reintroduced by forest officers to fill up the gap areas. It is a sub-climax community as many dominant species which would not be distributed and replaced by other new community.

Successional Trends in Ground Substratum

Substratum was usually heterogeneous. Litter is composed mainly of needles of conifers, leaves of *Qurecus* in patches dried of climbers and grasses. Mosses and Lichen are found as pioneer species and the succession is presumed to be allogenic type.

Undecomposed litter from *Pinus* is seen abundantly, soil with low pH, i.e., acidic. This shows that *Pinus* litter has phenolic compounds a high lignin confers which leads to acidic soils unsuitable for establishment of any other plant near it. However, mosses quickly invade such forest floor of the conifers. Minerals and humus content dissolves out being acidic as H_2O percolates. This is called podzolization and this ash coloured soil is Pod soil.

Ecology

Understanding tree ecology of an area, the most important aspects that help in suggesting some management steps include interspaces (Singh *et al.* 2008). Major forest forming species seen in forest were gymnosperms and angiosperms. *Pinus, Cedrus* and *Picea* are usually seen and is codominant species. Barren areas in the forest give an insight into the succession trends and different species that have inhabited that area. The plant species growing under the shade of tree in that particular area die due to sudden exposure to high sunlight. The flora and fauna particularly the microbes and lower plants inhabiting earlier, in that place are not able to adapt to sudden change in climatic condition. This wipes out the sensitive species and establishment and invasion of new plants in that particular area. These plants are not only different in their adaptability to new environment but also lead to the formation of totally new community in that area. As seen in the case of felling of *Cedrus* plant due to forest fires and anthropogenic activity, there is replacement of *Cedrus* with *Pinus* seedlings, which are able to survive under harsh conditions.

Soil Fertility

This refers to the ability of soil to provide plants with the nutrients. Nutrients required for plant growth are present in the soil in variety of forms. They may be dissolved in soil solution, from which they can be utilized directly. They may be absorbed on the exchange sites, from which they may enter soil solution or be directly utilized by tree roots of microorganisms that come in contact with exchange sites; alternatively they may be firmly fixed in clay (lattices) immobilized, in decomposition resistant organic matter, etc. Potential hydrogen ion concentration (pH) considered the most important factor in determining the availability of many elements because of its relationship with solubility, rates of decomposition and type of vegetation. In the present study the soil is found to be rich in nutrients.

Soil Analysis

The pH of soil samples was estimated by a rapid field test which indicated that the soil is acidic. Although it showed variation from being moderately acidic (pH = 5.5) to strongly acidic (pH = 4) areas where *Pinus*, *Cedrus* were in abundance, soil varied from pH 4-5, i.e., acidic.

At certain places forest too have added to low pH of soil, i.e. (high acidity). Plants species observed in fire prone area were *Oplismenus*, *Sanicula, Sarcococca, Fragaria, Ocimum* and also the canopy was open one. Soil samples showing pH = 6 (slightly acidic) had *Cedrus, Quercus* sp., *Oplismenus, Viola,* and *Fragaria.* Estimated by Purdue Method or Cobalti Nitrite test the status of K^+ in soil samples was found to be slightly variable from being deficient in some places to adequate in other areas. Regions from where soil samples showed deficiency of traces for the presence of K^+ had sparse vegetation. Topography showed steep slope with *Cedrus, Qurecus* as tree sp. Plant species found in these area were *Carex, Berberis, Viola,* and *Oplismenus.* Adequate amount of K^+ observed in gradual slopes with plant sp. *Oplismenus, Onychium, Dryopteris.*

Rapid Field Test for PO_4-P

Rapid test performed for PO_4-P in soil sample of Kanasar showed overall deficiency of PO_4-P. Although soil samples area have shown adequate to medium levels of PO_4-P, areas with medium amount of PO_4-P have plant species like *Viola, Oplismenus, Sanicula, Stellaria, Fragaria. Cedrus deodora* was a dominant species. Highly disturbed area with burnt logs and lumbering was observed in nearby region also.

Organic Matter

Using Walkely and Black's rapid titration method, the % soil organic matter present in soil samples showed variation from 1.52% to 19.463%. Regions with low percentage of organic matter had human disturbance. Tree species like *Cedrus, Pinus* were abundant. Plant species like *Oplismenus, Oxalis, Fragaria, Majus, Sanicula, Ranunculus*, etc. were present. Topography indicated steep slopes and open canopy cover.

High percentage of organic matter was in regions with gradual slopes, the abundant species were *Cedrus deodara, Quercus.* Bryophytes like *Pellia, Funaria hygrometrica* patches (fertile) were observed over leaves litter. Other plant species include *Onychium, Oplismenus, Fragaria* and *Rubia quadrifolia.* Most of communities observed were of *Cedrus deodara* forest at the range of 1700-2500 m. *Pinus wallichiana* forest 1700- 2500 m in small pockets. *Pinus roxburghii* forest ranges from 1500–2200 m. *Querus semecarpifolia* ranges from 2400–3000 m (Singh and Singh, 1987).

Anthropogenic Activities

Deforestation is going on a large scale in the Kanasar region (Fig. 1B). The chief reasons are:

1. Unscientific lopping, i.e., cutting the branches up to meristem was very common.
2. Labour is specially employed for cutting trees mainly *Cedrus.* The land that is already deforested is used by these people as agricultural land for their sustainance. In these areas mainly *Amaranthus* is grown.

3. Equipment were used for trimming timber in the forest, near the deforested area.
4. Clear cut fencing of *Cedrus* trees were done in which there were young trees too.
5. Presence of more of *Pinus* and less of *Quercus* trees also supports large-scale deforestation as *Pinus* grows in areas where plenty of sunlight is available. Gaps introduced by man, created favourable conditions for growth of *Pinus*.
6. Canopy cover (%) was too less in many sites indicating large-scale deforestation.

The deforestation is mainly at higher elevations at Site 1 indicating that forests were cleared from inside. From outside it appeared as if this is a dense forest. In the top hill large areas of forests are being converted into agricultural land.

Impact on Environment due to Anthropogenic Activities in Kanasar

During studies in and around Kanasar forest division, we found that the area is highly disturbed due to anthropogenic activities and the major causes of the depletion of forest cover are:

1. Commercial forest exploitation
2. Unscientific lopping
3. Illicit felling of trees
4. Forest fires
5. Conversion of forest land to agricultural land (land clearing)

The broad-leaved forests are repeatedly lopped for fodder and fire wood and for various other purposes leading to their gradual disappearance. Oaks form the main forest composition but the canopy gaps were found in the forest. Large-scale deforestation has led to the disturbance of catchments area of river Yamuna. Surface water is found in River Tone near Kanasar which finally falls in river Yamuna. Due to deforestation, the soil layer, deprived of biotic control by roots, gets detached easily leading to massive soil loss in the region. Absence of cover on soil leads to less percolation of water leading to lowering of water tables and thus subsoil surface system (aquifers) is disrupted. This ultimately leads to drying up of springs and hence water flow in major rivers. Source of CO_2: Forests are becoming source of CO_2 rather than sink because of over-exploitation. (Singh and Singh, 1987) especially Himalayas. This increase would lead to global warming leading to melting of ice caps and thereby reducing the land owing to submergence of lowlands.

Replacement of *Quercus sp.* by *Pinus roxburghii*

This change over of a single species brings about series of changes in the area. These changes bring about climatic changes not only in the surrounding areas but in other places too. Soil erosion is more in *Pinus* forests compared to *Quercus* forests. As *Quercus* supports herbaceous vegetation is rich in broad-leaved forests rather than a *Pinus* forest.

High tree canopy reduces the erosion potential of rainfall thereby there is a less erosion. Many species associated with *Quercus* does not grow along with *Pinus* thus the vertical stratification is also disrupted. *Pinus* forests have less of horizontal stratification and thus less percolation of rainwater. Over a period of time this would lead to drying up of aquifers.

1. CO_2 Source: *Pinus* is a fast growing compared to *Quercus* which is slow growing shade tolerant tree. The fast growing trees have less dense wood compared to slow growing ones and therefore with the increase in populations of fast growing species which also have high turnover rate, the forests become a net source of CO_2.
2. Landslides: *Pinus* does not support much of herbaceous vegetation and hence the area is not bonded much. Thus, this area is prone to a landslide which further brings about loss of species and disturbances in catchments areas.
3. *Pinus* forests are prone to fires. Of the *Pinus* litter, fires result in marked nitrogen losses through volatilization, thus augmenting the nitrogen shortage.
4. Intrusion of Pine has decreased the resilience of 'oak forests' in Himalayas as *Pinus* creates conditions that are unfavourable for regeneration of oak.
5. With the destruction of forests, habitat for wildlife is shrinking, resulting in loss of animal diversity.

There are the impacts when there is a changeover of single species (*Quercus* to *Pinus*). This change leads to replacement of herbaceous sp. which results in further changes. Thus, one can clearly visualize that even a single change brings about series of changes that deteriorates the community structure.

DISCUSSION

Community structure

Community structure and functions are manifestations of a complex array of interactions, directly or indirectly laying all members of a community together into an intricate web. It comprises of partially closed communities with not so rapid replacement of species along the gradient. Distinct soil types of north facing and south facing mountain slopes. North facing soil types was generally compact, wet, little partly decomposed. South facing soil type was comparatively drier and rocky and litter was abundant. Among the tree species *Cedrus, Oak, Phoebe, Pinus* are abundant. Lower slopes due to human disturbance and interference was of young *Pinus* seedlings but on higher slopes was in patches along with *Cedrus.*

Among a Quercus species, *Q. incana, Q. semecarpifolia* were abundant in the area covered on the basis of frequency, density and abundance, dominant plant species of ground vegetation are of *Oplismenus* (Poacee). *Fragaria* (Rosaceae), *Chenopodium embersosum, Viola pilosum, Galium aparine.*

Stratification

In the forest of the area the top storey made by *Cedrus* and *Pinus,* second storey by *Quercus* and *Picea* and third storey of shrubs and tall herbs. Uppermost layer of canopy receives the full sunlight gradually moving down the forest strata light intensity is progressively reduced to only a fraction of full sunlight. In Oak forest less amount of light reaches forest while in pine forest due to dense upper canopy lower strata does not develop at all. Various strata are determined by light moisture which modify one environment from forest canopy and forest floors. The forest formation is more or less a mixture of: (a) Low to mid mountains consisting of semi-sclerophyllous broad leaves forest with concentrated summer leaf drop (Singh and Singh, 1992); (b) Mid mountain, i.e., needle leaf evergreen forest.

Regenerative Capacity of Forest

In the studied area *Cedrus* and *Picea* species are co-dominant and add up to large biomass. These species form uppermost canopy. The saplings of same age were not occurring naturally but were planted by the Forest Department to counter the indiscriminate lumbering. Many oak trees and other coniferous trees are individually occupying an area. The distribution is almost in the range of 1,300–2,300m. *Pinus* species are most dominant and have distinct features. They are light demanding and are adapted to fire. Actually the seeds of *Pinus* remain dormant till fire occurs. They occur on driest and steepest of slopes even in rocky terrains. Its species occur in abundance due to cutting and burning of other forest trees. Due to thin stress and other adaptable characters, they are able to replace in particular the oak formation. *Pinus* species also have great nutrient conserving ability which makes soil unfit for other plants to grow. They do not allow oak and other trees to reoccupy the deforested area.

CONCLUSIONS

1. The trends of succession along almost all the ecological gradients studied show allogenic and secondary succession. Replacement of the existing communities seems to be caused largely by external conditions rather by the existing communities themselves. Also the new species inhabiting the area are formed on the already built up substratum.
2. Lichen and mosses could be the pioneering species followed by species of *Polytrichum, Dryopteris, Viola, Fragaria, Geranium, Galium, Rumex, Pinus, Cedrus* and *Picea.*
3. The sub-communities formed could be associated with certain woody climbers, i.e., lianas with thick stems. This may be an indication of well developed and an old forest formation. Replacement of oak forests by coniferous forests is a very serious and improper situation that needs immediate attention.
4. Based on our observations we have concluded that oak forest is being replaced by coniferous forests slowly. The reasons for the conclusion is evident from one observation, the large-scale deforestation resulting in *Pinus* take over. Frequency of oak is very low (20%). Average of both sites compared to frequency of *Pinus* (60%) and *Cedrus* (40%).
5. Regeneration potential of *Quercus* is mostly decreasing. Young saplings of *Quercus* are almost absent but young saplings of *Pinus* have been observed. Old trees of *Quercus* are present with height up to 15m but not in flowering stage, since top shoots were being several used as fodder by the local people. Unscientific lopping has led to the death of several *Quercus* trees. Thus, regeneration of *Quercus* is almost absent in the region. *Quercus* requires high nitrogen. Through soil tests we found that nitrogen is deficient in the soils thus hindering the growth of the genus adversely. Nitrogen deficiency is mainly due to encroachment of pines.
6. Heavy lopping of oaks has caused canopy gaps and reduction of leaf-fall. *Pinus* needles contain lignin and make the soil acidic, which fails to attract invertebrates. This results in slow immobilization of nutrients resulting in deficiency of nitrogen and other minerals.

7. Very few legumes have been observed in the area to fix the nitrogen at this altitude, e.g. *Desmodium*. It is inadequate for *Quercus* growth. *Cedrus* and *Quercus* can coexist and form a climax community but *Quercus* and *Pinus* can't coexist as their requirements are contrary to each other.
8. Care should be taken that the growing regions of the plants are not hurt. Awareness should be generated among locals so that they use their resources judiciously. Only dead trees should be cut and that too selectively, i.e., a large number of trees should not be cut from the same region.
9. Aforestation of the area should be done by native species of the area and not by alien species (e.g. *Pinus* in Kanasar). Illegal felling of trees should be banned.

RECOMMENDATIONS

The young Himalayan mountain belt is in a state of continuous dynamic shift, rendering itself very vulnerable to anthropogenic influences of all kinds. In view of the great importance of the stability of Himalayan ecology in the economic life of the nation, the prudent appreciation of geological principle through multi-layered checks and control is the primary requisite for human intervention.

The ecosystems here are highly fragile and their disturbances may lead to serious consequences, through indiscriminate human intervention in future. Following recommendations are being put forward for saving the further depletion of biodiversity from the Kanasar area:

1. Awareness and participation of the local people in processes of evolving and evaluation will ensure minimum damage to ecology, environment and sustainability of the whole esosystem.
2. The area being highly sensitive requires a body of professionally competent consultants independent of vested interest as well as of bureaucratic pressures, for monitoring and providing data base for long term sustainability and development.
3. Projects without prior approval of environmentalists should not be implemented. These need to be assessed by people who are specialized in the above fields especially in respect to Environmental Impact Assessment.
4. The policies made by authorities concerned should aim to balance the commercial and protective aspects taking into account conservation, resource renewal as the guiding principles.
5. In order to ensure better management of catchments areas, recharge of springs and stability of hill slopes, there should be complete prohibition on encroachment into forests and pasturelands for cultivation, cattle breeding and habitation.

REFERENCES

Dhar, U Rawal, RS & Samant, SS (1997). Structural diversity and representativenenss of forest vegetation in a protected area of Kumaon Himalaya, India, Implications for conservation Biodiv. *Conser.* **6:** 1045-1062

Gaur, RD (1999). *Flora of the District Garhwal North West Himalaya with Ethnobotanical Notes*. Transmedia Publication, Srinagar, Garhwal, Uttaranchal

Govindapyari, H, Bahuguna, YM, Semwal, DP & Uniyal PL (2010). Plant diversity status and conservation aspect in Kanasar Forest Ranges, Uttarakhand. *The Botanica* **58:** 29-32

Magurran, AE (1988) *Ecological diversity and its measurement.* Croom Helm, London

Misra, R (1968). Ecology *Work Book.* Oxford and IBH Publishing Company, New Delhi

Obute, GC (2010). Pollination: A threatened vital biodiversity service to humans and the environment. *International J. Biodiversity and Conservation* **2:** 001-013

Ram, J, Kumar, A & Bhatt, JP (2004). Plant diversity in six forest types of Uttaranchal, Central Himalaya, India. *Curr. Sci.* **86:** 638-647

Rawat, TS, Menaria, BL, Dugya, D & Kotwal, PC (2008) Sustainable forest management in India *Curr. Sci.* **94:** 996-1002

Semwal, DP, Saradhi, PP, Nautiyal, BP & Bhatt, AB (2007). Current status, distribution and conservation of rare and endangered medicinal plants of Kedarnath Wildlife Sanctuary, Central Himalaya, India. *Curr. Sci.* **92:** 1733-1738

Semwal, DP, Saradhi, PP & Bhatt, AB (2006). Comparing structure and composition of *Taxus baccata* sub-sp. *wallichiana* Zucc. (Himalayan Yew) with other associates in a part of Garhwal Himalaya, Uttarakhand, *Phytomorphology*, **56:** 127-131

Semwal, DP, Uniyal, PL & Bhatt, AB (2010). Structure, composition and dominance-diversity relation in three forest types of part of Kedarnath Wildlife Sanctuary, Central Himalaya. Notulae Scientifica Biologiae **2(3):** 128-132

Semwal, RL & Mehta, JP (1996). Ecology of forest fires in chir pine (*Pinus roxburghii* Sarg.) forest of Garhwal Himalaya. *Curr. Sci.* **70:** 426-427

Shannon, CE & Weaver, W (1963). *The Mathematical Theory of Communication,* University of Illinois Press, Champaign, Illinois, USA

Simpson, EH (1949). Measurement of diversity. *Nature,* **163:** 688

Singh, JS (2002). The biodiversity crisis: A multifaceted review. *Curr. Sci.* **82:** 638-647

Singh, JS (2006). Sustainable development of the Indian Himalayan region: Linking ecological and economic concerns. *Curr. Sci.* **90:** 784-788

Singh, JS & Singh, SP (1987). Forest vegetations of the Himalaya, *Bot. Rev.* **53:** 80-192

Singh, JS. & Singh, SP (1992). *Forests of Himalaya: Structure, Functioning and Impact of Man,* Gyanodaya Prakashan, Nainital

Singh, JS, Singh, SP & Gupta, SR (2008). *Ecology, Environment and Resource Conservation.* Anamaya Publishers, New Delhi

Suleiman, MK, Bhat, NR, Abdal, MS, Zaman, S, Jacob, S & Thomas, RR (2010). Propagation studies in *Farsetia aegyptia* Turra *International J. Biodiversity and Conservation* **2:** 030-032

Valdiya, KS (1980). *Geology of Kumaun Lesser Himalaya*, Wadia Institute of Himalayan Geology, Dehradun

9

Forest Plant Biodiversity in the State of Uttarakhand

AN Singh and Ajay Kumar

EIA Division, Directorate of Extension, ICFRE, PO. New Forest, Dehradun–248 006, Uttarakhand
E-mail: dransamit@rediffmail.com, singhan@icfre.org

ABSTRACT

The forest plant biodiversity provides a wide range of goods and services for all living beings. The goods include timber, fuelwood, fodder, food produce, medicinal produce, etc. Forests and trees play a significant role in the conservation of ecosystems, in maintaining quality of water, in preventing or reducing the severity of floods, soil erosion and draught. These also provide economic, social and livelihood benefits. The forest plants absorb carbon dioxide from atmosphere and store carbon in woods, leaves, litter, roots and soil by acting as 'carbon sink' and moderate local as well as global climate.

Uttarakhand has a geographic area of 53,483 km^2. Of these, about 19% is under permanent snow cover, glaciers and steep slopes. The recorded forest area of the State is 34,662 km^2, which constitutes 64.79% of its geographic area. Major forest types occurring in the State are Tropical Moist Deciduous, Tropical Dry Deciduous, Subtropical Pine, Himalayan Moist temperate, Himalayan Dry Temperate, Subalpine and Alpine Forests. The State has rich forests comprising over a thousand species of trees, shrubs, herbs and climbers. The floristic exploration of different regions of Uttarakhand has revealed that the region is very rich in vegetation. Sustainable use vs. conservation of plant biodiversity is warranted through regulatory regime and people's participation.

Keywords: Forest, goods and services, plant diversity, Uttarakhand, Western Himalaya.

INTRODUCTION

The term biodiversity implies variability among living organisms from all sources including, inter alia, terrestrial, marine and other aquatic ecosystems and the ecological complexes of which they are a part; this includes diversity within species, between species and of ecosystems (CBD, 1992). The biological diversity may be considered a unique and irreplaceable natural resource, which is being rapidly lost from the earth. This eventually constrains and counteracts economic development. The United Nations Convention on Biological Diversity (UNCBD) proposes to conserve biological diversity, promote sustainable use of its components and encourage suitable sharing of the benefits from the utilization of genetic resources.

The forest provides a variety of goods and services ranging from timber, food, fodder, NTFPs, medicinal produce, shelter for wildlife, various produce from wildlife, minerals, environmental amelioration, recreation and many more still to mention, besides offering livelihood to millions of people.

The National Forest Policy of India (1988) has rightly been focused on 'Conserving the natural heritage of the country by preserving the natural forests with the vast variety of flora and fauna, which represent the remarkable biological diversity and genetic resources of the county'. The Indian National Environment Policy, 2006 recognizes that the 'forests provide habitat for wild life and the ecological conditions for maintenance and natural evolution of genetic diversity of flora and fauna'. It also stresses that 'forests of high indigenous genetic diversity should be treated as entities with 'Incomparable Value'. It further aims to 'strengthen the protection of areas of high endemism of genetic resources (Biodiversity hotspots) while providing alternative livelihoods and access to the resources to local communities which may be affected thereby'.

An overview of the forest biodiversity scenario of India with emphasis on what is being done and what more needs to be done with regard to its conservation has recently been outlined (ICFRE, 2005; Singh and Kushwaha, 2008). In the present article, however, status and extent of forest cover and plant diversity with respect to the forests of Uttarakhand only has been provided, based on the relevant publications.

Demographic Details and Status of Forests in the Uttarakhand

The State of Uttarakhand forms the eastern most part of the Western Himalayas. This region is separated by Tons river from Himachal Pradesh in the west and Kali river in the east from Nepal. In this Sector, the Great Himalayas range from 4800 to 6000m above MSL. These areas are with a permanent snow cover. The Shiwalik ranges go up to an elevation of 3500 m above MSL. The State is situated in the northern part of India and shares an international boundary with China in the north and with Nepal in the east. This nascent State has an area of 5.35 million ha and lies between lat 28°43'N and 31 °28' N and long 77°34'E and 81°03'E. The State can be divided into three physiographic zones, namely, the Himalayas, the Shiwaliks and the Tarai region. The State has a temperate climate except in the plains where the climate is tropical with temperatures ranging from sub-zero to 43°C. The average annual rainfall is 1,550 mm.

The total population of the State is 8.48 million *(Census 2001)* of which rural population is 74.33% and urban 25.67%. The population density is 159 persons per km^2. Population of the Scheduled Tribes in the State is 3%. The livestock population of the State is 4.94 million (as per the Livestock Census, 2003).

Of the geographic area (53,483 km^2), about 19% is under permanent snow cover, glaciers and steep slopes where it is not possible to grow trees due to physical limitations. Uttarakhand comprises 13 districts spreading over an area of 53,082 sq km. The Himalayan Montane have rich forests comprising over a thousand species of trees, shrubs, herbs and climbers. The types of forests in a particular area depend on its climatic conditions, altitude, topography, soils, etc. Forests of the western Himalaya differ from those of the eastern Himalayas. The western Himalayan forests are diverse both in content and composition, whereas eastern Himalayan forests are very rich both in flora and fauna (Champion and Seth, 1968; Negi, 1990 and Dhar *et al.*, 1997).

Uttarakhand lies on the southern slope of the Himalayan range, and the climate and vegetation vary greatly with elevation, from glaciers at the highest elevations to tropical forests at the lower elevations. The highest elevations are covered by ice and bare rocks. The Western Himalayan Alpine Shrub and Meadows eco-region lies between 3,000-3,500 and 5,000m elevation; tundra and alpine meadows cover the highest elevations, transitioning to *Rhododendron*-dominated shrub lands below. The Western Himalayan subalpine conifer forests (Fig. 3) lie just below the tree line; at 3,000-2,600 meters elevation these exhibit transition to the Western Himalayan broadleaf forests, which lie in a belt from 2,600 to 1,500m elevation. Below 1,500 metres elevation is extended western end of the drier Tarai-Duars Savanna and grasslands belt, and the Upper Gangetic Plains moist deciduous forests. This belt is locally known as Bhabhar. These lowland forests have mostly been cleared for agriculture, although a few pockets still remain.

Indian National Parks in Uttarakhand include the Jim Corbett National Park (the oldest national park of India) at Ramnagar in Nainital District, Valley of Flowers National Park (Fig. 4) and Nanda Devi Biosphere reserve in the Chamoli District, which together are a UNESCO World Heritage Site, Rajaji National Park in Haridwar District, and Govind Pashu Vihar National Park and Gangotri National Park in Uttarkashi District.

Land Use Pattern of the State

Land use pattern of the State has been furnished in Table 1. It is evident from the Table that the entire geographic area is under varied land use (Fig. 2). However, the forests are distributed in about 61% of the entire geographic area.

Table 1: Land use pattern in the State of Uttarakhand

Land Use	Area in '000 ha	Percentage
Total Geographic Area	5348	—
Reporting area for land utilization	5689	100.00
Forests	3468	60.96
Not available for land cultivation	465	8.17
Permanent pastures and other grazing-lands	229	4.03
Land under misc. tree crops and groves	252	4.43
Cultivable Wastelands	386	6.79
Fallow land other than current fallows	71	1.25
Current fallows	41	0.72
Net area sown	776	13.64

Source: Land Use Statistics, Ministry of Agriculture, GOI, 2005.

Forest Cover in the State

The recorded forest area of the State is 34,662 km^2, which constitutes 64.79% of its geographic area. By legal status, Reserved Forests constitute, 71.08%, Protected Forests 28.51% and Unclassed Forests 0.41 % of the total forest area. Major forest types occurring in the State are Tropical Moist Deciduous, Tropical Dry Deciduous, Subtropical Pine, Himalayan Moist temperate, Himalayan Dry Temperate, Subalpine and Alpine Forests. Forests are largely distributed throughout the State with conifers and sal being the major forest formations. District-wise forest cover in the state is given in the Table 2 (FSI, 2005).

Table 2: District-wise forest cover in Uttarakhand (Number of Districts: 13, area in km^2)

SN.	Districts	Geographic area	2005 Assessment			Total	Per cent of G. A.	Change	Scrub
			Very dense forest	Mod. dense forest	Open forest				
1.	Almora	3,139	168	969	440	1,577	50.24	0	10
2.	Bageshwar	2,246	159	875	346	1,380	61.44	0	4
3.	Chamoli	8,030	406	1,558	734	2,698	33.60	0	7
4.	Champawat	1,766	327	605	230	1,162	65.80	0	9
5.	Dehradun	3,088	487	664	442	1,593	51.59	0	27
6.	Pauri Garhwal	5,329	450	2,065	756	3,271	61.38	0	64
7.	Haridwar	2,360	29	327	274	630	26.69	-8	0
8.	Nainital	4,251	548	1,936	604	3,088	72.64	-3	17
9.	Pithoragarh	7,090	470	1,229	378	2,077	29.29	0	69
10.	Rudraprayag	1,984	179	605	336	1,120	56.45	0	5
11.	Tehri Garhwal	3,642	227	1,255	656	2,138	58.70	0	85
12.	Udhamsingh Nagar	2,542	144	246	174	564	22.19	-7	0
13.	Uttarkashi	8,016	408	2,062	674	3,144	39.22	0	23
	Total	**53,483**	**4,002**	**14,396**	**6,044**	**24,442**	**45.70**	**-18**	**320**

Source: State of Forest Report, FSI, 2005

Biodiversity Measurement and Monitoring

There are numerous standard methods for assessment of plant biodiversity of an area. Some of them are Plot Scale measurement, Temporal Assessment, Geospatial Assessment through satellite imageries and Participatory Assessment through people's biodiversity registers. The influence of forest flora coupled with the climatic and geographical characteristics has resulted into varied ecosystems and reservoirs of natural resources. The forests are sustaining the population of region, both ecologically and economically.

The restricted distribution of various floristic components has resulted in evolution of several intra-specific variabilities, thus enriching the genetic diversity. There are numerous endemic species reported from this region (Samant, 1999). Plant diversity in various forest types of Uttarakhand has been studied and reported (Jeet Ram *et al.*, 2004; Priti Kumari and Tewari, 2009).

The number of forest floras published in respect to various localities have been enlisted in the Table 3 (Biswas, 2005).

Table 3: Floristic richness of some parts of the region explored

S. No.	Titles of the Floras	No. of Species described
1.	Forest Flora of Chakrata and Dehradun	620
2.	Forest Flora of Tehri Garhwal Circle	507
3.	Forest Flora of Kumaon	723
4.	Herbaceous Flora of Dehradun	1230
5.	Flora of Mussoorie	1125
6.	Flora Nainitalensis	870
7.	Flora of district Uttarkashi	507
8.	Flora of Chamoli	2022
9.	Flora of Corbett National Park	594
10.	Flora of the District Garhwal, North-West Himalaya (With Ethnobotanical Notes)	2150

The Forest vegetation of Uttarakhand can be divided into the following 03 major zones:

1. Subtropical zone of Uttarakhand

The subtropical zone has pure as well as mixed forests of *Shorea robusta* (Sal), the others being *Lannea coromandelica* (Jhingan), *Buchanania lanzan* (Chironji), *Dalbergia sissoo* (Shisham), *Haldina cordifolia* (Haldu), *Syzygium cumini* (Jamun), *Mallotus philippinensis* (Rohini), *Mitragyna parviflora*, *Terminalia* spp. (Myrobalans), *Ficus* spp. (Figs), *Macaranga pustulata*, *Callicarpa arborea*, *Diopoknema butyracea* (Chyura), *Bauhinia variegata* (Kachnar), *Bombax ceiba* (Semal), *Lydia claying* (Pula), *Schleichera oleoso* (Kokum), *Holoptelea integrifolia Cassia fistula* (Amaltas), *Nyctanthes arbor-tristis* (Parijat, Harsingar), *Anogeissus latifolia* (Bakli, Dhaura), etc. The shrub vegetation is represented by *Bergera koenigii*, *Carissa opaca*, *Clerodendrum viscosus*, *Adhatoda vasica*, *Jasminum multiflorum*, *Solanum erianthum*, *Calicarpa macrophylla*, *Eranthemum nervosum*, *Phlogacanthus thyrsiformis*, *Jatropha curcas*, *Rhus parviflora*, *Dodonaea viscosa*, *Woodfordia fruticosa*, and many others.

A. *Saussurea obvallata* (Brahmakamal) — State Flower of Uttarakhand.

B. *Rhododendron arboreum* (Burans) — State Tree of Uttarakhand.

Fig. 1. A–B.

2. Temperate zone of Uttarakhand

This is marked by the presence of *Quercus leucotrichophora* (Banj oak), *Rhododendron arboreum* (Burans) (Fig. 1. B), *Myrica esculenta* (Kaphal), *Leonia ovalifolia* (Aynor), *Ilex dipyrena*, *Quercus semecarpifolia* (Kharsu Oak), *Q. dilatata* (Moru Oak), etc. The coniferous forests in this zone are unique. Pure stands of *Pinus roxburghii* (Chir Pine) and *Cedrus deodar* (Deodar, Cedar), *Abides pindrow* (Raga), *Pinus wallichiana* (Kail), *Taxus wallichiana* (Thunera, Himalayan Yew) still visible at some places give a pristine look to the slopes. These also have insectivorous plants like *Drosera peltata* and species of *Utricularia*. Another such species is *Pinguicula alpina* seen in Martoli Bugyal, Kumaon. The saprophytes and parasites are also well represented by *Monotropa uniflora*, *Dendrophthoe falcata* (Banda), *Balanophora involucrata*, and species of *Viscum, Korthalsella, Arceuthobium, Scurrula*, etc. The zone has a variety of useful plants, some of them have been well-known for centuries. These include *Cedrus deodara*, *Pinus* spp., *Abies pindrow*, *Quercus* spp., *Aconitum heterophyllum*, *Paeonia emodi*, *Swertia chirayta, Bergenia ciliata, Dioscorea deltoidea,* and *Angelica glauca* (Choru), etc.

A. Typical mixed forest with cultivated land at Chopta, Ukhimath

B. Mixed oak forest in Alaknanda Valley

Fig. 2. A–B.

3. Sub-alpine and alpine zones of Uttarakhand

The altitude above 3,000 meters is generally considered a zone of sub-alpine and alpine vegetation. The tree species are represented by *Pinus wallichiana*, *Abies pindrow, Prunus cornuta, Acer caesium, Betula utilis* (Bhojpatra) and *Salix* sp. Species of the genera *Cotoneaster, Rosa, Berberis, Ribes, Juniperus, Rhododendron anthopogon, Rhododendron campanulaium* are the shrubby components of the zone. The herbaceous vegetation is represented by a number of species of genera *Potentilla, Primula, Aster, Saxifraga, Achaean, Delphinium, Polygonum, Corydalis, Pleurospermum, Meconopsis, Pedicularis, Saussurea,* (Fig. 1. A) *Rheum, Silene*, etc. The Bugyals of this zone are well-known for a rich and diversified flora. Plant species like *Nardostachys grandiflora* (Jatamansi), *Podophyllum hexandrum* (Himalayan May-Apple, Papri), *Picrorhiza kurroa* (Kutki), *Gentiana kurrooa, Armenia benthonica* (Bal char), *Rheum* sp. (Dole), *Ephedra gerardiana, Dactylorrhiza hatagirea* (Hatthajari), etc., common in this zone, are of immense medicinal value.

A. Deodar forest at Chakrata

B. Chirpine forest at Uttarkashi

Fig. 3. A–B.

The floral diversity is further exhibited by the species that grow in the rain shadow areas of Uttarakhand. Such species develop adaptive characters to survive the wrath of adverse climate. *Labium* sp., *Thylacospermum caespitosum, Acantholimon iycopodioides,* and *Dracocephalum heterophyllum*, etc. are a few such examples.

Orchids

Uttarakhand has more than 225 species of this charming group, well-known for ornamental flowers with great horticultural potential and long shelf-life. *Aerides*, *Coelogyne*, *Cymbidium*, *Dendrobium*, *Thunia* and *Rhynchostylis*, for which Mandai, Baram, Shandev, Kaflani, etc. are rich. Among the global orchids, species of *Calanthe*, *Habenaria*, *Anoectochilus*, *Satyrium*, etc are fairly common whereas the Lady's Slipper Orchid—*Cypripedium* is scarcely distributed. Many small river valleys offer wonderful experience to nature lovers and hikers. The vast open hay fields, above the tree line present endless views of the variously coloured Himalayan flowers. The most interesting of them, aesthetically or botanically are seen in the higher altitudes, from 2,450 metres and above. The arrival of spring brings forward an uprising of colours, when the Semal and Palash put the lower altitude forests on fire with their blazing red flowers. It is also

(A)

(B)

Fig. 4 (A–B). Valley of Flowers, Chamoli District.

the time for Burans to spread its fire at a height of or above 2,450 metres adding colour to the blue and white panorama of snow. The flowers not only grow in the 'Valley of Flowers' but are also found on different treks habitually up to great heights as also on the hay field and even in rock cracks or moraines.

Fig. 5. Forest cover map of Uttarakhand. (FSI Report, 2005)

Some economically important forest tree species, shrub species and herbaceous species are enlisted in the Tables 4, 5, and 6. These are recorded from different regions of Uttarakhand. Ethnobotanical usages, local and trade names of the enlisted plants have also been provided.

Table 4: Some economically important forest tree species in the regions of Uttarakhand

Botanical Names	Families	Trade Names	Local Names	Uses
Acacia catechu (L.f.) Willd.	Mimosaceae	Cutch Tree	Khair	M, Ag, Tn.
Acacia nilotica (L.) Dl. ssp.	-do-			
India (Benth.) Brenan		-	Babul	Tm, Ag, Gu, M
Haldina cordifolia Roxb.	Rubiaceae	Haldu	Haldu	Tm, Tn, Fo
Aegle marmelos (L.) Corr.	Rutaceae	Bael Tree	Bel	M, Gu
Albizia lebbek (L.) Benth.	Mimosaceae	Siris Tree	Siris	T, Ag, Gu, Or
Albizia odoratissima Benth.	Mimosaceae	Black Siris	Kala Siris	Tm, Fo
Anogeissus latifolia Wall. ex Guill.	Combretaceae	Axle Wood	Bakli	Tn, Gu, Ag, Fo
Bombax ceiba L.	Bombacaceae	Silk Cotton Tree	Semul	M, Fb, Ag, Sf, Oil
Bridelia retusa (L.) Spreng	Euphorbiaceae	Kasi	Ekdania	T,Tn,Fo
Casearia elliptica Willd.	Flacourtiaceae	-	Chilla	Tn, M
Casimiroa edulis Llave	Rutaceae	White Sapota	–	Ed., Or, Sf.
Cassia fistula Linn.	Caesalpiniaceae	Indian Laburnum	Amaltas	Or,Tn, M, Ag
Cassine glauca (Roth.) Kuntz.	Celastraceae	-	Dhebri	T, Gu, Tn.
Catunaregam spinosa (Thunb.) Tirv.	Rubiaceae	Common Emetic Nut	Madana	M, Ed, Ag, Or
Cedrus deodara (Roxb. ex D. Don) G. Don	Pinaceae	Deodar	Deodar	Tm, M, Oi, Or.
Celtis australis L.	-do-	Nettle-Wood	Kharik	Fo
Cupressus torulosa D.Don	Cupressaceae	Himalayan cypress	Surai	Tm, Oil, Sf.
Dalbergia sissoo Roxb.	Fabaceae	Sissoo	Shisham	Tm
Ehretia laevis Roxb.	Boraginaceae	-	Chamror	Fo
Emblica officinalis Gaertn.	Euphorbiaceae	Indian Gooseberry	Amla	M
F. auriculata Lour.	Moraceae	-	Timla	Ed
F. racemosa L.	-do-	-	Gular	M, Fo, Tn
F. religiosa L.	-do-	Peepal	Pipal	M, Tn
F. rumphii Bl.	-do-	-	Kabar	Fo
F. virens Ait.	-do-	-	Pilkhan	Fo
Grewia asiatica		Tiliaceae	-	Dhaman Fb
G. optiva Drumm.	Tiliaceae	-	Biul	Fo, Fb, Ag
Grevillea robusta Cunn. & R.Br.	Protaceae	Silver Oak	–	Gu, Tn, Or.
Holarrhena pubescens (Buch.-Ham.) Wall. ex G. Don	Apocynaceae	-	Kora, Kura	Tm, Oil
Hymenodictylon excelsum Wall.	Rubiaceae	Kuthan	Borang	Tm, Fo, Dy
Jacaranda mimosaefolia D.Don	Bignoniaceae	–	–	Or, Sf.
Kydia calycina Roxb.	Malvaceae	Pula	Pula	Tm, Ag, Fb
Lannea coromandelica (Houtt.) Merr.	Anacardiaceae	Jhingan	Jhingan	Tm, Gu, Dy.
Lagerstroemia parviflora Roxb.	Lythraceae	Lendia	Dhaura	Tm, Fb, Gu
Litsea chinensis Lam.	Lauraceae	–	Maida-Lakri	M, Tm, Oil
Leucaena leucocephala (Lam.) de Wit	Mimosaceae	Lead Tree, White Popinac	Lucuna Subabul	Fo, Tn, Oil, Or, Sf.

Abbreviations

Ag. = Agricultural implements; Dy. = Dye; Ed. = Edible; Fb. = Fiber; Fo. = Fodder; Gu. – Gum; M. = Medicinal; Oil = Oil; Or. = Ornamental; Sf = Tm. = Timber; Tn. = Tannin.

Table 4: Contd.

Botanical Names	Families	Trade Names	Local Names	Uses
Mallotus philippensis (Lam.) Muell.-Arg.	Euphorbiaceae	Kamla Tree	Rohini	Dy, Fo, M, Tn.
Mangifera indica L.	Anacardiaceae	Mango	Aam	El, M, Tm
Melia azedarach L.	Meliaceae	Persian Lilac, Bead Tree	Dainkan, Bakain	M, Gu, Ag, Or, Sf.
Moringa oleifera Lam.	Moringaceae	Horse Radish Tree Drumstick Tree	Shajana	Ed, M, Gu, Oil
Myrica esculenta Buch.-Ham.	Myricaceae	Box Myrtle	Kaphal	Ed, Tn, M
Nyctanthes arbor-tristis L.	Oleaceae	Night Jasmine, Coral Jasmine	Harsinghar	M, Oil, Tn, Or.
Ougeinia oojeinensis (Roxb.) Hochr.	Fabaceae	Sandan	Sandan	Tm, Tn, Fo, M
Parkinsonia aculeata L.	Caesalpiniaceae	-	-	Or, Sf.
Phoenix humilis Royle	Arecaceae	Dwarf Date Palm, Hill Date Palm	Khajur	Ed
Pinus roxburghii Sargent	Pinaceae	Chir Pine	Chir, Kulain	Tm, Oil, Fo
Pittosporum eriocarpum Royle	Pittosporaceae		Raduthia Meda-tumri	Fo
Prosopis juliflora (Sw.) DC.	Mimosaceae	Mesquite	Vilayti babool	Tm, Ed, Fo
Prunus cerasoides D. Don.	Rosaceae	Himalayan Wild Cheory	Painya	M, Tn, Gu, Or.
Pyrus pashia Buch.-Ham. ex Don	-do-	Mehal	Malu	Ed, Ag.
Quercus leucotrichophora Camus	Fagaceae	Banj Oak	Banj.	Tm, M, Tn
Quercus serrata Thunb.	Fagaceae	Manipuri Banj.		Oil, Sf.
Rhododendron arboreum Sm.	Ericaceae	Rose Tree	Burans	Ed.
Schleichera oleosa (Lour.) Oken	Sapindaceae	Lac Tree	Kusum	Tm, Ag, Oil, Dy, M.
Shorea robusta Roxb. ex Gaertn.	Dipterocarpaceae	Sal	Sal	Tm, Oil
Sterculia villosa Roxb.	Sterculiaceae	Udal	Udala	Ag, Gu, Fb.
Syzygium cumini (L.) Skeels	Myrtaceae	Black Plum Java Plum	Jamun	Ed, M, Tm, Dy, Fo
Tectona grandis L.f.	Verbenaceae	Teak	Sagun	Tm, Oil, M, Tn, Sf.
Terminalia alata Heyne ex Roth.	Combretaceae	Laurel	Sain	Fo, Tm, Gu
Terminalia bellirica (Gaertn.) Roxb.	Combretaceae	Belliric Myrobalan	Bahera	M, Oil
Terminalia chebula Retz.	-do-	Chebulic, Myrobalan	Harar	M, Oil
Thevetia peruviana (Pers.) Merr.	Apocynaceae	Yellow oleander	Pili kaner	M, Or.
Toona ciliata Roem.	Meliaceae	Red Cedar, Moulmein Cedar	Tun	Tm, Dy, Oil
Wendlandia heynei (R.andS.) Sant. and Merch.	Rubiaceae	Terchunia, Udashu	–	Fo, Ag.
Ziziphus mauritiana Lam.	Rhamnaceae	Indian Jujube Common Jujube	Ber	Ed, Fo, M, Tn

Table 5: Some economically important Shrub species in Uttarakhand

Botanical Names	Families	Trade Names	Local Names	Uses
Acalypha indica Linn.	Euphorbiaceae	Indian Acalypha	Kuppi	M
Adhatoda zeylanica Med.	Acanthaceae	Vasaka	Basinga	M, Oil
Aerva sanguinolenta (L.) Bl.	Amaranthaceae	-	Safedphulia	M
Agave cantala Roxb.	Agavaceae	Bombay Aloe, Cantala, Magney	Rambans	Fb.
Anisomeles indica (L.) Ktze.	Lamiaceae	-	Goplya	M, Oil
Artemisia nilagirica	Asteraceae	Indian wood Fleabane	Kunjaa	Oil
Asclepias curassavica L.	Asclepiadaceae	Curassavian Swallow	Lalma	Or, M, Fb.
Asparagaus adscendens Buch.-Ham. ex Roxb.	Liliaceae	Blood flower		M
Barleria prionites L.	Acanthaceae	Yellow Hedge	Kala-bansa Barleria	M, Or.
Berberis lyceum Royle	Berberidaceae	-	Kilomora, Kirmora	M
Boehmeria platyphylla D.Don	Urticaceae	-	Khagsa	Fb.
Callicarpa macrophylla Vahl	Verbenaceae	Beauty Berry	Priyangu, Daiya	M
Calotropis procera (Ait.) Dry.	Asclepiadaceae	-	Aak	M
Cannabis sativa L.		True Hemp, Soft Hemp, Marijuana	Bhang, Bhanglu	M,Fb.
Carissa opaca	Apocynaceae	-	Karaunda	Ed, M
Cassia occidentalis L.	Caesalpiniaceae	Negro Coffee	Taror	M
C. tora L.	-do-	Sickle Senna	Chakunda	M, Gu
Clerodendrum viscosum Vent.	Verbenaceae	Glory Bower	Bhant	M
Colebrookia oppositifolia Smith.	Lamiaceae	-	Bhirmoli, Binda	M
Cotinus coggygna Scop.	Anacardiaceae	Smoke Tree, Indian Sumach	Gad-tungla	Tn
Datura stramonium L.	Solanaceae	Stramonium, Mad Apple	Dhatura	M
Debregeasia longifolia (Burm.f.) Wedd.	Urticaceae	Wild Rhea	Tusara, Siaru	Fo
Duranta repens L.	Verbenaceae	Golden Dew Drop, Pigeon Berry,	Durenta	Or.
Euphorbia royleana Boiss.	Euphorbiaceae	-	Sulla, Surao	M
Ficus palmata Forsk.	Moraceae	Fig	Baru	Ed, M
Glycosmis arborea (Roxb.) DC.	Rutaceae	-	Padera	M
Helicteris isora L.	Sterculiaceae	East Indian Screw Tree	Marorphali	Fb, M
Hypericum oblongifolium Choisy	Hyperaceae	-	Chitroi	Or
Indigofera cassioides Rottl. ex DC.	Fabaceae	-	Sakena	Ed, M
Inula cuspidata (DC.) Cl.	Asteraceae	-	Jhuri	Or
Jatropha curcus L.	Euphorbiaceae	Purging Nut, Barbedos Nut	Phari Arand	Oil
Lantana camara L.	Verbenaceae	Lantana, Wild Sage	Kuri	Oil
Lantana indica Roxb.	-do-	-do-	-do-	Oil
Martynia annua L.	Martyniaceae	Devil's Claw, Tiger Claw	Bagnakha, Bichhwa	M, Oil
Bergera koenigii (L.) Spr.	Rutaceae	Curry Leaf Tree	Kurry patta	Ed, Oil, ag.
Nicandra physaloides (L.) Gaertn.	Solanaceae	Apple of Peru	Tambukya	M

Table 5: Contd.

Botanical Names	Families	Trade Names	Local Names	Uses
Opuntia vulgaris Mill.	Cactacee	-	Nagphani	Ed, Fb, M
Pavetta indica L.	Rubiaceae	White Pavetta, Indian Pellet Shrub	Angari	M
Pogostemon benghalense (Burm.f.) Kuntz.	Lamiaceae	-	Lujra, Kala-basan gu	M, Oil
Rhus parviflora Roxb.	Anacardiaceae	-	Tungla	Ed, M
Ricinus communis L.	Euphorbiaceae	Castor Plant	Erandi, Arandi	M
Rubus ellipticus Sm.	Rosaceae	Himalayan Yellow Raspberry	Hinsar, Hisalu	Ed.
Rubus niveus Thunb.	-do-	Mysore Raspberry	Anchu, Kala Hisnar	Ed.
Rumex hastatus D.Don	Polygonaceae	-	Almora	Tn.
Solanum erianthum D.Don	Solanaceae	Potato Tree	Akra	M
S. torvum Sw.	-do-	-	Bhurat	M, Ed, Oil
S. viarum Dunal	-do-	-	-	M
Trema politoria (Planch.) Bl.	Urticaceae	-	Khagshi,	M
Urtica dioica Linn.	Urticaceae	Stinging Nettle	Kandali,	Ed.
Vitex negundo L.	Verbenaceae	-	Shimalu, Nirgundi	M, Oil
Woodfordia fruticosa (L.) Kurz.	Lythraceae	Fire Flame Bush, Shiranji tea	Dhaula	Dy. Ag, Med.

Table 6: Some economically important Herbaceous species in Uttarakhand

Botanical Names	Families	Trade Names	Local Names	Uses
Achyranthes aspera L.	Amaranthaceae	Prickly chaff flower	Apamarga	M
Ageratum conyzoides L.	Asteraceae	Goat weed	Gandela	M
Amaranthus viridis L.	Amaranthaceae	–	Cholai	Ed
Anagallis arvensis L.	Primulaceae	Scarlet Pimpernel	–	M
Anisochilus carnosis (L.f.) Wall. ex Benth.	Lamiaceae	–	–	M
Argemone mexicana L.	Papavaraceae	Prickly Poppy	–	M
Boerhavia diffusa L.	Nyctaginaceae	Spreading Hog-Weed	Punarnava	M
Capsella bursa-pastoris (L.) Medik.	Brassicaceae	Shepherd's Purse	–	M
Celosia argentea L.	Amaranthaceae	Qual Grass	Sufaid margha	M
Chenopodium album L.	Chenopodiaceae	Lamb's Quarters	Bathua	Ed.
Cleome viscosa L.	Capparidaceae	–	Jakhia	Ed.
Clinpodium umbrosum (Bab.) Koch.	Lamiaceae	–	–	–
Coleus forskohlii (Willd.) Briq.	-do-	–	Patharchoor	M
Colocasia esculenta (L.) Schott.	Araceae	Cocoyam	Pindalu	Ed.
Commelina benghalensis L.	Commelinaceae	–	–	Ed.
Curculigo orchioides Gaertn.	Hypoxidaceae	–	Kalimusli	M
Desmodium gangeticum (L.) DC.	-do-	–	Salparni	M

Table 6: Contd.

Botanical Names	Families	Trade Names	Local Names	Uses
Eclipta lebeian (L.) L.	-do-	–	Brangraj	Dy
Euphorbia hirta L.	Euphorbiaceae	–	Dudhi	M
Evolvulus alsinoides (L.) L.	Convolvulaceae	–	Sankhpush pi	M
Fumaria indica (Haussk.) Pugs.	Fumariaceae	–	Kurri	M
Launaea procumbens (Roxb.) Ram. and Raj.	Asteraceae	–	–	M
Malva verticillata L.	Malvaceae	–	–	M
Oxalis dehradunensis Raizada	-do-	Indian Sorrel	Khatti Melhi	M
Perilla frutescens (L.) Britt.	Lamiaceae	–	Bangjeera	Oil
Phyla nudiflora (L.) Green.	Verbenaceae	–	–	M
Phyllanthus amarus Sch. and Thonn.	Euphorbiaceae	–	Bhuinmala	M
Plantago erosa Wall.	Plantaganaceae	–	–	M
Portulaca oleracea L.	-do-	Common Purslane	Luanya	Ed.
Rhabdosia rugosa (Wall. ex Benth.) Hara	Lamiaceae	–	–	M
Scoparia dulcis L.	Scrophulariaceae	Sweet Broom Weed	–	M
Senecio nudicaulis Buch.-Ham. ex D.Don	Asteraceae	–	Neel Kanthi	M
Sesamum orientale L.	Pedaliaceae	Gingelly	Til	Oil
Sida cordifolia L.	-do-	Country-Mallo	Bela	M
Solanum nigrum L.	Solanaceae	Black Nightshade	Makri	M
S. surattense Burm.f.	-do-	Yellow Berried	Kateli	M
Stellaria media (L.) Vill.	-do-	Chick weed	–	M
Tagetes minuta L.	Asteraceae	Stinking-Roger	Jangli gainda	Oil
Thalictrum foliolosum DC.	Ranunculaceae	–	Mamira	M
Uraria lagapodioides (L.) Desv.	Fabaceae	–	–	M
Vernonia cinerea (L.) Less.	Asteraceae	–	Sahdevi	M
Verbascum thapsus L.	-do-	Cow's Lungwort	Akelbeer, Bantamkhu	M
Viola canescens Wall.	Violaceae	Sweet violet	Banafsa	M
Xanthium indicum Koen	Asteraceae	Burweed	–	M

Abbreviations

Ag. = Agricultural implements; Dy. = Dye; Ed. = Edible; Fb. = Fiber; Fo. = Fodder; Gu. – Gum; M. = Medicinal; Oil = Oil; Or. = Ornamental; Sf = Tm. = Timber; Tn. = Tannin.

CONCLUSIONS

For the State like Uttarakhand which is very rich in forest cover and plant species diversity, the assessment and conservation of biodiversity is of paramount importance. There is a need to set up a coordinated programme for systematic biodiversity measurement and monitoring, at least in major forest types or eco-regions. It is also desirable to keep the regulatory regime in place and concerted efforts need to be made to attract people's participation in a big way. Education should be provided and capacity should be built in such a way that people may derive intangible as well as economic benefits from conservation efforts to improve their livelihood.

REFERENCES

Biswas, S (2005). Developing a floristic model for restoration of threatened biodiversity of North Western and Central Himalaya, pp.148-151. In: *Proceedings of the Workshop on Conservation of Biodiversity in India-Status, Challenges and Efforts.* Publ. Indian Council of Forestry Research and Education, Dehradun, Uttarakhand, India

CBD (1992). Convention on Biological Diversity: text and annexes, article 2. Secretariate of the Convention on Biological Diversity, Montreal, Canada

Champion, HG & Seth, SK (1968). A Revised Survey of the Forest Types of India. Govt. of India Publication, Delhi

Dhar, U, Rawal, RS & Samant, SS (1997). Structural diversity and representativeness of forest vegetation in a protected area of Kumaun Himalaya, India: Implications for conservation. *Biodivers. Conserv.* **6:** 1045-1062

FSI (2005). *State of Forest Report*, Forest Survey of India, Dehradun, Uttarakhand (India)

Gaur, RD (1999). Flora of the District Garhwal, North-West Himalaya (With Ethnobotanical Notes) Transmedia, Srinagar, Pauri, Uttarakhand

ICFRE (2005). Proceedings of the Workshop on Conservation of Biodiversity in India- Status, *Challenges and Efforts.* Publ. Indian Council of Forestry Research and Education, Dehradun, Uttarakhand (India)

Jeet Ram, Kumar, A & Bhatt, J (2004). Plant diversity in six forest types of Uttaranchal, Central Himalaya, India. *Curr. Sci.* **86:** 975-977

Priti Kumari, & Tewari, LM (2009). Biodiversity in Uttarakhand Himalaya region. *Nature and Science* **7:** 113-124

Negi, SS (1990). *Himalayan Forests and Forestry.* Indus Publishing Company, New Delhi, India

Samant, SS (1999). Diversity, nativity and endemism of vascular plants in a part of Nanda Devi Biosphere Reserve in West Himalaya I. Himalayan Biosphere Reserve (Biannual Bulletin) **1:** 1-28

Singh, JS & Kushwaha, SPS (2008). Forest biodiversity and its conservation in India. *Internat. Fore. Rev.* **10:** 292-304

10

Economics of Crop Diversification: Uttarakhand State

HARJEET SINGH, AND AK SHARMA[1]

Agricultural Finance Corporation Ltd., B-1/9, Janakpuri, New Delhi-110 058
[1]Additional PCCF, Department of Forests, Van Bhawan, 17, Rana Pratap Road, Lucknow-226 001 (U.P.)
E-mail: h.harjeet@gmail.com

ABSTRACT

Since agriculture is not as productive in the hills as compared to the plains due to lack of sufficient manpower and poor economic returns, the decision of the farmers to go in for crop diversification is influenced by factors like weather conditions, price support, market intervention, risk factor and relative profitability of these crops vis-a-vis the other crops. As such, any economic impact of this process would be a function of a host of relevant variables. Diversification can be a useful means to increase crop output under different situations. The integration of medicinal and aromatic plant species on the farmland paves the way for the improvement and diversification of existing systems. It offers an opportunity to explore the possibility of growing commercial species with high value crops (HVCs) of medicinal and aromatic plant species such as kesar (*Crocus sativus*), Kutki (*Picrorhiza kurroa*), sarpgandha (*Rauwolfia serpentina*) and lemongrass (*Cymbopogon flexuosus*) on the farmland in different parts of the State.

INTRODUCTION

Modern agriculture is a complex and demanding profession requiring inputs from a series of institutions and individuals. In essence, the technology developed by various research institutions is transferred to the farm community through extension. However, once the technology has reached the field, the farmer requires integrated services from a series of technical units. How to grow a new variety in the field with the use of appropriate mix of recommended inputs to realize its potential yield, offers continuous challenges for the farmers.

The diversification of existing cropping patterns coupled with development of suitable technology packages are required to cope up with the increasing demand for an array of agricultural, horticultural, medicinal and aromatic plant cropping systems. The wealth of information available on optimizing traditional agricultural and forestry systems has undoubtedly, helped to maximize the productivity, but the rising population coupled with falling productivity, land degradation, soil erosion, loss and overuse of cultivable land are the major constraints for attaining sustainable production. The existing land use systems with separate allocation to agriculture and forest are not adequate to meet these demands. The shift to agri-horticulture (agriculture + fruit trees), agro-forestry (growing trees + crops), horti-medicinal (fruit trees + medicinal crop) cropping models provides a viable option to these systems with opportunities to diversify as well as increase overall land productivity. It is necessary to develop viable, acceptable, diversified and sustainable cropping systems, which would ensure enhanced crop production by maximizing utilization efficiency of the available resources.

Economic feasibility provides the main rationale to bring a species under cultivation unless of course sufficient quantities of the material can still be obtained at a lower price from wild. The cultivated material will be competing with the material harvested from the wild that arrives in the market by commercial gatherers who have incurred no input costs for cultivation. However, there is need for domesticating and varietal upgrading of the wild collected species for commercial production to meet the demand of the local and international market and industry.

A partnership approach needs to be developed between banks, buyers, R&D institutions and the farmers to achieve this objective. The institutes should take lead in providing the technical expertise for making plant inventories, developing agro-technologies and processing techniques. Banks should design marketing strategies based on studies launched by them, and the farmers should form community groups for cultivation and primary processing.

Considering this, crop diversification has been recognized as an important instrument through which the farmers can grow the best profitable commodities on their land and earn sufficient money. For this, they should be aware as to which commodities are suited to their locality and will bring more profits for them. They should also know as to how these commodities can be produced efficiently and sold in the markets, both internal and external.

STUDY AREA AND METHODOLOGY ADOPTED

(i) Study Area

The sample for the study was representative of variations in agro-climatic situations in the State and four districts were selected for undertaking the survey. In each district three blocks and four villages were selected. Twenty-four farmers were selected at random from the sample villages for in-depth discussion.

(a) Chamoli;
(b) Dehra Dun;
(c) Nainital; and
(d) Udhamsingh Nagar.

A number of checklists were designed for data collection and discussion with the farmers to understand the reasons for shift/diversification from traditional crops to new crops. The data thus collected have been tabulated and analyzed with appropriate statistical techniques.

DIVERSIFICATION TOWARDS NON-TRADITIONAL CROPPING PATTERN

There is a growing awareness about the cultivation of non-traditional crops, viz medicinal and aromatic plant species over the traditional ones. The villagers are taking keen interest in this direction with the assistance provided by the Central and State Governments, NGOs and financial institutions in some areas and are successfully cultivating *Picrorhiza kurroa* (Kutki), *Crocus sativus* (Saffron), *Rauwolfia serpentina* (Sarpgandha), *Cinnamomum tamala* (Tejpat), *Emblica officinalis* (Amla), *Acorus calamus* (Bach) and *Cymbopogon flexuosus* (lemongrass) in their agriculture fields and on wastelands. The economics of medicinal and aromatic plant species is reasonably good. Regretfully, the current cultivation practices of medicinal and aromatic plant species in the State are still primitive due to lack of suitable agro-technology and poor marketing coupled with poor processing facilities.

The district-wise information about trends of crop diversification by the sample farmers during the years 2000-2001 and 2005-2006 has been given in Table 1 below:

Table 1: District-wise information about trends of crop diversification area (acre)

Crops	2000-01	2005-06	Area increase/ decrease	Percentage
I - District Chamoli				
(a) Agriculture Crops				
Wheat	25	26.5	+ 1.5	6.0
Paddy	11.5	10.5	- 1	- 8.69
Potato	2.5	4.5	+ 2	80
(b) Horticulture Crop				
Apple	4.0	4.0	0	0
(c) Medicinal Plants				
Kutki	0	6.20	+ 6.20	+ 100
Saffron	0.62	0.62	0	0
II - District Dehradun				
(a) Agriculture Crops				
Wheat	31.9	24	- 7.9	- 24.76
Paddy	22	19.5	- 2.5	- 11.36
(b) Horticulture Crop				
Litchi	27.60	27.60	0	0
c) Medicinal Plant				
Sarpgandha	0	2.9	+ 2.9	+ 100
(d) Aromatic Plant				
Lemongrass	0	7.5	+ 7.5	+ 100
III - District Nainital				
(a) Agriculture Crops				
Wheat	18.5	20.5	+ 2	+ 10.81
Paddy	12.5	10.5	- 2	-16
(b) Horticulture Crop				
c) Litchi	19.0	19.0	0	0

Table 1: Contd.

Crops	2000-01	2005-06	Area increase/ decrease	Percentage
(d) Medicinal Crop				
Aonla	7.5	7.5	0	0
IV - District Udhamsingh Nagar				
Agriculture Crop				
Wheat	215	192	- 23	-10.69
Paddy	170	143	- 27	-15.88
Medicinal Crops				
Aonla	14.5	14.5	0	0
Bach	0	2	+ 2	100
Aromatic Crop				
Lemongrass	0	48	+ 48	100
Forest Tree				
Poplar	14	14	0	0

In the Chamoli district, four villages have been selected from three blocks, viz. Joshimath (Baragoan, Parsari), Ghat (Ramni) and Dewal (Ghes) whereas in Dehradun district Vikasnagar (Badripur), Doiwala (Simlat) and Sahaspur (Rajawala and Sahaspur). Similarly, as regards the Nainital district Bhimtal (Dogora), Ramnagar (Kotabagh and Bhelpora) and Okhal Kanda (Remal Gaon). Sitarganj (Loka and Nenkuliya), Rudrapur (Bajpur) and Khatima (Chanda) have been selected from the district Udhamsingh Nagar.

During the span of one decade, the area under horticulture in all the four districts has grown rapidly especially under Litchi (*Litchi chinensis)* in Dehradun and Nainital districts. Horticulture is emerging as a major component in the cropping pattern. The limited marketing and agro-processing facilities are some of the important problems particularly for *Malus pumila* (apple), *Litchi chinensis* (litchi) and *Emblica officinalis* (aonla).

The major constraints that tend to limit the crop diversification are high cost of inputs and low quality of produce owing to the use of archaic technologies, often resulting in low benefits. It is necessary to invest in research for developing not only improved varieties, but also better agro-techniques that can enhance the yield potential of varieties. More attention needs to be paid to techniques that do not require expensive inputs. Farmers should also be encouraged to adopt measures to reduce crop duration besides raising off-season crops.

It is necessary to reduce the area under paddy in order to enable farmers diversify towards high value crops. To motivate the farmers in this regard, there is a need for public investment in: (a) marketing infrastructure to increase the trade value of alternative crops and (b) value addition through improved processing.

ECONOMICS OF CROP DIVERSIFICATION

Table 2 below indicates the benefit-cost ratio (BCR) and internal rate of return (IRR) for various one acre crop models.

Table 2: The BCR and IRR for various one acre crop models

Sr. No	Districts	Crop Models	BCR	IRR (%)
(A)	**District Chamoli**			
1.	**Traditional Crops**	Potato	1:3.82	NA
		Wheat	1:2.00	NA
		Paddy	1 : 1.92	NA
	Non-traditional Crops	Kutki	1:2.79	39.23
		Apple + Saffron	1:5.62	19.98
(B)	**District Dehradun**			
2.	**Traditional Crops**	Wheat	1:3.65	NA
		Paddy	1 : 1.84	NA
	Non-traditional Crops	Sarpgandha	1:3.70	39.0
		Lemongrass	1: 3.41	NA
		Litchi + Wheat	1:7.49	21.11
(C)	**District Nainital**			
3.	**Traditional Crops**	Wheat	1:1.94	NA
		Paddy	1 : 1.85	NA
	Non-traditional Crops	Tejpat	1:4.93	19.76
		Litchi + Wheat	1:8.51	20.55
		Anola + Wheat	1:6.64	29.0
(D)	**District Udhamsingh Nagar**			
4.	**Traditional Crops**	Wheat	1:4.13	NA
		Paddy	1 : 1.84	NA
	Non-traditional Crops	Lemongrass	1:5.63	51.59
		Bach	1:2.61	NA
		Poplar + Wheat	1:7.50	29.42

Viability analysis indicates that B:C ratio is as high as 1 : 8.51 for the litchi + wheat cropping pattern. The benefit/cost ratio for Tejpat crop is 1:4.93 and the internal rate of return 19.76 per cent.

The production costs for enterprises of individual agriculture, horticulture, medicinal and aromatic plant species vary due to factors such as location, crop (annual or perennial), size, machinery, labour use and marketing activities.

In district Chamoli, BCR of traditional crops is 1: 3.82 for potato and 1: 2.00 for wheat. However, in non-traditional cropping pattern, it is 1: 5.62 for the apple + saffron cropping model.

In district Dehradun, the benefit/cost ratio is available for 5 crops and IRR for 3 crops. The litchi + wheat cropping model shows the highest Benefit/Cost Ratio (BCR), i.e. 1 : 7.49 and the Internal Rate of Return (IRR) is 21.11 per cent.

In district Nainital, under the medicinal plant sector, Tejpat cultivation, on the boundary of the agriculture land is eminently suited. The climate of this region allows vigorous growth of a variety of medicinal and aromatic crops also. Moreover, a related advantage for district Nainital is its relative proximity to North Indian Urban Centres that have an increasing demand of medicinal plant species and their products.

Similarly in case of district Udhamsingh Nagar, lemongrass introduced as a new aromatic crop has shown high degree of promise over the traditional cropping pattern. The Benefit/Cost Ratio (BCR) for lemongrass is 1 : 5.63 and Internal Rate of Return (IRR) is 51.59 per cent.

SUGGESTIONS AND RECOMMENDATIONS

There is good potential for high value low volume crop in the State and a strong case is made out for diversification of crops from cereals to horticulture, medicinal and aromatic plants for improving economic condition of the farmers of the hills.

The climate and geography of the State are cumulatively favourable and make it possible to grow a variety of high value crops suited to different regions which offer a wide scope for diversification leading to all round agricultural development.

As would be expected, the decision of the farmers to go for crop diversification is influenced by factors like weather conditions, price support, market intervention, risk factor and relative profitability of these crops vis-a-vis the others. As such, any positive and negative impact of this process would be a function of a series of related factors. According to the information gathered through interaction with the State officials and group discussions with the farmers during the survey in selected sample districts coupled with statistical data provided by the State, the process of diversification of crops has already been initiated though on a moderate scale.

(i) Agriculture

Currently, more than three-fourth of the population in the hilly areas has its livelihood dependent on agricultural sector. Thus, any attempt at inclusive growth has to increase the income from this sector. Agricultural produce consists largely of cereals like rice, wheat and millets, although horticultural produces including fruits and vegetables also constitute a significant part of the net sown area.

The yield from the field crops like rice and wheat is not very high in the hilly areas of the State. This is largely due to the mountainous terrain that makes it difficult to adopt mechanized, modern agriculture. Thus, these crops are produced in the hills mostly to fulfil the subsistence needs of the people.

(ii) Horticulture

Apple, walnut, guava, peach, pear, apricot, citrus varieties, kaphal, strawberry, and banana are the major fruits grown in the hilly localities. Apple orchards are located mainly in Joshimath in Chamoli and Ramgarh-Mukteshwar belt of the district Nainital. These alongwith the litchi of Dehradun and Ramnagar make Nainital an important fruit-growing belt.

Farmers in the hilly regions should be encouraged to shift from the cultivation of cereals to the cultivation of horticulture crops as well as value-added horticulture based products since this sector has vast potential. Regretfully, despite the advantages, horticultural produce is still not providing the farmers with higher incomes as expected, due to the absence of necessary market infrastructure, institutional support and incentives.

The development of horticulture also depends on the emphasis being given to transport and power sectors. The perishable nature of these crops leads to the necessity of a network of warehouses and cold storages for which assured power supply is needed. The high input costs and the risk of crop failures in this sector makes it necessary to develop financial institutions that can provide credit as well as crop insurance.

The incentives that the State can offer to the small and marginal farmers to shift to horticultural production are also bound to encourage the development of this sector. The most important justification for providing such incentives is that the small and marginal farmers who are dependent on subsistence agriculture have no savings of their own and are not creditworthy. This makes it very difficult for them to switch over to horticulture due to the higher input costs involved. Thus, the State needs to provide them with subsidized seeds, fertilizers, insecticides, etc. The second reason for providing incentives is to take care of the uncertainties due to market failure and natural calamities. The most common form of market failure is the lack of competition among the purchasers of these products leading to a low price and profit for the farmer. The State should provide minimum support prices in order to make sure that the poor farmers get a remunerative price. Finally, provision has to be made to minimize farmer's losses due to natural calamities through crop insurance. If the premium for such a insurance is too high for the small farmers, then the State should subsidize this to make it affordable. There is a lot of potential for the development of horticulture in the State and this needs to be tapped efficiently to bring it at par with Himachal Pradesh.

The returns earned by the farmers from horticulture crops are also low due to the lack of sufficient agro-processing units and agro-industries in the State with the result, the farmers try to sell off low quality fruits alongwith those of the better quality. Consequently, they get an average price for all the produce and this is usually very low. It must be pointed out that the practice of producing horticultural crops is not new in the mountains of Uttarakhand, although in the past, a large part of it was for domestic consumption rather than for commercial purposes. Evidently, for changing the scenario through increasing the incomes of farmers in the rural mountainous regions, the thrust has to be on modern commercial horticultural practices. The problem, however, is that the small and marginal farmers lack adequate knowledge about the most suitable and remunerative crops or the scientific practices that can ensure realization of high yields of these crops. Even when this knowledge and information barrier has been bridged, they are handicapped by the high cost of acquiring high-yielding seeds, fertilizers, insecticides, pesticides, etc., partly due to their high prices and partly due to their outlets being far off from the villages and farms.

The policy package for inclusive growth must ensure that the above-mentioned forward and backward linkages for this sector are provided to ensure growth in the sector.

(iii) Medicinal and Aromatic Plant Species

The existing marketing structure for medicinal plants in the country is unorganized to a great extent. The market is dominated by middlemen. Government is trying to improve the marketing system through regulations and creation of infrastructure. However, the measures being taken still seem to be inadequate. Furthermore, since the question of livelihood is associated with the decision to reduce the dominance of the middlemen from the market, it is desirable that a proper balance is achieved. In this regard the following approaches deserve to be considered :

1. Establishment of Regulated Markets: A regulated market is a place where transactions are governed by various rules and regulations framed by local bodies through State Legislation. There are certain rules and regulations in the medicinal plant market also but these are not implemented properly. As such it is desirable that the Govt. makes more efforts to monitor the trade of medicinal plants.

2. Buy-back Policy: This may prove to be one of the most effective measures to reduce the dominance of the middlemen in the market as well as exploitation of the farmers or primary collectors of medicinal plants. The main problem farmers face in cultivation of medicinal plants is that they get very poor return from selling their products. Due to no direct contact with the market, they have to sell their product to the village agents at a rather low price. In this way the medicinal plant cultivation has become less attractive for the farmers. As long as they are not given any assured market, their problems remain unapprised. In this context, the buy-back policy will be of great help for the farmers. In buy-back policy what normally happens is that the Govt. or a private organization signing a contract with the farmers undertakes that the farmers will grow medicinal plants on their land and these organizations will buy their produce at the price prevailing in the market. In several parts of the country this buy-back policy has been adopted with good results. In Gujarat, the Zhandu Pharmaceuticals have taken initiatives in this regard. The company is providing propagules of medicinal plants to the farmers and buying from them the ready product at a fixed price. Wimco has done the same with regard to the tree plantation in a number of districts of Uttar Pradesh and Uttarakhand including Bareilly, Badaun, Shahjahanpur and Pilibhit. Now the scenario is changing in Uttarakhand as well. In the village Gheas, the High Altitude Plant Physiology Research Centre (HAPPRC), Srinagar, has taken initiative in this direction and provided buy-back guarantee to the cultivators through Delhi-based pharmaceuticals. Under the technical guidance of HAPPRC, the villagers are cultivating two prominent herbal plants, viz *Picrorhiza kurrooa* (Kutki) and *Saussurea lappa* (kuth) in their barren lands. These efforts should be encouraged to streamline the marketing system of medicinal plants in the State. In this context the scheme for the development of commercial cultivation of MAPs being implemented by National Medicinal Plants Board is a commendable initiative since it links production with assured buy-back arrangements.
3. Improvement in Transport Facilities: Being a hilly State, transport is a great problem in Uttarakhand. People of remote villages cannot bring their collections to the town. Firstly, it is time-consuming and secondly they have to sell their products under unfavourable conditions in markets located in remote towns. Chamoli, Uttarkashi, Pauri and Pithoragarh are the main districts where economically important medicinal plants are available due to high altitude in these regions. However, the main markets for crude medicinal plants are Dehradun, Ramnagar, Tanakpur and Haridwar which are far away from these districts. This disadvantage affects them greatly and leads to dependence on the middlemen. Thus, the improvement of road facilities is a long-term strategy and it should take into account a lot of parameters. However, it leads to numerous advantages that can be derived by shortening and speeding up the marketing channel. The development of road infrastructure will lead to threefold advantages, viz:
   (i) Cheaper transports enable rural people to reach larger centres to sell their products.
   (ii) Better price realization is possible through direct contact with the market dealers.
   (iii) Minimum middlemen dependence in the market channel.

The lack of marketing institutions and infrastructure has to be dealt with immediately by the Government. The objective of this policy should be to provide farmers with alternative options to sell their products, so that the part of the profit now going to the middlemen is minimized and the farmer gets a better price for his produce. There are three types of institutions that the Government needs to create or strengthen for this purpose.

1. Herbal Research and Development Institute (HRDI) and Uttarakhand Forest Development Corporation (UKFDC) that will help, particularly the small and marginal farmers, to grow medicinal and aromatic crops and market them at remunerative prices.
2. Strengthen the farmers through cooperative associations and encourage them to corporatise themselves so that they can employ professionals to help them market their products.
3. Allow and encourage contract farming between the farmers and processors of medicinal plant-based pharmaceutical industries so that the role of the middlemen can be minimized.

It must be understood that whereas the first institution, i.e., the marketing board, should primarily target the poor and marginal farmers, the medium and large farmers can use the second and third category of institutions more effectively. It is pertinent to point out that although the State has been declared an **"Organic State"**, it will be useful to the farmers only when the produce can be marketed at a high price with the help of organic certification.

REFERENCES

Anonymous, (2000). National Agriculture Policy. Department of Agriculture and Co-operation, Ministry of Agriculture, Government of India, New Delhi

Anonymous, (2001). The Diversification of Punjab's Agriculture. Pamphlets issued by the Managing Director, Punjab Agro-Food Corporation Ltd., Chandigarh

Anonymous, (2002). Conservation, Cultivation and Marketing of Medicinal Plants in Uttarakhand. Agricultural Finance Corporation Ltd., NRO, New Delhi

Abdul Kalam, APJ (2003). Prosperous, Peaceful and Safe Human Habitat. Inaugural Address, D.S.D.S 2003, New Delhi, February 06, 2003

Arnold, JEM (1983). Economic considerations in agroforestry projects. *Agroforestry Systems* **1**: 299- 311

Arnold, JEM (1987). Economic Considerations in Agroforestry, pp.173-190. In : Stepper H A & Nair PKR (eds) Agroforestry: A Decade of Development. International Centre for Research in Agroforestry, Nairobi, Kenya

Arnold, JEM & Dewees, P (1998). Rethinking Approaches to Tree Management by Farmers, ODI Natural Resource Perspective, No. 26, Overseas Development Institute, London, UK

Atul Punam, & Kholsa, PK (1990). Classification of Traditional Agro-forestry Systems, pp. 24-27: In Proc. IUFRO 19th World Forestry Congress, Montreal, Canada

Bajaj, YPS (1995). Biotechnology in Agriculture and Forestry: Medicinal and Aromatic Plants, VII. 1-8, Springer-Verlag, Berlin

Bhakuni, DS (1999). Medicinal Plants of Uttarakhand and Problems of Pharmaceutical Industries. pp. 35-44. In: Nautiyal AR, Nautiyal MC & Purohit AN (eds). In: Harvesting Herbs. Bishan Singh Mahindra Pal Singh, Dehradun

Bhalla, GS & Singh, G (1997). Recent Developments in Indian Agriculture: A State of Level Analysis. Economic and Political Weekly, Ludhiana, March 29

Biswas, K (1956). Common Medicinal Plants of Darjeeling and the Sikkim Himalayas. Govt. Press, Calcutta

Chauhan, NS (1996). Business of medicinal and aromatic plants. *Sci. Tech. Entrep.* **7**: 34 38

Dadhwal, KS, Narain, P. & Dhyani, S.K. (1989). Agro-forestry Systems in the Garhwal Himalayas in India. *Agrofor. Syst.* **7**: 213-225

Dhar, U, Rawat, RS & Jyoti, U (2000). Setting priorities for conservation of medicinal plants: A case study from Indian Himalayas. *Biol. Conser.* **95**: 57-65

11

Medicinally Important Endangered Orchid *Habenaria intermedia* D. Don and *H. edgeworthii* Lindl.: A Phenological Study

CHETNA BISHT*, ANOOP BADONI** AND AR NAUTIYAL*

**High Altitude Plant Physiology Research Center*
***Department of Seed Science and Technology*
H.N.B. Garhwal Central University, Srinagar–246 174, Pauri, Uttarakhand, India
E-mail: bishtchetna@yahoo.com and annabadoni@yahoo.co.in

ABSTRACT

The Orchidaceae is a cosmopolitan family distributed throughout the world, barring a few isolated islands and the frozen continent of Antarctica, which grows in almost all kinds of habitat except the aquatic and marine ecosystems. The family exhibits lesser diversity in its vegetative parts, but shows enormous diversity in its highly specialized floral morphology. The North-Western Himalayas, comprising the states of Jammu and Kashmir, Himachal Pradesh, and Uttarakhand, have 255 species of orchids of which 217 are common with the Eastern Himalayas. The orchids represent a group of botanically significant and commercially important plants which are still in an evolutionary flux; these account for nearly one-tenth of the total flowering plant species of the world. Several orchids are frequently mentioned in ancient Indian literature and some of them including *Eria muscicola* (Jivanti), *Eulophia campestris* (Salem), *Habenaria edgeworthii* (Vridhi), *Habenaria intermedia* (Ridhi), *Malaxis mucifera* (Jeevak), *Orchis latifolia* (Salampanja), *Vanda cristata* (Rasna), and *Zeuxine strateumatica* (Shwethuli) are still extensively used in ayurvedic preparations. In the Ayurvedic system of medicine a group of eight drugs, known as "Ashtawarga" – a group of 8 drugs in Ayurvedic system which are used for preparation of tonics, such as 'Chyavanpras', consist of 4 orchid species, viz *Malaxis muscifera* (Lindl.) Kuntze, *Malaxis acuminate* D.Don, *Habenaria intermedia* D.Don and *Habenaria edgeworthi* Hook. F. *Habenaria* is a large genus of orchids found in tropical and temperate parts of the world. About 8 species are reported from India, of which two are very important species, namely, *Habenaria intermedia* and *H. edgeworthii.* In the present work the phenology (plant emergence to plant senescence and tuber dormancy stages) of both the species have been presented.

Keywords: Diversity, endangered orchids, floral morphology.

INTRODUCTION

Orchidaceae comprising about 750 genera and 18,000 species (Heywood, 1993), or 788 genera and 18,500 species (Mabberley, 1988), is the second largest family of flowering plants in the world. The great geographic expense of the country encompassing a variety of bioclimatic zones and the enormous diversity of ecosystems account for the bewildering array of orchid species in India. The wide range of phyto-climatic conditions met across the country have definitely influenced the distribution and composition of orchid flora in different parts of the country. The North-Western Himalayas, comprising the states of Jammu and Kashmir, Himachal Pradesh, and Uttarakhand, has 255 species of orchids (Seidenfaden & Arora, 1982; Deva & Naithani, 1986) of which 217 are common with the Eastern Himalayas. The Eastern Himalayas, together with the North-Eastern states, the Western Ghats, and the Western Himalayas, therefore, represents three major centers of diversity of orchids in India.

The Indian orchids are paradoxically victims of their own beauty and popularity; the result is that their natural population is rapidly declining. However, commercial exploitation of orchids has remained rather negligible in India due to low priority and lack of awareness about its commercial significance and poor marketing, packaging and transport strategies. The orchids in the region are fast depleting from their natural habitats due to deforestation and over-exploitation for commercial purposes. The people of the region are fond of growing orchids besides other flowering plants. As a result, the orchids are collected indiscriminately in large numbers from nature and sold in the market place, and also house to house. Therefore, there is a need for both conservation and sustainable utilization of orchids (Kumaria & Tandon, 2001). Many of them particularly in genera *Aerides, Arachnis, Cattleya, Cymbidium, Dendrobium, Epidendrum, Oncidium, Paphiopedilum, Phalaenopsis, Renanthera, Vanda,* etc., have been extensively used to progenate high profile and internationally acclaimed hybrids. The Indian orchids are paradoxically victims of their own beauty and popularity; the result is that their natural population is rapidly declining because of unbridled commercial exploitation on the one hand and deterioration of their habitats on the other. More then 150 species are thus in danger of becoming extinct while some other like *Calanthe pachystalix, C. whiteana, Coelogyne albolutea, C. assamica, and Paphiopedilum charlesworthii,* have already become so (Pathak, 2001).

Several such species are on the verge of extinction due to evolutionary trends of man-made threats, of which the latter is predominant factor during past few decades. *Habenaria* is a large genus distributed in tropical to temperate regions of the world. It is represented by approximately 500 species of which eight have been reported from India. *H. intermedia* and *H. pubescens* are medicinally important among them, the former is rapidly disappearing due to indiscriminate digging for its edible tubers, is known as 'Riddhi' in ayurveda and occurs in grassy slopes between 2,000-3,000m in the Himalayan region. Its tender leaves as well as tubers are edible and cooked as vegetables. The mature tubers are also useful to prepare salep (Chauhan *et al.*, 2007).

COMMERCIALLY POTENTIAL ORCHIDS OF THE WESTERN HIMALAYAS

The orchids represent a group of botanically significant and commercially important plants which are still in an evolutionary flux; these account for nearly one-tenth of the total flowering plant species of the world. They rank very high in the evolutionary hierarchy of monocots and are still in an active state of speciation; weak barriers of reproductive isolation promote free gene flow across the taxonomic limits, hybrids (interspecific to plurigeneric) are easily formed and their number far exceeds the number of natural species (Vij, 1995). They stand distinct in having intricately fabricated and colorful flowers with suppressed development of endosperm and microscopic seeds and highly reduced embryos. Many of them are rich in alkaloids and other phytochemical contents which have been effectively used, in local medicines, for their curative and aphrodisiac properties (Lawler, 1984). Several orchids are frequently mentioned in ancient Indian literature and some such including *Eria muscicola* (Jivanti), *Eulophia campestris* (Salem), *Habenaria edgeworthii* (Vridhi), *H. intermedia* (Ridhi), *Malaxis mucifera* (Jeevak), *Orchis latifolia* (Salampanja), *Vanda cristata* (Rasna), and *Zeuxine strateumatica* (Shwethuli) are still extensively used in ayurvedic preparations. More recently, anti-viral and anti-cancerous properties of *Vanda parviflora* have been positively tested. The most important commercial product of the orchids, however, is 'Vanillin' an odorous constituent obtained by sweating of cured beans of *Vanilla planifolia* Andr. which is used for flavouring chocolates, beverages, confectionery, cakes, custards, puddings and ice creams and also used as deodorant in soaps, perfumes and powders.

History of orchids probably started with their use as medicinal plants. As early as 200 BC the Chinese pharmacoepoia - "*the Sang Nung Pen Tsao Ching*", mentions *Dendrobium* as a source of tonic, astringent, analgesic, and anti-inflammatory substances. In Europe, in the medieval belief of the doctrine of signatures, tubers of some of these plants were thought to confer virility and potency. In India also orchids have been in use in indigenous system of medicines since Vedic period. (Handa,1986; Kaushik, 1983; Kirtikar & Basu, 1935; Trivedi *et al.*, 1961; Vij *et al.*, 1997). In the Ayurvedic system of medicine a group of eight drugs, known as "Ashtawarga" is used for preparation of tonics, such as 'Chyavanpras', consist of 4 orchid species, viz *Malaxis muscifera* (Lindl.) Kuntze, *M. acuminata* D. Don, *Habenaria intermedia* D. Don and *H. edgeworthi* Hook. F. Around 40 other species are being used in indigenous medicine system. The important orchid genera that are being used as medicinal sources are: *Acampe, Bulbophylum, Calanthe, Coelogyne, Cymbidium, Cypripedium, Dactylorhiza, Dendrobium, Epipactis, Eria, Eulophia, Filckingeria, Habenaria, Liparis, Luisia, Malaxis, Pecteilis, Pholidota, Rhynchostylis, Satyrium and Vanda.*

GENERAL DESCRIPTION

Habenaria edgeworthii; Lindl, (2000-3000 m):

Stem	:	60-70 cm. high, robust.
Leaves	:	7-15 cm. oblong, 3-nerved, distributed chiefly towards the upper portion of the stem.
Flowers	:	3-4 cm. across, green.
Petals & Sepals	:	Petals forming a hood with the dorsal sepal, narrowly and falcately oblong and acute.

H. intermedia D. Don (2000-3000m):

Stem	:	30-70 cm. high.
Leaves	:	4-10 cm. long, ovate, acute and fleshy.
Flowers	:	1-5 cm. across, crowded, deflexed in bud.
Sepals	:	Green, pubescent, the dorsal erect, concave, lateral pairs longer, deflexed.
Petals	:	Yellow, erect, thick, obliquely triangular, from the broad base, apex curved inwards and forming a hood with the dorsal sepal. Lip yellow, longer than the lateral sepals, concave.

Both the species are terrestrial in habitat and distributed in Dhanolty, Chopta, Ghimtoli, Mussoorie, Rudranath, Rambara, Madhyamaheshwar of Garhwal Himalayas at altitudes of 2000-3000m asl (Table 1). Phenological data (Table 2) reveals that these species complete all the phenophase in a short period, between the months of March to October. During the month of March-April when the temperature increases, *H. intermedia* starts to emerge whereas in *H. edgeworthii* emergence begins in the month of April-May. The month of June-July is the season of flowering in *H. intermedia* (Fig. 1a) and July-August in *H. edgeworthii* (Fig. 2a). During the months of July-August and August-September pod formation and development of seeds occurs in *H. intermedia* (Fig. 1b) and in *H. edgeworthii* (Fig. 2b), respectively. The seeds (Fig. 3 a,b) mature between August-September and September-October, respectively. The aerial parts of the plants dry up whereas tubers enter into a dormant phase (Fig. 3c) until the return of favorable conditions in the next growing season.

Table 1: Distribution of *Habeneria intermedia* and *H. edgeworthii*

Taxon	Habit	Altitude (in m.)	Distribution in Garhwal Himalaya
H. intermedia	Terrestrial	2000-3000	Dhanolty, Chopta, Ghimtoli, Mussoorie, Rudranath, Rambara, Madhyamaheshwar
H. edgeworthii	Terrestrial	2000-3000	Dhanolty, Raithal, Ghimtoli, Mussoorie

Table 2: Phenology of *Habeneria intermedia* and *H. edgeworthii*

Taxon	Emergence	Flowering	Pod Formation and Seed Development	Seed Maturation	Senescence of Plant and Tuber Dormancy
H. intermedia	March-April	Jun-July	July-Aug.	Aug-Sep.	Sep.-Oct.
H. edgeworthii	April-May	July-Aug.	Aug.-Sep.	Sep-Oct.	Oct. last or first week of November

Although, the species are listed as endangered, there is no management plan for conservation due to the lack of related information. Collection of these species continues from the wild

(a)  (b)

Fig. 1 (a-b): *Habenaria intermedia* D. Don.

(a)  (b)

Fig. 2 (a-b): *Habenaria edgworthii* Landl.

through illegal means. Which alongwith the destructive harvesting has also been reported earlier for a few other medicinal plants (Chauhan, 2004; Kala, 2005). Domestication and cultivation of such endangered medicinal orchids may be encouraged to fulfil the market demand, which will increase the income of local people on the one hand and reduce pressure on the natural habitat on the other. It is pertinent to point out that sale of medicinal herbs generates more income than the traditional crops in Uttarakhand (Silori and Badola, 2000).

(a) (b)

(c)

Fig. 3. (a) Dried spikes with seed pods of *H. edgeworthii* (b) mature seeds of *H. intermedia* (c) Senescence of aerial parts of Plant and Tuber Dormancy in *H. intermedia*.

Besides the most beautiful flowers, the orchids exhibit an incredible range of diversity in size, shape, and color. Large scale *in-vitro* multiplication of these plants using tissue culture techniques has opened a large number of possibilities. In fact, the production of quality plants in large quantities and propagation of exquisite and rare hybrids through these techniques have catapulted them among the top ten cut flowers in the international market.

The schematic representation of Fig. 4 shows different stages of *in-vitro* propagation of orchids. The suitable explants (shoot apex, stem, leaf, seed and root) may be cultured in suitable medium with or without any plant growth regulators. After a few weeks the protocorm-like bodies develop in the culture, which are then subcultured for shoot and root proliferation. Hardening is the important and last step of *in-vitro* propagation after the seedling development.

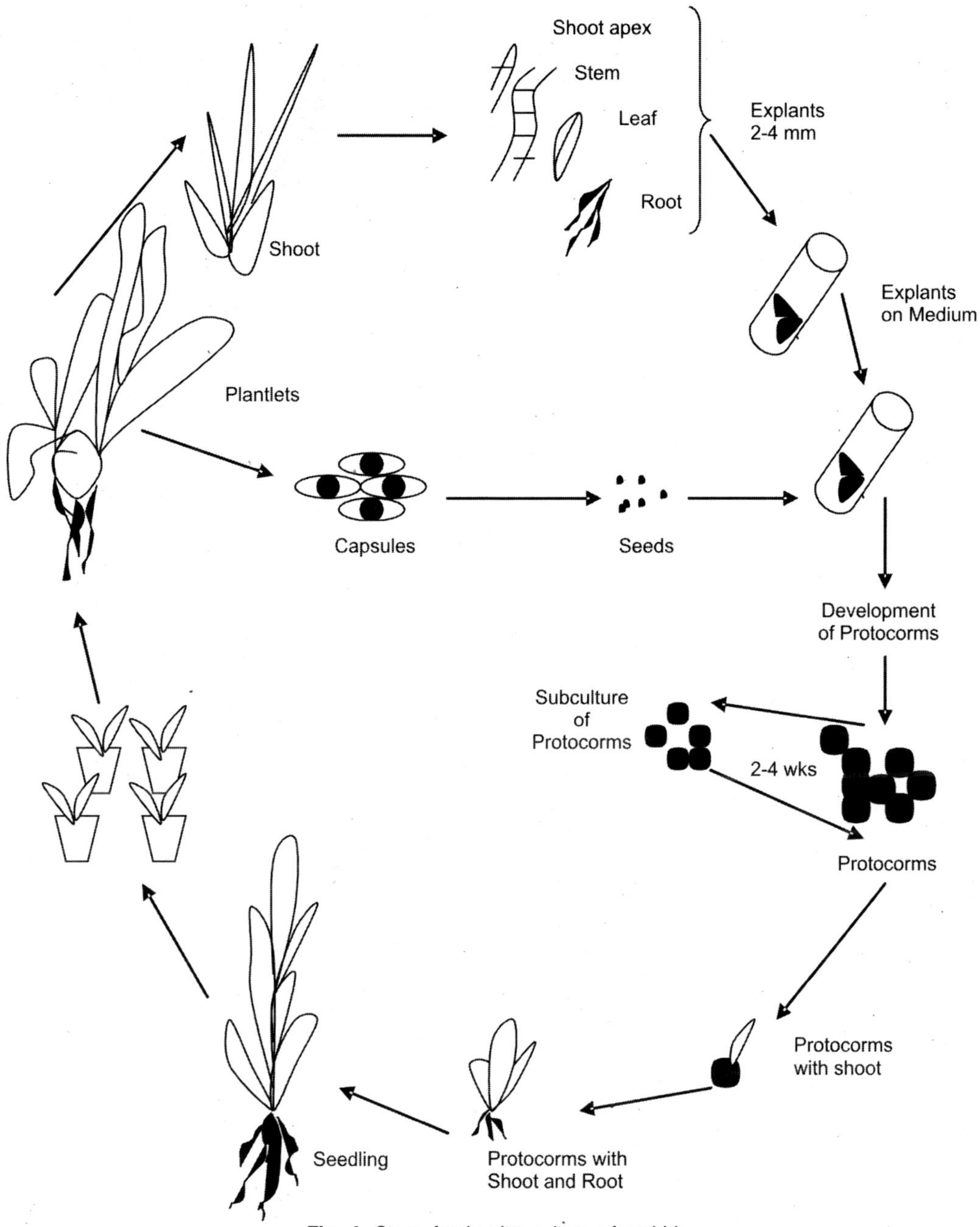

Fig. 4. Steps for *in vitro* culture of orchids.

During recent years, the technique of 'embryo culture'/'green pod culture'/'ovule culture' has gained wide acceptance. In several commercial laboratories, this technique has been replaced dry seed culture. Embryo culture refers to "culturing of immature embryos procured from green capsules prior to their dehiscence". It reduces the period between pollination and culturing of seeds. Embryos from complex crosses, which otherwise abort in the capsule, can be rescued using this technique. Moreover since only the surface of the fruit is exposed to sterilizing agent, the death of embryos from over-exposure to sterilizing agent is minimized. Hence, it results in increased germination percentage of seeds. The difference in harvesting time between dry seed and the green pod culture may be as much as 6-8 months in some genera. This reduction in harvesting time decreases the time required for flowering.

The orchids are fast depleting from their natural habitats due to deforestation and over-exploitation for commercial purpose. The people are fond of growing orchids besides other flowering plants. As a result, the orchids are collected indiscriminately in large numbers from nature and sold in the markets, and also house to house. Therefore, there is a need for both conservation and sustainable utilization of orchids.

Save Orchids……Save Natural Beauty!

ACKNOWLEDGMENTS

The authors are thankful to Prof. M.C. Nautiyal and Dr P. Prasad, for their valuable suggestions and guidance. Financial assistance from Uttarakhand State Council for Science and Technology, Department of Science and Technology, Dehradun, Uttarakhand, is gratefully acknowledged.

REFERENCES

Atwood, JT (1986). The size of Orchidaceae and the systematic distribution of epiphytic orchids, *Selbyana* **9:** 171-86.

Chauhan, RS (2004) Eco-physiology, Agro-technology and Trade of an Endangered Medicinal herb *Nardostachys jatamansi* DC. *Ph. D. Thesis*, H.N.B. Garhwal University, Srinagar, Garhwal

Chauhan, RS, Nautiyal, MC & Prasad, P (2007). *Habenaria intermedia D. Don*-an endangered medicinal orchid, *The McAln In. Orchid Soc. J.* **8:** 15-20

Deva Som & Naithani, HB (1986). The Orchids Flora of North West Himalaya, *Print and Media Associates,* New Delhi, India

Handa, SS (1986). Orchids for drugs and chemicals, pp 89-100. *In:* Vij, SP (Ed.) *Biology Conservation and Culture of Orchids. Affiliated East West Press,* New Delhi, India

Heywood, VH (Ed.) (1993). Flowering Plants of the World, *B.T. Batsford Ltd.* Oxford

Kala, CP (2005). Indigenous use, population density and conservation of threatened medicinal plants in protected areas of the Indian Himalayas. *Conser. Biol.* **19:** 368-378

Kaushik, P (1983). Ecological and Anatomical Marvels of the Himalayan Orchids. *Today and Tomorrow's Printers and Publishers,* New Delhi, India

Kirtikar, KR & Basu, BD (1935). Indian Medicinal Plants Ed. 2, (Reprinted, 1975). *Bishen Singh Mahendra Pal Singh,* Dehra Dun, India

Kumaria, Suman, & Tandon, Pramod, (2001). Orchids: The World's Most Wondrous Plant, pp. 17-28. *In:* Pathak Promila (Ed.) *Orchids Science and Commerce*

Lawler, LJ (1984). Ethnobotany of the Orchidaceae: A Manual, pp. 27-149 *In:* Arditi, J (Ed.) *Orchid Biology, Reviews, and Perspectives Vol. III Comell Univesity Press, Ithaca,* New York, USA

Mabberley, DJ (1998). The Plant-Book ed. 2, *Cambridge University Press*, Bath, UK

Seidenfaden, G & Arora, CM (1982). An emumeration of orchids of North-Western Himalaya, *Nord J. Bot.* **2:** 7-27

Silori, CS & Badola, R (2000). Medicinal plants cultivation and sustainable development: a case study in buffer zone of the Nanda Devi Biophere Reserve, western Himalayas, India. *Mountain Res. & Dev.* **20:** 272-279

Trivedi, KP, Garg, DS, Agarwal, JP & Garg, DD (1961). Ridhi-Vridhi, Jeevak and Rishbhak, *Dhanvantari Vanaushadhi Visheshank,* Part-I pp. 529-533

Vij, SP, Kaur, S & Manjari, (1997). Orchids in medicine. *Orchid News* **13:** 19

Vij, SP (1995) Genetic resources of orchids, pp. 153-181. *In:* Chadha, KL & Bhattacharjee, SK (eds). *Advances in Horticulture (**12**)-Ornamental Plants I. Malhotra Publishing Houses,* New Delhi, India

12

Ichthyodiversity of River Ramganga of Western Uttar Pradesh

JK Pathak*, Mohd. Alam and Shikha Sharma

Hydrobiology Laboratory, Department of Zoology, Hindu College, Moradabad- 244 001, U.P., India
E-mail: jugal28@yahoo.com

ABSTRACT

Biodiversity and its conservation are the major issues of enabling sustainable use of aquatic resources. The present investigation deals with the ichthyofaunal diversity of Ramganga river, in Uttarakhand and Uttar Pradesh states. The investigation was undertaken for a period of three year, monthly collections were made from June 2006-April 2008. The results of present investigation confirm a total of 38 fish species belonging to 10 families, is identified from the study area. Cyprinidae was dominant with twelve species followed by Bagridae with six species. Results reveal that maximum fish species inhabit upstream sampling stations mainly attributed the good water quality whereas less distribution in downstream sampling stations was mainly due to the high BOD and low dissolved oxygen level.

Keywords: Fish diversity, ichthyofauna, Kalagarh dam, pollution, Ramganga river.

INTRODUCTION

In India, the fishery is sector immensely contributing to the economy of the country by providing valuable foreign exchange and employment to millions of people. It is an instrument of livelihood of a large number of economically backward populations of the country. India is the fourth largest fish producers in the world and second in inland fish production (Ayyappan and Dewan 2006). Fishery sector has played a pivotal role in the socio-economic development of the country. Uttar Pradesh is one of the main states of India which is blessed with vast water resources in the form of rivers, canals reservoir, (constructed for various purposes such as irrigation, power generation, flood relief and recreational) lakes, ponds, etc. Rivers and canals are the main flowing water resources of the State having an area of about 720,000 hectare offering an enormous scope of fisheries development in Uttar Pradesh.

Ganga river system is the main flowing water resource of the State of Uttar Pradesh. River Ganga alongwith its tributaries Yamuna, Ramganga, Gomti, Ghaghra and Gandak, pass through the State. As many as 111 different fish species have been reported in 1980 from the freshwaters of Uttar Pradesh (Sultan, 2006), which may have decreased in recent times due to the long-term over-exploitation, pollution and man-made obstructions. The average production of the fish from the Indian rivers has been estimated by the Central Inland Fisheries Research Institute (CIFRI) to one ton/square km/year but regarding the rivers of Uttar Pradesh, no detailed study has been carried so far.

Food scarcity and population explosion are the major problems in India. Fish is an important food which provides cheap and best quality proteins and other by-products; therefore, information on the fish fauna of open water ecosystems such as rivers is of paramount importance. Besides enhancing our knowledge on fish diversity of rivers, such information would help in optimize exploitation of this important resource. Riverine environment provides imperative habitats for several species of fish serving as spawning grounds, nurseries and feeding grounds.

Extensive work has been done by various workers to investigate icthyofauna of riverine resources such as Chandra *et al.* (1990); Tamubi *et al.* (1992); Kumar (1995); Pathak *et al.* (2001); Srivastava (2002); Ugurlu & Polat (2002); Sinha *et al.* (2004); Sukumaran *et al.* (2005); Ahmed *et al.* (2005); Ugurlu & Polat (2008) and Shahnawaz *et al.* (2009).

STUDY AREA

River Ramganga is one of most important tributaries of Ganga. It is a perennial river with high discharge of water during monsoon months but extremely low during summer. It is a spring-fed river which originates in the southern slopes from Dudhatoli (3, 110 masl) hill of middle Himalaya of the Uttrakhand State. During its course, the river flows through a mountainous terrain and has a number of falls and rapids. The river enters the plains at Kalagarh near the border of Uttar Pradesh where a famous hydroelectric dam has been constructed. The dam has altered the natural habitat of the river and is also responsible for the obstruction of continuous flow of water in the downstream. The river has a torrential flow of water at its upstream sampling station and passes through canyon at Kalagarh while it is calm and lethargic in flow at downstream segment. The substratum of the river at upstream and somewhat middle segment consists of bedrocks, boulders and gravels, being composed of sand, silt and clay at downstream. The river traverses about 158 km before it meets the reservoir and continue to downstream for about 370 km before joining river Ganga at Kannauj in Uttar Pradesh. During its travels upto Kalagarh dam, the river is joined by the main tributaries viz. Binoo, Khatraun, Nair, Badangad, Mandal, Helgad and Sona Nadi.

At Kalagarh, a barrage has been constructed and the major portion of the river is diverted through the canal for irrigation purposes and only a small portion is left for downstream areas. The water discharge is regulated by the Irrigation Department. Beyond Kalagarh, the river flows southeasterly direction and finally joins the river Ganga at Kannauj. The important tributaries that join the river Ramganga after Kalagarh (Downstream) are Kho, Gangan, Aril, Kosi and Deoha.

METHODOLOGY

Due to the lack of any commercial fish landing centre in the entire stretch of the study area, casual to professional fishermen were followed for the assessment of fish fauna and fish production of the river. During the study period, survey and collection of fish fauna were made from all the six sampling stations, with the help of local fisherman. Fishes were harvested during entire study period 2005-2008 and these were brought to laboratory for further study. Identification was done as outlined in Talwar & Jhingran (1991), and Sunder *et al.* (1995). The parameters were analyzed as given in instrument manual and APHA (2001), Trivedi *et al.* (1986), Golterman *et al.* (1976), Goldman & Horne (1983), Welch (2003) and Weaver *et al.* (2007).

RESULTS AND DISCUSSION

The faunistic survey conducted during the study period revealed that the principal fish fauna, which support, the fishery of river Ramganga, are *Tor* sps, *Schizothorax* sps. *Labeo* sp., *Mystus* sp., *Channa* sp., *Barilius* sp., *Nemachelius* sp. and other miscellaneous groups. A total of 38 species belonging to 10 families were identified from the study area. The icthyofauna is primarily dependents on various hydro-chemical characteristics of the river water. Various anthropogenic activities in the study area influence the distribution of fishes, as the river flows towards downstream a sharp decline in the fish species is noticed. In general, the following species were found inhabiting the river Ramganga from upstream of hills to downstream of plains upto Moradabad.
Systematics of fishes:

Grade- Pisces
Subgrade- Teleiostomi
Class- Osteichthyse
Subclass- Actinopterygii
Infraclass- Teleostei
Division III – Euteleostei

Family – Cyprinidea

1. *Labeo rohita*
2. *L. dero*
3. *L. calbasu*
4. *Barillius barna*
5. *Barillius bola*
6. *Barilius vagra*
7. *Tor tor*
8. *Tor puititora*
9. *Garra gotyla*
10. *G. lamta*
11. *Schizothorax plagiostomus*
12. *S. richrdsonii*

Family - Bagridae

13. *Mystus seenghala*

14. *Mystus vittatus*
15. *M. punctatus*
16. *Mystus cavasius*
17. *Rita rita*
18. *Walago attu*

Family - Belontiidae

19. *Colisa fasciatus*
20. *Barbus amphibious*
21. *Barbus ticto*
22. *Barbus conchonius*

Family - Clariidae

23. *Clarius batrachus*

Family - Centropomidae

24. *Chanda ranga*
25. *Chanda nama*
26. *Gobius cristalus*

Family - Channidae

27. *Channa punctatus*
28. *Channa gachua*
29. *Channa striatus*

Family - Cobitidae

30. N*oemacheilus rubidipini*
31. *N. Capirat*
32. *N. botia*
33. *N. montanus*
34. *Botia almorhae*

Family - Mestacembelidae

35. *Mastacembelus armetus*

Family - Belonidae

36. *Xenenthodon cancila*

Family – Sisoridae

37. *Glyptothorax pectinoptrus*
38. *Pseudoecheneis sulcatus*

Tyagi & Chaturvedi (1978) had reported 65 species of the fishes belonging to the 19 families from Moradabad region. A sharp decline in the fish diversity from this region can be noticed during the recent years. The base line data generated from this study may be useful for the better management of the fisheries resources in river Ramganga and may be used by the authorities in making strategies for the rational exploitation of the fishery resources of the river.

Species distribution of the fishes in river Ramganga among various sampling stations from Chaukhutiya of upstream to downstream of Katgharh is presented in Table 1. Results reveal that a total 19 species inhabit at Sampling Stations 1 and 2, whereas Sampling Station 3 and 4 harbour 23 species each, Sampling Station 5 is inhabited by 21 species, and at Sampling Station 6 and 7 only 15 and 12 species have been recorded, respectively. Sampling Station 1 to 3 have been found to be richest in number of species due to good water quality such as high dissolved oxygen, low BOD, optimum pH and total dissolved solids and availability of food in abundance as against Sampling Stations 6 and 7 which have lowest species distribution attributed to the high pollution load at these two stations.

In downstream stations of river Ramganga, the level of BOD, COD, DOM and heavy metals was found to be very high and the value of dissolved oxygen has been recorded to be below minimum required level (5 mg/l), during most part of the year. The low oxygen value at these stations was mainly due to the high amount of organic matter discharged alongwith the sewage and effluent poured into the river directly which influence the survival of the fishes at these stations. Kumar (1995) has mentioned that the availability of diverse fishes in river Mayurakashi was mainly due to the high oxygen content. According to Shivprakasha *et al.* (1990), the low production of the fishes in river is directly related to its pollution level. Similar issues have been put forward by Sunder *et al.* (1995). The average value of hydro-chemical parameters of three years analysis has been illustrated in Table 2.

Table 1: Ichthio-diversity of River Ramganga from hilly region Uttarakhand to plains of U.P.

	Fishes	SS1	SS 2	SS 3	SS 4	SS 5	SS 6	SS 7
1	*Labeo rohita*	-	+++	+++	+++	++	-	-
2	*L. dero*	-	+++	++	+++	+++	-	-
3	*L. calbasu*	-	+++	+++	++	++	+	-
4	*Barillius barna*	-	+	++	+++	+++	+	+
5	*Barillius bola*	-	+++	+++	+++	++	++	+
6	*Barilius vagra*	++	-	-	-	-	-	-
6	*Tor tor*	+++	+	-	-	-	-	-
7	*Tor puititora*	+++	+	-	-	-	-	-
8	*Mystus seenghala*	-	+	+++	+++	+++	++	++
9	*Mystus vittatus*	-	-	+++	+++	+++	+++	++
10	*Mystus punctatus*	-	++	++	++	+++	+	+
11	*Mystus cavasius*	-	++	+++	+++	+++	+	++
12	*Rita rita*	-	-	++	++	+++	++	++
13	*Walago attu*	-	++	+++	+++	+++	+	++
14	*Colisa fasciatus*	-	-	+	++	++	+	++
15	*Barbus amphibious*	-	++	++	++	-	-	-
16	*Barbus conchonius*	++	-	-	-	-	-	-
17	*Barbus ticto*	++	-	-	-	-	-	-
18	*Clarius batrachus*	-	-	++	++	++	++	+
19	*Chanda ranga*	-	-	+++	+++	+++	+	-
20	*Chanda nama*	-	+	++	+++	++	-	-
21	*Gobius cristalus*	-	-	++	+++	++	-	-
22	*Channa punctatus*	+	+++	++	+++	+++	+++	+++
23	*Garra gotyla*	++	-	-	-	-	-	-

Table 1: Contd.

	Fishes	SS1	SS 2	SS 3	SS 4	SS 5	SS 6	SS 7
24	*Garra lampta*	+++	-	-	-	-	-	-
25	*Channa gachua*	++	++	+++	+++	+++	+++	++
26	*Schizothorax plagiostomus*	+++		-	-	-	-	-
27	*Schizothorax richardsonii*	+++		-	-	-	-	-
28	*Channa striatus*	-	+	++	++	++	+	+
29	*Channa gachua*	+++	+	-	-	-	-	-
30	*Channa punctatus*	+++	-	-	-	-	-	-
31	*Noemacheilus rubidipinis*	-	++	++	+	++	-	-
32	*Nemachelius botia*	++	-	-	-	-	-	-
33	*Nemachelius montanus*	++	-	-	-	-	-	-
34	*Nemachelius capirat*	-	-	++	++	-	-	-
35	*Mastacembelus armetus*	++	+++	++	++	+	-	-
36	*Glyptothorax pectinoptrus*	++	-	-	-	-	-	-
37	*Pseudoecheneis sulcatus*	++	-	-	-	-	-	-
38	*Botia almorhae*	+++	-	-	-	-	-	-

Hill stream fishes confined to Sampling Station 1, *Labeo* species have been found to be dominant at all the stations except 6 and 7. Only one species, *Labeo calbasu* has been recorded from Station 6 during high flow period. *Tor, Garra, Nemachelius* and *Schizothorax* species were found absent from all the sampling station except Sampling Station 1, perhaps due to cold water at this station. *Barillius* species and *Channa* species have been recorded from all the station throughout the study period but at Sampling Stations 6 and 7 these were not in abundance except *Channa punctatus.* This type of distribution is closely related to the high pollution status at these sites. Sampling Stations 6 and 7 were inhabited mostly by catfishes and murrells since they are hardy fishes and can survive in extreme environmental conditions (Table 2). *Colisa fasciatus* and *Clarias batrachus* have been recorded at all the stations except Stations 1 and 2 *Noemacheilus* species, *M. armatus* and *Xenentodon cancila* are distributed at all the stations except Stations 5 and 6; they contribute little to the total catch. The absence of fishes and other aquatic life forms in Ubeji creek in Niger delta of Nigeria due to high concentration of heavy metals in water was reported by Achudume (2009).

Fig. 1: Fishing Method: Collecting fishes from sampling station by drag net and by caste net.

Table 2: Average values of various hydro-chemical parameters of River Ramganga.

Parameters	N	Minimum	Maximum	Mean	±SE	Std.	Deviation Variance
AT	324	3.00	40.00	26.49	.391	7.04	49.61
WT	324	5.00	34.40	23.65	.354	6.37	40.62
HUMD	324	31.00	96.00	75.72	.636	11.44	131.06
TRANS	324	.00	105.00	19.07	1.099	19.788	391.57
TURB	324	1.00	1408.00	133.95	8.051	144.92	21004.50
V	324	.06	2.63	.548	.025	.456	.20
D	324	.10	4.32	1.09	.026	.477	.22
W	324	12.00	76.00	29.37	.713	12.84	164.89
Q	324	.84	452.11	28.99	3.005	54.10	2927.55
C	324	.19	1239.65	37.17	6.782	122.08	14905.85
TS	324	120.00	9520.00	605.05	43.902	790.24	624489.45
TDS	324	35.00	425.00	148.59	3.669	66.04	4362.39
TSS	324	33.00	9360.00	456.46	43.642	785.56	617112.42
EC	324	160.00	1800.00	621.32	15.598	280.77	78834.72
PH	324	6.20	9.60	7.72	.027	.492	.24
DO	324	.00	14.95	6.42	.200	3.60	12.97
FCO_2	324	.00	14.54	7.63	.230	4.14	17.14
BOD	324	.25	29.00	6.17	.354	6.37	40.68
COD	324	5.00	268.00	44.93	2.276	40.97	1679.14
ALK	324	32.00	480.00	180.51	4.763	85.73	7350.88
HCO_3	324	39.04	585.60	226.55	6.180	111.24	12375.55
SAL	324	5.16	296.99	47.63	1.975	35.54	1263.78
CL	324	2.84	164.52	28.12	1.141	20.54	422.01
HARD	324	29.00	374.00	181.43	4.026	72.47	5252.01
CA	324	16.82	107.65	44.92	.917	16.51	272.82
MG	324	2.04	57.44	16.82	.552	9.95	99.04
NA	324	2.00	96.00	29.12	1.265	22.78	518.96
K	324	1.00	150.00	20.49	1.267	22.82	520.87
SO_4	324	2.00	85.00	22.98	.717	12.90	166.64
PO_{4-} P	324	.04	3.56	.777	.035	.630	.39
NO_{3-} P	324	.06	2.68	.641	.020	.365	.13
SiO_2	324	4.00	30.00	13.13	.201	3.62	13.14

Following conclusions have been drawn from the present study:

- Anthropogenic environmental degradation primarily due to water abstraction, construction of dams and barrage siltation, soil erosion in the catchment areas and pollution from industrial, agricultural and municipal wastes have all caused devastating effects on the fish stock.
- The biologically and economically more important species have started giving way to low value species, and the symptoms of 'Pauly's Ecosystem overfishing' are discernible.
- The faunistic diversity is unfavourably affected by various anthropogenic activities in the catchment area from upstream to downstream.
- Size related population dynamics revealed that immature fishes constitute 80% of the total fish haul in this region; as a result the recruitment of new fish crop will go down sharply in coming years.

Fig. 2. Indiscriminate killing of the fingerlings has been noticed at various sampling stations during study period.

RECOMMENDATIONS

The following recommendations are being put forward for effective management of river Ramganga in order to conserve fish germplasm.

- Factual information about the water quality status at regular intervals of time and the related effects are needed for an effective control of water quality.
- The river pollution control programmes should be achieved by adopting any of the two approaches, the defensive control method and the offensive control method.
- To reduce the wasteload and protect the river against overloading – it enhances its self purification capacity.
- The industrial effluents into the Ramganga should be regulated in such a way so as to equalize the waste assimilating capacity of the river.
- It is not possible to increase the water discharge from the dam in order to dilute the effluents entering it; the wastewater may be stored in ponds and released according to the variation of the river flow. In downstream areas to increase the dissolved oxygen in the river, it should be agitated artificially by surface aerators.
- Culture fishes by completely blocking of small channels draining water from depressions. Fishes should be conserved in the pools thus created and are fished later in the seasonal cycle at times when fish from other sources are becoming scarce.
- The knowledge of breeding period of fishes can be spread among the rural people, with a total ban on fishing during pre-spawning and spawning period of major fish species.
- Capacity building- facilities should be created to educate and train personnel in river quality management (holistic approach).

ACKNOWLEDGMENTS

The author (JKP) is thankful to Department of Science and Technology, Government of India, New Delhi for providing financial assistance.

REFERENCES

Achudume, AC (2009). The effect of petrochemical effluent on the water quality of Ubeji Creek in Niger delta of Nigeria. *Bull. Environ. Contam. Toxicol.* **83:** 410-415

APHA, AWWA, WEFC, (1995). Standard methods for the examination of water and waste water. Washington DC

Ayyappan, S & Dewan, AD (2006). Fisheries research and development in India. *Fishing Chimes.* **26:** 19-23

Chandra, R, Yadava, YS, Kolekar, V & Choudhury, M (1990). A checklist of fishes from Dighali beel district Kamrup, Assam. *J. Inland Fish. Soc. India,* **22:** 92-95

Goldman, CR & Horne, AJ (1983). Limnology. McGraw-Hill Inc. Book Co. Tokyo, 464

Golterman, H, Clymo, RS & Ohnstad, MAM (1975). Methods for physical and chemical analysis of freshwater. I.B.P. Handbook, Blackwell

Kumar, A (1995). Fish fauna and hydrological conditions of the river Mayurakashi in Santhal Pargangs (India). *J. Inland. Fish. Soc. India,* **27:** 97-100

Pathak, V, Sarkar, A, Mahavar, LR & Bhattacharya, BK (2001). Ecological status and fish production potential of Siang, Diband and Lohit, the three forerunners of river Brahmaputra. *J. Inland. Fish. Soc. India,* **33:** 23-28

Shivprakasha, SM, Shambhogue, SL & Srikanth, GK (1990). Riverine fishery resources of Karnataka–An overview. *J. Inland Fish. Soc. India.* **22:** 1-6

Srivastava, NP (2002). Phytoplankton, primary production and fish production potential of Tawa reservoir. *J. Inland. Fish. Soc. India,* **34:** 23-29

Sukumaran, PK, Srivastava, NP, Singh, DN & Das, AK (2005). Status of fish and fisheries of river Godavari. *J. Inland. Fish. Soc. India,* **37:** 15-20

Sultan, S (2006). Intensification and diversification strategies for aquaculture promotion in Uttar Pradesh. *Fishing Chimes.* **26:** 223-230

Sunder, S, Raina, HS, Mohan, M & Singh, B (1995). Ecology and fisheries potential of the Gaula river with special reference to proposed impoundment (Jamrani dam) on the system. *J. Inland. Fish. Soc. India,* **27:** 35-45

Talwar, PK & Jhinagaran AG (1991). Inland fishes of India and adjacent countries. Vol. I and II. Oxford and IBH Publishing Co. Pvt. Ltd. New Delhi, India

Trivedi, RK, Goel, PK & Trishul, CL (1987). Practical methods in ecology and environmental science. Environmental Publication, Karad

Tyagi, VK & Chaturvedi, LD (1978). Fishes of Moradabad. *Proc. All India Sem. Ichthyol.* pp. 50-55

Welch, PS (2003). *Limnological Methods.* Narendra Publishing House, New Delhi. pp. 371

13

Mapping of Forest Plantations in Forest Research Institute, Dehradun Using Aerospace Remote Sensing

RAJAN KUMAR GUPTA, HIMANSHU RAI, ANAMIKA TRIPATHI[1] AND SK KATIYAR[2]

Pt. LMS Govt. PG Autonomous College, Rishikesh–249 201 (Dehradun), Uttarakhand
[1]Hindu Collage, Moradabad–244 001, Uttar Pradesh
[2]Govt. P.G. Collage, Lansdowne–246 155, (Pauri), Uttarakhand

ABSTRACT

Continuous growth in human and livestock population have exerted pressure on forest vegetation leading to deforestation and degradation of existing forests. In addition, forest areas are being cleaned for alternative uses like rehabilitation, agriculture, mining, etc. causing environmental problem. The disturbance of forest and other vegetation in any form not only reduce the forest/vegetation cover but also threaten the environment. The aerospace remote sensing provided an important technique for mapping and monitoring the forest resource on real time basis. Increased spatial response has brought in significant improvement in the mapping technology. Initial application of satellite remote sensing provided information about forest cover use in visual and digital interpretation technique. The satellite imagery of the seventies and eighties revealed that the forest was loosing tree cover at the staggering rate of one to three million hectare a year. Out of seventy-five million hectare of forest under forest management in India, forty million hectares are now without tree cover. Essentially the present study focuses on methodological aspect followed with two main procedures such as: Preparation of forest cover and landuse mapping by visual interpretation on aerial photograph of 1976 and IKONOS image of Dec. 2001, and secondly creation of various theme layers in the GIS domain. An attempt has been made to classify the loss in the plantation cover as well as the appearance of new plantations in twenty-five years in Forest Research Institute into 21 forest cover and land use classes. This study has achieved a number of results and outputs based upon well-defined objectives. A brief discussion on the findings of this study has been listed and mentioned in the paper.

Keywords: Aerospace remote sensing, forest cover map, landscape mapping, remote sensing.

INTRODUCTION

The word 'forestry' has been derived from Latin word 'foris' meaning outside the village boundary or away from habitat land and the 'forest' is referred to as an area occupied by different kinds of trees, shrubs, herbs and grasses maintained for productivity of wood and non-wood materials. They are the natural strands of woody vegetation in which trees predominate. The forests play a vital role in the environmental conservation and forest ecosystem is most complex, self-perpetuating and magnanimous from the point of view of flora and fauna. It is the most diverge gene pool. Because of the forest the natural regulatory processes excel producing one of the most stable ecosystems. They have an important role in socio-economic progress of country by providing a wide range of services essential for human welfare. India is one of the few countries in the world, which adopted a forest policy as early as 1894 AD, that has been revised in 1952 and 1988. The prime objectives of the National Forest Policy, 1988 are protection, conservation and development of forest. According to the National Forest Policy of India, one third of the country's land is required to be maintained under forest cover, but according to SFR Report, 1997, the forest cover is only 19.27% of total geographic area. India, the second most populous and seventh largest country in the world has an area of 328.8 million hectare with sizable human (950 million) and livestock (about 500 million) populations (Porwal, 1997). The continuous growth in human and livestock population have exerted pressure on forest vegetation leading to deforestation and degradation of existing forests. In addition, forest areas are being cleared for alternative uses like rehabilitation, agriculture, mining, etc., causing environmental problems. The disturbance of forest and other vegetation in any form not only reduce the forest/vegetation cover but also threatens the environment itself.

REMOTE SENSING AND GIS IN FORESTRY

The main objective of the Forest Department is "Caring for the land and serving the people", depends mainly upon the quality and quantity of information about land and the resources upon the land. The mapping of forest resources is necessary for rational planning. At present, it is being done through conventional method, which has many limitations. In the modern era of scientific development and technological advancement with periodic changes in the forest, require closer and vigilant monitoring. A number of studies in forest resource mapping have been conducted in India in recent years (Roy, 1985; 1987; Roy *et al.*, 1991; Sudhakar *et al.*, 1992; and Madhavan *et al.*, 1991). Large variation, in respect to climate, physiography and soil markedly influence their occurrence, nature, structure, composition and density (Rao, 1990). Thus to maintain and develop adequate vegetation of sustainable bases, it is essential to map the forest vegetation cover regularly and at relatively short time intervals.

The aerospace remote sensing provides an important technique for mapping and monitoring the forest resource on real time basis. Increased spatial response has brought in significant improvement in the mapping technology. Initial application of satellite remote sensing provided information about forest cover use in visual and digital interpretation technique. The satellite imagery of the seventies and eighties revealed that the forest was loosing tree cover at the staggering rate of one to three million hectare a year. Out of seventy-five million hectare of forest under forest management in India, forty million hectares are now without tree cover (Chaudhary, 1992).

Some of the benefits of remotely sensed data are:

- These provide larger views of the forest and can also show neighbouring and regional land use, which is less possible with ground surveys.
- These enable to detect change in land cover, forest fires, insect infestation, etc., and repeated observations of the same area.
- These provide general information about forest location and distribution.
- These provide accurate information about forest health through assessment of forest canopy and
- These are effective for large area inventories.

The role of Remote Sensing and GIS in ecology and forestry are as follows:

- Vegetation mapping
- Slope/aspect mapping
- Soil mapping
- Monitoring biotic pressure on land degradation
- Mapping and monitoring of biodiversity
- Change detection in vegetation cover

The geographical information system (GIS) is a computer based technology characterized by a specific hardware and software that permits the simultaneous storage, management and analysis of data along with its geographic location. It is a spatial data base management system (Jordan, 1989). The information is contained in map (Graphics) or tabular formats. Database is normally geographically referred using a map projection. The role of GIS in forest management is considerable.

IKONOS

The advent of Remote Sensing and GIS greatly complemented forest mapping, forest change detection and monitoring activities. The state-of-the art sensors on board IKONOS satellite provide the opportunity to accomplish this globally, nationally, regionally, or locally. It is now possible to look at the forest once a day to three times in a day, using IKONOS data. IKONOS is the world's first one meter resolution commercial imaging satellite, launched on September, 24, 1999, from Vandenberg Airforce base, California, USA. The spatial resolution of an IKONOS image is 82 cm in Nadir viewing. It is just like an aerial photograph (Tanka and Sugimura, 2001). The IKONOS satellite system was initiated as the commercial remote sensing system which provide good quality data of Globe (Ramanathan, 2001) The data frequencies can encompass all degree of forest dynamics anywhere on the earth. The IKONOS satellite/sensors have spectral range 450-750 mm with 4 MS 1 PAN Band and 1-4 m GSD with 1-3 day revisit alongwith 13 km swath width.

STUDY AREA

Established in 1906, FRI, Dehradun is one of the oldest institutions of its kind and is acclaimed the world over (Plate 1, Fig. 1). The institute's history is virtually synonymous with the evolution and development. The area is situated between 30°19′55′ and 30°21′16′′ N latitude and 77°58′40′′ and 78°01′00′′ E latitude. The main township of the area is the Dehradun city about

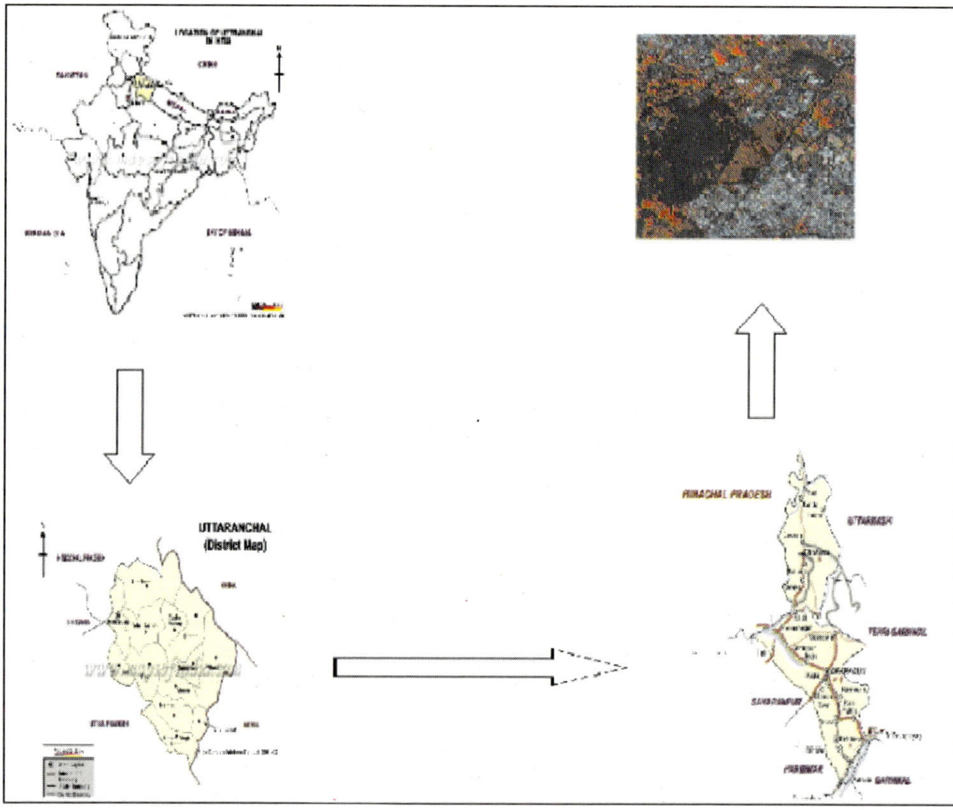

Fig. 1. Map showing their site of study.

5 km from the main campus, Panditwari, Kaulagarh, Ballupur and Premnagar. On the south-west side of the tract, Indian Military Academy is situated and there is the military component of Premnagar. On the north-eastern side is situated the ONGC complex. Different blocks of FRI and their areas have been given in Fig. 2.

Problem stuttering and hypothesis with the help of study area information and its problem is as below. We have tried to structure the problem and made assumptions on some of the hypotheses for testing. The hypotheses are:

- Different types of forest will exhibit different spectral signatures
- Biotic factors do have impact on phytodiversity
- Different climatic, biotic and edaphic factors govern the biodiversity of plants
- The environmental impact change must result in some kind of physical manifestation

Plate 1a. Forest Research Institute, Dehradun

Plate 1b. Chir-pine plantation (Champion Block)

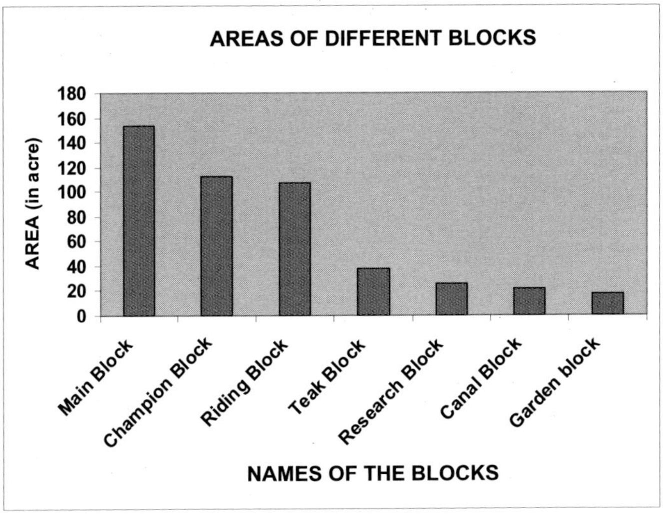

Fig. 2. Different FRI blocks showing their areas.

Based on the above hypotheses, the project objectives have been formulated. These are:

- Vegetation mapping of the study area
- Structural analysis of the vegetation
- To see the changes in vegetation cover with the help of RS and GIS technique
- Stratify forest into homogenous units
- Compare changes within the forest crop, using historical data.
- Suggest updating of map using satellite remote sensing technique.

The entire work has been divided into the following parts.

A. PRE-FIELD WORK

STUDY MATERIALS

Following primary and secondary data have been used to accomplish the task:

(a) Primary data

1.	Satellite	IKONOS 2
	Sensor	Optical Sensor Assembly (OSA)
	Data Product	Standard IKONOS FCC
	Date	4 December 2001
	Spatial resolution	4 meter
	Swath Width	11 km
	Pointing capacity	± 30°
	Revisit time	1-3 days
2.	Aerial photograph	1976
	Scale	1:5000
	Sensor	Panchromatic B/W

Size of Photograph	18 × 18 cm
Number	706A/15-12, 706A/14-13, 706A /14-15, 706A/15-10, 06A/14-11, 706A/16-5, 706A/16/6

(b) Secondary data

S.O.I. Toposheet	
Number	53/J/3
	53/F/15
SCALE	1:50,000

(i) Reconnaissance survey

The survey aimed to collect various categories of information regarding the area. A birds' eye view of the whole of the study area has been acquired during (11/06/02, 13/06/02 and 14/06/02). The training sets were generated completely during the reconnaissance survey and ground truthing.

(ii) Analytical tools and computer hardware/software

The following analytical tools and computer facilities were used in the plantation mapping.

GIS Software. ARCVIEW 3.2 A, ERDAS IMAGINE 8.4, PC based computer system, MS WORD 2000.

Methodology for computer Analysis:

All the spatial data in the form of maps were scanned using hp. HEWWLETT PACKARD (FED, IIRS) and transferred to computer readable digital maps. The digital maps were imported to ERDAS IMAGINE and rectification process have been carried out using this software.

The image to image registration was done by obtaining a transformation with all points on one image to points on the other image. To accomplish this GCPs pairs of GCP which are known to be some physical locations on both images are located. The polynomial transform which describes as to how the uncorrected image data must be wrapped over the raster data is based on ground control point (Figs. 3, 4).

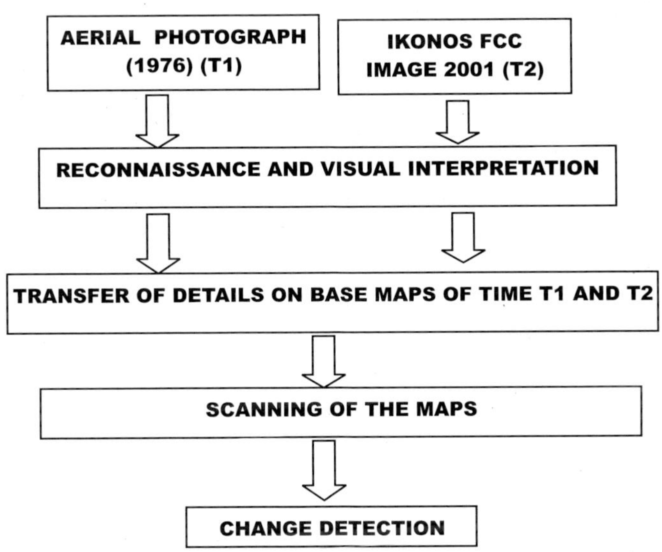

Fig. 3. Flow diagram showing methodology.

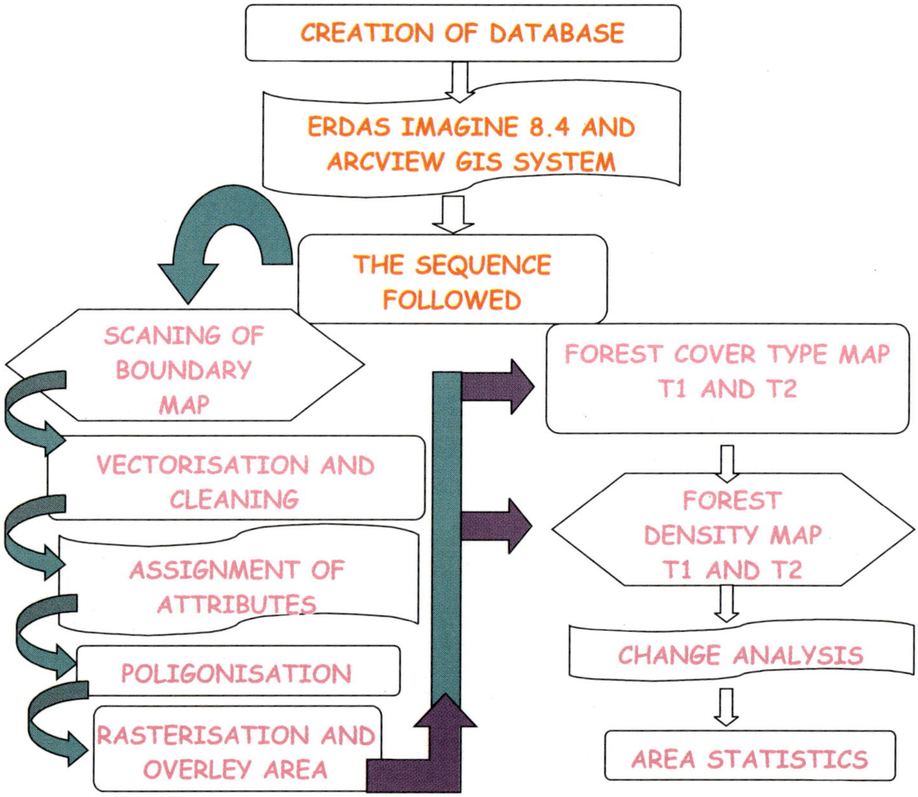

Fig. 4. Flow diagram outlining GIS analysis.

At the IKONOS and aerial photograph scenes of the projected area were rectified separately. The rectified digital maps were consequently imported to ARC VIEW in order to digitize and generate various theme layers visualized forest cover type map, forest density type map, landuse map, etc. The on screen digitization done in ARC VIEW was used for process of cleaning and editing. Subsequently, all the concern attributes were assigned to each cover class. Transformation processes, which are important for analysis operations, were carried out in ARC VIEW using geographical Lat. Long projection system. All the theme layers were assigned proper colour and legend and stored in GIS for further analysis.

B. POST-FIELD WORK

PRELIMINARY VISUAL INTERPRETATION

Based on the preliminary interpretation keys, the forest has been classified into various forest cover types, density and height class categories on the IKONOS satellite imagery and aerial photograph through visual interpretation technique. The interpretation of the image is an essentially deductive process and the identification of certain key features leads to recognition of others. There are nine elements of image interpretation (shape, size, tone, colour, shadow, pattern, texture, location and association) regarded as being of general significance irrespective of the précised nature of the imagery. In practice, these elements assure a variety of rank of importance. However, the tone or colour is the most important paremeter than the others, in broad delineation of forest cover type in any forest area. Simultaneously, the shape and pattern are also helpful to identify the plantation areas and texture for density classification. Based on the final interpretation key, a comprehensive visual interpretation on IKONOS satellite imagery has been carried out and the discrimination of various forest cover type and landuse types have been mainly by tracing the details on tracing undertaken sheets. The discrimination of density, height and crown diameter classes has also been carried out which is being presented in Table 1.

Table 1: Image interpretation key for landuse mapping

S.No.	Forest Cover Types	Symbols
1.	SCRUB	Sc
2.	BAMBOO	B
3.	CHIRPINE	CP
4.	NURSERY	N
5.	*DALBERGIA*	D
6.	*EUCALYPTUS*	Eu
7.	*POPULAS* MEDIUM	Pm
8.	*POPULAS* YOUNG	Py
9.	SAL FOREST	S
10.	TEAK	T
11.	TEAK YOUNG	Ty
12.	MIXED PLANTATION	MP
13.	*AGATHIS ROBUSTA*	Ar
14.	*TERMINALIA* SP.	Ts
15.	*SYZIGIUM* SP.	Sz
	Non-forest Cover Types	Symbols
16.	AVENUE TREES	AT
17.	DEER PARK	DP
18.	LAWN and PARK	LandP
19.	BUILDING	S
20.	ORCHARD	O
21.	BOTANICAL GARDEN	BG
	Crown Density class	
	Density classes	Codes
	Less than 20%	1
	21 to 40%	2
	41 to 60%	3
	61 to 80%	4
	More than 81%	5

Table 1: Contd.

S.No.	Forest Cover Types	Symbols
	Crown diameter classes	
	Crown diameters	Class codes
	Less than 5m	1
	6 to 10 m	2
	11 to 15 m	3
	16 to 20 m	4
	21 to 25 m	5
	26 to 30 m	6
	More than 30 m	7
	Stand Height classes (For aerial photograph)	
	Height classes	Codes
	Less than 10 m	1
	11 to 15 m	2
	16 to 20 m	3
	More than 20 m	4

RESULTS AND DISCUSSION

Essentially the present study focuses on methodology aspect accompanied with two main procedures such as preparation of forest cover and landuse mapping by visual interpretation on aerial photograph of 1976 and IKONOS image of December 2001, and secondly, creation of various theme layers in the GIS domain (Figs. 5, 6, 7, 8 and 9). This study has achieved a number of results and outputs based upon well-defined objectives. A brief discussion on the findings of this study has been listed and mentioned below.

LANDUSE MAPPING AND FOREST COVER TYPE

An attempt has been made to classify the FRI forest, into 21 forest cover and landuse classes (Fig. 10), the major chunk of forest consists of Chirpine plantations which occupied 748.4 hectare and 162.84 hectare of the total forest area during 1976 and 2001, respectively, followed by *Eucalyptus* (103.1 hectare during 1976 and 103.7 during 2001), Sal (14 hectare in 1976 and 14.3 hectare in 2001) and bamboo (24.4 hectare to 16.2 hectare in 2001).

The mixed plantation was observed to be 84.4 hectare and 88.97 hectare in 1976 and 2001, respectively.

The analysis of the common selected area for the change detection showed that the area of chirpine plantation has abruptly been reduced from 784.4 hectare to 162.84 hectare within 25 years (Table 2; Fig. 11), consequently the area of scrub was observed to have increased from 3.2 hectares to 243.11 hectares during 25 years whereas the mixed plantation area has increased only marginally from 88.4 hectare in 1976 to 88.97 hectare in 2001.

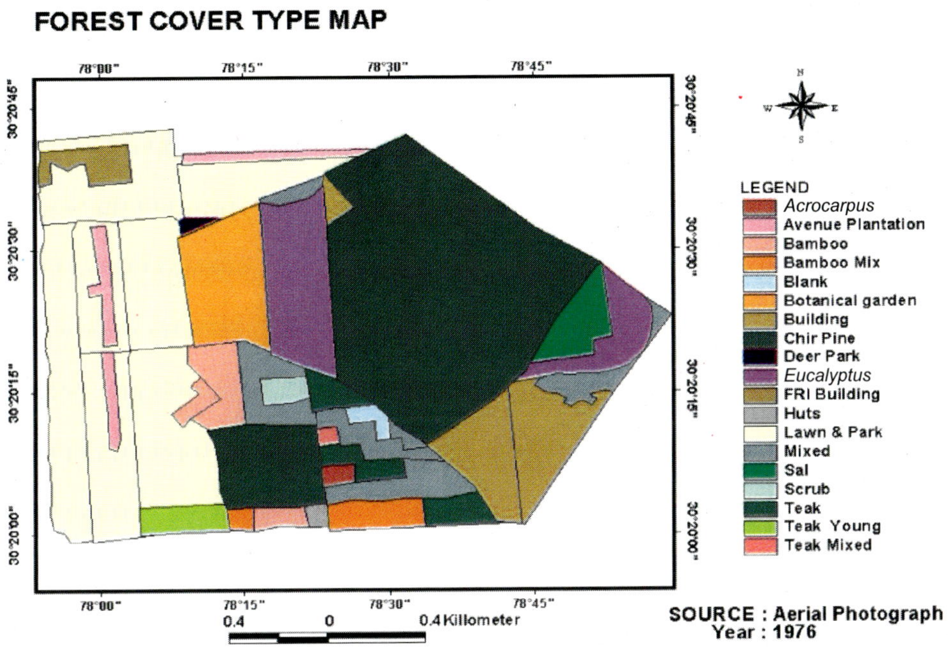

Fig. 5a. Forest cover type map (Year 1976).

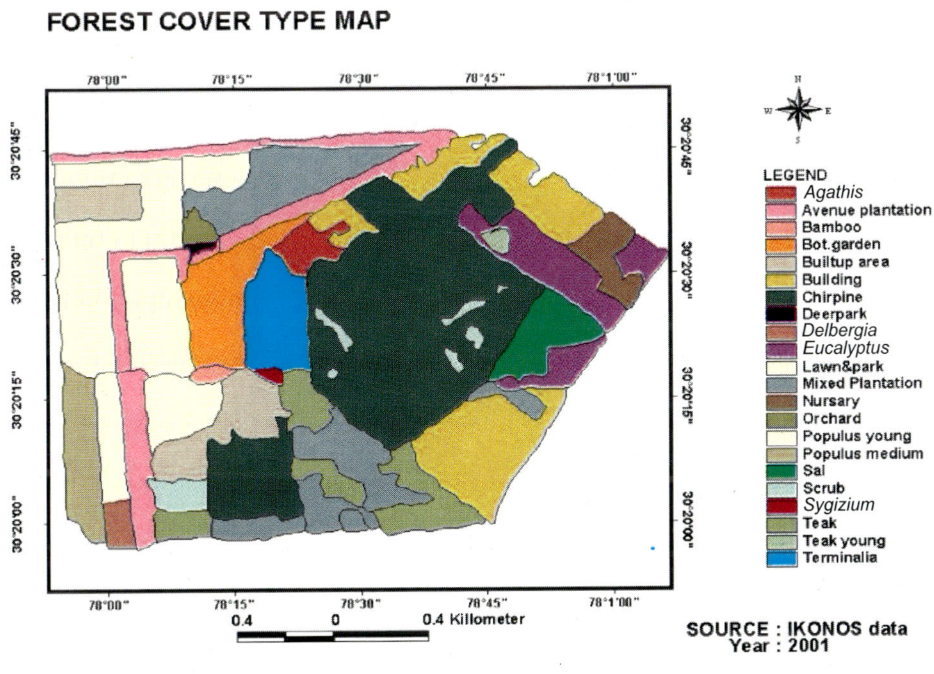

Fig. 5b. Forest cover type map (Year 2001).

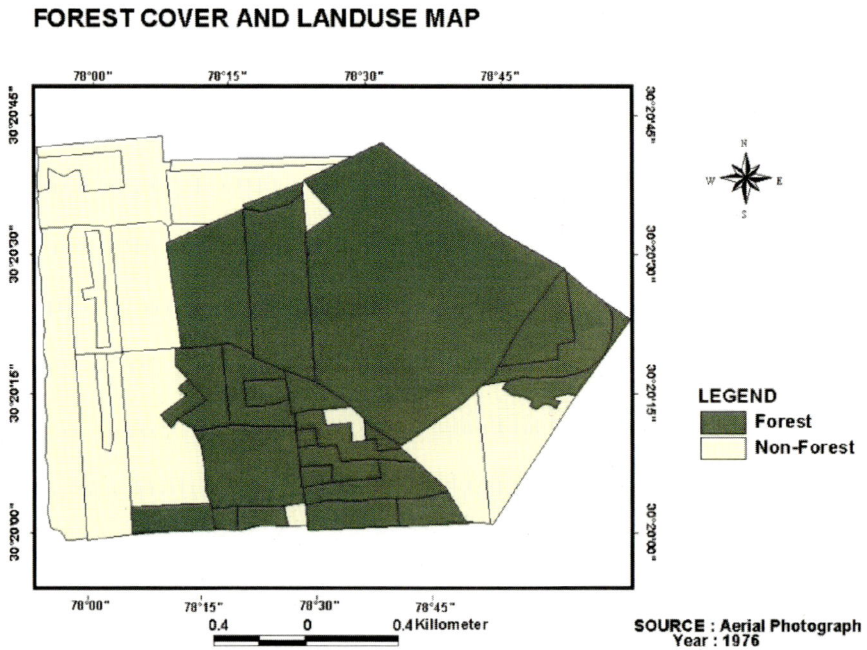

Fig. 6a. Forest cover and landuse map (Year 1976).

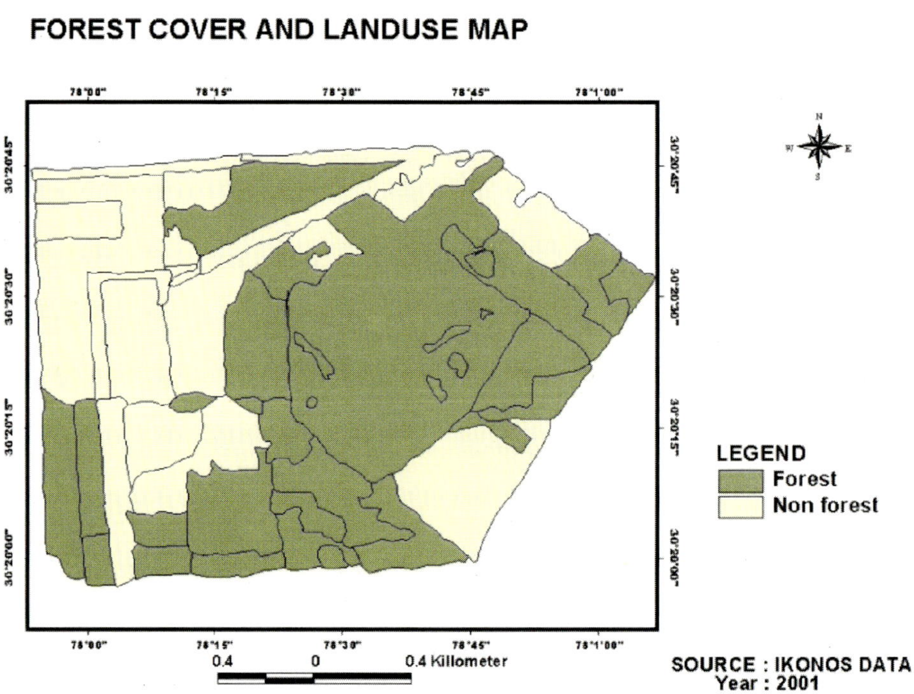

Fig. 6b. Forest cover and landuse map (Year 2001).

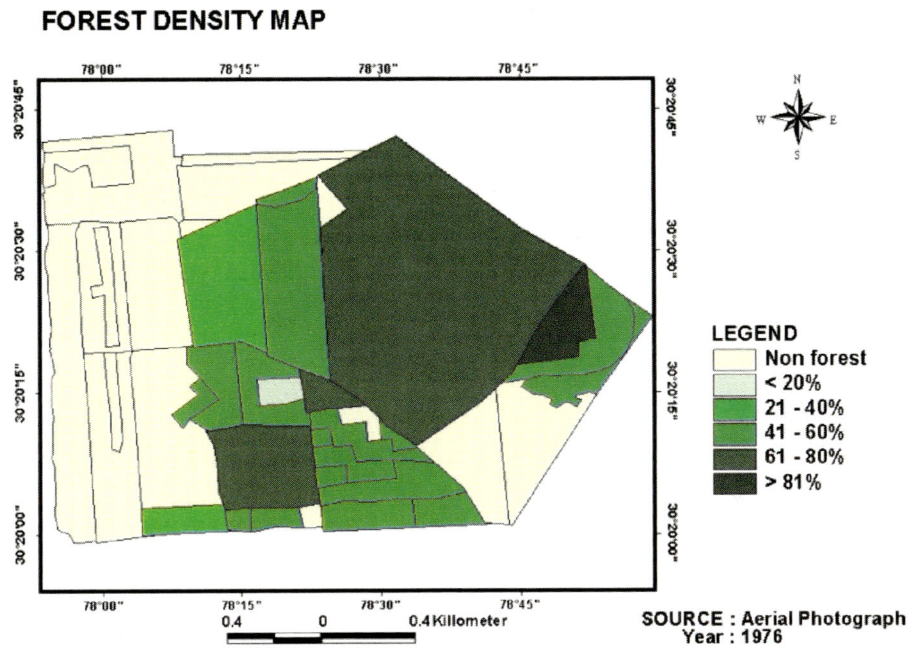

Fig. 7a. Forest density map (Year 1976).

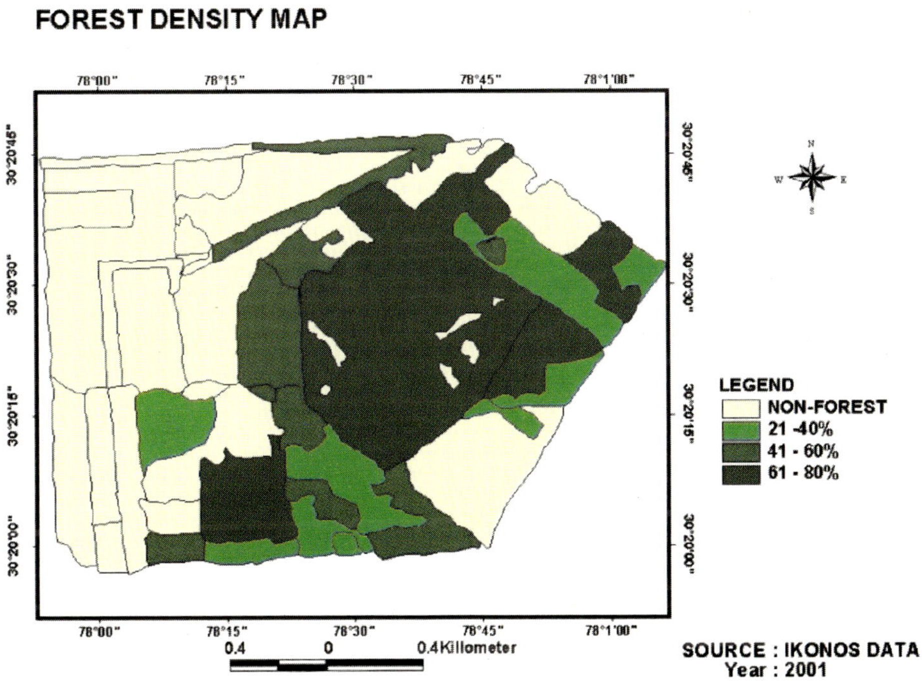

Fig. 7b. Forest density map (Year 2001).

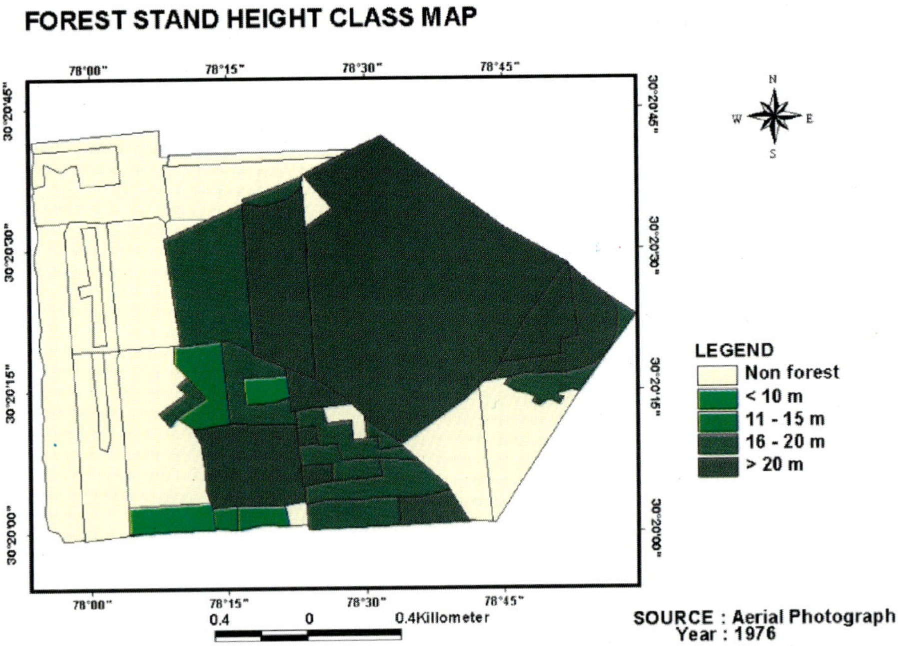

Fig. 8. Forest strand height class map (Year 1976).

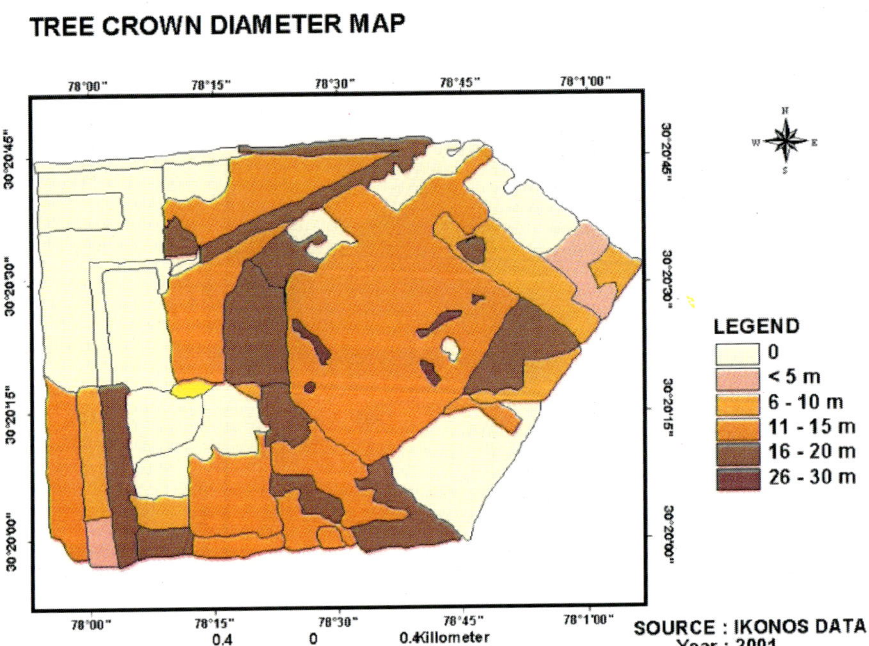

Fig. 9. Tree crown diameter map (Year 2001).

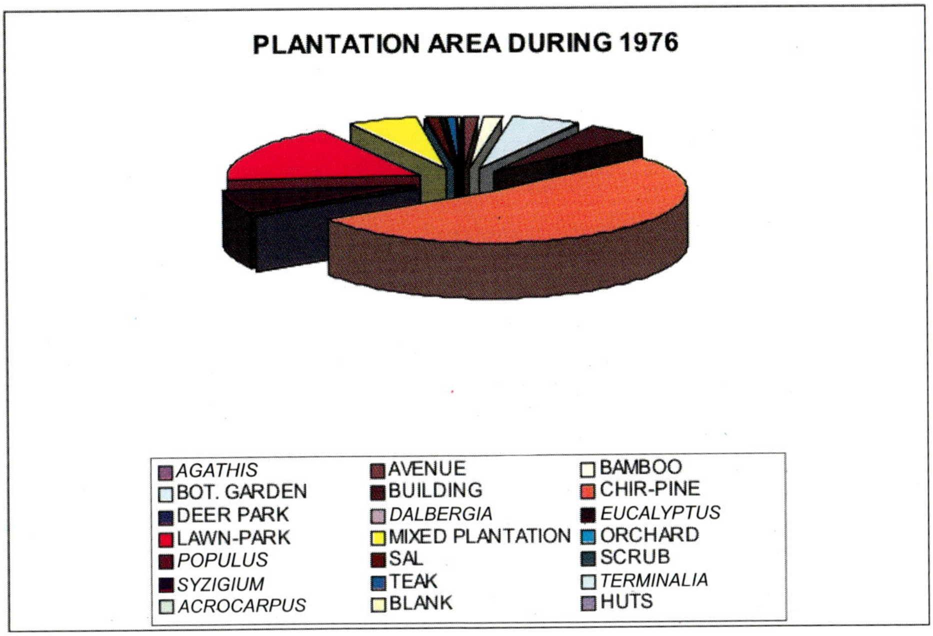

Fig. 10a. Plantation area (Year 1976).

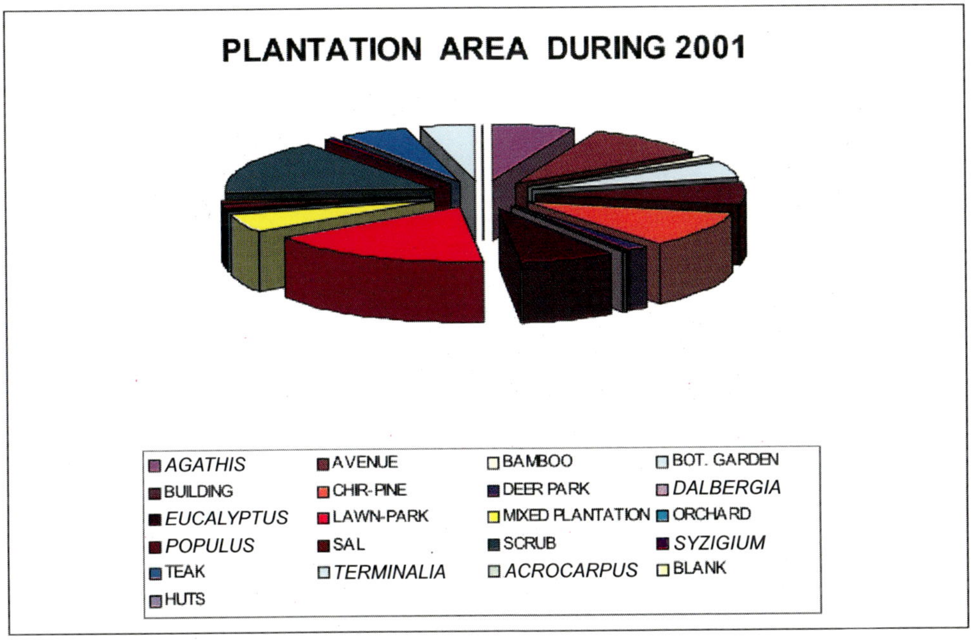

Fig. 10b. Plantation area (Year 2001).

The new plantations noticed in 2001 IKONOS imaginary work are of *Agathis, Syzigium, Populus* and *Terminalia* (Table 2) whereas the *Acrocarpus* plantation lost its unique identity in 2001 and has been incorporated in mixed plantation.

Table 2: Comparative area figures for time 1 and time 2

TIME 1 AERIAL PHOTOGRAPH 1976 CLASS	AREA (T1) HACTARE	TIME 2 IKONOS 2001 CLASS	AREA (T2) HACTARE	CHANGE DETECTION (T2-T1)
AGATHIS	0	*AGATHIS*	104.06	104.06
AVENUE PLANTATION	16.68	*AVENUE PLANTATION*	155.5	138.82
BAMBOO	22.66	*BAMBOO*	16.32	-6.34
BOT. GARDEN	81.63	BOT. GARDEN	73.35	-8.28
BUILDING	75.87	BUILDING	114.88	39.01
CHIR-PINE	748.41	*CHIR-PINE*	162.84	-585.57
DEER PARK	1.82	*DEER PARK*	24.05	22.23
DALBERGIA	0	*DALBERGIA*	7.12	7.12
EUCALYPTUS	103.15	*EUCALYPTUS*	113.77	10.62
LAWN-PARK	255.12	*LAWN-PARK*	281.81	26.69
MIXED PLANTATION	84.83	MIXED PLANTATION	88.97	4.14
ORCHARD	0	*ORCHARD*	5.91	5.91
POPULUS	0	*POPULUS*	25.29	25.29
SAL	14.02	*SAL*	14.34	0.32
SCRUB	3.25	*SCRUB*	246.36	243.11
SYZIGIUM	0	*SYZIGIUM*	13.93	13.93
TEAK	11.65	*TEAK*	82.37	70.72
TERMINALIA	0	*TERMINALIA*	68.75	68.75
ACROCARPUS	1.19	*ACROCARPUS*	0	-1.19
BLANK	2.28	BLANK	0	-2.28
HUTS	1.29	HUTS	0	-1.29

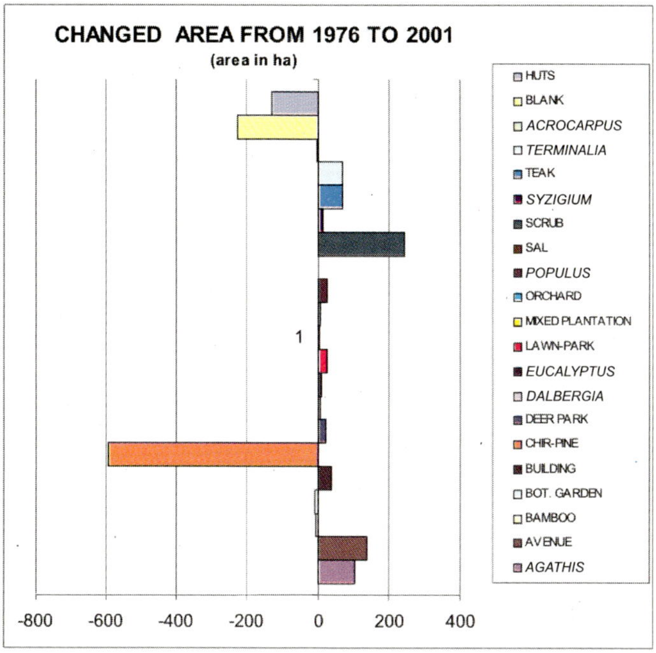

Fig. 11. Changed area between 1976 – 2001.

Fig. 12. Imagery of FRI area.

ACKNOWLEDGEMENTS

I express my sense of deep gratitude to Dr P.S. Roy, Dean, IIRS for providing us an opportunity to undergo this course. I am indebted to Dr. S.P.S. Kushwaha, Head, Forestry and Ecology Division for providing all the facilities and making the training successful. I extend my sincere thanks to my supervisors, Dr. I.J. Singh and Shri K.K. Das for providing useful inputs and suggestions. I am also thankful to UGC, New Delhi and UCOST, Dehradun for financial assistance.

REFERENCES

Chaudhary Kamala, (1992). Structural change for better. *The Hindu Survey of Environment* pp. 31-33

Madhawan, Un, NV, Jadhar, PS, Tiwari, RN, Sudhakar, AKS, Ranganath, BK & Dabral, SK (1991). RS-14 Applications in Forestry. *Curr. Sci.* **63:** 188-192

Porwal, MC (1997). Priortisation of probable sites for Forest Plantation – A Remote Sensing and Geographical Information System Approach. Remote Sensing for Natural Resources

Ramnathan Sugimaran, (2001). Forestland cover classification using statistical and artificial neural network approaches applied to IRS LISS-III sensor. *Geo Carto International* **16:** 37-42

Rao, UR (1990). Space and forest management. Proc. 41 IAE Cong. Dresden, Germany

Roy, PS (1985). Forest type mapping and monitoring in Varatang Forest Division (Andaman Island)

Roy, PS (1987). Montane vegetation stratification through digital processing of Landsat MSS data. *Geo Carto International* 19-26

Roy, PS, Ranganathan, BK, Diwakar, PG, Vohra, TPS, Bhan, SK, Singh, IJ & Pandian, VC (1991). Tropical Forest Type Mapping and Monitoring using Remote Sensing. *Internat. J. Remote Sensing* **12:** 2205-2225

Sudhakar, S, Krishnan, N, Ramana, IV, Pal, DK, Das, PK & Raha, AK (1992). Forest Cover mapping using IRS-1A LISS -II data. *Asian Pacific Remote Sensing J.* **4**

Tanka, & Sugimura, (2001). A New Frontier of Remote Sensing from IKONOS image. *Internat. J. Remote Sensing* **22:** 1-5

14

Biodiversity and Conservation of Mangroves in the State of Orissa with Special Reference to Bhitarkanika

MIHIR KUMAR DAS

Department of Botany, G.M. Post-Graduate College (Autonomous), Sambalpur–768 004, Orissa, India
E-mail: mihir_gmc@rediff.com

ABSTRACT

Orissa has a large mangrove habitat due to nutrient rich soil formed in the Mahanadi delta and the Brahmni-Baitarani delta. The coast has a smooth and gradual slope which provides large area, for colonization of mangroves. Bhitarkanika, the hot-spot of rich biological diversity, harbours several species of mangroves and their associates. It is also blessed with huge populations of resident birds, estuarine crocodiles, largest rookery of Olive Ridley Sea turtles, large populations of King cobra, spotted dear, lizard, python, fishing cat and other reptilian fauna. This ecosystem supports a range of inter-connected food-webs within it to maintain the balance of nature. The livelihood patterns of the villagers surrounding Bhitarkanika directly or indirectly influence the very existence and survival of the flora and fauna in this ecosystem. There is much anthropogenic and biotic pressure on the biodiversity. The present paper deals with the diversity of mangrove flora and fauna, their socio-economic aspects and the conservation measures to be taken to preserve the rich biodiversity.

Keywords: Bhitarkanika, biodiversity, conservation, mangroves.

INTRODUCTION

"Forest is a peculiar organism of unlimited kindness and benevolence which makes no demands for its sustenance and extends generously the products of its life activity, it affords to all beings, offering shade even to the axe men who destroyed it" – Buddha.

Orissa, due to its specific geographical position, with a broad spectrum of edaphic factors coupled with climatic variations harbours very rich forest cover. However, its luxuriance and gregariousness have been immensely interfered by various anthropogenic factors. Consequently, at present, it has come down to around 56,060 sq.kms, which is only 36% of the geographical area of the State (Mishra, 1999). This is according to the satellite imageries of 1995. It is apprehended, there is further shrinkage of this forest cover as a result of which various climatic hazards have seriously affected the general populace (*cf.* Super Cyclone, 1999); specially, the germplasm of wildlife is threatened and the rich biodiversity is on the brink of graverisk. Hence, there is urgent need for the effective conservation of various forms of wild lives extant in the State.

In this context, it is significant to mention that the International Union for Conservation of Nature and Natural Resources (IUCN) indicates that one of the chief strategies is to create and maintain a series of "protected areas" on national as well as international basis like " Wild life sanctuaries; 'National Parks' and "Biosphere Reserve". Despite some nomenclatural confusion pertinent to the above said protected areas in different parts of the world, ten categories have been suggested at the 25th working sessions of the Commission held at Corbett National Park (India) in February 1985, with different levels of management objectives. These are:

1. Scientific Reserve/Strict Nature Reserve
2. National Park
3. National Monument/National Landmark
4. Nature conservation Reserve/Manage Nature Reserve/Wild life Sanctuary
5. Protected Landscape
6. Resource Reserve
7. National biotic Area/Anthropological Reserve
8. Multiple use Management Area/Managed Resource Area
9. Biosphere Reserve
10. World Heritage Site (Natural)

The management objectives of Wild life Sanctuaries, National Parks and Biosphere Reserves as approved by the Commission on Natural Parks and Protected Areas [CNPPA] in Orissa are indicated below (Padhi, 1985):

- (A) National Park: Bhitarkanika
- (B) Biosphere Reserve: Similipal
- (C) Wild life Sanctuary: Nandan Kanan

Out of these, Bhitarakanika, the national park in the State of Orissa deserves special mention.

Bhitarakanika: National Park in the State of Orissa

The State of Orissa harbours a varied and rich vegetational wealth owing to its unique geographical location and congenial climatic conditions coupled with varied topography and edaphic factors. Champion and Seth (1968) had recognized as many as 32 forest types within the State, of which five main categories are: (i) Tropical moist deciduous forests, (ii) Orissa semi-evergreen forests, (iii) Central Indian sub-tropical hill forests, (iv)Tropical dry-deciduous forests and (v) Littoral and Tidal Swamp forests. The Littoral and Tidal swamps are distributed along the coastal belt of Bay of Bengal in discontinuous patches, mostly in the estuaries of the rivers like the Mahanadi, Brahmani, Baitarani, Devi, Jambu, Budhabalanga, etc. Next to Sunderbans of West Bengal, probably the mangroves of Orissa occupy the 2nd position in India at least in respect of species diversity. However, the various aspects of mangrove vegetation of Orissa have not yet been properly studied.

The rich heritage of forest cover in Orissa was believed to be about 43.42 per cent of the total landmass, which has now said to have degraded to 23.21 per cent only in some quarters. Consequently, the Littoral and Tidal swamps also exhibit considerable amount of shrinkage at places (where they exist) and they are totally absent at other places. This plant community is now experiencing a high degree of hazard and devastation due to large-scale destruction of habitat, change of optimum conditions of the habitats besides the operation of pestilent biotic factors. Therefore, the rate of depletion of the mangroves is rather fast. According to an estimate (Samal and Patnaik, 1989), there is a depletion of about 20 sq.kms of mangrove forests in Orissa over a period of one decade. The mangroves of Orissa are now restricted to an area of 150 sq.kms. only as per an estimate given in the status report: "Mangroves in India" (Anonymous, 1987). This figure is pertinent to the mangroves of Mahanadi delta only since the remote sensing survey report conducted in 1984 indicates the existence of 214.58 sq. kms. of mangrove forests in the State of Orissa. However, this plant community occurring in the coastal tracts of Balasore, Puri and Ganjam are now in highly degraded form and are often ignored. On the contrary, Bhitarkanika, in the district of Kendrapara harbors a rich and luxuriant mangrove vegetation although in other regions (of this district) like Paradeep, Hukitola, Kujanga, Jambu, etc. they are either in highly denuded form or are totally absent due to construction of port, factory, network of transport system, reclamation of forest land for the human settlement, paddy cultivation and prawn culture etc.

Geographical Location and Topography

Orissa is a maritime State located in the eastern coast of Indian Peninsula, having a coastline of 480 km. The Protected Area Network of Orissa covers 6611.12 km^2 of land area which is 4.25% of the geographical area and 11.37% of the forest area of the state of Orissa. Of the 17 notified wildlife sanctuaries of Orissa Source: Office of the Principal, Chief conservator of Forests (Wildlife) Orissa, Bhitarkanika Wildlife Sanctuary lies between 20°4′ and 20°8′ North latitude to 86°45′ and 87°5′ East longitude covering an area of about 650 kms^2. (Anonymous, 1986) of which, about 380 sq.kms. is under forest cover, whereas Bhitarkanika core spreads over an approximate area of 141.44 kms^2 including about 115.5 km^2 of mangrove forest. It is bound by the river Dhamra (confluence of Brahmani and Baitarani) and Maipara (part of Dhamra) and about 35 km. long stretch of seashore. The landmass is about 0-8 m. above the sea level. A large number of meandering creeks and channels, sand dunes and sand bars near Gahirmatha, estuaries at the mouths of Brahmani, Baitarani, Patsala (distributaries of the Brahmani) provide peculiar ecological niches for the growth and development of the mangroves and sandy vegetation alongwith their associates.

Geology and Soil

The core area of the sanctuary extends from Dangmal to Thakurdian river a mouth close to Sataveya and Gahirmatha coastal region which is mostly formed by the alluvial filling up of the littoral zones composed of sandy and muddy beaches along the fringes of the coast of Bay of Bengal including the rivers, meandering creeks and network of channels. Due to regular inundation by tidal ebb and flow, the soil is mostly clayey loam and highly slushy in nature. Through constant deposition of humus and other organic matters, it is rather fertile. Surface soil is composed of silt loam and clayey loam which is about 3-4 meters in thickness. It is more or less saline in nature. Sodium chloride is found in high quantity whereas calcium carbonate is rarely found. However, in the elevated area away from the creeks and channels the soil is less slushy and less saline. Often it is mixed with coarse sand. These places are occasionally flooded with the saline water through narrow canals connected to the main creeks during rainy season and high tide. In general, the soil pH varies from 6.8 to 7.3.

Environment

Bhitarkanika wildlife sanctuary experiences mainly three seasons namely (a) summer (b) rainy and (c) winter. Summer starts from mid-February and continuous up to mid-June. The temperature rises up to 45°C. April is the hottest month in the Bhitarkanika Mangrove Swamps. Occasional rains and cyclones are met with towards the end of the summer. After summer, rainy season begins which extends up to the end of October. Due to influence of the South-West monsoon maximum amount of rainfall is received during this period. The mean annual rainfall is 1,700 mm and the wind velocity is about 40 kms per hour. Winter starts from November and continues up to February. January is the coldest month. The temperature comes down to 12°–10°C, while mean maximum temperature is 27°C. Storms and oceanic cyclones are usually experienced during the transitional period of rainy and winter season. This causes heavy rainfall. The wind velocity ranges from 180-250 km/hr. In fact, this terrain is highly prone to severe cyclones, which bring about catastrophic damage to the plant community and various life forms. The relative humidity of the driest month is 75% and that of the wettest 82%.

The core area of the sanctuary is influenced by the freshwater flow from the river Brahmani and salt water from the river Patsala. It enjoys a macro-tidal range. Water salinity of the rivers varies seasonally. During summer it increases and in rainy season it decreases considerably due to admixture of freshwater. Similarly, salinity variation is also observed at different depths of the water bodies. It increases from the surface water towards bottom whereas it decreases from the river mouth to inland area. Water salinity ranges from 20-28 gm/lit. The tidal height varies from 3.6 to 5.5 metres.

Floristic Studies: Past and Present

Despite the rich heritage of mangrove flora, very scanty and fragmentary information has been provided by Haines (1921-1925), the pioneer plant explorer of the Orissa State, for this plant community. Haines in his treatise "The Botany of Bihar and Orissa" has provided an account of 45 mangrove species (including some of their associates) under 39 genera belonging to 29 families. Of these, as many as 29 taxa have been collected only from the tidal forests of Mahanadi delta while the localities of collections in respect of other species are Chilika Lake (in Puri and Ganjam districts) and Chandipur (in Balasore district). It appears therefore that the estuaries of other rivers like Devi, Subarnarekha, Budhabalanga have not been explored by Haines. Paradoxically, Bhitarkanika with its magnificent and luxuriant mangroves remained untouched probably due to lack of communication facilities to this impenetrable forest, packed with many savage animals both in land and water.

Mooney (1950), the later plant explorer also ignored this mangrove tract. Probably, this is due to the fact that he mainly concentrated his exploration activities in the hilly tracts of Western Orissa, which were not botanized by Haines. However, he could collect as many as 12 mangrove taxa through a short botanical trip to the Mahanadi delta in the regions of false point during the 1st week of June 1949, when the manuscript of his book entitled "Supplement to the Botany of Bihar and Orissa" was in press. In the appendix III of this treatise Mooney has provided description of these taxa including some new reports for the State of Orissa. A few sporadic reports on the occurrence of mangrove species at Bhitarkanika have been made by Banerjee and Das (1972), Rao and Sastry (1974). However, in recent times Choudhury (1984, 1986, and 1987), Patnaik and Choudhury (1989) have provided some information regarding the floristic composition, distributional pattern, sociability, vulnerability, socio-economic importance, management and conservation strategies of the mangroves of Bhitarkanika Wildlife Sanctuary.

General Aspects of the Flora

The vegetation of Bhitarkanika can broadly be classified into (a) mangrove formation, (b) salt-bush formation. The former is mostly zoned in the fringes of estuarine banks near the funnel-shaped river mouths of Dhamra and Maipara. This also stretches along the banks of creeks and channels of the inner estuarine regions and elevated banks of upper riverain systems as well as islets. The environmental factors prevailing in these habitats favor the luxuriant growth of the typical mangrove taxa. This plant community enjoys high salinity of soil (which is silty clay type), strong wind velocity as well as tidal effects. On the other hand, salt bush formation is found along the littoral tract of Satavaya and Gahirmatha sea shore where soil is not slushy rather sandy in nature, less saline and is not subjected to inundation or active intertidal forces but enjoys salt-loaded wind and other physical forces of the ocean.

Mangrove vegetation

The mangrove taxa occurring in Bhitarkanika wildlife sanctuary constitute two-storey system of the mangrove forest namely top storey and middle storey whereas the ground flora is practically absent or very poor. The top canopy is mostly composed of *Sonneratia apetala, Avicennia officinalis, A. alba, Heritiera fomes, Xylocarpus moluccensis, Excoecaria agallocha*, etc. of which, *S. apetala* occurs along the banks and creeks of the river Bhitarkanika (a distributary of the Brahmani) gregariously. Of course, occasionally this taxon is also found in the interior part having very wide girth of the trunk indicating its antiquity. These places are more or less swampy and soil is saline in nature. Such habitats are not directly influenced by tide and ebb. However, during high tide, saline water spreads over these places. Along the fringes of some creeks which are under the direct spell of inundation and high degree salinity *S. apetala* is found almost in pure formation whereas *A. officmalis*, though less gregarious, is also found as dominant elements in the banks of the other creeks and channels. *Ex. agallocha, H. fomes* are less abundant in similar habitats. On the contrary, these taxa are found almost in pure formation in the drie and elevated areas. Especially adjacent to the Dangmal forest-shed *Ex. agallocha* occurs ir thick patches. It is to be noted that *X. granatum* exhibits a tangible degree of shrinkage in it population.

Besides these trees (constituting the top storey), *Pongamia pinnata*, a deciduous element i also found in abundance away from the water bodies. All these taxa exhibit different degrees o xerism. Whereas some of them are true mangroves having vivipary, pneumatophores, knee roots etc., others fall under semi-mangroves type.The dominant species are: *Kandelia candel Brownlowia tersa, Aegiceros corniculatus, Rhizophora apiculata, Hibiscus tiliaceus, Bruguier gymnorrhiza, B. cylindrica, Phoenix paludosa, Tamarix troupii, Acanthus ilicifolius, Ceriop decandra, Clerodendrum inerme, Cynometra iripa, Flagellaria indica, Lumnitzera racemosa* and *Salvadora persica,* etc. These form the second storey of the forest. Of these, *Kandeli candel, Rhizophora mucronata, Brownlowia tersa, Aegiceras corniculatus, Intsia bijuga*, etc enjoy the direct influence of tide and ebb as well as high degree of water salinity. *Rhizophor* community gives a distinctive appearance due to profuse knee roots which remain partiall inside water during tide. The muddy beach is studded with innumerable pointed pneumatophores

Although *Hibiscus tiliaceus* and *Ceriops decandra* are found in the vicinity of the wate bodies in association with other taxa, these are mostly met with in the interior parts, which ar less inundated. The former is rather gregarious whereas the latter is less abundant. In simila habitat, *Clerodendrum inerme, Lumnitzera racemosa, Phoenix paludosa, Bruguiera gymnorrhiza Tamarix troupii, Flagellaria indica*, etc. are found. Majority of these taxa fall under the categor of semi-or transitory mangals.

It is interesting to note that long stretches of *Phoenix paludosa* and *Tamarix troupii* ar extant at Dangmal. Former is also very gregarious in the Kalibhanjadian Island which is situate at the bifurcation of the river Dhamra. Along the fringes of the defunct creeks usually thicket of *Acanthus ilicifolius* in association with *Caesalpinia nuga, Acrostichum aureum, Phoeni paludosa* form ideal niches for the crocodile to lay eggs therein. *Pandanus tectorius* is als occasionally met with in similar habitat forming thick patches. The common climbers or associate of these shrubby elements are *Derris trifoliata, D. scandens, Caesalpinia nuga, Mucuna gigantea Dalbergia spinosa, Finlaysonia obovata, Sacrolobous carinatus*, etc. *Drynaria quercifolia* an *Hoya parasitica* are the only epiphytes of this mangrove forest.

Mangrove swamps

The ground flora close to the creeks and channels is rather poor. *Cyperus malaccensis* an *Myriostachya wightiana* are generally gregarious in the muddy flats enjoying full inundation *Porteresia coarctata* is also found in similar habitat but is comparatively less abundant in drie areas where the soil is less saline and clayey in nature. The ground flora is composed of element like *Laggera aurita, Crotolaria retusa, Tylophora tenuisama, Hemidesmus indicus, Fimbristyli ovata, Leucas sp., Ageratum conyzoides, Salicornia branchiata, Carissa spinarum, Suaed maritima, S. nudiflora, Synostemon bacciformis, Ischaemum barbatum, Eleusine indica, Eugeni bracteata, Cynodon dactylon*, etc. The typical mangrove forests gradually pass into scrub jungl or semi-deciduous forests in the drier and elevated area where *Lannea coromandelica, Ixor pavetta, Suregada multiflora, Bridelia spp., Strychnos nux-vomica, Amoora cucullata, Ficu* spp., *Manilkara hexandra, Ziziphus oenoplia, Pterospermum xylocarpum, Dodonaea viscosa*, etc. are generally found.

It is interesting to note that in Bhitarkanika, *Cerbera managhasa*-a small tree, *Acanthus volubilis*-a robust climber are rarely met with. These are not traceable in other mangrove forests of Orissa. Hence, these taxa fall under the vulnerable category. Further, Majumdar and Banerjee (1985) have identified a rare taxon, which is named as *Heritiera kanikensis* by them.

Bhitarkanika harbours a good number of species belonging to the genus *Heritiera*, which also occur abundantly. However, these are disappearing from Sunderbans of West Bengal at faster rate (Mukherjee and Mukherjee,1978). On the other hand, *Nypa fruticans* is singularly absent from Bhitarkanika which is rather gregarious in Sunderbans. It has been suggested that the species of *Heritiera* might have migrated to other coastal regions of Indian Sub-continent from the Orissa Sea Coast. Distribution of *Xylocarpus granatum* in Bhitarkanika also deserves special attention since this taxon is not found in the other mangrove forests of Orissa.

Saltbush formation

Along the littoral tract of Satavaya and Gahirmatha the scattered sand dunes as well as sand bars mostly sustain xeric herbs of which extensive patches of *Ipomoea pes-caprae* attract one's attention. Other common xeric elements are: *Hydrophylax maritima, Bulbostylis barbata, Launaea sarmentosa, Spinifex littoreus, Cyperus arenarius, Sesuvium portulacastrum, Pedalium murex, Gisekia pharnaceoides, Trianthema portulacastrum*, etc. These elements act as sand binders as well as sand stabilizers. Soil erosion is considerably checked by the extensive spread of these taxa.

Fauna

Bhitarkanika wildlife sanctuary, harboring abundant mangrove elements and their associates, innumerable water bodies, alongwith sea-beach provides an ideal niche for both aquatic as well as terrestrial organisms of diverse categories. The Olive Ridley sea turtle (*Lepidochelys olivacea*) is an annual migrant reptile to Gahirmatha beach during its breeding season (December – March). Visit of about 100,000 to 200,000 sea turtles every year for mass nesting provides a spectacular sight. Salt water crocodile (*Crocodylus porosus*) is another important reptile found in the rivers and creeks of Bhitarkanika. They attain a size of 6-7 meters in length. It has been reported that the world's largest crocodile was killed in the year 1926 from the Bhitarkanika river system. Partial albino variety, locally known as 'Sankhua' is also found in the water bodies. Due to large-scale hunting *Crocodilus perosus* was fast disappearing and was on the threat of extinction. However, after 1975 this endangered species has been enjoying adequate protection and care due to declaration of wildlife sanctuary. At Dangmal, a Crocodile Research Centre was established for research and conservation of this salt water crocodile where active research works are being conducted and artificially hatched juveniles are being regularly released into the creeks and channels of Bhitarkanika. As per the Report (Kar, 1985), 550 crocodiles consisting of 27 males and 523 females have been released into the water bodies of this sanctuary. Their movement, behavior as well as survival rates are being regularly monitored. It has been observed that they prefer hypo-salinity condition of water as well as areas which are thickly populated with mangrove elements and are less interfered by human activities. Further, from time to time juvenile salt-water crocodiles are being sent to various research centers, Zoological/Biological parks of India for research and display. Captive breeding programme is now a fascinating facet of crocodile research, which will provide some interesting data on breeding behaviour of this species. As a matter of fact, due to adequate care and protection its population has attained stability and therefore, this species is now out of danger.

However, rehabilitation of mangrove taxa in the partially degraded areas within this terrain will provide better environment for the growth and development of these crocodiles. In addition to these sea turtles and crocodiles, other aquatic animals are water monitor lizards, mudskippers, limulus, red crabs and varieties of fishes. Migratory birds like open-billed storks, white ibis, grey pelicans, bar-headed goose, Bramhiny ducks, etc. come in large number during winter season of every year and these are special attractions for the ornithologists as well as nature lovers. Besides these migratory birds, the resident birds also add to the diversity of wildlife. Night herons, cattle egrets, cormorants, grey herons, three varieties of king fishers, water fowls, common sand piper, collard dove, Alexandrian parakeet, etc. come under this category. The terrestrial wild animals such as python, king cobra, spotted deer, wild boars, chital, hyena, porcupine, fishing cat, mongoose, etc. are notable in this sanctuary. In comparison to the national status, the composition of vertebrate fauna of Bhitarkanika ecosystem represents 8% mammals, 17.7% birds, 9.40% reptiles and 2.5% amphibians (Kar and Kar, 2008).

Socio-Economic Aspects of the Mangroves

This mangrove forest possesses manifold socio-economic potentialities. Local people make use of mangrove wood for fuel, charcoal, agricultural implements, house building materials etc. The woods of *Avicennia spp. Bruguiera cylindrica, Ceriops decandra, Excoecaria agallocha, Tamarix* spp. having high caloric value are used as fuelwood. Wood for furniture is generally available from timber of *Xylocarpus moluccensis*. Stems of *Phoenix paludosa, Heritiera fomes* are used as poles for the thatched houses whereas the leaves of *Phoenix paludosa* and those of *Myriostachya wightiana* are used as roofing materials. The latter one is also a good fodder for the buffaloes. For basket making, the leaves of *Flagellaria indica* and *Myriostachya wightiana* are extensively used. The mangrove elements provide food for the aquatic life forms. High percentage of tannin is present in the bark of *Rhizophora mucronata, R. apiculata, Sonneratia caseolaris, S. apetala, Aegiceras corniculatum, Bruguiera* spp., *Avicennia* spp., *Candelia candel* etc. Saponin, flavonoids etc. are available from the species like *Excoecaria agallocha, Sonneratia apetala, Xylocarpus granatum, Aegiceros corniculatum, Phoenix paludosa* etc. Fruits of *Cynometra iripa* and *Sonneratia caseolaris* are edible. Similarly, *Suaeda maritima* and *S. nudiflora* are used as pot herbs. Different parts of *Xylocarpus granatum* are used to cure elephantiasis, swelling of the breast and other abdominal troubles. The bark of *Xylocarpus moluccensis* is used to cure dysentery and diarrhoea. *Salvadora persica, Pongamia pinnata, Sonneratia caseolaris, Pandanus tectorius, Acanthus ilicifolius*, etc. are potential sources of medicine to cure rheumatism, fever, joint pain, joint swelling etc. Similarly, *Lumnitzera racemosa, Salicornia brachiata, Tamarix spp.* possess curative properties for skin diseases of various categories. The mangrove forest provides breeding ground and nursery for majority of commercial fishes and prawn, etc. The forest meadows and swampy places support a large number of resident and migratory birds. Good quality honey and other minor forest products are also available from this mangrove forest. In addition to such economic importance, the mangrove forests check the oceanic cyclones considerably and also encroachment of the sea acting as shoreline stabilizer.

Management Aspects of Mangrove Vegetation

The various life-forms extant in the Bhitarkanika wildlife sanctuary is now enjoying adequate protection since 1975. Formerly, Bhitarkanika consisting of 81 villages alongwith extensive forest cover was under the control of the Kanika Raja. However, with the enforcement of Zamindari Abolition Act, 1951, the Kanika Zamindari was vested in the Revenue Department of Orissa with effect from 26th November, 1952. Thereafter, this was transferred to Athagarh Forest Division in 1957 and finally in 1980 the Kanika Range Forest came under a separate wildlife Division i.e. Chandbali Wildlife Conservation Division in the Balasore District which is under the direct supervision of a Divisional Forest Officer. During the management of Kanika Raja the forestlands were given on lease, which was a potential source of revenue. Indiscriminate cutting of the tree species for fuel, timber and other useful materials devastated these forestlands. Forestlands were reclaimed for human habitation and paddy cultivation. This was a good hunting spot for the high officials as well as nearby inhabitants who killed spotted deers, crocodiles, sambars, leopards, pitted sterks, giant water monitor lizards, heron of various categories on large scale. This affected the overall equilibrium of the ecosystem and many wild lives were on the threat of extinction.

Realizing the importance of conservation of this vulnerable plant community and some endangered life forms therein, the Government of Orissa (Forest, Fisheries and Animal Husbandry Department) declared the forestland and wasteland, known as Kanika forest in Ali, Rajanagar, Pattamundai and Mahakalpada police station within the Kendrapada Civil Sub-division in the district of Cuttack including all creeks, channels, nalas, wetlands and riverbeds situated within the above-said protected forest as Bhitarkanika Wildlife Sanctuary (vide Notification No.6958-4F(W)-34/75-FFAH/dt.22.4.1975). However, recently this sanctuary has been upgraded to a national park (vide Notification No.8F(W)-53/88-22904/FFAH/dt. 3.10.88) which spreads over an area of 367 kms^2 as against the previous coverage i.e. 650 kms^2. Consequently, the wild lives were well protected and are thriving well. Still, there are some illegal poaching activities and encroachment of the forest lands for prawn culture and paddy cultivation by the nearby villages.

It has been experienced that the destruction of mangrove forests has brought about various types of environmental hazards as well as genetic erosion of many useful life-forms present therein. Hence, in view of their high socio-economic importance, genetic diversity and significant role in maintaining environmental stability there is an urgent need for the conservation of this mangrove vegetation. Adequate attention should also be paid for the germplasm conservation of the vulnerable and rare taxa like *Cerbera manghas*, *Acanthus volubilis*, *Heritiera kanikensis*, etc. Judicious utilization of the mangrove elements may be adopted. Concurrently, adequate awareness should also be generated among the common people especially those living nearby, about the importance of this mangrove forest and not to cut down the trees for fuel purpose. Alternative source of energy like gobar gas plants or solar cookers should be popularized among these people.

Attempts should be made to rehabilitate the mangrove taxa in the places where these have been partially or totally destroyed. Development of germplasm bank and propagation of vulnerable mangrove taxa by different methods should also be initiated to restore the rich heritage of mangrove vegetation of Orissa.

Strategies for Conservation of Biodiversity

As indicated earlier, the biodiversity of Orissa mangroves at present exhibits tangible shrinkage and is still in a declining state. This has posed several environmental hazards especially, in the coastal belt. Ultimately it has affected the socio-economic status of populace of Orissa in general and that of coastal terrain in particular. The fury of the environmental hazards has reached its climax in the end of Oct.1999 causing widespread devastation of various life-forms, loss of human habitations as well as properties. This supercyclone also affected the neighboring regions. Restoration of habitats, avenue plantation and croplands has not yet been completed (Choudhury, 2002) and the imbalance of coastal environment is still continuing. As a matter of fact, frequent oceanic cyclones are still occurring in the State of Orissa.

Hence, it is high time for conservation of coastal zone biodiversity of Orissa at least to improve the lifestyle of coastal dwellers and to strengthen the socio-economic status of the State in general.

(i) In the first place, although Bhitarkanika mangroves are well protected due to its status of being a National Park, germplasm conservation of some endangered / vulnerable / endemic taxa should be attempted on priority basis. The population of mangrove species such as *Xylocarpus granatum, Cerbera manghas, Acanthus illicifolius, Bruguiera gymnorrhiza, Heritiera kanikensis*, etc. which are restricted to Bhitarkanika only exhibiting shrinkage of population (Choudhury *et al.*, 1995) should be proliferated by modern biotechnological methods.

(ii) Regeneration, rehabilitation as well as adequate protection of rare / threatened mangrove taxa should be attempted. Prawn culture and conversion of Mangrove forests for paddy cultivation and human habitation as practiced by the immigrants and inhabitants especially in the Mahanadi delta, Paradeep, Astaranga, Ramnagar etc. should be prohibited.

(iii) Simultaneously, felling of woody elements by the local people for fuel purpose and to meet the other commercial usages especially in the mangrove denuded area (of Mahanandi delta, Jambu, Hukitola Astaranga, etc.) should be checked strictly.

(iv) Alternative fuel energy should be implemented by popularizing the biogas chula and plantation of fast-growing trees/shrubs in order to provide fuelwood to the general public inhabiting the mangrove forest regions.

(v) Mangrove nursery should also be developed in Hukitola Island and Bhitarkanika where under adequate protection and care, mangrove taxa and associates such as *Excoecaria agallocha, Intsia bijuga, Cynometra isipa, Clerodendrum inerme* should be planted. After successful stabilization of population of these species, other typical mangroves may be planted in the denuded areas, so that the gene pools of the mangrove species will be ideally maintained.

(vi) The inhabitants, especially school / college students should be educated for conservation of mangrove species through organization of eco-friendly camps / seminars / symposia / workshops using audio-visual aids.

(vii) Government as well as Non-Government Organizations (NGO) should play an active role in developing awareness towards judicious exploitation and proper conservation of mangrove species. Unless otherwise grass-root level attempts are made in this direction, conservation of mangroves will not be successful.

(viii) Finally, development of artificial mangrove forests in the mangrove denuded areas is the need of the hour. In this connection man-made mangrove forest in Hochiminch City is a bright example (Choudhury, 1999). By spray of poisonous chemicals a vast mangrove forest was destroyed. However, through successful plantation programme over an area of 20,000 ha. in south of Hochiminh City (Vietnam), a pure stand of *Rhizophora apiculata* has come into existence which provides charcoal of high calorific value, honey, woody elements which are being exploited by the nearby inhabitants. This type of artificial mangrove forests is also met within India. Development of such forests selecting some mangrove elements may be followed in Chilika area, Mahanadi delta, Paradeep, Hukitola etc. This will promote eco-friendly tourism which will enhance the economic conditions of the inhabitants.

In recent time, "M.S. Swaminathan Researh Foundation (Chennai) in collaboration with "International Tropical Timber Organisation (Japan)" have provided the final report on the "Travelling workshops to mangrove forests in South and Central Africa, Oceania, West and Central Africa". Out of 23 sites recommended from 9 countries, Bhitarkanika (India) has been selected as one for the establishment of Mangrove Genetic Resource Conservation Centre (MGRCC).

The summary of Evaluation Report for four sites in India has been highlighted. Four mangrove chunks such as (i) Choras, (ii) Pichavarum, (iii) Bhitarakanika and (iv) Coringa have been studied by the mangrove experts. Various parameters like (a) mean temperature (b) mean annual rainfall (c) mean relative humidity (d) tidal conditions (e) water pH (f) salinity and (g) total available area have been taken into account. Other salient features of Bhitarakanika have also been critically evaluated. From such a study it has been concluded that of the four sites in India, Bhitarkanika is the most suitable one for establishment of MGRCC.

Adequate steps should be taken up for implementation of this Report given by experts of the above-said organizations so that the germplasm as well as luxuriance of Orissa mangrove will be restored. This will provide salubrious environment in the Orissa coast and the lost glory of mangroves will be revived.

ACKNOWLEDGEMENTS

The author is grateful to late Dr. Bhabani Prasad Choudhury, former Professor of Botany, Utkal University, Bhubaneswar (Orissa) for his inspiration and providing necessary information to write this article before his sad demise on September 08, 2008.

REFERENCES

Anonymous, (1986). Wildlife of Orissa. Forest Department, Government of Orissa, Bhubaneshwar

Anonymous, (1987). Mangroves in India–Status Report. Government of India, Ministry of Environment and Forests, New Delhi

Banerjee, LK & Das, GC (1972). New distributional records from Orissa Coast. *Bull. Bot. Surv. Ind.* **14(1-4):** 184-186

Champion, HG & Seth, SK (1968). A Revised Survey of the Forest Types of India, Govt. Press, New Delhi

Choudhury, BP (1984). A glimpse into the vegetation of Bhitarkanika Wildlife Sanctuary in the state of Orissa. *Ind. Bot. Rept.* **3:** 121-124

Choudhury, BP (1986). Nomenclature and distribution of some mangrove elements in the Bhitarkanika Wildlife Sanctuary. News letter, *Nature and Wildlife Cons. Soc.* **4:** 58-60

Choudhury, BP (1987). Socio-economic aspect of Bhitarkanika Wildlife Sanctuary in the State of Orissa. *Orissa Review* **43:** 45-47

Choudhury, BP (1999). Mangroves of Bhitarkanika. *In*: Tiwari, AN (Ed.), Reference Orissa, Enterprising Publishers, New Delhi, 26-32

Choudhury, BP (2002). Probable solutions of environmental problems arising after super cyclone. *Orissa Review* **56(6):** 32-34

Choudhury, BP, Biswal, AK & Rath, SP (1995). Biodiversity in the Bhitarkanika Wildlife Sanctuary in the State of Orissa pp. 53-60 *In*: Mohanty, RC (ed.), Environment: Change and Management, Kamal Raj Enterprises, New Delhi

Haines, HH (1921-1925). The Botany of Bihar and Orissa (6 parts), London; reprinted ed. 1961, BSI, Howrah

Kar, SK (1985) A decade of Salt water crocodile (*Crocodylus porosus* Schneider) conservation programme in Bhitarakanika Wildlife Sanctuary, *Orissa Proc. Nat. Conf. Biosphere Reserve*, Similipal, Orissa pp. 103-108

Kar, SK & Kar, CS (2008). Biodiversity resources of Bhitarkanika mangrove ecosystem: It's significance and threats pp. 13-16 *In*: *Proc. Symp. Wetland and Mangrove Biodiversity in Orissa Coast*, Ministry of Environment and Forests, Govt. of India, New Delhi

Majumder, NC & Banerjee, LK (1985). A new species of *Heritiera* dry and Sterculiaceae from Orissa, *Bull. Bot. Surv. Ind.* **27(1-4):**150-151

Mishra, GC (1999) Green Forest of Orissa pp. 16-20. *In*: Tiwari, A.N. (ed.), Reference Orissa, Enterprising Publishers, New Delhi

Mooney, HF (1950) Supplement to the Botany of Bihar and Orissa. Catholic Press, Ranchi

Mukherjee, BB & Mukherjee, J (1978). Mangroves of Sunderbans. *Phytomorphology* **28(2):** 192

Padhi, SC (1985). The management aspect of Similipal as a Biosphere Reserve pp. 4-7. *In*: Conservation of Similipal in its wilderness. *Orissa Env. Soc.*, Berhampur

Patnaik, SN & Choudhury, BP (1989) Present status and future development of mangrove vegetation in Bhitarkanika pp. 89-95 *In:* INDO-US workshop on Wetlands, Mangrove and Biosphere Reserve. Government of India, Ministry of Env and Forestry, New Delhi

Rao, TA & Sastry, RK (1974). An ecological approach towards classification of coastal vegetation of India II. Estuarine border vegetation. *Indian For.* **100:** 438-452

Samal, RC & Patnaik, JM (1989). Management strategies for Bhitarkanika Mangrove Forest. *In*: INDO-US workshop on Wetlands, Mangroves and Biosphere Reserves, Govt. of India, Ministry of Environment and Forests, New Delhi, 96-99

15

Invasive Alien Plant Species: A Threat to Native Biodiversity

AK PALIWAL, BEENA KUMARI[1] AND TARIQ HUSAIN[2]

Department of Botany, Government Post-Graduate College, Bageshwar–263 642 (Uttarakhand)
[1] Department of Botany, Hindu College, Moradabad–244 001, (Uttar Pradesh)
[2]Biodiversity & Angiosperm Taxonomy Division, National Botanical Research Institute, Rana Pratap Marg, Lucknow–226 001, (Uttar Pradesh)

ABSTRACT

Invasive alien species, introduced and/or spread outside their natural habitats, have affected native biodiversity in almost every type of ecosystem on the earth and constitute one of the greatest threats to biodiversity. These often coexist with native species for an extended time, and gradually the superior competitive ability of an invasive species becomes apparent as its population grows larger and denser and it adapts to its new location. The problem continues to grow at a great socio-economic, health and ecological cost around the world. Gradually, these exacerbate poverty and threaten development through their impacts on agriculture, forestry, fisheries and natural systems, which are important bases of peoples' livelihoods in the developing countries. This damage is further aggravated by climate change, pollution, habitat loss and human-induced disturbances.

Keywords: Alien invasive plant species, conservation of nature, native biological diversity, disturbed ecosystem, vulnerable to invasions.

INTRODUCTION

According to World Conservation Monitoring Centre (WCMC), 16,04,000 species have been described at the global level. India accounts for 8% of the global biodiversity existing in only 2.4% land area of the world. According to Nayar (1989), the number of flowering plant species endemic to the present political boundaries of this country is 4900 out of a total of 15000, i.e.

33%. Hajra & Mudgal (1997) have reported 5,400 endemics in 17,000 angiospermous species of India, which comes to 31.76%. India is an important center of agri-biodiversity having contributed 167 species to the world agriculture and being homeland for 320 species of wild relatives of crops. The flora of India shows close affinity with the flora of Indo-Malayan and Indo-Chinese region. This fact prompted Hooker (1904) to arrive at an erroneous conclusion that "India has no flora as a separate entity but is an admixture of the floras from adjacent countries". According to Nayar (1977), Indian flora has south-east Asian (35%), Malayan (8%), steppe (1%), African (2%), and Mediterranean-Iranian (5%) elements. The cultivated and naturalized aliens constitute only 18%. Chatterjee (1940) reported 6,850 species as endemic to India for the first time. The subsequent phytogeographers after critical analysis of flora have convincingly concluded that India has a flora of its own and as many as 5,400 (33%) taxa of Indian flora are endemic to the present Indian boundaries.

Plant invasions have been recognized as one of the most serious global processes impacting the structure, composition, and function of natural and semi-natural ecosystems (Mooney & Hobbs 2000). International Union for Conservation of Nature and Natural Resources (IUCN) defines Invasive Alien Species as a foreign (alien) species coming from outside which becomes established in natural or semi-natural ecosystems or habitats, acting as an agent of change, and thereby threatening native biological diversity. These invasive species are widely distributed in all kinds of ecosystems throughout the world, and include all categories of living organisms. Invasive species are good at exploiting bare soil and empty niches. These species are generally non-natives that infest natural ecosystems, including forest, rangelands, and pastures. One way to avoid these species is to choose the ones that are native to the area. Natives often are adapted to a specific environmental niche, and have natural controls that keep them in balance. Invasive species, on the other hand, are tolerant against environmental extremes and possess greater flexibility for survival in wide range of conditions with high water, light and nutrient use efficiencies. Some of the invasive species also exhibit fire resistance quality, besides better competitive ability and allelopathy.

Distribution of notorious weeds has caused a strong effect on the native flora, since several of these have replaced much of the indigenous species, probably because of their strong harmful effects (Nayar 1977). According to the Convention for Biological Diversity, invasive exotic species are the second largest cause of biodiversity loss in the world and impose high costs to agriculture, forestry, and aquatic ecosystems. In fact, introduced species are a greater threat to native biodiversity than pollution, harvest, and disease combined altogether. It is very difficult to choose exotic species, from all over the country, that really are "invasive" or "worse" than any others. The interactions of these species with ecosystems are very complex.

INVASIVE SPECIES OF INDIA

Invasive alien species of plants are as equally ancient as human civilization, and are chronologically indistinguishable by man. However, biological invasions have lately become one of the major global issues of concern since June 1992 soon after the UNCED's Earth Summit held in Rio de Janeiro, Brazil. It was estimated that about 40% of the species in the Indian flora are alien, of which 25% are invasive in nature. *Parthenium hysterophorus*, first reported in 1951 from Maharashtra was introduced from Tropical America and presently has stretched throughout the length and breadth of the country. *Lantana camara* which was introduced in 1809-1810 from Tropical and subtropical America has reached the submontane regions of the outer Himalaya to southernmost parts of the country. *Eichhornia crassipes* which was introduced in 1914-1916 from Brazil has extended to a large area almost all over the water bodies of India. *Mikania micrantha* has a rather neotropical origin but has extended natural forests and plantations. A total of 173 species of 117 genera belonging to 44 families have been documented as Invasive alien plant species, representing 1% of the Indian flora (Reddy *et al.* 2008).

Introduced Plants from Australia

Acacia dealbata, A. decurrens, A. holosericea, A. longifolia, A. mangium, A. melanoxylon, Araucaria bidwilli, A. columnaris, A. cunninghamii, A. excelsa, Callistemon lanceolatus, C. cupressiformis, Callitris glauca, Castenospermum australe, Casuarina equisetifolia, Eucalyptus alba, E. camaldulensis, E. citriodora, E. globulus, E. grandis, E. paniculata, E. punctata, E. saligna, E. tereticornis, E. torelliana, E. viminalis, Grevillea banksii, G. robusta, Melaleuca leucodendron, Stenocarpus sinuatus, Thuja occidentalis, and *T. orientalis.*

Introduced Species from Western Asia and Europe

Allium cepa, A. sativum, Althaea rosea, Asparagus officinalis, Avena sativa, Brassica oleracea, Cicer arietinum, Coriandrum sativum, Ficus carica, Lactuca sativa, Lens culinaris, Medicago sativa, Morus nigra, Nerium oleander, Papaver somniferum, Pisum sativum, and Spinacea oleracea.

Introduced Species from the Mediterranean Region and Africa

Aloe barbadensis, Antirrhinum majus, Asparagus plumosus, Avena fatua, Briza maxima, Calendula bicolor, Capsella bursa-pastoris, Carthamus tinctorius, Cassia didymobotrya, Catharanthus roseus, Chenopodium album, C. murale, Chrysanthemum coronarium, Cichorium endivia ssp. divaricatum, Convolvulus arvensis, Cyperus flabelliformis, Foeniculum vulgare, Haworthia coarctata, Hordeum vulgare, Ipomoea batatas, I. cairica, Lathyrus sativus, Nasturtium officinale, Nerium indicum, Nigella damascena, Oxalis purpurea, Pelargonium inquinans x zonale, Pelargonium x hybridum, Pennisetum typhoides, Plantago major, Poa annua, Raphanus sativus, Ricinus communis, Salvia officinalis, Sansevieria trifasciata, Senecio vulgaris, Sisymbrium officinale, Spergula arvensis, Stellaria media, Taraxacum officinale, Verbena officinalis, V. supina, Veronica arvensis, and *Vicia benghalensis.*

Introduced Species from Europe/America/Mexico/West Indies/Brazil

Agave americana, A. sisalana, Allamanda cathartica, A. violacea, Anacardium occidentale, Arundo donax, Angelonia biflora, Annona muricata, A. reticulata, A. squamosa, Arachis hypogaea, Asclepias physocarpa, Averrhoa bilimbi, A. carambola, Bauhinia blakeana, Begonia manicata, Beta vulgaris, Bignonia alliacea, B. megapotamica, Bixa orellana, Bougainvillea glabra, B. spectabilis, Brassica rapa, Browallia viscosa, Brunfelsia americana, Caesalpinia coriaria, C. pulcherrima, Caladium bicolor, Capsicum annuum, C. frutescens, Carica papaya, Cassia grandis, Catesbaea spinosa, Ceiba pentandra, Centaurea cyanus, Cereus caesius, Cestrum diurnum, C.

nocturnum, Chenopodium ambrosioides, Chorisia speciosa, Citharexylum subserratum, Cleome spinosa, Clitoria ternatea, Cocos nucifera, Colvillea racemosa, Conyza bonariensis, Coreopsis lanceolata, C. stillmanii, Coriandrum sativum, Cosmos bipinnatus, Couroupita guianensis, Crescentia cujete, Cuphea lanceolata, Dahlia variabilis, Datura stramonium, Daucus carota, Delonix regia, Delphinium ajacis, Dianthera secunda, Dianthus barbatus, Dieffenbachia seguina, Duranta repens, Eruca sativa, Eschscholtzia californica, Euphorbia pulcherrima, Fragaria vesca, Furcraea foetida, Gaillardia pulchella, Godetia amoena, Gossypium herbaceum, G. hirsutum, Graptophyllum pictum, Guazuma tomentosa, Gustavia augusta, Hamelia patens, Helianthus annuus, Hibiscus elatus, H. liliflorus, H. sabdariffa, Iberis amara, Ipomoea indica, Iresine herbstii, Jacaranda mimosaefolia, Jacquemontia pentantha, Jacquinia ruscifolia, Jatropha multifida, J. panduraefolia, Kigelia africana, Lathyrus aphaca, Lychnis coeli-rosa, Lycopersicon esculentum, Magnolia grandiflora, Malpighia coccigera, M. glabra, Malva parviflora, Malvaviscus arboreus, Manihot esculenta, Manilkara achras, Medicago lupulina, M. polymorpha, Mentha piperita, Millettia peguensis, Monstera deliciosa, Morus nigra, Musa acuminata, Nicotiana alata, N. tabacum, Ochroma pyramidale, Oenothera rosea, Opuntia elatior, O. imbricata, O. microdasys, O. vulgaris, Oxalis latifolia, Papaver rhoeas, Parkinsonia aculeata, Passiflora edulis, P. suberosa, Pavonia hastata, Persea americana, Petrea volubilis, Petunia axillaris, Phoenix dactylifera, Physalis peruviana, Pinus caribaea, P. gregii, P. merkusii, P. michoacana, P. occidentalis, P. oocarpa, P. patula, P. pringlei, P. pseudostrobus, P. rudis, P. taiwanensis, Pisonia grandis, Pithecellobium dulce, Plumeria alba, P. rubra, Portlandia grandiflora, Portulaca grandiflora, Portulacaria afra, Prunus domestica, Psidium guajava, Punica granatum, Pyrostegia venusta, Rauvolfia tetraphylla, Ravenia spectabilis, Reseda odorata, Rhoeo discolor, Rivina humilis, Russelia equisetiformis, Salvia splendens, Solanum grandiflorum, S. tuberosum, Solidago canadensis, Spergula arvensis, Stellaria media, Stigmaphyllon ciliatum, Tabebuia chrysantha, T. pentaphylla, T. rosea, Tagetes erecta, Talinum calycinum, Tamarindus indica, Tecoma gaudichaudii, T. stans, Tecomaria capensis, Tecomella undulata, Theobroma cacao, Thevetia peruviana, Thunbergia alata, T. erecta, Thyrallis glauca, Tithonia rotundifolia, Tropaeolum majus, Vaccaria pyramidata, Verbena peruviana, Vernonia elaeagnifolia, Vicia hirsuta, Viola odorata, V. tricolor, Yucca gloriosa, Zea mays, Zebrina pendula, Zephyranthes grandiflora, Zinnia elegans which are affecting the agricultural systems in northeast and southwest India.

ECOLOGICAL IMPACTS OF INVASIVE ALIEN SPECIES

Biological species invasions alter ecosystems in a multitude of ways. Worldwide, an estimated 80% of endangered species suffer losses by competition with, or predation by, invasive species. Reports indicate that introduced species, such as corn, wheat, rice, cattle, and poultry, provide more than 98% of the US food system at a value of approximately $800 billion per year. Since highly adaptable and generalized species are introduced to environments already impacted by human activities, some native species may be put at a disadvantage to survive whereas the survival of other species is enhanced. Land clearing and human habitation put significant pressure on local species. This disturbed habitat is prone to invasions that can and do have adverse effect on local ecosystems and changing ecosystem functions. Multiple successive introductions of different non-native species can have interactive effects like the introduction of a second non-native species can enable the first invasive species to flourish. These species can also change the functions of ecosystems. For example, invasive plants can alter the fire regime, nutrient cycling, and hydrology in native ecosystems. The invasive species which are closely related to the rare native species have even the potential to hybridize with them. Harmful effects of hybridization have resulted in a decline and even extinction of native species. Natural, wild species can be threatened with extinction through the process of genetic pollution.

Genetic pollution is uncontrolled hybridization and introgression which leads to homogenization or replacement of local genotypes as a result of either a numerical or fitness advantage of the introduced species. It can bring about a form of extinction either through purposeful introduction or through habitat modification, bringing previously isolated species into contact. Such phenomena can be especially detrimental for rare species coming into contact with more abundant ones where the latter can interbreed with them, creating hybrids and swamping the entire rare gene pool, thus driving the native species to extinction. Attention has to be focused on the gravity of this problem, as it is not always apparent from morphological observations alone. Some degree of gene flow may be a normal, evolutionarily constructive process, and all constellations of genes and genotypes cannot be preserved. However, hybridization with or without introgression may, nevertheless, threaten the very existence of a rare species (Singh & Singh 2009).

Though weeds cause an overall reduction in yield, they often provide essential nutrients for subsistence of farmers as some deep-rooted weeds can "mine" nutrients from the subsoil and bring them to the topsoil, whereas the others may provide a habitat for beneficial insects and/or alternative food for pest species. Several weed species are accidental introductions with crop seeds and imported plant material. A lot of introduced weeds in pastures compete with native forage plants, proving toxic to young cattle (older animals will avoid them) or non-palatable because of their thorns and spines. A number of the plant species intentionally introduced in various countries in the past have become major threats to the biodiversity and their control measures consume substantial financial resources. A few countries have formulated/are formulating strategies and establishing institutions to exclusively deal with these. Realizing this serious threat to forest ecosystems, the Asia Pacific Forestry Commission (APFC) has formally launched a 'Asia Pacific Forest Invasive Species Network' (APFISN) at the 20th session of the APFC held at Nadi, Fiji, from April 17-21, 2004 with the following objectives:

- To promote exchange and sharing of information on forest invasive species among the member countries;
- To facilitate access to expertise and resources such as research and education and training opportunities;
- To strengthen capacities of the member countries to conduct research and management of forest invasive species;
- To increase coordination and cooperation among the member countries by developing regional strategies for forest invasive species; and
- To raise awareness of invasive species as a significant issue throughout Asia and the Pacific Region.

Anthropogenic factors also facilitate the invasion through land clearings and human habitation, which put significant pressure on native species. It is clear, however, that the changing climate is favouring the expansion and survival of invasive alien species, but still it is difficult to predict the degree to which the changing climate will benefit invasive species. Studies are urgently needed to minimize the benefits that the invasive species might draw from a changing climate. The common Indian invasive species are: *Ageratum conyzoides* L., *Argemone mexicana* L., *Bombax ceiba* L., *Calotropis procera* (Ait.) R. Br., *Cannabis sativa* L., *Cymbopogon martinii* (Roxb.) Wats., *Eucalyptus X hybrid, Ipomoea fistulosa Mart. ex Choisy, Lantana camara* L., *Parthenium hysterophorus* L., *Prosopis juliflora* (Sw.) DC., *Solanum nigrum* L., *Tectona grandis* L., etc. The aquatic weeds falling in this category are: *Eichhornia crassipes* Solms., *Hydrilla verticillata* (L.f.) Royle, *Nelumbo nucifera* Gaertn., and *Pistia stratiotes* L. etc.

Invasive species have the capacity to make the environment friendly for themselves allowing these species to thrive while making it less favourable for competitor species. These might be able to use resources previously unavailable to native species, such as the deep-water sources accessed by long taproots or an ability to live on uninhabited soil or making the soil toxic for the competing species by secreting chemicals (allelopathy). For example, *Parthenium hysterophorus* L. commonly known as the 'Congress grass' is a good example whose root extract secreted at intervals of 120 and 180 hours, has been identified as one of the most harmful extracts for Chickpea (Oudhia *et al.,* 1997).

On the other hand, allelopathic effect of common weed, *Lantana camara* has been shown to be beneficial for rice, as it increases its chlorophyll content and grain yield at the same time significantly reducing the growth of other associated weeds of the rice crop e.g., i.e. *Panicum psilodium* Trin. and *Commelina benghalensis* L. (Sharma *et al*., 1981). Disturbances such as forest fire change the fundamental nature of the ecosystem, whereas the normal succession favours those species which can multiply and spread faster, having traits similar to the invasive species, which also alter the fire regimes. These spread rapidly after burning and increase the frequency and intensity of fires. *Lantana camara* is one such example which is favoured by fire. It has several characteristics that might give it an advantage under conditions of frequent fire. In Biligiri Rangan Hills of South India, it has been observed that the initial spread of *Lantana* sp. coincided with the last 'mass flowering' and dieback of bamboo, which was followed by widespread fires. Changes in the ecosystem also affect the dynamics of species interactions and niche establishment. These features favour the growth of alien species that would not have been able to colonize the area and niches. Evidences clearly indicate that a more disturbed ecosystem is highly vulnerable to invasions by alien species than the less disturbed ones. Ecosystem invasibility is also related to species diversity. It is thought that 'species poor biotas' and "empty niches" of an area/habitat easily make it vulnerable to colonization by alien species.

In addition to negative impact on indigenous flora and economy, some exotic plants are quite useful for local people. Prominent examples of this group are: *Borassus flabellifer, Leucaena leucocephala and Prosopis juliflora.* These plants have the advantage of good adaptability, rapid growth and several other multiple uses. *P. juliflora* is used for coal extraction, firewood, erosion control and fencing. *Borassus flabellifer* is also an economically important species, which had been introduced in India long ago. Its cut flower stalks yield sugar and toddy, the fruits are roasted and eaten and the leaves are used for thatching. *Leucaena leucocephala* is used for firewood, erosion control, fodder, green manure, and providing shade and shelter for the other crops. Several exotic weedy species like: *Argemone mexicana, Cassia tora, Chamaesyce hirta, Cleome viscosa, Croton bonplandianum, Eclipta prostrata, Ipomoea carnea, Malachra capitata,*

Mimosa pudica, Physalis angulata, Tridax procumbens are also used in several of native medicines. *Alternanthera philoxeroides and Portulaca oleracea* are used as leafy vegetables. Although India is mainly a tropical country, due to marked altitudinal variations, variable climatic conditions-from hot to cold deserts are introduced here. India's rich vegetational wealth and diversity is undoubtedly due to the vast climatic and altitudinal range coupled with varied ecological habitats.

GLOBAL ACTION AGAINST INVASIVE ALIEN SPECIES

Since 1996, it has been repeatedly realized that the invasive species might be one of the most significant "negative externalities" of globalization. As such seventy-eight countries, numerous international and inter-governmental organizations participated at the 'Trondheim Conference.' This meeting, sponsored by the Government of Norway and the United Nations, was the first global effort to assess the impact of invasive species on the environment. The major conclusions arrived at are:

- The impact of invasive species is "immense, insidious, increasing, and irreversible." In other words, every country has been impacted by these, the patterns and trends follow that of globalization, and as long as we are engaged in international trade, travel, and transport we will need to manage this problem;
- Apart from the climate change, invasive species are the most significant threat to the environment worldwide. Developing countries will be severely impacted, particularly the Small Island Developing States (SIDS); and
- A global plan and strategy is urgently needed to address the problem. In 1997, three international organizations came together with a commitment to share their expertise and other resources in order to address the scientific and technical aspects of the problems identified in Trondheim, *viz*. International Union of Conservation of Nature (IUCN), CAB International (CABI), and the Scientific Committee on Problems of the Environment (SCOPE) formed the Global Invasive Species Programme (GISP)—a coalition of scientific and technical experts. The CBD and its members (there are 191 Parties, as on October 2008) recognize that there is an urgent need to address the impact of invasive alien species. Article 8(h) of the CBD states that, "Each contracting Party shall, as far as possible and as appropriate, prevent the introduction of control or eradicate those alien species which threaten ecosystems, habitats or species". The CBD sets global priorities, guidelines, collects information and helps to coordinate international action on invasive alien species.

The CBD has adopted guidance on prevention, introduction and mitigation of impacts of alien species that threaten ecosystems, habitats or species, which can be accessed on the CBD website (Decision VI 23). The website also provides further information on invasive species and relevant decisions of the Parties to the organization.

THREATS TO AGRICULTURE AND BIODIVERSITY MANAGEMENT

The significant threat from alien weed species is being increasingly recognized both in agricultural and non-agricultural lands. In India, over 60 weed species have been introduced from other parts of the world and found their route in plant materials/seeds of economically important crops. A few examples of invasive alien weed species which have a significant impact on agri-ecosystem and also on human health are: *Ageratum conyzoides, Argemone mexicana, Chromolaena odorata, Cuscuta campestris, Eichhornia crassipes, Ipomoea fistulosa, Lantana camara, Mikania micrantha, Mimosa spp., Parthenium hysterophorus* and *Salvinia molesta.* Threats posed by these alien weeds to the agriculture and biodiversity are being discussed and management strategies are highlighted below:

Ageratum conyzoides L.

Ageratum conyzoides, an alien, obnoxious weed is a native of tropical America. It outcompetes the native associated vegetation for nutrients, moisture and light thereby suppressing them. The invasion by the species is leading to shrinkage of grazing area for animals and reduction in the productivity of grasslands by 90 per cent. The growth of newly planted trees in forests interferes with successions of natural forests besides threatening the biodiversity. Though the weed provides shelter to the wild animals, farmers working in fields infested with the species suffer from nausea, giddiness, headache and skin as well as eye irritation. Similarly, cases of poisoning in farm animals have also been reported. In crops like maize, it interferes in the harvesting and establishment of crops in rotation.

Management: Mechanical and manual weeding is limited due to its allergic effects and strong foul smell which often results in giddiness, vomitting, fever and headache. Puddled rice-wheat rotation brings about lower weed infestation than maize-wheat system. Seeding of *Cassia tora* prior to the onset of monsoons has been found to be effective in minimizing this weed in wastelands. Alachlor at 1.5 kg/ha applied at the pre-emergence stage coupled with hand weeding has been found to be very effective in controlling its growth in the field of maize. Application of atrazine at 1.5 kg/ha (PE) once during May or October in pasturelands or glyphosate at 1.5 kg/ha in May-June and/or September-October prior to flowering also helps in controlling the weed.

Cuscuta campestris Yuncker

Cuscuta campestris, commonly known as "dodder" is an invasive obnoxious parasite weed that attaches itself to the stems and leaves of a wide variety of host plant species. It draws nutrients from the host plant for sustenance and provides a dense barrier that drastically retards growth and vigour of the host-plant and reduces grain yield by 35-100 per cent. It parasitizes legumes, clovers, niger, linseed, onion, sugarbeet and many other crops. The yield reductions due to *Cuscuta* are reported to the tune of 70% in chillies, 31-34% in green gram and black gram, 60-65% in niger and 60-70 per cent in alfalfa. *Cuscuta* infestation is also known to bring down the quality besides enhancing the cost of harvesting in agricultural crops.

Management: *Cuscuta* falls under the category of those few weeds that can be controlled completely by rotating host crop with non-host crop. The shade from dense crop foliage suppresses *Cuscuta* effectively to control it almost completely. Several soil and foliage applied herbicides such as, 2,4-D, glyphosate, trifluralin, pendimethalin, ethofumesate, etc, are known to achieve moderate to good control of *Cuscuta* most of which are either not available to the farmers or can not be used selectively in rice-fallow cropping system. Extreme sensitivity of crop seedlings to foliage applied herbicides prohibits their use as post-emergence treatment in host crops.

Lantana camara L.

Lantana camara is an invasive weed affecting pastures and native forest in over 60 countries worldwide. A native of tropical America, it was introduced into Netherlands by the Dutch explorers from Brazil in the late 1600s, before its spread to other countries as an ornamental. *Lantana* invades both the natural forest and agricultural ecosystems in disturbed native forests. It can even become the most dominant under-storey species, disrupting succession and decreasing biodiversity. The taxon is widely reported to be poisonous to cattle, buffalo, sheep, goats and horses as has been reported from numerous states like Himachal Pradesh, Maharashtra Uttarakhand and Uttar Pradesh in India.

Management: *Lantana* cannot withstand frequent soil disturbances. This is the major reason because of which our cultivated areas under crops and orchards have so far been free from this weed. A three phase integrated strategy developed for its management can be practiced, as outlined below:

Phase I: Cut the *Lantana* bushes up to 5-8cm above the ground level between August and October months.

Phase II: Spray glyphosate (0.75-1%) solution on regenerated foliage during the months of September to November.

Phase III: Plant competitive species such as guinea grass (*Panicum maximum*) or *Setaria siratro* NB-37 or any other fast growing fodder trees like *Robinia* or poplar can be grown as per the land suitability.

Parthenium hysterophorus L.

Parthenium hysterophorus is currently considered as one of the seven most dreaded weeds of the world. It is suspected to have sneaked into India alongwith wheat imported from the USA under PL-480 scheme sometime in early fifties. It has high reproductive capacity, with light weight seeds which disseminate fast by wind, rain and traffic activity. Lack of seed dormancy also enables the weed to complete 3-4 life cycles in a year. Chemicals like parthenin, hysterin, hymenin and abborsin secreted by this weed exert strong allelopathic effects on different crops grown in association manifested by poor germination and crop growth. The weed also affects nodulation in legumes due to inhibition of the activity of nitrogen-fixing bacteria. Direct contact with plant causes contact dermatitis in human-beings. Even the presence of pollen in the air is allergic to some and may result in diseases like fever and respiratory ailments including asthma.

Management: Post-emergence application (before flowering) of glyphosate (0.5%), 2, 4-D, (0.5%) or metribuzin (0.3%-0.5%) provides effective control of *Parthenium*. The Biological Control of *Parthenium* using *Zygogramma bicolorata,* a Mexican beetle has been found to be very promising. Growing competitive crops (fodder *Sorghum*, sunflower and maize) which are self-perpetuating and species like *Abutilon indicum, Cassia sericea, C. tora,* and *Tagetus erecta* in non-crop areas are also recommended for the control of the weed.

Eichhornia crassipes Solms.

Water hyacinth (*Eichhornia crassipes* Solms-Laubach) originating from South America, is ranked among the top ten worst weeds world-wide. It is one of the most successful colonizers in the plant kingdom with beautiful lavender blue flowers which produces seed prolifically, an important factor contributing to its distribution. In India, the plant was first introduced in Bengal, most probably in early 1780s. The explosive growth rate of this weed is due to a large extent, because of eutrophicatiion in water bodies. In addition, the absence of natural enemies of the plant contributes to the rapid colonization of the weed. Water hyacinth is a big menace severely affecting navigation, fishing, recreational use of aquatic bodies and hydro-electric generation. It also plays a major role in human health hazards by harbouring harmful insects and vectors of diseases like malaria, encephalitis, filariasis, etc. and by providing a safe hideout for wild animals like crocodiles and snakes. By increasing the evapo-transpiration losses of water, the weed is responsible for drying up of large quantities of water from tanks and ponds which are the mainstay of agriculture in the rain-fed areas.

Management: Manual removal of water-hyacinth using mechanical mowers, dredgers or manual extraction methods, is probably the most widely practiced method of control in the developing countries. Several herbicides have been found promising in controlling the weed. A single application of 2, 4-D at 4 kg/ ha is found to give better control than two split applications of 2 kg each. However, control with glyphosate at 2.20 kg/ha has been found to be more effective with least effect on water quality and fish mortality. Three exotic bio-control agents, two coleopteran weevils viz. *Neochetina bruchi* and *N. eichhorniae* and one mite viz. *Orthogalumna terebrantis* have been introduced in India in 1982. About 3,000 weevil would be required in one hectare for control of this weed.

Ipomoea fistulosa Mart. ex Choisy

Ipomoea fistulosa, a native of South America has spread fast in India during the past three decades and has become an obnoxious weed in diverse ecosystems. It is a highly invasive plant, which does not allow the establishment of other vegetation, posing a great threat to biodiversity.

Management: Because of the semi-woody and deep-rooted nature of this weed, manual removal is not practicable. Cutting this weed using knife/ sickle leads to re-growth and hence is not effective. A foliar application of 2, 4-D Na salt (0.2%), soap oil (1 ml/l of water) is quite effective in drying up the plants.

Mikania micrantha Kunth.

Mikania micrantha H.B.K. (Asteraceae), a perennial fast-growing weed of neotropical origin is a major menace in natural forests, forest plantations and agricultural systems in northeast and southwest India. The apparent negative impact of the weed includes reduction in yield of subsistence and cash crops, loss of native biodiversity and prevention of forest regeneration. Among commercial crops, tea, coffee, pineapple, banana and ginger are under its serious threat.

Management: Of the herbicides tested for the control of *Mikania*, triclopyr @ 500 ml/ha, glyphosate @205-5 1ml/ha and diuron @ 1-1.5 kg/ha gave good control if applied before flowering/seed setting stage of the weed (August-September). However, the treatment lasts only for 9-12 months since their application is carried out during monsoon season which gets easily washed away. Therefore, integrated approach involving mechanical, chemical and biological methods is required to manage the control of this weed.

CONCLUSION

The threats posed by the invasive alien weed species are enormous in dimensions and intensity. Measures and guiding principles for dealing with invasive alien weeds are broadly known including the prevention of their entry, early detection, control and management practices, restoration effort, research support, information management and public awareness. However, their fruitful and effective integration is required to address the emerging threat posed by invasive alien species to the agriculture and native biodiversity in India.

REFERENCES

Chatterjee, D (1940). Studies on the endemic flora of India and Burma. *J Asiat Soc Bengal.* **5:** 19-67

Hajra, PK & Mudgal, V (1997). Plant diversity hotspots in India – *An overview. BSI,* India

Mooney, HA & Hobbs, RJ (eds.). (2000). Invasive Species in a Changing World. *Island Press,* Washington, DC

Nayar, MP (1977). Changing patterns of the Indian Flora. *Bull. Bot. Surv. India* **19:** 145-155

Nayar, MP (1989). *In-situ* conservation of Wild flora resources. *Bull. Bot. Surv. India* **29:** 319-333

Oudhia, P, Kolhe, SS & Tripathi, RS (1997). Allelopathic effects of white top (*Parthenium hysterophorus* L.) on chickpea. *Legume Research* **20:** 327-329

Sharma, OP, Makkar, HP, Dawra, RK & Negi, SS (1981). A review of the toxicity of *Lantana camara* in animals. *Clin. Toxico.* **18:** 1077-1094

Singh, RP & Singh, RK (2009). Invasive Alien Weeds: A threat to agriculture and Biodiversity. National Conference on Invasive Alien Species, UP State Biodiversity Board, Lucknow, pp. 118-121

Reddy, S, Bagyanarayana, G, Reddy, KN, & Vatsavaya, S Raju (2008). Invasive Alien Flora of India, National Biological Information Infrastructure, USGS, USA

16

Pollen Diversity of Higher Vascular Plants Growing in and Around Firozabad District of Uttar Pradesh: A Taxonomic View

SHALINI PALIWAL, SP PALIWAL AND ANUBHA JOSHI[1]

Department of Botany, Narain P.G. College, Shikohabad–205 135, Firozabad, U.P., India
[1]Department of Botany, Guest Faculty, Sri Venkateswara College, University of Delhi, New Delhi–110 021

ABSTRACT

An ecofloristic survey in the Firozabad district was carried out between July, 2007 and June, 2009 to prepare a Pollen Calendar of different angiospermous plants with their local names, habit, flowering period, mode of pollination, prevalence and nature. All these data are essential pre-requisites for aerobiological investigations. During the survey, pollen from flowering plants were collected in different seasons of the year to find out the morphology and exine sculpture which may prove helpful in identification of the diverse vegetation of the area. The survey work in Firozabad district revealed a total of 291 angiospermous species under 222 genera and 73 families. Out of the total families 66 belong to dicots and 7 to monocots. One Gymnospermous family Cupressaceae was represented by an ornamentally cultivated *Platycladus orientalis* Syn. *Thuja orientalis*. On the basis of mode of pollination, 220 species were found to be entomophilous, 42 anemeophilous and 29 amphiphilous from the total 291 species studied. The vegetation cover of the district Firozabad showed hydrophytes, mesophytes and some xerophytic plants which grow in heterogeneous assemblages of both monocots and dicots. Maximum plants bloom, during spring and rainy season followed by winter and summer. Aerobiologically the pollen of species of *Achyranthes, Bauhinia, Brassica, Cassia, Euphorbia, Lindenbergia, Mangifera, Ocimum, Parthenium*, etc are abundant, *Azadirachta, Boerhaavia, Calotropis, Eclipta, Ipomoea, Melia, Nerium, Quisqualis, Ricinus, Solanum, Tabernaemontana, Tagetes, Withania*, etc. are common, *Bombax, Celosia, Crataeva, Digera, Holoptelia, Michelia, Parkinsonia, Tectona, Terminalia*, etc. are frequent and *Alangium, Alanizia, Averrhoa, Jacaranda, Manilkara, Tecoma*, etc. have been rarely observed. During the annual cycle the frequency of pollen in the air was different in different seasons. It might be due to local climatological influence on flowering of plants. Overall 118 pollen types have been recorded from the air. Of these 62.71% occur in spring, 58.47% in rainy, 36.44% in winter, and 56.77% in summer season of an annual cycle.

INTRODUCTION

Aerobiology is understood to be the study of airborne bioparticulates, their identity, behaviour, movements and survival (Boral and Bhattacharya 1999). The concentration of pollen in the atmosphere is subject to variation with time because of the reproductive cycles of plants. Variation in pollen counts also depends on the distribution of vegetation and pollen dispersal mechanisms. On the other hand, the air flow and geography of the area which cause geographical variation are responsible for pollen abundance (Latorre and Bianchi 1977, Latorre 1996c, Latorre *et al.* 2001). The flora of towns differs from that of rural or natural areas. In a town the biodiversity is reduced and plants form characteristic synantrophic communities. The content of herbaceous pollen in the atmosphere depends on the vegetational cover, climate, weather and geographical conditions (Kasprzyk 2006). The aim of the study conducted here was to compare the aerobiological data obtained from pollen monitoring stations located at sites different with respect to their flora and microclimate i.e., a town and a rural area.

Firozabad district of Uttar Pradesh is situated in the western part of the State at an altitude of nearly 163m above mean sea level and between 27°7N' and 79E' longitude. It experiences a tropical monsoon type of climate having the year divided into winter, spring, summer and rainy seasons. In summers the temperature ranges between 26 and 42°C and maximum rains are recorded in July and August. In winters the temperature ranges from 7 to 24°C. The relative humidity has been observed to be more during July and August (72%).

METHODOLOGY

In order to evaluate the aerobiological assemblage of three different sites of Firozabad district viz., Shikohabad, Firozabad and Jasrana, (Map 1) a good knowledge of its ground flora and their pollen grains were of prime importance. Accordingly, a detailed and systematic ecofloristic survey was also carried out. The frequency of occurrence is based on visual observations carried out during the survey period from July 2007 to June 2009, covering agriculture fields, scrub jungle, road sides, barren lands, etc. For aerobiological study, a modified Durham Gravitational Settling Sampler was used. It was installed at a height of 10m above the ground level on the terrace of a house. Two microscopic slides coated with stained adhesive glycerine jelly were exposed daily in starting position at 8.00 am for 24 hours. These slides were carried to the laboratory for further observations and identification of pollen types. Reference slides of pollen were also made for identification purpose.

OBSERVATIONS AND DISCUSSIONS

The district Firozabad experiences a tropical monsoon type of climate and four distinct seasons viz., summer, rainy, winter and spring conditions prevaile in the area. The ecological factors such as climate, soil and biotic factors directly influence the vegetation of the area studied. All types of vegetation comprsing hydrophytic, mesophytic and certain xerophytic plants grow in ponds, wastelands, along roadsides, canals and railway tracks in the territorial limits of the district.

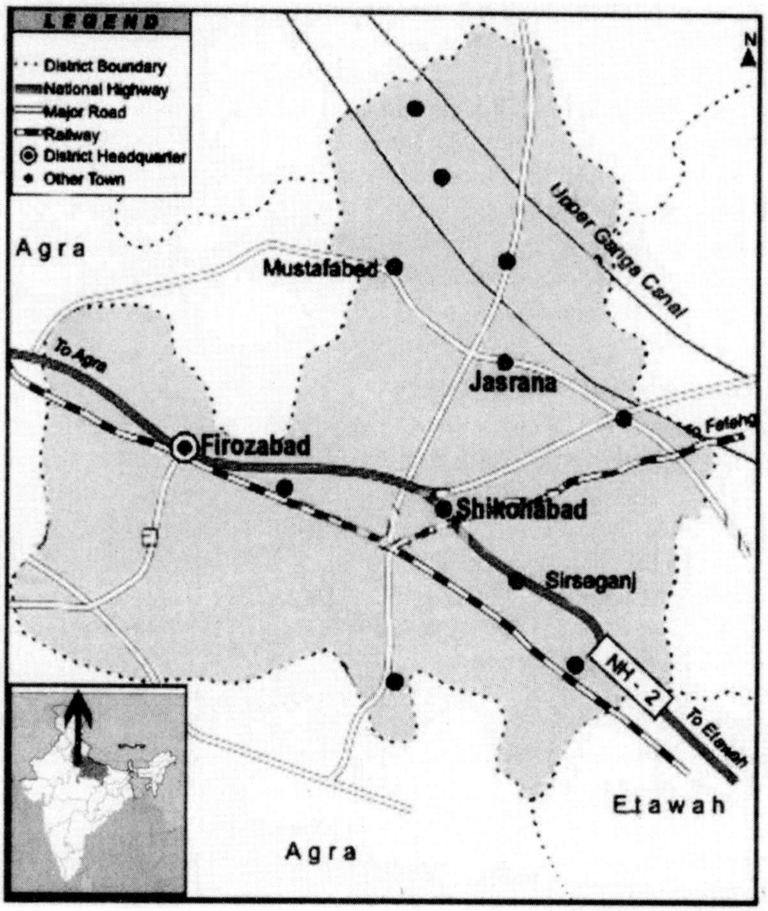

Map of district Firozabad showing different sites.

[A] Vegetation of aquatic and marshy habitats

The common habitats of the hydrophytes in district Firozabad are the canals and ponds nearby Jasrana, Shikohabad, and Sirsaganj, etc. On the basis of their contacts with soil, water and air, the hydrophytes of the area can be placed under the following categories:

(a) Free-floating hydrophytes- These are in contact with water and air only being commonly represented by *Eichhornia crassipes, Trapa natans* (Syn. *T. bispinosa*), *Wolffia* sp., etc. These occur abundantly during the months of September, October and November.

(b) Attached submerged hydrophytes- These plants are in contact with soil and water. Two common taxa are: *Hydrilla verticillata* and *Vallisneria spiralis*. The vegetative shoots of these plants are under water, but the flowers are raised to or are slightly above the surface of the water. These species grow at different depths forming tangled masses of vegetation.

(c) Attached hydrophytes with floating leaves- These hydrophytes are in contact with soil, water as well as with air. Among these are *Aponogeton natans,* and *Sagittaria guayanensis*.

(d) Emergent amphibious hydrophytes- The roots, lower part of the stems and in some instances even the lower leaves of these hydrophytes are usually submerged in water. These are represented by *Polygonum barbatum* and *Sesbania bispinosa.*

(e) Submerged hydrophytes- These are rootless submerged hydrophytes which remain in contact with water only, e.g. *Ceratophyllum demersum.*

(f) Reed swamp hydrophytes- These are attached to the soil which is usually saturated with water at least in early part of their life. They occur on soft wet mud forming a reed swamp association of *Alternanthera sessilis, Bacopa monnieri, Cyperus difformis, C. iria, Dichanthium annulatum, Eclipta alba, Fimbristylis dichotoma, Phyla nodiflora, Polygonum plebeium* and *Typha angustata*, etc.

[B] Vegetation on the wastelands

The characteristic ruderal flora occurs in open places where the soil is distributed from time to time, in fallow fields, trails and roads, waste places, in and near settlements, old garden sites and cultivated lands. All such man-made habitats are very susceptible to the invasion of weed flora which quickly occupies such areas. Some common weeds of such lands are as follows:

(a) Trails, roadsides and railway line weeds- *Adhatoda vasica, Ageratum conyzoides, Argemone mexicana, A. ochroleuca, Calotropis procera, Cassia tora, Cleome viscosa, Croton bonplandianum, Lantana indica, Solanum surattense, Xanthium strumarium, Erianttens munja, S. spontaneum* and *Tribulus terrestris*, etc.

(b) Weeds of cultivated fields- These occur in different seasonal crops, e.g. Kharif (rainy season) and Rabi (winter season). The dominant taxa are as follows:

- **(i)** The common weeds associated with the kharif (rainy season) crops are: *Borreria hispida, Celosia argentea, Cleome viscosa, Convolvulus arvensis, Corchorus aestuans, C. tridens, Crotalaria medicaginea, Cynodon dactylon, Cyperus iria, C. rotundus, Dactyloctenium aegyptium, Digera muricata, Echinochloa colonum, Eragrostis tenella, Evolvulus alsinoides, Lindernia crustacea, Oldenlandia corymbosa, Phyllanthus virosus, Polygonum glabrum, Rhynchosia minima, Sorghum halepense, Sporobolus diander, Trianthema portulacastrum* and *Tridax procumbens*.
- **(ii)** The common weeds associated with the rabi (cold season) crops are: *Alternanthera pungens, Amaranthus spinosus, Anagallis arvensis, Argemone mexicana, Asphodelus tenuifolius, Boerhaavia diffusa, Coronopus didymus, Cynodon dactylon, Eclipta alba, Euphorbia dracunculoides, E. hirta, Lathyrus aphaca, L. sativus, Launaea procumbens, Medicago denticulata, Melilotus alba, Orobanche aegyptiaca, Rumex dentatus, Solanum nigrum, Sonchus oleraceus, Stellaria media, Vernonia cinerea* and *Vicia hirsuta.*

[C] Vegetation of walls, crevices and rocky places

Several plants are found on the crevices of old semi-demolished walls of buildings. Seedlings of *Ficus benghalensis* and *F. religiosa* frequently occur on the walls of the buildings. *Aristida adscensionis, Boerhaavia diffusa, Cyperus triceps, Euphorbia hirta, Heliotropium strigosum, Kikxia ramosissima, Lindenbergia indica, Phyllanthus amarus* are found growing in the crevices and on the rocky slopes.

[D] Escape plants

A large number of plants are originally under cultivation in fields and gardens for food, vegetable and ornamental purposes found along roadsides, canals and in wastelands. These are escape plants and common among them are *Barleria cristata, Crotalaria juncea, Cucumis melo, Mimordica charantia, Portulaca grandiflora* and *Sesamum indicum*.

[E] Vegetation of saline and alkaline lands

Saline and alkaline soils are encountered at Jasrana, Makkhanpur, Dabrai and some other parts of the district. It is not suitable for agricultural purposes. The most common plants of such soils are: *Alhagi pseudalhagi, Cyperus rotundus, Heliotropium curassavicum, Phyla nodiflora, Salsola baryosma, Sporobolus marginatus, Suaeda fruticosa* and *Trianthema portulacastrum*.

[F] Roadside trees

Common roadside trees which are generally planted as avenue trees are *Albizia lebbek, Azadirachta indica, Cassia fistula, C. siamea, Dalbergia sissoo, Delonix regia, Ficus racemosa, F. religiosa, F. virens, Holoptelea integrifolia, Mangifera indica, Pongamia pinnata, Syzygium cumini, Tamarindus indica, Terminalia arjuna* and *Neolamarckia cadamba*.

[G] Climbers creepers and twiners

These are very common during the rainy and winter season. The common climbers are *Abrus precatorius, Antigonon leptopus, Cardiospermum halicacabum, Cocculus hirsutus, Cuscuta reflexa, Ipomoea nil* and *Quisqualis indica*.

[H] Seasonal vegetation

This includes the herbs and undershrubs of rainy, winter and summer seasons growing in different habitats of the district.

(a) Rainy season annuals- Commencement of the rainy season in July revives the germination of seeds and fruit-stocks which have been lying dormant in the soil. Most of the species of this season flower and fruit in September when sky gets clear and provides long hours of sunshine. The common monsoon species are *Alysicarpus vaginalis, Aristida adscensionis, Borreria hispida, Bulbostylis barbata, Cassia occidentalis, C. pumila, C. tora, Celosia argentea, Cenchrus ciliaris, Cleome viscosa, Commelina benghalensis, C. forskohlii, Corchorus capsularis, Crotalaria medicaginea, Cynodon dactylon, Cyperus rotundus, Dactyloctenium aegyptium, Desmodium triflorum, Eragrostis pilosa, Euphorbia hirta, Evolvulus alsinoides, Heliotropium strigosum, Indigofera linifolia, Polygala erioptera, Sida cordifolia, Tephrosia purpurea, Trianthema portulacastrum, Triumfetta rhomboidea* and *Zornia diphylla*.

(b) Winter season annuals- The winter season vegetation starts to appear from the month of October since the temperature almost which initiates the germination of winter plants. These produce flowers and fruits at the middle or towards the end of the winter. Common winter annuals encountered in the area are: *Anagallis arvensis, Argemone mexicana, A. ochroleuca, Asphodelus tenuifolius, Bergia ammanioides, Coronopus didymus, Fumaria officinalis, Gnaphalium luteo-album, Lathyrus odorata, L. sativus, Lindenbergia indica, Medicago* *denticulata, Melilotus indica, Oxalis corniculata, Potentilla supine, Rumex dentatus, Sonchus oleraceus, Spergula arvensis, Stellaria media, Vernonia cinerea, Vicia hirsuta*, and *V. sativa.*

(c) Summer season annuals- Seeds of several annual herbs start germinating in the month of March. The temperature gradually rises in March and continues till the end of June. These months are dry and plants show various xerophytic characteristics such as thorns, woolly tomentum or stellate hairs, thick cuticle, etc. A few summer annuals are enumerated below: *Alhagi pseudalhagi, Alternanthera pungens, Boerhaavia diffusa, Citrullus lanatus, Dichanthium annulatum, Echinops echinatus, Glinus lotoides, Gomphrena globosa*, and *Heliotropium eichwaldi.*

[I] Cultivated plant communities

The cultivation of crops is determined by climate, soil, and water facilities. The agricultural year of the area is divided into two crop seasons i.e., Kharif (rainy season) and Rabi (winter season) crops.

(i) Kharif (Rainy season) crops- The pure or mixed crops with several edible legumes are grown in the district. The common ones are as follows–*Abelmoschus esculentus* (Bhindi), *Cajanus cajan* (Arhar), *Capsicum annuum* (Mirch), *Crotalaria juncea* (Sanai), *Dolichos lablab* (Sem), *Oryza sativa* (Dhan), *Pennisetum typhoides* (Bajra), *Vigna mungo* (Urd), *V. radiata* (Mung), *Sesamum indicum* (Til), *Sorghum vulgare* (Jowar) and *Zea mays* (Makka).

(ii) Rabi (Cold season) crops- The important field crops of this season are- *Avena sativa* (Jai), *Brassica campestris* (Peeli sarson), *Cicer arietinum* (Chana), *Hordeum vulgare* (Jau), *Linum usitatissimum* (Alsi), *Pisum sativum* (Matar), *Triticum aestivum* (Gehun), *Saccharum officinarum* (*Eekh* or *Ganna*), etc.

(iii) Vegetable crops- A number of vegetable crops are sown in December, January and February. Some of them are *Allium cepa* (Piyaz), *A. sativus* (Lehsun), *Anethum graveolens* (Sowa), *Brassica oleracea* var. *botrytis* (Phoolgobhi), *B. oleracea* var. *capitata* (Bandgobhi), *B. rapa* (Shaljam), *Coriandrum sativum* (Dhaniya), *Daucus carota* (Gajar), *Raphanus sativus* (Muli), *Solanum melongena* (Baigan), *S. tuberosum* (Alu), and *Spinacia oleracea* (Palak).

During the summer months, i.e. April, May and June, principal cucurbits are grown. The common ones are- *Citrullus* sp. (Tarbooz, tinda), *Cucumis* sp. (Kharbooz, kakri, khira), *Cucurbita maxima* (Sitaphal), *C. pepo* (Kaddu), *Lagenaria siceraria* (Lauki), *Luffa acutangula* (Kali torai), *L. cylindrica* (Ghia torai), *Momordica charantia* (Karela), *Trichosanthes anguina* (Chachinda), and *T. dioica* (Parwal).

[J] Fruit plantation and their wild flora

Most of the fruit orchards producing summer and rainy season fruits are owned by farmers and some of them are present along the roadsides and village sides. The common fruits grown in the district are: *Annona squamosa* (Sharifa), *Carica papaya* (Papita), *Citrus aurantifolia* (Kagazi nimbu), *C. limon* (Bara nimbu), *Cordia dichotoma* (Lasoora), *Ficus racemosa* (Gular), *Mangifera indica* (Aam), *Manilkara hexandra* (Khirni), *Musa paradisiaca* (Kela), *Psidium guajava* (Amrood), *Punica granatum* (Anar), *Tamarindus indica* (Imli), and *Ziziphus mauritiana* (Ber).

This survey work was undertaken to identify the pollen types found in the atmosphere of Firozabad district alongwith the preparation of a Pollen Calendar of plants in this locality. Pollen calendars are useful for the clinical management of respiratory disorders caused by airborne pollen and fungal spores in order to identify the taxa and season of acute sensitivity and immunological responses (Bianchi and Olabuenaga 2006).

In the Pollen Calendar, we have furnished botanical and local names, flowering period, habit, pollination period, mode of pollination, prevalence and nature of different species which have been observed to disseminate large quantities of pollen in the atmosphere (Table 1).

Table 1: Pollen Calendar of plant species of various families observed with their flowering period, mode of pollination, prevalence and nature from July 2007 to June 2009 in the district Firozabad, Uttar Pradesh

Taxa	Local Names	Habit	Flowering Periods	Mp	P	N
ANGIOSPERMS						
DICOTYLEDONS						
DIVISION-POLYPETALAE						
AIZOACEAE						
Trianthema portulacastrum L.	Santhi, Biskhapara	Herb	Jul-Oct	Am	A	We
ANACARDIACEAE						
Mangifera indica L.	Aam	Tree	Feb-Jul	En	A	Cu,Wi
ANNONACEAE						
Polyalthia longifolia (Sonner.) Thw.	Ashok, Ashapala	Tree	May-Jun	An	Co	O
APIACEAE						
Anethum graveolens L.	Sowa	Herb	Feb-Apr	En	Co	Cu
Coriandrum sativum L.	Dhaniya	Herb	Nov-May	En	Co	Cu
Daucus carota L.	Gajar	Herb	Jan-Mar	En	R	Cu
Foeniculum vulgare Mill.	Saunf	Herb	Feb-Mar	En	R	Cu
BALSAMINACEAE						
Impatiens balsamina L.	Gulmehndi	Herb	Aug-Nov	En	R	O
BOMBACACEAE						
Bombax malabaricum DC.	Semal	Tree	Feb-Apr	En	F	O
BRASSICACEAE						
Brassica campestris L. Prain	Peeli sarson	Herb	Jan-Feb	En	A	Cu
B. juncea (L.) Czern and Coss	Rai	Herb	Dec-Mar	En	A	Cu
B. nigra (L.) Koch.	Kali rai	Herb	Nov-Jan	En	A	Cu
B. oleracea L. var. *botrytis* L.	Phoolgobhi	Herb	Nov-Apr	En	A	Cu
B. oleracea L. var. *capitata* L.	Bandgobhi	Herb	Sep-Nov	En	A	Cu
B. rapa L.	Shaljam	Herb	Dec-Mar	En	A	Cu
Iberis amara L.	Candytufft	Herb	Feb-May	Am	Co	O
Raphanus sativus L.	Muli	Herb	Dec-Jun	Am	Co	Cu
CACTACEAE						
Opuntia dillenii Haw.	Nagphani	Herb	Jan-Sep	Am	R	We
CAESALPINIACEAE						
Bauhinia acuminata L.	Safed Kachnar	Shrub	Mar-May	En	A	O
B. purpurea L.	Kachnar	Tree	Nov-Apr	En	F	O

Contd.

Table 1: Contd.

Taxa	Local Names	Habit	Flowering Periods	Mp	P	N
B. racemosa Lamk.	Kachnal	Tree	Mar-Apr	En	F	O
B. tomentosa L.	Kachnar	Shrub	Aug-Oct	En	Co	O
B. variegata L.	Kachnar	Tree	Feb-Apr	En	F	Cu
Caesalpinia crista L.	Uru, Karanju	Shrub (Cl)	Aug-Jan	En	F	O
Cassia fistula L.	Amaltas	Tree	Apr-Sep	En	Co	Cu,O
C. obtusifolia L.	Pawad	Herb	Sep-Feb	En	Co	We
C. occidentalis L.	Bara pawar	Shrub	Jul-Oct	En	Co	We
C. siamea Lamk.	Kassod	Tree	May-Aug	En	F	O
C. sophera L.	Kasaundi	Herb	Jul-Feb	En	Co	O
C. tora L.	Pawaar	Herb	Aug-Nov	En	A	We
Casuarina equisetifolia J. R. and G. Forst.	Vilayati jhau	Tree	Mar-May	An	R	O
Delonix regia (Boj.) Raf.	Gulmohar	Tree	Apr-Dec	En	Co	O
Parkinsonia aculeata L.	Vilayati Kikar	Tree	Oct-Mar	En	F	O
Saraca asoca (Roxb.) de Wilde	Ashok, Sita ashok	Tree	Mar-Apr	En	Co	O
Tamarindus indica L.	Imli	Tree	Jul-Nov	En	R	Cu,O
CAPPARACEAE						
Capparis decidua (Forsk.) Edgew.	Kareel	Shrub	Mar-Jul	En	R	Wi
Cleome viscosa L.	Hulhul	Herb	Jul-Sep	En	Co	We
Crataeva religiosa auct. (Non Forst.)	Barna	Tree	Apr-Jun	Am	F	Cu,O
CARICACEAE						
Carica papaya L.	Papita	Tree	Jan-Dec	En	Co	Wi,Cu
CARYOPHYLLACEAE						
Dianthus caryophyllus L.	Carnation	Herb	Dec- Feb	An	R	O
Spergula arvensis L.	Muchmuchia	Herb	Dec-Apr	An	Co	We
COMBRETACEAE						
Quisqualis indica L.	Malti	Shrub (Cl)	Jan-Dec	En	Co	O
Terminalia arjuna (Roxb.) Wt. and Arn.	Arjun	Tree	May-Aug	En	F	Cu
CUCURBITACEAE						
Benincasa hispida (Thunb.) Cogn.	Petha	Herb (Cl)	Jul-Oct	En	Co	Cu
Citrullus lanatus (Thunb.) Mansf.	Tarbooz	Herb (Cl)	Apr-Jun	En	R	Cu
C. fistulosus Stocks	Tinda	Herb (Cl)	Jul-Oct	En	R	Cu
Coccinia grandis (L.) Voigt	Kundru	Herb (Cl)	Jun-Nov	En	A	Cu,Wi
Cucumis melo L.	Kharbuza	Herb (Cr)	Jul-Sep	En	Co	Cu
C. momordica Duthie and Fuller	Phunt	Herb (Cr)	Jun-Sep	En	Co	Cu
C. sativus L.	Khira	Herb (Cl)	Apr-Jul	En	F	Cu
C. utilissimus Roxb.	Kakri	Herb (Cr)	Apr-Sep	En	Co	Cu
Cucurbita maxima Duch.	Sitaphal	Herb (Cr)	Oct-Nov Mar-Jun	En	R	Cu,Wi
C. pepo L.	Kaddu	Herb (Cr)	Jan-Dec	En	R	Cu
Lagenaria siceraria (Molina) Standley	Lauki	Herb (Cl)	Jul-Oct	En	Co	Cu
Luffa acutangula (L.) Roxb.	Torai	Herb (Cl)	Jul-Sep	En	Co	Cu
L. aegyptiaca Mill.	Ghiatorai	Herb (Cl)	Jul-Sep	En	Co	Cu
Momordica charantia L.	Karela	Herb (Cl)	Jul-Oct	En	F	Cu
Trichosanthes anguina L.	Chachinda	Herb (Cl)	Jul-Oct	En	R	Wi
T. dioica Roxb.	Parwal	Herb (Cl)	Jul-Oct	En	Co	Cu

Contd.

Table 1: Contd.

Taxa	Local Names	Habit	Flowering Periods	Mp	P	N
FABACEAE						
Arachis hypogaea L.	Moongphali	Herb	Nov-Jan	En	R	Cu
Butea monosperma (Lamk.) Taub.	Tesu, Dhak	Tree	Mar-May	En	R	Wi
Cajanus cajan (L.) Millsp.	Arhar	Shrub	Nov-Dec	En	R	Cu
Cicer arietinum L.	Chana	Herb	Nov-Feb	En	R	Cu
Crotalaria juncea L.	Sanai	Shrub	Sep-Nov	En	R	Cu
C. medicaginea Lamk.	Gulali, Jhojhru	Herb	Aug-Oct	En	F	We
Dalbergia sissoo Roxb.	Shisham	Tree	Mar-Jul	En	Co	Cu
Erythrina variegata L.	Raktamadar	Tree	Mar-May	En	R	O
Indigofera linifolia Retz.	Neel	Herb (Cr)	Apr-Sep	En	Co	We
Lablab purpurea (L.) Sweet	Sem	Shrub (Cl)	Oct-Dec	En	Co	Cu
Lathyrus odoratus L.	Jangli matar	Herb (Cl)	Dec-May	En	Co	We
Lens esculenta Moench	Masur	Herb	Nov-Feb	En	R	Cu
Melilotus indica (L.) All.	Banmethi	Herb	Dec-Apr	Am	Co	Cu,We
Pisum sativum L.	Matar	Herb (Cl)	Jan-Mar	En	R	Cu
Pongamia pinnata (L.) Pierre	Papri, Karanj	Tree	Mar-Jul	En	R	Cu,O
Tephrosia purpurea (L.) Pers.	Sarphonka	Herb	Apr-Sep	En	Co	Wi
Trifolium alexandrium L.	Berseem	Herb	Nov-Feb	En	Co	Cu
Trigonella foenum-graecum L.	Methi	Herb	Jan-Mar	Am	Co	Cu,Wi
Vigna mungo (L.) Hepper	Urd	Herb	Jul-Nov	En	F	Cu
V. radiata (L.) Welczek	Moong	Herb	Jul-Nov	En	F	Cu
Vicia faba L.	Bakla	Herb	Dec-Mar	En	A	Cu,We
FUMARICACEAE						
Fumaria officinalis L.	Gajari	Herb	Nov-Feb	En	Co	We
HYDROCOTYLACEAE						
Centella asiatica (L.) Urban	Mandukparni	Herb (Cr)	Feb-May	En	Co	Wi
LINACEAE						
Linum usitatissimum L.	Alsi	Shrub	Jan-Mar	En	R	Cu
LYTHRACEAE						
Lagerstroemia speciosa (L.) Pers.	Jarul	Tree	Apr-Oct	En	R	O
Lawsonia inermis L.	Mehndi	Shrub	Jul-Oct	En	F	O
MAGNOLIACEAE						
Michelia champaca L.	Champa	Tree	Apr-May	En	F	Cu
MALVACEAE						
Abelmoschus esculentus (L.) Moench	Bhindi	Herb	Jun-Oct	En	Co	Cu
Abutilon indicum (L.) Sweet	Atibala, Kanghi	Shrub	Aug-Mar	En	Co	Wi
Althaea rosea (L.) Cav.	Gulkhera	Herb	Mar-Jun	En	R	O
Gossypium herbaceum L.	Bari	Shrub	Oct-Feb	En	R	Cu
Hibiscus cannabinus L.	Patsan	Shrub	Sep-Nov	En	R	Cu
H. rosa-sinensis L.	Gulhar	Shrub	Jan-Dec	En	Co	O
Malva parviflora L.	Panirak	Herb	Dec-Jan	En	Co	We
M. sylvestris L.	Gulkhair	Herb	Nov-Feb	En	R	We
Malvastrum coromandelianum (L.) Garcke	—	Herb	Aug-Apr	En	Co	We
Sida cordifolia L.	Kharenti	Herb	Sep-Nov	En	Co	Wi
S. rhombifolia L.	Mahabala	Shrub	Sep-Oct	En	Co	Wi
Thespesia populnea (L.)						

Contd.

Table 1: Contd.

Taxa	Local Names	Habit	Flowering Periods	Mp	P	N
Soland. ex Corr.	Paraspipal	Tree	Mar-Jun	En	Co	O
MELIACEAE						
Azadirachta indica A. Juss.	Neem	Tree	Mar-May	Am	Co	Cu
Melia azedarach L.	Bakain	Tree	Mar-Jun	En	R	Cu
MENISPERMACEAE						
Tinospora cordifolia (Willd.) Miers ex. Hook. f. and Thoms.	Giloe, Gudich	Shrub (Cl)	Mar-May	En	Co	O, Cu
MIMOSACEAE						
Acacia auriculiformis A. Cunn.	Babul	Tree	October	En	Co	Cu
A. nilotica (L.) Willd. ex Del subsp	Desi Babul	Tree	Jul-Feb	En	Co	O, Wi
Albizzia lebbeck (L.) Benth.	Siras	Tree	Apr-Oct	An	R	O, Cu
A. procera (Roxb.) Benth.	Safed Siras	Tree	Jul-Aug	An	R	Cu
Leucaena leucocephala (Lamk.) Wit	Subabul	Shrub	Jul-Oct	Am	F	Cu
Mimosa pudica L.	Lajwanti	Shrub	Aug-Nov	An	F	Cu, Wi
Pithecellobium dulce (Roxb.) Benth.	Jangaljalebi	Tree	Mar-Aug	Am	F	Cu
Prosopis chilensis Stuntz	Kabuli Kikar	Tree	Feb-Sep	An	A	Wi
MORINGACEAE						
Moringa oleifera Lamk.	Sahjan	Tree	Mar-Jul	En	F	Cu
MYRTACEAE						
Callistemon lanceolatus DC.	Bottle brush	Tree	Oct-Feb	Am	A	C,
Eucalyptus hybrida Labill.	Safeda	Tree	Nov, Feb-Mar	Am	A	Cu, O
Psidium guajava L.	Amrood	Tree	Jul-Sep	An	A	Wi, Cu
Syzygium cumini (L.) Skeels	Jamun	Tree	Apr-Jul	En	F	Wi, Cu
NYMPHYACEAE						
Nymphaea nouchali Burm. f.	Chhota kamal	Herb (Aq)	Jul-Nov	An	Co	O
OXALIDACEAE						
Oxalis corniculata L.	Khat-mitthi	Herb (Cr)	Dec-Apr	En	Co	We
O. martiana Zucc.	—	Herb	Dec-Apr	En	Co	We
Averrhoea carambola L.	Kamrakh	Tree	Aug-Dec	En	R	Cu
PAPAVERACEAE						
Argemone mexicana L.	Satyanashi	Herb	Oct-May	En	Co	We
A. ochroleuca Sweet. Brit.	Kateli	Herb	Dec-May	En	F	We
Papaver somniferum L.	Afim, post	Herb	Jan-Apr	En	Co	O, Wi
PORTULACACEAE						
Portulaca oleracea L.	Kulfa	Herb	May-Jun	En	Co	Cu, We
PUNICACEAE						
Punica granatum L.	Anar	Shrub	Apr-Jul	En	R	Cu
RANUNCULACEAE						
Ranunculus sceleratus L.	Jaldhania	Herb	Jan-Jun	En	Co	Wi

Contd.

Table 1: Contd.

Taxa	Local Names	Habit	Flowering Periods	Mp	P	N
Thalictrum foliolosum DC.	Pinjari	Herb	October	En	R	Wi
RHAMNACEAE						
Ziziphus jujuba (L.) Lamk.	Ber	Tree	Sep-Feb	En	F	Cu,Wi
Z. nummularia (Burm. f.) Wt and Arn	Jharbery	Shrub	Jul-Dec	En	Co	Wi
ROSACEAE						
Rosa damascena Mill.	Gulab	Herb	Jan-Dec	En	Co	O,Cu
RUTACEAE						
Aegle marmelos (L.) Correa	Bel, Bael	Tree	Mar-Jul	En	Co	Cu,Wi
Bergera koenigii L. (Syn. *Murraya koenigii* (L.) Spreng.)	Meetha neem	Shrub	Apr-Jun	En	R	Cu
Citrus aurantifolia (Christm.) Swingle	Kaghazi nimbu	Shrub	Jan-Dec	En	Co	Cu
C. limon (L.) Burm. f.	Bara nimbu	Tree	Feb-Apr	En	Co	Cu
Limonia acidissima L. (Syn. *Feronia elephantum* (L.) Swingle)	Kaith	Tree	May-Sep	En	F	Wi,O
Murraya paniculata (L.) Jack	Kamini	Shrub	May-Oct	En	R	O
SIMAROUBACEAE						
Ailanthus excelsa Roxb.	Arru	Tree	Feb-Jun	En	Co	O
TAMARICACEAE						
Tamarix aphylla (L.) Karst.	Farash	Tree	Jul-Nov	An	R	Cu,Wi
T. dioica Roxb.	Jhau	Shrub	Jul-Oct	An	R	Wi
TILIACEAE						
Corchorus capsularis L.	Kharenti	Shrub	Aug-Oct	En	R	Cu,Wi
TRAPACEAE						
Trapa natans L. (Syn. *T. bispinosa* Roxb.)	Singhara	Herb (Aq)	Sep-Nov	An	R	Cu
ZYGOPHYLLACEAE						
Tribulus terrestris L.	Gokhru	Herb	Jul-Nov	En	Co	Wi
DIVISION-GAMOPETALAE						
ACANTHACEAE						
Adhatoda vasica Nees	Adusa	Shrub	Oct-Mar	En	A	Wi
Hygrophila auriculata (Schum.) Heine	Talmakhana	Herb	Oct-Dec	En	Co	Wi
Justicia diffusa Willd.	—	Herb	Aug-Nov	En	Co	We
Peristrophe bicalyculata (Retz.) Nees	Missi	Herb	Aug-Nov	En	Co	We
Rungia repens (L.) Nees	Kharmor	Herb	Oct-Feb	En	Co	W
ALANGIACEAE						
Alangium salviifolium (L.f.) Wang.	Akorha	Tree	Feb-Aug	En	R	Wi
APOCYNACEAE						
Alstonia scholaris (L.) R. Br.	Chitwan, Saptparni	Tree	Nov-Feb	Am	F	O

Contd.

Table 1: Contd.

Taxa	Local Names	Habit	Flowering Periods	Mp	P	N
Carissa carandas L.	Karaunda	Shrub	Apr-Jun	En	F	Cu
Nerium indicum Mill.	Kaner	Shrub	May-Jul	En	Co	O
Plumeria acutifolia (Poir.) Woodson	Sonchampa	Tree	Jun-Sep	En	R	O
P. rubra L.	Lalchampa	Tree	Jun-Sep	En	R	O
Tabernaemontana divaricata (L.) R. Br. ex Roem. and Schult.	Chandni	Shrub	May-Oct	En	Co	O
Thevetia peruviana (Pers.) Merril	Pili Kaner	Shrub	Jun-Nov	En	R	O
Catharanthus roseus (L.) G. Don	Sadabahar	Shrub	Jan-Dec	En	F	O,We
ASCLEPIADACEAE						
Calotropis procera (Willd.) Dryand ex W. Ait.	Akaua	Shrub	Dec-Sep	En	Co	Wi
C. gigantea Ait.	Bada aakha	Shrub	Jan-Dec	En	F	O,Wi
ASTERACEAE						
Ageratum conyzoides L.	Uchunti	Herb	Jan-Apr	Am	Co	We
Blumea lacera DC.	Kakranda	Herb	Nov-May	En	Co	We
Calendula officinalis L.	Zergul	Herb	Aug-Nov	Am	Co	O
Carthamus tinctorius L.	Kusum	Shrub	Mar-May	Am	Co	Cu
Chromolena odorata (L.) King and Robin	—	Shrub	May-Jul	Am	Co	Wi
Chrysanthemum coronarium L.	Yellow Guldaudi	Herb	Nov-Feb	En	Co	O
C. indicum L.	Guldaudi	Herb	Nov-Feb	En	Co	O,Cu
Eclipta prostrata (L.) L.	Mochkand	Herb	Jan-Dec	En	Co	Wi
Helianthus annuus L.	Surajmukhi	Herb	Apr-Jun	En	F	O
Launaea procumbens L.	Jangaligobi	Herb	Aug-Apr	En	C	Wi
Parthenium hysterophorus L.	Congress grass, Gajar ghas	Herb	Jan-Dec	An	A	We
Sonchus oleraceus L.	Pilidudhi	Herb	Dec-Feb	En	A	We
Sphaeranthus indicus L.	Mundi	Herb	May-Jul	Am	A	We
Tagetes erecta L.	Genda	Herb	Jan-Dec	En	Co	O
Tridax procumbens L.	Shavanti	Herb	Dec-Apr	An	A	Wi
Vernonia cinerea (L.) Less.	Phoolni	Herb	Sep-Mar	An	A	Wi
Xanthium strumarium L.	Chhota gokhru	Herb	Mar-May	En	A	We
BIGNONIACEAE						
Bignonia atrovirens Roth.	—	Tree	Jan-Jun	En	F	O
Jacaranda mimosifolia D. Don	Nili Gulmohor	Tree	Mar-May	En	R	O
Kigelia afriema (Lam.) Benth.	—	Tree	May-Jul	En	Co	O
Millingtonia hortensis L. f.	Akashneem, Neemchameli	Tree	Nov-Dec	En	R	O
Tabebuia spectabilis Nichols.	Basant rani	Tree	Dec-Mar	En	R	O
Tecoma stans (L.) H. B. and K.	Yellow bells	Shrub	Jul-Jan	En	R	O
Tecoma undulata G. Don	—	Shrub (Cl)	May-Oct	En	Co	O
BORAGINACEAE						
Heliotropium indicum L.	Nilkatti	Herb	Apr-Sep	En	C	We
CONVOLVULACEAE						
Convolvulus arvensis L.	Hirankhuri	Herb (Cr)	Aug-Jan	En	Co	We
C. microphyllus L.	—	Herb	Mar-Nov	En	Co	Wi

Contd.

Table 1: Contd.

Taxa	Local Names	Habit	Flowering Periods	Mp	P	N
Evolvulus alsinoides L.	Shankhpushpi	Herb	Aug-Nov	En	Co	Wi
Ipomoea aquatica Forssk.	Nari	Herb (Cl)	Sep-Dec	En	Co	Wi
I. fistulosa Mart. ex. Choisy	Besharam	Shrub	Jul-Mar	En	Co	Wi
I. nil (L.) Roth	Nilkalmi	Shrub (Cl)	Aug-Oct	En	Co	Wi
I. palmata Forssk.	Railway creeper	Shrub (Cl)	Aug-Nov	En	R	O
I. pes-tigridis L.	Ghiabati	Herb (Cl)	Aug-Nov	En	Co	O
I. purpurea (L.) Roth	Morning glory	Herb	Dec-Mar	En	Co	O
CUSCUTACEAE						
Cuscuta reflexa Roxb.	Amarbel	Herb (Cl)	Nov-Feb	En	R	W
LAMIACEAE						
Leucas aspera (Willd.) Spreng.	Chhota halkusa, Gopha	Herb	Sep-Feb	En	Co	We
Mentha piperita L.	Podina	Herb	Mar-Jun	En	F	We
Ocimum americanum L.	Kalitulsi	Herb	Oct-Feb	En	Co	O
O. basilicum L.	Bankasturi	Herb	Sep-Mar	En	A	Wi
O. tenuiflorum L.	Tulsi	Herb	Feb-May	En	Co	O
OLEACEAE						
Jasminum auriculatum Vahl	Juhi	Shrub (Cl)	Aug-Oct	En	F	O,Cu
J. grandiflorum L.	Chameli	Shrub	May-Sep	En	F	O,Cu
J. sambac (L.) Ait.	Bela, Mogra	Shrub	Apr-Jun	En	F	O
PEDALIACEAE						
Sesamum indicum L.	Til	Herb	Aug-Oct	En	R	Cu
PRIMULACEAE						
Anagallis arvensis L.	Neel	Herb	Dec-Apr	En	A	We
RUBIACEAE						
Hamelia patens Jacq.	Scarlet bush	Shrub	May-Oct	En	R	O
Ixora arborea Roxb. ex Sm.	Jilpai	Shrub	Mar-Sep	En	R	O
I. coccinea L.	Rookmini	Shrub	Apr-Aug	En	R	O
Neolamarckia cadamba (Roxb.) Bosser Syn. *Anthocephalus cadimba* (Roxb.) Miq.	Kadamb	Tree	May-Nov	En	F	O
SAPOTACEAE						
Madhuca indica Gmel.	Mahua	Tree	Mar-Apr	En	F	Wi
Manilkara hexandra (Roxb.) Dub.	Khirni	Tree	Oct-Jun	En	R	Cu,Wi
Mimusops elengi L.	Maulshari	Tree	May-Jan	En	Co	Cu
SCROPHULARIACEAE						
Lindenbergia indica (L.) Vatke	—	Herb	Jan-Apr	Am	A	Wi
Verbascum thapsus L.	Jangali Tambaku	Herb	Jan-Dec	En	F	Wi
SOLANACEAE						
Capsicum annuum L.	Mirch	Herb	Oct-Nov	En	R	Cu
Cestrum diurnum L.	Din-ka-raja	Shrub	Jul-Sep	En	F	O
C. nocturnum L.	Rat-ki-rani	Shrub	Jul-Sep	En	F	O
Datura metel L.	Dhatura	Shrub	Dec-Feb	En	Co	Wi

Contd.

Table 1: Contd.

Taxa	Local Names	Habit	Flowering Periods	Mp	P	N
Lycopersicon esculentum Mill.	Tamatar	Herb	Nov-Feb	En	Co	Cu
Solanum melongena L.	Baigan	Herb	Jul-Sep	Am	Co	Cu
S. nigrum L.	Makoi	Herb	Jan-Mar	Am	Co	Wi
S. tuberosum L.	Alu	Herb	Jan-Mar	Am	Co	Cu
S. xanthocarpum Schrad. and Wendl.	Berkateli	Herb	May-Sep	Am	Co	Wi
Withania somnifera (L.) Dunal	Asgandh	Shrub	Jan-May	En	Co	Wi
VERBENACEAE						
Clerodendrum indicum (L.) Kuntze	Bhatkateri	Shrub	Jul-Nov	En	Co	O
C. inerme (L.) Gaertn.	Lanjai	Shrub (Cl)	Jun-Dec	En	Co	O
C. splendens G. Don.	—	Shrub (Cl)	Jan-Mar	En	R	Cu
Lantana camara L.	Kurri	Shrub (Cr)	Jun-Sep	En	Co	Wi
L. indica Roxb.	Tulsidal	Shrub	Aug-Nov	En	Co	Wi
Tectona grandis L.	Sagwan, Sagon	Tree	Sep-Oct	En	F	Cu
DIVISION-MONOCHLAMYDEAE						
AMARANTHACEAE						
Achyranthes aspera L.	Latjeera, Apamarg	Herb	Jan-Dec	En	A	We
Alternanthera sessilis (L.) DC.	Kanchari	Herb	Aug-Dec	En	Co	We
Amaranthus oleracea Duthie	Chaulai	Herb	Aug-Dec	An	Co	We
A. spinosus L.	Kantewali chaulai	Herb	Aug-Jan	An	A	We
Celosia argentea L.	Cilmil, Safed murga	Herb	Oct-Jan	An	F	We
Digera muricata (L.) Mart.	Latmahuria	Herb	Aug-Dec	En	F	We
Gomphrena globosa L.	Bachelor's button	Herb	Jul-Jan	En	A	We
CANNACEAE						
Canna orientalis Rose.	Sabbajaya	Herb	Jan-Dec	En	F	O
Cannabis sativa L.	Bhang	Herb	Feb-Apr	An	A	We,Cu
CHENOPODIACEAE						
Chenopodium album L.	Bathua	Herb	Dec-Mar	An	Co	We
C. murale L.	Khartua	Herb	Dec-Mar	An	Co	We
Spinacia oleracea L.	Palak	Herb	Jun-Jul	Am	Co	Cu
EUPHORBIACEAE						
Acalypha indica L.	Kuppi	Herb	Sep-Feb	En	Co	We
Croton bonplandianum Baill.	Kala bhangra	Herb	Jul-Oct	An	Co	We
Emblica officinalis Gaertn.	Amla	Tree	Mar-May	En	Co	Cu
Euphorbia hirta L.	Lal dudhi	Herb	Jan-Dec	An	A	We
Jatropha curcas L.	Ratanjot	Shrub	Jul-Sep	En	R	O
J. gossypifolia L.	Bherenda	Shrub	Aug-Oct	En	R	We
Ricinus communis L.	Arandi	Shrub	Nov-Feb	An	Co	Wi
MORACEAE						
Artocarpus heterophyllus Lamk.	Kathal	Tree	May-Jun	Am	Co	Cu
Ficus benghalensis L.	Bargad	Tree	Mar-Jun	En	Co	Cu,Wi
F. carica L. C	Anjir	Tree	Sep-Nov	En	Co	Cu
F. glomerata Roxb.	Gular	Tree	Mar-Jun	En	Co	Cu

Contd.

Table 1: Contd.

Taxa	Local Names	Habit	Flowering Periods	Mp	P	N
F. virens Ait.	Pilkhan	Tree	Oct-Dec	En	Co	Cu
F. religiosa L.	Pipal	Tree	Mar-May	En	Co	Cu,Wi
F. rumphii Blume C	Pakar	Tree	May-Jun	En	Co	Cu
Morus alba L.	Shahtoot	Tree	Feb-May	En	Co	Cu,O
NYCTAGINACEAE						
Boerhavia diffusa (L.) Druce.	—	Herb	Aug-Dec	En	Co	Cu
B. erecta L.	Santhi, Punarnava	Herb	Jul-Dec	En	Co	Wi
Bougainvillea glabra chois.	—	Shrub (Cl)	Sep-Nov	En	Co	Cu
Bougainvillea spectabilis Willd.	Baganbelias	Shrub (Cl)	Jan-Dec	En	Co	O
Mirabilis jalapa L.	Gulabbas	Herb	Aug-Dec	En	Co	Wi
Nyctanthes arbor-tristis L.	Harsingar	Shrub	Sep-Nov	En	Co	O
POLYGONACEAE						
Antigonon leptopus Hook. and Arn	Coral creeper	Shrub (Cl)	Dec-Apr	En	R	O
Polygonum barbatum L.	Jalbahar	Herb	Sep-Jan	En	Co	Wi
ULMACEAE						
Holoptelea integrifolia (Roxb.) Planch.	Chilbil	Tree	Feb-Apr	Am	F	Cu
MONOCOTYLEDONS						
ARECACEAE						
Phoenix dactylifera L.	Khajoor	Tree	Feb-Mar	An	F	Cu,W
P. sylvestris (L.) Roxb. ex Buch. Ham	Khajeria	Tree		An	F	Cu,W
Livistona chinensis R. Br. ex Mart	Fan Palm	Tree	Dec	Am	R	O
COMMELINACEAE						
Commelina benghalensis L.	Kankaua	Herb	Jul-Nov	Am	Co	Wi
C. forskalii Vahl	Kankawwa	Herb	Dec-Mar	Am	Co	Wi
Cyanotis axillaries Roem. and Schult. f.	Soltraj	Herb	Aug-Nov	En	F	W
Rhoeo spathacea (Swartz) Stearn	—	Herb	Jan-Dec	En	R	O
CYPERACEAE						
Cyperus esculentus L.	Chichoda	Herb/Grass	Oct-Jun	An	Co	Wi
C. rotundus L.	Motha, Nagarmotha	Herb	Jul-Dec	An	A	We
LILIACEAE						
Allium cepa L.	Piyaz	Herb	Nov-Feb	En	R	Cu
A. sativum L.	Lehsun	Herb	Mar-May	En	R	Cu
Aloe barbadensis Mill.	Gwarpatha, Ghrit Kumari	Herb				
		Herb	Mar-Apr	En	R	O
Asparagus racemosus Willd.	Shatawar	Herb (Cl)	Jul-Oct	En	Co	O
Asphodelus tenuifolius Cav. An.	Piazi	Herb	Dec-Apr	En	A	We
MUSACEAE						
Musa paradisiaca L.	Kela	Tree	Apr-Jun	En	F	Cu

Contd.

Table 1: Contd.

Taxa	Local Names	Habit	Flowering Periods	Mp	P	N
POACEAE						
Avena sativa acut. (non Linn.) L.	Jai	Herb/Grass	Jan-Mar	An	F	Cu
Cynodon dactylon (L.) Pers.	Doobghas	Grass (Cr)	Jan-Dec	An	A	We
Dichanthium annulatum (Forssk.) Stapf.	Zarga, Apang	Grass	May-Dec	An	A	We
Digitaria setigera Roth apud Roem. and Schult.	Kiwai	Grass	Jul-Oct	An	Co	We
Eragrostis pilosa (Schrad.) Nees	Kiwa, Chiriya	Grass	Jul-Dec	An	Co	We
Hordeum vulgare L.	Jaun	Grass	Jan-Mar	An	F	Cu
Oryza sativa L.	Chawal, Dhan	Herb	Sep-Dec	An	Co	Cu,Wi
Pennisetum typhoides (Burm. f) Stapf and C.E. Hubb.	Bajra	Grass	Sep-Oct	An	R	Cu
Erianthms munja (Roxb.) Tesw.	Munj	Grass	Oct-Dec	An	F	Wi
Saccharum officinarum L.	Ganna	Grass	Sep-Dec	An	F	Wi
Sorghum vulgare Pers.	Jowar	Grass	Sep-Oct	An	Co	Wi
Triticum aestivum L.	Gehun	Grass	Feb-Apr	An	A	Cu
Zea mays L.	Makka	Grass	Oct-Feb	An	Co	Cu
TYPHACEAE						
Typha angustata Bory and Chaub.	Patera, Hathighas	Herb	Sep-Mar	An	R	Wi
GYMNOSPERM						
CUPRESSACEAE						
Platycladus orientalis L. Franco Syn. *Thuja orientalis* L.	Morpankhi	Shrub	Dec-Mar	An	F	O

A-Abundant, **Am**-Amphiphilous, **An**-Anemophilous, **Aq**-Aquatic, **Cl**-Climber, **Co**-Common, **Cr**-Creeper, **Cu**-Cultivated, **En**-Entomophilous, **F**-Frequent, **Mp**-Mode of Pollination, **N**-Nature, **O**-Ornamental, **P**-Prevalence, **R**-Rare, **Wi**-Wild, **We**-Weed.

On the basis of flowering and pollen types a total of 291 species belonging to 222 genera, representing 73 families have been surveyed, of these 66 belong to dicots and 7 to monocots (Plate I; Table 2; Fig. 1). The plants of varied habit and habitats, e.g. aquatic, climber, herbs, weeds, grasses, shrubs and trees form distinct communities in the particular area. Statistically the area is dominated by herbs and weeds occupying 49.83% followed by trees 24.39%, shrubs 21.31% and grasses 4.47% (Table 3 and Fig. 2). In the present survey 79.60% plants were found to be entomophilous whereas 14.43% are anemophilous and 9.47% amphiphilous (Table 4 and Fig. 3). Most of the plants flower during spring and rainy season (243) whereas least percentage of flowering was noticed in the plants of summer season (106). Preliminary aerobiological survey for two years (July, 07-June, 09) revealed that the air of Firozabad district has never been free of pollen grains. A total of 119 pollen types have been identified and recorded. The present investigations point that poaceous pollen was found to be dominant (42.92%) followed by those of number of Asteraceae (31.49%), Amaranthaceae/Chenopodiaceae (16.75%), Euphorbiaceae (4.74%) and Caesalpiniaceae (4.10%) which occur round the year (Table 5; Fig. 4). Pollen grains of some of the species have been reported to be allergic (*See* Shivpuri *et al.*, 1960a and 1960b Anand *et al.*, 1987, Singh *et al.*, 1993, Chakraborty *et al.*, 1996) which may cause dermatitis, rhinitis, bronchitis and asthma to the local people.

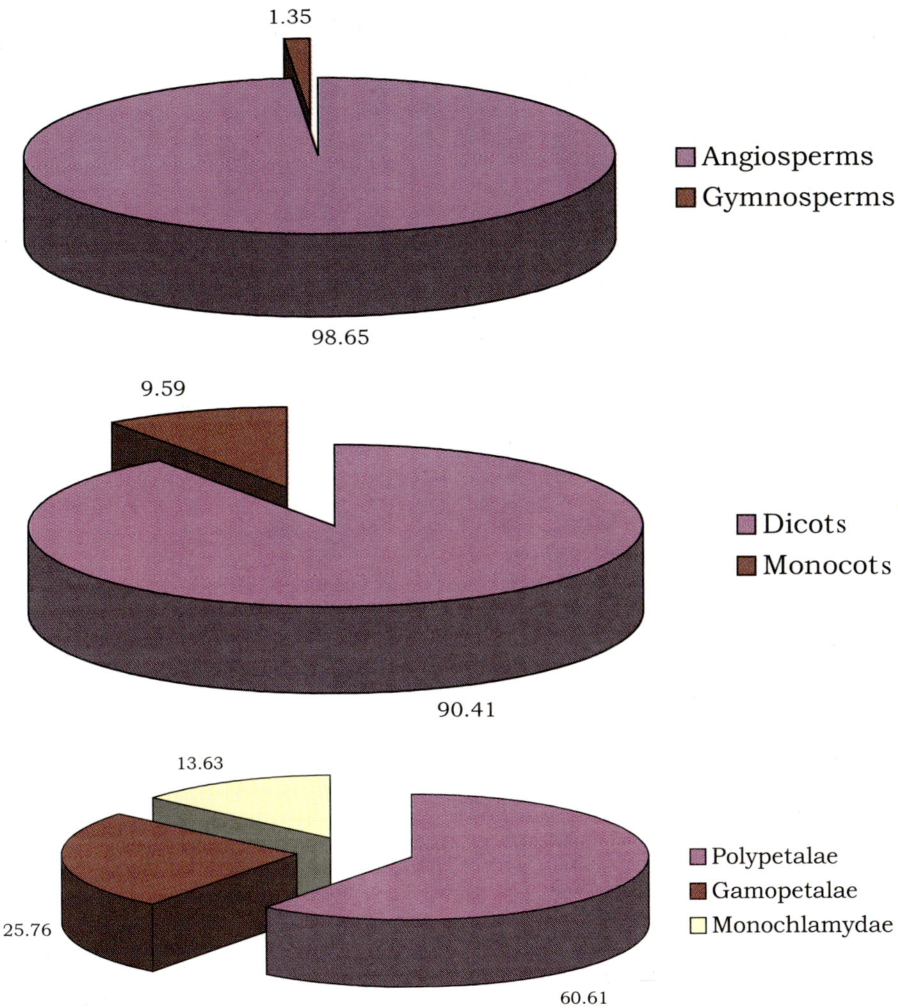

Fig. 1. Spectrum of different plant groups in the district Firozabad.

Table 2: Taxonomic plant groups observed in the district Firozabad

Groups	Families		Genera		Species	
	No.	%	No.	%	No.	%
A. Angiosperms	73	98.65	222	99.55	291	99.66
1. Dicots	66	90.41	198	89.19	262	90.03
(a) Polypetalae	40	60.61	107	54.04	140	53.44
(b) Gamopetalae	17	25.76	64	32.32	85	32.44
(c) Monochlamydae	09	13.63	27	13.64	37	14.12
2. Monocots	07	9.59	24	10.81	29	9.97
B. Gymnosperms	01	1.35	01	0.45	01	0.34

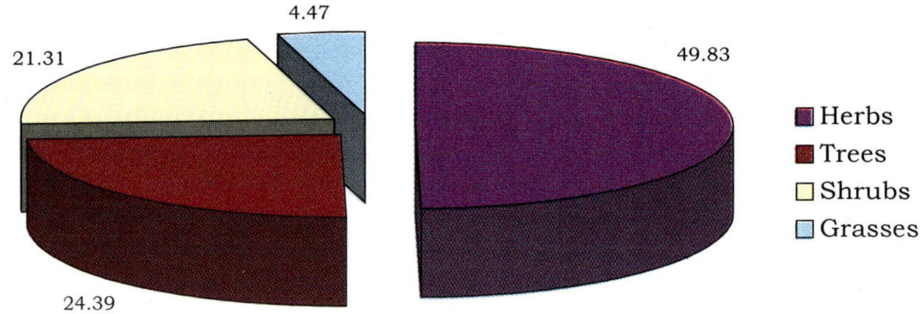

Fig. 2. Spectrum of observed angiospermous species based on their habit.

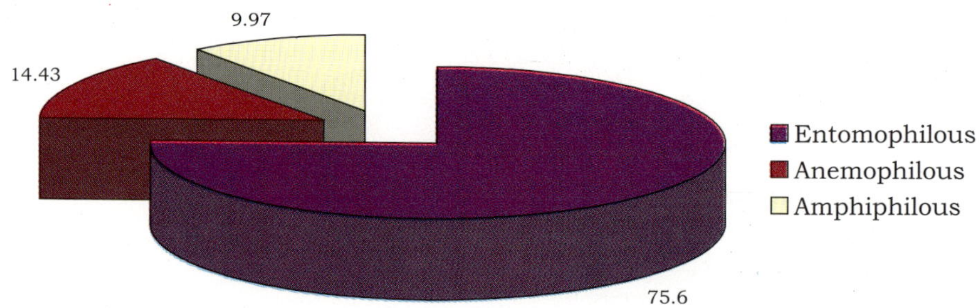

Fig. 3. Spectrum of observed angiospermous species based on their mode of pollination.

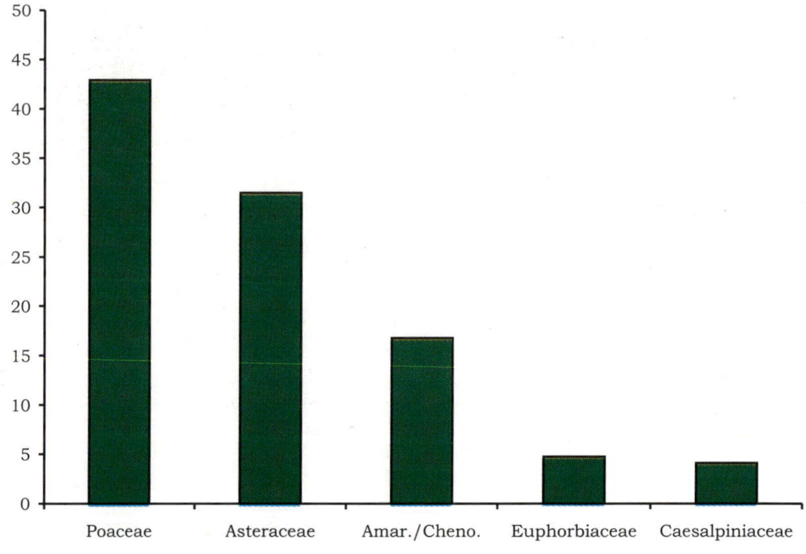

Fig. 4. Dominant families where pollen grains are represented in the atmosphere.

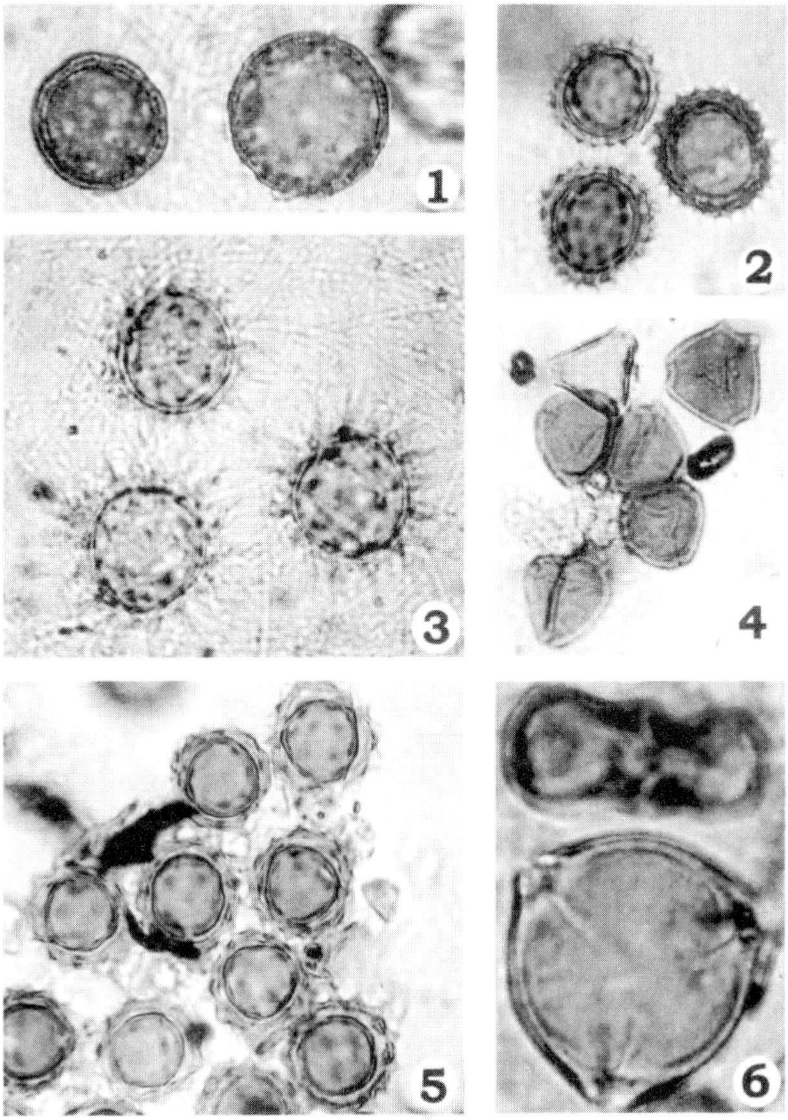

PLATE-1: Microphotographs of dominant pollens grains.

Fig. 1 Poaceaeous pollen (x750)
Fig. 2 *Eclipta prostrata* (x750)
Fig. 3 *Pennisetum typhoides* (x1000)
Fig. 4 *Cynodon dactylon* (x1000)
Fig. 5 *Tridax procumbens* (x750)
Dichanthium annulatum (x750)
Fig. 6 *Dichanthium annulatum* (x1000)
Celosia argentea (x1000)

Table 3. Analysis of observed angiospermous plant species based on their habit

Habit	No. of Species	Percentage
Grasses	13	4.47
Herbs and weeds	145	49.83
Shrubs	62	21.31
Trees	71	24.39
Total	**291**	**100.00**

Table 4. Analysis of observed angiospermous taxa based on their modes of pollination

Habit Categories	No. of Species	Anemophilous (An)	Entomophilous (En)	Amphiphilous (Am)
Grasses	13	13	—	—
Herbs and weeds	145	17	111	17
Shrubs	62	03	56	03
Trees	71	09	53	09
Total	**291**	**42**	**220**	**29**
Percentage	**100.00**	**14.43**	**75.60**	**9.97**

Table 5: Dominant families of pollen grains in the air of district Firozabad during July, 2007 to June, 2009

Dominant Components of Aerobiota	Shikohabad	Firozabad	Jasrana	Total	Percentage
A. Families					
1. Poaceae	2154	2147	2483	6784	**42.92**
2. Asteraceae	1550	1599	1828	4977	**31.49**
3. Amaranthaceae / Chenopodianceae	823	769	1056	2648	**16.75**
4. Euphorbiaceae	233	198	318	749	**4.74**
5. Caesalpiniaceae	204	197	247	648	**4.10**

ACKNOWLEDGEMENTS

S.P. Paliwal is highly thankful to the UGC, New Delhi for financial assistance.

REFERENCES

Anand, P, Agashe, SN & Manjunath, K (1981). Aerobiological studies of Bangalore city and their scope in Immunotherapy. *Proc. Nat. Conf. Env. Biol.* Aurangabad, pp. 117-124

Bianchi, MM & Olabuenaga, SE (2006). A 3-year airborne pollen and fungal spores record in San Carlos de Bariloche, Patagonia, Argentina. *Aerobiologia*, **22:** 247-257

Boral, D & Bhattacharya, K (1999). Ecofloristic survey of Behrampur town with reference to Aerobiology. *Indian J. Aerobiol,* **12:** 11-13

Chakraborty, P, Gupta-Bhattacharya, S & Chanda, S (1996). Comparative aerobiology, allergenicity and biochemistry of three palm pollen grains in Calcutta. *Aerobiologia* **12:** 47-50

Kasprzyk, I (2006). Comparative study of seasonal and intradiurnal variation of airborne herbaceous pollen in urban and rural areas. *Aerobiologia* **22:** 185-195

Latorre, F & Bianchi, MM (1977). Relación cntre la connentración arbórea de Mar del Plata (Argentina). *Pollen* **8:** 43-59

Latorre, F (1999c). El pollen atmosférico como indioator de la vegetación y de su fenologia floral. Doctoral Thesis, Universidad de Buenos Ares. pp. 244

Latorre, F, Romero, EJ & Mancini, MV (2001). Representatividad de la vegetación en el espectro de pollen atmosférico de la ciudad de mar del Plata (Argentina). *Publicación Especial Ameghiniana,* **8:** 119-124

Shivpuri, DN, Vishwanathan, R & Dua, KL (1960a). Studies in pollen allergy in Delhi area. Part I: Pollination calendar. *Ind. J. Med. Res.*, **48:** 15-20

Shivpuri, DN, Vishwanathan, R & Dua, KL (1960b). Studies in pollen allergy in Delhi area. Part II: Survey of atmospheric pollens. *Ind. J. Med. Res.*, **48:** 21-30

Singh, AB, Malik, P, Prakash, D & Gangal, SV (1993). Identification of specific IgE binding fractions of castor bean (*Ricinus communis*) pollen obtained from different source materials. *Grana* **32:** 376-380

17

Response of Higher Plant Biodiversity to SO_2 Exposure

VINITA KATIYAR[1*] AND S BHANUMATI[2]

IEMPS (SS in Botany), Vikram University, Ujjain-456 010
[1]Present Address: Civil Engineering Department, Indian Institute of Technology New Mehrauli Road, New Delhi-110 016
E-mail: vinita.katiyar@gmail.com
[2]Department of Chemistry, Gargi College, University of Delhi, New Delhi-110 049

ABSTRACT

Air pollution is one of the severe problems world is facing today. Adverse effects of air pollution on biota and ecosystems have been demonstrated worldwide. Much experimental work has been conducted on the analysis of air pollutant effects on crops and vegetation at various levels ranging from morphological to molecular processes. Study of air pollutants on tree saplings in plant exposure chamber system may be a significant approach in order to assess the impact of air pollutants on individual plant life forms.

Sulfur dioxide a major air pollutant acts as precursor of fine particulate and acid rain. It poses a significant threat on health and vegetation. Assessment of total chlorophyll contents, leaf area dry weight ratios (LDR), nitrate reductase activity (NRA), foliar protein contents, free amino acids, proline contents and residual sulphate contents of diverse forms of wooden tree saplings named *Azadirachta indica* (A. Juss), *Cassia siamea* (L.), *Dalbergia sissoo* (Roxb), *Eucalyptus rostrata* (L.), *Mangifera indica* (L.), *Syzygium cumini* (L.) Skeels were observed as indicator parameters of sulphur dioxide (SO_2) exposure under open top chamber system (OTC) system. Effects of SO_2 on plant biodiversity can be assessed in two ways: first direct effect of SO_2 in the gaseous form and indirect form of SO_2 accumulated as residual sulphate contents in the leaves.

Declined protein contents, free amino acids, total chlorophyll contents, free nitrate reductase activity and enhanced sulphate, proline contents and leaf area - dry weight ratios were the general responses elucidated as evidence of adverse effects of SO_2 while compared to unexposed saplings of diverse life forms of wooden trees saplings. However, the magnitude of quantitative changes in the contents of different parameters and their statistical analysis revealed that *Azadirachta indica*, *Cassia siamea* and *Eucalyptus rostrata* had appeared as moderately tolerant tree species among the tested diversity of higher plants. Residual sulphate contents were found to have significant effect on the total chlorophyll contents, free amino acids and NRA of plant diversity followed by foliar protein contents. Diversity of plants responded to SO_2 exposure irregularly, thus studied physiological and biochemical activities can serve as good indicators of exposure of SO_2, in SO_2 prone areas.

Keywords: Residual sulphate, open top chamber (OTC), plant biodiversity, SO_2 fumigation, tree saplings.

INTRODUCTION

Biodiversity is the variability among living organisms from all sources, including inter alia, terrestrial, marine, and other aquatic ecosystems, and the ecological complexes of which they are part; this includes diversity within species (genetic diversity), between species and of ecosystems. India is one of the 12 'mega diverse' countries of the world with a large array of environmental conditions by virtue of its tropical location, varied physiographic features and climatic types. India has constituted approximately seven per cent part of world's bio-diversity in the form of angiosperms (higher plants) (UNDP/UNEP/GEF 2001). The degradation of tropical forests and destruction of habitat due to anthropogenic activities are the major causes of decline in the global biodiversity. The air pollution is one of the important factors that have contributed to staggering loss of biodiversity.

Atmospheric gases like SO_2, NO_2 and O_3 have been considered damaging pollutants and their phytotoxicity are well-known (Ozolincius *et al.* 2005, Paoletti *et al.* 2010). Sulphur dioxide, which is a well known recognized air pollutant and a phytotoxic agent, enters in leaves via stomata; it is metabolized into sulphite, bisulphate and sulphate, which in turn affect several biochemical and cellular processes, and plant characteristics (Horsman and Wellburn 1975). In this process, SO_2 ultimately induces the release of several free radicals. SO_2 toxicity involves either sulphite or free radicals formed during photo-oxidation of sulphite to sulphate (Peiser *et al.*, 1982) and their reaction with biological molecules, especially protein and lipids (Navari-Izzo *et al.,* 1992). At the same time, the toxicity of SO_2 reduces through various detoxifying mechanisms. A plant would finally be left over with a certain degree of sulphate accumulation. This kind of accumulation of sulphate, termed as "residual sulphate", is toxic enough to bring about chlorosis and nutritional imbalances (Katiyar and Dubey 2001).

Effects of SO_2 on biodiversity of wooden plants have been observed in two ways i.e. effect of gaseous form of SO_2 on the physiological process of plants and indirect effect of residual sulphate content on the various internal mechanisms of the plants.

Experimental Set-up

Indicator parameters named total chlorophyll content (Arnon 1949), leaf area dry weight ratio (Rao and Dubey 1988), nitrate reductase activity (Shrivastava and Mathur 1980), free proline content (Bates *et al.,* 1973), foliar protein content (Lowry *et al.*, 1951), free amino acid (Jayaraman 1981), foliar (residual) sulphate content (Patterson 1958) were determined after one-year old saplings of *Azadirachta indica* (Neem), *Cassia siamea* (Amaltas), *Dalbergia sissoo* (Sheesham), *Eucalyptus rostrata* (Eucalyptus), *Mangifera indica* (Aam), *Syzygium cumini* (Jamun) had been exposed to 495 $\mu g/m^3$ of SO_2 for 40 days 4 hd^{-1} in Open Top exposure chamber system.

METHODOLOGY

Wooden plants were grown in earthen pots of 35 cm in diameter containing black cotton soil (sand, peat and compost) kept in $1m^3$ open top chambers. The environmental conditions were $20.4 \pm 1.6\%$, 10.8 ± 1.8^oC day/night temperature and relative humidity 50-60 ± 18%/60-70 ± 15%, respectively.

One year old tree saplings were exposed to SO_2 gas in open top chamber system (Figure 1). Sulphur dioxide generated by bubbling air through known speed in an impinger containing 1% aqueous metabisulphite (Hashimato and Tanaka 1980). The generated and available SO_2 concentration at the plant canopy height was determined using Toxic Gas Monitor (TGM) CEA Inst, USA by sucking air into pararosaniline reagents for 30 min. intervals of 1 h. The plants were exposed to SO_2 for 40 days 4 h d^{-1}. Plants considered to be control received a similar handling in a parallel exposure chamber without SO_2.

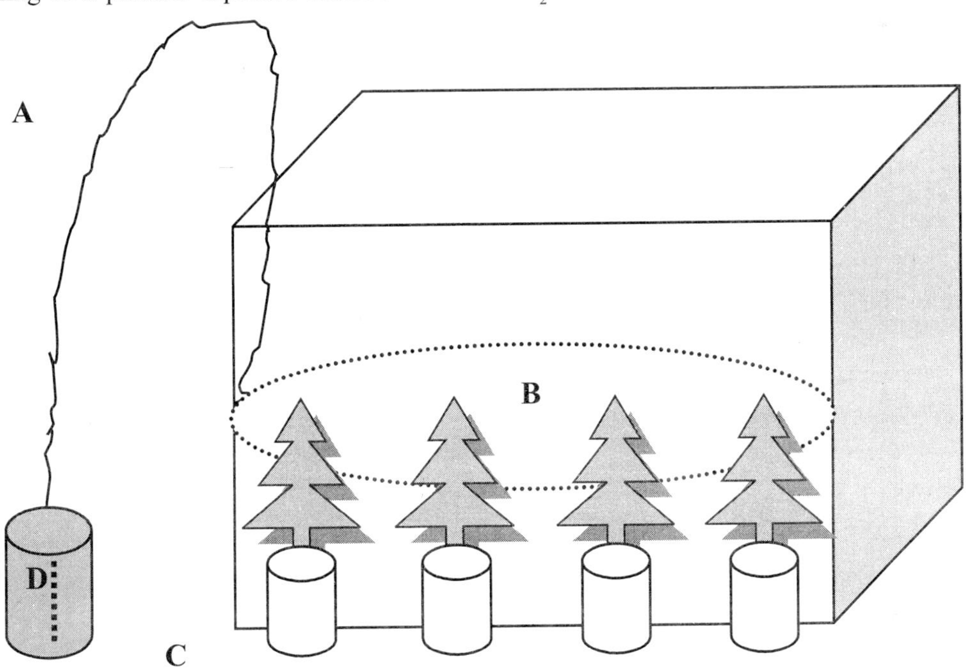

A:Teflon/non-perforated tubes, **B:** Perforated teflon tubes, **C:** Generation of SO_2, gas **D:** Aeration

Fig. 1. Gaseous Exposure Chamber.

Pearson correlation coefficient r' between residual sulphate contents and other parameters was established in order to estimate the effect of accumulated foliar sulphate content on the activity of plant physiology. Student t' test had facilitated the analysis of effect of SO_2 exposure on various physiological mechanisms in diversity of wooden tree saplings.

RESULTS AND DISCUSSION

India is one of the 12 'mega biodiversity' countries in the world, by virtue of having high species richness and more number of endemic species. With only 2.4% of the land area, India accounts for 7-8% of the recorded species of the world. Over 45,000 species of plants and 89,000 species of animals have been recorded from our country so far (India's 3rd National Report to CBD, 2006).

Threats of air pollution on the diversity of plants are really a great challenge to all over the world. It deteriorates ecological conditions and can be defined as the fluctuation in any atmospheric constituent from the value that would have existed without human activity (Tripathi and Gautam 2007). The atmospheric SO_2 adversely affects various morphological and physiological characteristics of plants such as reduction in dimension of leaf blade of five tree species had been observed as the effect of heavy dust and SO_2 pollution (Jahan and Iqbal 1992).

Total chlorophyll contents, nitrate reductase activity, protein contents and amino acids were decreased in response to SO_2 exposure whereas foliar sulphate, leaf area-dry weight ratios and free proline content were the parameters, which increased due to SO_2 exposure in a variety of wooden tree saplings (Table 1-4). Significant loss in total chlorophyll contents was distinguished of tree saplings named *E. rostrata and S. cumini.* Chlorophyll pigments are highly sensitive to SO_2 and this is the reason they are affected first by short-term exposure of SO_2 and have been frequently used as indicators of plant response to air pollutants (Panigrahi *et al.* 1992, Katiyar and Dubey 2001).

Table 1: Magnitude of total chlorophyll content (mg g^{-1} fresh wt.) and leaf area dry weight ratio (sq cm g^{-1}) due to SO_2 exposure

S.No.	Experimental Tree Species	Total chlorophyll content Unexposed	Exposed	r-value	Leaf area dry weight ratio Unexposed	Exposed	r-value
1.	*A. indica*	2.347	2.019^{ns}	0.353^{ns}	74.35	77.43^{ns}	0.517^{ns}
	Sr	0.571	0.067		0.871	0.689	
2.	*C. siamea*	1.787	1.652^{ns}	0.976*	105.65	111.11^{ns}	0.537^{ns}
	Sr	0.036	0.040		0.563	1.449	
3.	*D. sissoo*	1.580	1.442^{ns}	0.998*	76.07	83.08^{ns}	0.779^{ns}
	Sr	0.098	0.022		0.852	0.597	
4.	*E. rostrata*	2.090	1.808*	0686^{ns}	73.67	76.95^{ns}	0.371^{ns}
	Sr	0.050	0.030		0.136	0.265	
5.	*M. indica*	1.997	1.779^{ns}	0.927*	67.79	78.88^{ns}	0.596^{ns}
	Sr	0.028	0.086		1.317	0.141	
6.	*S. cumini*	1.720	1.526*	0.996*	66.27	69.59^{ns}	0.883^{ns}
	Sr	0.076	0.022		0.868	0.257	

Table 2: Magnitude of free amino acid (mg g^{-1} dry wt.) and free proline content (ì mol g^{-1} fresh wt.) due to SO_2 exposure

S. No.	Experimental Tree Species	Free amino acid Unexposed	Exposed	r-value	Foliar proline content Unexposed	Exposed	r-value
1.	*A. indica*	0.117	0.106^{ns}	0.847^{ns}	13.55	15.56*	0.829^{ns}
	Sr	0.005	0.002		0.926	0.513	
2.	*C. siamea*	0.330	0.309^{ns}	0.967*	8.68	10.10^{ns}	0.927*
	Sr	0.003	0.040		0.101	0.300	
3.	*D. sissoo*	0.393	0.359*	0.990*	7.60	9.00*	0.489^{ns}
	Sr	0.003	0.006		0.378	0.225	
4.	*E. rostrata*	0.309	0.286*	0.965*	5.07	5.86^{ns}	0.443^{ns}
	Sr	0.030	0.003		0.175	0.225	
5.	*M. indica*	0.213	0.178^{ns}	1*	5.50	6.25*	0.381^{ns}
	Sr	0.003	0.004		0.476	0.327	
6.	*S. cumini*	0.419	0.374*	0.618^{ns}	7.21	8.03^{ns}	0.989*
	Sr	0.005	0.001		0.325	0.02	

Table 3: Magnitude of nitrate reductase activity (ì mol g^{-1} h^{-1} fresh wt.) and free protein content (mg g^{-1} fresh wt.) due to SO_2 exposure

S. No.	Experimental Tree Species	NR activity Unexposed	Exposed	r-value	Foliar protein content Unexposed	Exposed	r-value
1.	*A. indica*	6.670	6.12*	0.974*	98.70	88.62	0.562^{ns}
	Sr	0.208	0.08		3.292	3.865	
2.	*C. siamea*	9.12	8.50^{ns}	0.482^{ns}	113.65	98.61	0.389^{ns}
	Sr	0.568	0.04		3.650	2.165	
3.	*D. sissoo*	7.20	6.43*	0.277^{ns}	66.83	54.17	0.9818*
	Sr	0.100	0.15		3.090	3.222	
4.	*E. rostrata*	6.76	6.26*	0.993*	165.02	155.06	0.293^{ns}
	Sr	0.152	0.05		3.842	4.777	
5.	*M. indica*	2.23	1.90*	0.998*	195.86	159.73	0.993*
	Sr	0.305	0.11		4.761	4.182	
6.	*S. cumini*	2.60	2.16*	0.981*	129.81	107.96	0.971*
	Sr	0.200	0.15		3.099	7.098	

ns = Not significant
* = Significant at 5% level
r = Correlation-coefficient
Sr = Standard error

Table 4: Magnitude of foliar sulphate content (mg g^{-1} dry wt.) due to SO_2 exposure

S.No.	Experimental condition	Foliar sulphate content *A. indica*	*C. siamea*	*D. sissoo*	*E. rostrata*	*M. indica*	*S. cumini*
1.	Unexposed	7.09 ±0.049	3.33 ±0.02	5.44 ±0.03	4.09 ±0.03	6.16 ±0.02	5.10 ±0.02
2.	Exposed	8.93 ±0.04	3.86 ±0.05	6.06 ±0.04	4.49 ±0.06	6.86 ±0.04	5.76 ±0.03

± = Standard deviation

Sulphur dioxide induced a decrease in photosynthetic pigments supposed to remove Mg^{++} from tetrapyrol ring of the chlorophyll molecules degrading it into phaeophytin, a brown non-photosynthetic pigment (Khan and Malhotra 1982). Figures in Table-I illustrate a decline in photosynthetic pigments in terms of total chlorophyll contents among every tested tree seedlings, and significant 'r' value expresses that diminution in photosynthetic pigments attributed to more accretion of foliar sulphate content that reflect the plant physiology. Similar results have been reported by Katiyar and Dubey (2000).

Reductions in leaf area and leaf number may be due to decrease in leaf production rate and enhanced senescence. The reduced leaf area results in reduced photosynthetic rate of leaves (Tiwari *et al.*, 2006). Impairment of metabolic activities is reflected in dry matter production. The dry weight of foliage represents the plant growth and its physiological conditions. Reduced dry matter production can be expressed with the increase in leaf area - dry weight ratios (Rao 1989, Jain 1993, Katiyar and Dubey 2001). Dry weight of the leaves of plants in response to SO_2 can be severely decreased even in the absence of visible injury (Kumawat 1990, Jain 1993, Katiyar and Dubey 2001). See also Table-1.

The free proline accumulation in SO_2 stress is a known fact but other soluble amino acids did not increase to the same extent (Erickson and Dashek, 1982). Proline is catabolic amino acid that accumulates due to protein degradation in stress leaves is the hypothesis of inhibition of protein synthesis and protein degradation (Raggi 1994,). Proline accumulates within the leaves of a variety of plants when these are subjected to the environmental stresses (Erickson and Dashek 1982), like SO_2 and NO_2 (Katiyar and Dubey 2000). Table-II elucidates that significant changes are observed in *A. indica, D. sissoo* and *M. indica* in proline contents due to SO_2 exposure.

However, amino acid pool enriched with protein depletion at severe injury levels due to SO_2 exposure had been reported by Beg and Farooq (1988). SO_2 might have altered the composition of individual amino acids by affecting nitrate reductase activity under SO_2 fumigation. Reduction has been reported in total amino acids along with amines, asparagines and glutamines by means of SO_2 (Cowling and Bristow 1979).

Nitrate reductase is an essential enzyme in higher plants that catalyses the conversion of nitrate to nitrite which is essential for synthesis of vital compounds (Shrivastava and Mathur 1980). Any alteration in the NRA leads to change the amino acid composition and protein synthesis pathway. Kumawat (1990) has observed inhibition of nitrate reductase activity in plants exposed to SO_2. Degradation of protein under SO_2 exposure was confirmed by Beg and Farooq (1988), and Kumawat (1990) as accounted in this study. Significant correlation-coefficient 'r' demonstrates that residual sulphate content appeared to have a significant influence on the activity of nitrate reductase enzyme and the content of free amino acids (Table III).

The pollutants like sulphur dioxide (SO_2^-), nitrogen oxides (NO_2) and CO_2 when absorbed by the leaves may cause a reduction in the concentration of photosynthetic pigments viz., chlorophylls and carotenoids, which directly affect the plant productivity (Joshi and Swami 2009). Subsequently pointed out, earlier SO_2 absorbed through stomata into a leaf, is oxidized to sulphite or sulphate ions and then reduced to assimilate with a carbon skeleton to form cysteine (Garsed 1985). At sufficiently low concentrations, bisulfite and sulfite are effectively detoxified by plants, and SO_2 then provides a sulfur source for the plants. First, SO_2 might have followed the normal reduction pathway to form sulphur compounds, including the formation of insoluble materials and thiols. The formation of sulphate results from the extensive desulphuration of amino compounds especially cysteine which can be degraded rapidly to yield sulphate (Harrington and Smith 1980). It might be the cause of accumulation of sulphate fraction within the leaves. Higher concentration cannot be detoxified rapidly enough to avoid the injury (Eduardo Zeiger 2002). Regarding the plant responses in relation to SO_2 exposure, Dubey (1990) has established the high leaf area-dry weight ratios in spite of low stomatal conductance. It means that plant response to SO_2 and accumulation of sulphate contents in the leaves cannot be predicted by conductance only.

Sulphur concentration in scot pine needles was correlated with S emissions (Haapala *et al.*, 1996). Manninen and Huttunen (2000) reported that greater increase in the sulphate suggested more abundant stomatal uptake of SO_2. Accumulation of sulphate-sulphur takes place at low concentrations of SO_2 (Murray *et al.*, 1993, Katiyar and Dubey 2001) resulting in impairment of the plant physiology due to decrease in the photosynthetic pigments as well as dry matter production. Among the tested tree saplings, higher accumulation of sulphate contents appeared in *A. indica, C. siamea* and *E. rostrata* (Table 4).

CONCLUSION

Air pollution has played a key role in changing the distribution of plant biodiversity, and the ecology of susceptible plant communities in polluted areas. Different air pollutants have a range of effects on a single species or variety of plant life forms. Some pollutants can appear to be initially beneficial to a particular species, but later become harmful. Air pollution does not respect the boundaries of nature reserves and conservation areas. It is therefore a significant contributory factor, which is responsible for decline of global biodiversity. Figure 2 shows the

Many air pollutants combine to affect trees....

Sulphur oxides	Nitrogen oxides	Hydrocarbons	Heavy metals
Sulphuric acid	Nitric acids	Ozone	

...and act in a variety of ways...

Acid rain, mist and snow

↓

Dry deposition of ozone, sulphur and nitrogen oxides

↓

Increased pest numbers through SO_2 pollution

↓

Nitrogen fertilization of soil
Soil acidification, and release of metal ions

↓

Depression of mycorrhizal fungi

↓

...in concert with a range of other factors including climate, pests, diseases, management systems etc.

Fig. 2. Combined air pollution impacts on forest trees
(*Source:* Air pollution and biodiversity: a review Nigel Dudley & Sue Stolton
http://www.equilibriumresearch.com/upload/document/airpollutionandbiodi4f9.pdf)

effects of some air pollution stresses on biodiversity of higher plants in forest. In most situations pollutants are acting in the presence of other factors which themselves have an impact on ecosystems. Contributory factors fall into three main categories:

- anthropogenic factors: such as forest management systems
- natural factors: such as landform and soil types
- factors, which appear to be natural but have been influenced to some extent by human activities: such as climate changes induced by global warming, and introduced plant diseases.

Air pollution effects are thus both more complex and more wide-ranging than simply assessment of the damage to a few individual species might suggest (Nigel Dudley and Sue Stolton 2001). Urban air pollution is a serious problem in both developing and developed countries (Li 2003).

Biodiversity of saplings described above may be helpful in selection of indicators of SO_2 exposure and can be grown at SO_2 prone areas. The description of study reported here has enabled to draw a probable conclusion that diverse forms of wooden trees might have been developed the ability to resist the SO_2 toxicity through active antioxidant system to some extent, as revealed by the statistical analysis, but long term exposure leads to accumulation of sulphate within the foliage of saplings which might have been enough in quantity to reduce the tolerance potential of saplings as shown by significant r-value. Tree species *A. indica, C. siamea* and *E. rostrata* were observed to have good scavenging potential to overcome the SO_2 toxicity either in the gaseous form or in the form of residual sulphate content.

Biodiversity of higher plants served as good indicator of environmental stresses including air pollution. In-depth analysis of morphological, physiological and biochemical parameters require to be done to screen out a susceptible and tolerant nature of diversity of wooden trees at sapling stage.

ACKNOWLEDGEMENTS

First author is thankful to The University Grants Commission, New Delhi (India) for providing financial assistance.

REFERENCES

Arnon, DI (1949). Copper enzymes in isolated chloroplasts polyphenoloxidase in *Beta vulgaris. Plant Physiol.* **24:** 1-15

Bates, LS, Waldren, RP & Teare, ID (1973). Rapid determination of free proline for water stress studies. *Plant and soil* **39:** 205-207

Beg, MU & Farooq, MH (1988). Sulphur dioxide resistance of Indian trees-2, Experimental evaluation of metabolic profile. *Water, Air and Soil Pollut.* **40:** 317-326

CBD, India Report (2006). India's third national report to the Convention on Biological Diversity. MoEF, Govt. of India

Cowling, DW & Bristow, AW (1979). Effects of SO_2 on sulphur and nitrogen fractions and on free amino acids in perennial ryegrass. *J. Sci. Food Agric.* **30:** 354-360

Dubey, PS (1990). Study and Assessment of Plant Response against Air Pollution in Industrial Environments. Final Technical Report, All India Coordinated Program on Air Pollution & Plants, Ministry of Environment, Forests & Wild life Govt. of India, New Delhi

Eduardo Z (2002). The Effect of Air Pollution on Plants. Online *Plant Physiology* Essay **25:** 1

Erickson, SS & Dashek WV (1982). Accumulation of foliar soluble proline in sulphur dioxide stressed *Glycine max* cv. Procotor and Essex and *Hordeum vulgare* cv. excelsior seedlings. *Environ. Pollut.* Series A **28:** 89-103

Garsed, SG (1985). SO_2 uptake and Transport. In *Sulphur dioxide and Vegetation.* Winner, W.E. *et al.* (Eds), pp. 85-99, Stanford University Press, Stanford. CA

Haapala, H, Goltsova, N, Seppala, R, Huttunen, S, Kouki, J, Lampp, J & Popovichev, B (1996). Ecological condition of forests around the eastern part of the Gulf of Finland. *Environ. Pollut.* **91:** 253-265

Harrington, HM & Smith, IK (1980). Cysteine metabolism in cultured tobacco cells. *Plant Physiol.* **65:** 151-155

Horsman, DC & Wellburn, AR (1975). Synergistic effects of SO_2 and NO_2 polluted air upon enzyme activity in pea seedlings. *Environ. Pollut.* **8:** 123-133

Jain, N (1993). Assessment of Few Tropical Tree Species against Air Pollutants. Ph.D. Thesis, Vikram University Ujjain, India

Jayaraman, J (1981). Laboratory Manual of Biochemistry Western Wiley Ltd. Bombay, India

Joshi, PC & Swami, A (2009). Air pollution induced changes in the photosynthetic pigments of selected plant species. *J. Environ. Boil.* **30:** 295-298

Katiyar, V & Dubey PS (2000). Growth behavior of two cultivars of maize in response to SO_2 and NO_2. *J. Environ. Biol.* **21:** 317-323

Katiyar, V & Dubey PS (2001). Residual sulphate content correlated with chlorophyll loss in *Zea mays* (L). *Indian J. Environ. Toxicol.* **11:** 38-39

Khan, AA & Malhotra, SS (1982). Ribulose biphosphate carboxylase and glycolate oxidase from Jack Pine: Effects of sulphur dioxide fumigation. *Phytochem.* **21:** 2607-2612

Kumawat, DM (1990). Plant Response Against Air Pollution with Special Reference to Heavy Metals And Sulphur Dioxide, Ph.D. Thesis, Vikram University, Ujjain, India

Li, MH (2003). Peroxidase and superoxide dismutase activities in fig leaves in response to ambient air pollution in a subtropical city. *Arch. Environ. Contam. Toxicol.* **45:** 168-176

Lowry, O, Rosebrough, N, Farr A & Randall, R (1951). Protein measurements with the Folin phenol reagent, *Boil. Chem.* **193:** 265-275

Manninen, S & Huttunen, S (2000). Response of needle sulphur and nitrogen concentrations of Scots pine versus Norway spruce to SO_2 and NO_2. *Environ. Pollut.* **107:** 421-436

Murray, F, Wilson, S & Qifu, Mo (1993). Effects of SO_2 and NO_2 on growth and nitrogen concentrations in lucerne and barrel medic. *Environ. Expt. Bot.* **34:** 201-205

Navari–Izzo, F, Quartacci, MF, Izzo, R & Pinzino (1992). Degradation of membrane lipid component and antioxidant levels in *Hordeum vulgare* exposed to long term fumigation with SO_2. *Physiol. Pl.* **81:** 73-79

Ozolincius, R, Stakenas, V & Serafinaviciute, B (2005). Meteorological factors and air pollution in Lithuanian forests: possible effects on tree condition. *Environ. Pollut.* **137:** 587-595

Paoletti, E, Contran, N, Bernasconi, P, Günthardt-Goerg MS & Vollenweider, P (2010). Erratum to structural and physiological responses to ozone in Manna ash (*Fraxinus ornus* L.) leaves of seedlings and mature trees under controlled and ambient conditions. *Sci. Total Environ.* **408:** 2014-2024

Panigrahi, NC, Mishra, BB & Mohanty, BK (1992). Effects of SO_2 on chlorophyll content of two crop plants. *J. Environ. Bio.* **13:** 201-205

Patterson, J (1958). Colorimetric Analysis of Non Metals. ***In***: David Glick (Ed) Methods in Biochemical Analysis (Vol. II), Interscience Publ Inc., New York, USA

Peiser, GD, Liza MCC & Yang, SF (1982). Sulphite induced lipid peroxidation in chloroplasts as determined by ethane production. *Pl. Physiol.* **70:** 994-998

Raggi, V (1994). Changes in a free amino acid and osmotic adjustments in leaves of water stressed bean. *Physiol. Pl.* **91:** 427-434

Rao, MV & Dubey, PS (1988). Plant response against SO_2 in field conditions. *Asian Environ.* **3:** 1-9

Rao, MV (1989). Air Pollution Stress and Correlated Plant Response in Dewas Area. Ph.D. Thesis, Vikram University, Ujjain, India

Shrivastava, HS & Mathur, SN (1980). Nodulation and NR activity in nodules and leaves of black gram (*Vigna mungo* L.) as affected by varying day length. *Ind J. Expt. Biol.* **18:** 300-307

Tripathi, AK & Gautam, M (2007). Biochemical parameters of plants as indicators of air pollution. *J. Environ. Biol.* **28:** 127-132

UNDP/UNEP/GEF (2001). The Integration of Biodiversity into National Environmental Assessment Procedures. National Case Studies India. Produced for the Biodiversity Planning Support Programme. http://www.unep.org/bpsp/EIA/Case%20Studies/India.pdf

18

Diversity of *Jatropha* in India

KSHITIJ MALHOTRA, DINESH KUMAR, VIDYA DHAR PANDEY[1] AND SANJAY SINGH

Silviculture Division, Forest Research Institute, Dehradun–248 006.
[1]Department of Botany, Government Post-Graduate College, Rishikesh–249 201
E-mail: rishipop@gmail.com

ABSTRACT

Growing energy requirements, fast depleting petroleum resources and increased pollution require alternative sources of energy. In this scenario biofuel is one of the most potential renewable alternatives. The Planning Commission of India has selected *Jatropha* for large-scale planting besides other tree borne oil species. This priority given to *Jatropha* is due to its crude oil, which has high free fatty acid contents, more similar to fossil diesel in chemical composition and easy conversion into biodiesel. It has been grown in almost all the states of India as a live fence for many years. However, there is lack of information about ecological diversity of different species of *Jatropha* in India. As far as species diversity of this genus is concerned, twenty-six species have been variously reported in the Indian literature but some of them are synonyms. Investigation by the authors revealed that twelve species of *Jatropha* occur in India with internationally accepted botanical names. Genomic studies showed that there is high variation among species. However, *Jatropha curcas* diversity has been evaluated with SSR and AFLP analysis, very little diversity was found among most of the samples from Asia, South America and Africa. Despite limited genetic diversity within *J. curcas*, appreciable variability exists in important phenotypic, physiological and biochemical traits such as seed size, water use efficiency and seed oil content. Therefore, such variability can be used for the genetic improvement of *Jatropha*.

Keywords: Biofuel, *Jatropha*, petroleum resource.

INTRODUCTION

India is the second fastest growing developing country in the world after China. If the process of development will continue at the same pace it is inevitable that the energy consumption would increase manifold. As per Planning Commission estimates, the requirement of petrol is expected to grow from approximately 10 million tonnes in 2006-07 to 12.5 million tonnes in 2011-12. Similarly, the demand for diesel is likely to increase from about 52 million tonnes in 2006-07 to 66 million tonnes in 2011-12. Growing energy requirements, fast depleting petroleum resources and increased pollution require societal response so as to minimize ill effects of global warming and climate change. Contrary to the demand situation, the domestic supply is in a position to cater to only about 22% of the total demand. Therefore, attempt needs to be made to reduce dependence on imports and seek better alternatives. Biofuels are promising alternatives and the country has a ray of hope for the same. Biofuels are renewable and come from biological raw materials. Among biofuels i.e. ethanol and biodiesel are gaining worldwide acceptance as a solution to energy crisis.

The Planning Commission of India has selected *Jatropha* [The genus name *Jatropha* is derived from the Greek words jatrós (doctor) and trophé (food), which implies medicinal uses] as a potential plant species for large scale plantation in India besides other tree borne oil species like *Pongamia* (karanja), *Simmondsia chinensis* (jojoba), *Madhuca indica* (mahua), *Sapindus mukorossi* (reetha), *Azadirachta indica* (neem), etc. Today *Jatropha curcas* is the preferred species for biodiesel production in India among others tree borne oil seeds. Millions of *Jatropha* seedling have been planted in the country since 2003. This priority given to *Jatropha* is due to its crude oil, which has high free fatty acids content, more similar to fossil diesel in chemical composition and easy conversion into biodiesel. Its conversion efficiency is about 95%. *Jatropha* is one of the best alternatives for energy plantation due to the following additional characteristics:

- It can be grown in areas of low rainfall (600 mm per year) and in problematic soils. In high rainfall and irrigated areas too it can be grown with much higher survival and yields.
- *Jatropha* is easy to establish, grows very quickly and is hardy by nature.
- *Jatropha* lends itself to plantation on low fertility marginal, degraded, fallow, waste and other lands such as along the canals, roads, railway tracks, on borders of farmers' fields as a fence or live hedge in the arid/semi-arid areas and even on alkaline soils. As such it can be used to reclaim wastelands in the forests and outside.
- *Jatropha* is not browsed by animals.
- *Jatropha* can be established from seed, as well as from cuttings.
- Its gestation period is short. The bush can be expected to start bearing fruit within two years and in some cases after one year of planting.
- One hectare of *Jatropha* plantation can produce about 3 to 4 tonne seeds (upto yield 12 tonne have also been reported) with upto 40% oil content on seed weight basis as reported by various researchers.
- Its seed cake can be used as an organic manure as it is rich in nitrogen and phosphorous.
- Various parts of the plant have medicinal value, its bark contains tannin, the flowers attract bees and thus the plant has honey production potential.
- The latex of *Jatropha* contains the alkaloids such as 'jatrophine', 'jatropham', 'jatrophone' and 'curcain', which are believed to have anti-cancerous properties.
- The seed fat is rich in palmitic, oleic and linoleic acids. The toxicity of the seeds is because of 'curcin' alkaloids.
- *Jatropha* removes carbon from the atmosphere, stores it in the woody tissues (carbon sequestration) and assists in the build up of soil carbon. It is thus environment-friendly.

- The plant is undemanding in soil type and does not require tillage.
- It can meet a number of objectives such as meeting domestic needs of energy services, including cooking and lighting; as an additional source of household income and employment through markets for fuel, fertilizer, animal feed, medicine, and industrial raw material for soap, cosmetics, etc., creating environmental benefits – protection of crops or pasturelands, or as a hedge for erosion control, or as a windbreak.

On the basis of the above merits, following hectares of lands have been identified for *Jatropha* cultivation in different states of India (Table 1):

Table 1: State-wise distribution of wastelands suitable for *Jatropha* cultivation in India under National Mission on Bio-diesel

S. No.	States	Potential area for *Jatropha* cultivation (Lakh ha.)	Percentage of total waste/ land of state
1.	Andhra Pradesh	43.96	84.95
2.	Arunachal Pradesh	9.97	54.42
3.	Assam	14.56	72.76
4.	Bihar/Jharkhand	18.60	88.61
5.	Goa	0.40	65.14
6.	Gujarat	28.71	60.19
7.	Haryana	2.62	70.33
8.	Karnataka	17.89	85.85
9.	Kerala	1.00	68.84
10.	Madhya Pradesh/Chhattisgarh	66.20	94.96
11.	Maharashtra	48.55	90.78
12.	Manipur	12.62	97.49
13.	Meghalaya	9.37	99.85
14.	Mizoram	4.07	100.00
15.	Nagaland	8.40	100.00
16.	Orissa	18.88	88.47
17.	Punjab	1.06	47.41
18.	Rajasthan	56.88	53.84
19.	Sikkim	2.13	59.77
20.	Tamil Nadu	17.95	78.00
21.	Tripura	1.28	99.99
22.	Uttar Pradesh/Uttarakhand	12.14	31.31
23.	West Bengal	2.58	45.16
	Grand Total	**400.37**	**73.51**

Source: http://www.indiastat.com

Although a lot of information is already available in the scientific publications as well as print and electronic media about the prospects of *Jatropha* cultivation, the information about its biodiversity is still scarce.

LEVELS OF DIVERSITY

Biodiversity (variation of life at all levels of biological organization) is a portmanteau word, originating from biology and diversity, and used interchangeably with "biological diversity".

This term was first used by wildlife scientist and conservationist Raymond F. Dasmann (1968) advocating nature conservation.

Biologists often define "biological diversity" or "biodiversity" as the "totality of genes, species, and ecosystems of a region". An advantage of this definition is that it seems to describe most circumstances and present a unified view of the traditional three levels at which biological variety has been identified:

I. Ecological Diversity
II. Species Diversity
III. Genetic Diversity

I. Ecological diversity of Jatropha

Jatropha is believed to be a native of South America but it later spread to other continents of the world by Portuguese settlers. The Arabs have been using this plant for medicinal purpose. Today it is found scattered in all the tropical and sub-tropical regions of the world such as Brazil, Fiji, Honduras, India, Jamaica, Panama, Puerto Rico and Salvador. There are more than 200 different names for it all over the world which indicates its great significance to man and various possibilities of its uses. Several properties of the genus stemming from its hardiness, rapid growth, easy propagation and wide utility have resulted in the spread of *Jatropha* far beyond its original distribution. *Jatropha* exhibits a wide environmental tolerance and is found in both seasonally dry tropics as well as equatorial humid regions and is well adapted for cultivation in vast areas of marginal and degraded land in the semi-arid and arid tropics.

The current distribution shows that its introduction has been most successful in drier regions of the tropics with annual rainfall of 300-1000 mm. It occurs mainly at lower altitudes (0-500 msl) and can grow upto 1000 msl in areas with average annual temperatures well above 20°C. Kumar *et al.* (2009) have found *J. curcas* to be sensitive to frost. Frost is, therefore, expected to be a crucial limiting factor affecting its survival and growth in the Himalayas.

Though it has been grown in almost all the states of India as a live fence for many years. Gujarat, Rajasthan, Maharashtra, Madhya Pradesh, Chhattisgarh and Tamil Nadu are the main states cultivating this crop. It can grow on a wide range of soils including poor degraded soils and in low rainfall conditions with minimal inputs. However, there is still lack of information about the ecological diversity of different species of *Jatropha* and most of the information on this genus is based on *J. curcas* alone.

There are three indices which are used by ecologists:

- Alpha diversity refers to diversity within a particular area, community or ecosystem, and is measured by counting the number of taxa within the ecosystem (usually species).
- Beta diversity is species diversity between ecosystems; this involves comparing the number of taxa that are unique to each of the ecosystems.
- Gamma diversity is a measurement of the overall diversity for different ecosystems within a region.

Information on above indices of *Jatropha* needs to be collected.

II. Species diversity of Jatropha

A member of family Euphorbiaceae, *Jatropha* is a large genus comprising over 200 species in the world; including herbs, shrubs and trees. Twenty six species have been variously reported in

India in the literature (Hooker, 1887; Blatter, 1923; Dalzell *et al.*, 1861) However, some of those are synonyms (synonym is *an erroneous or improperly applied scientific name, as a new name applied to a taxon previously appropriately named or a specific name previously applied to another taxon*), homotypic synonyms (*homotypic or nomenclatural synonyms are synonyms that come about when a name is nomenclaturally incorrect or sometimes when a species is moved from one genus to another*) or basionyms (*the basionym (base name) is the first name validly published ever given to a biological species or genus which has priority over other names later given to the same species by different authors and the author of this name is called the basionym author*). In view of this, Table 2 has been prepared and twelve species of *Jatropha* have been listed that occur in India with internationally accepted botanical names. Their corresponding synonyms, homotypic synonyms and basionyms have also been mentioned. It must be pointed out that *J. heudelotii* Baill., *J. moluccana* L. and *J. manihet* L. were also published in India's literature but these plants do not belong to genus *Jatropha*. They are synonyms for *Ricinodendron hevdelotii* (Baill.) Heckel, *Aleurites moluccana* (L.) Willd. and *Manihot esculenta* Crantz, respectively. Furthermore *J. gossypifolia* has two varieties, *viz. J. gossypifolia* var. *elegans* (Pohl) Muell. Arg and *J. gossypifolia* var. *gossypifolia.* It may be mentioned here that *J. villosa* var. *ramnadensis*. Ramam is referred to as a variety in some publications but it is not an accepted variety of *J. villosa* (http://www.kew.org).

III. Genetic diversity of Jatropha

Gene is the fundamental unit of natural selection. For geneticists, *biodiversity* is the diversity of genes and organisms. They study processes such as mutations, gene exchanges, and genome dynamics that occur at the DNA level and generate evolution. Consistent with this, it is apt to state that genes are the ultimate source of biological organization at all levels of biological systems.

In India an investigation was carried out to assess the genetic diversity of 12 species/ accessions of *Jatropha* based on random amplified polymorphic DNA markers. From 26 random primers used, 18 primers gave reproducible amplification banding patterns of 112 polymorphic bands out of 134 bands scored accounting for 80.2% polymorphism across the genotypes. Three primers viz., OPA 4, OPF 11, and OPD 14 generated 100% polymorphic patterns. The polymorphic information content was highest for the primer OPD 14 (0.50) followed by the primers OPF 11 and OPAD 11 (0.48). Jaccard's coefficient of similarity varied from 0.00 to 0.85, indicative of high level of genetic variation among the genotypes studied (Ram *et al.*, 2007). Thus there is high variation amongst different species of *Jatropha*.

UPGMA (Unweighted Pair Group Method with Arithmatic Mean) cluster analysis indicated three distinct clusters, one comprising all accessions of *J. curcas,* whereas second included six accessions viz. *J. ramanadensis, J. gossypiifolia, J. podagrica, J. tanjorensis, J. villosa and J. integerrima. J. glandulifera* formed third cluster indicating its higher genetic distinctness from other species. The overall grouping of clusters confirmed the patterns of genetic diversity observed among the species (Ram *et al.*, 2007).

When *Jatropha* diversity was evaluated with SSR analysis, very little variability was noticed and most of the samples (192) of Asia, South America and Africa, were homozygous and general AFLP analysis showed that low genetic diversity (polymorphic loci <5%, mean heterozygosity <1%) (Yan, 2008).

Table 2: Species of *Jatropha* with accepted botanical names in India

S.No.	Accepted botanical names with publications	Synonyms* with publications	Homotypic synonyms* with publications	Variety with publications	Distribution		Life form/ Remark
					Major regions of the world	Major regions of India	
1.	*Jatropha curcas* L., Sp. Pl.: 1006 (1753).	-	-	-	Mexico to Trop. America	Naturalised in many parts of India particularly in Uttarakhand, Haryana, Rajastnan Gujarat, Chhattisgarh, Madhya Pradesh, Andhra Pradesh, Tamil Nadu, Kanyakumari, Coromandel, Konkan and Malabar coatal areas.	Nanophane rophytes or Phanerophytes/ Wild, cultivated and naturalized
2.	*Jatropha glandulifera* Roxb., Fl. Ind. ed. 1832, 3: 688 (1832).	-	-	-	S. India, Sri Lanka, Myanmar	Naturalised in Bengal, Deccan, Carnatic, and rarely found in Oudh and Punjab.	Succulent Nanophane-rophytes/ Wild, culti-vated and naturalized
3.	*Jatropha glauca* Vahl, Symb. Bot. 1: 78 (1790).	*Jatropha lobata* var. *glauca* (Vahl) Pax ex Blatter, Rec. Bot. Surv. India 8: 435 (1923).	# *Jatropha glauca* Vahl, Symb. Bot. 1: 78 (1790). *Adenoropium glaucum* (Vahl) Pohl, Pl. Bras. Icon. Descr. 1: 15 (1826). *Jatropha lobata* subsp. *glauca* (Vahl) Pax in H.G.A.Engler, Pflanzenr., IV, 147, III: 32 (1910).	-	N.E. Trop. Africa, S.W. Arabian Pen.	S. India	Nanophanerophytes/ Wild

Table 2: Contd.

S.No.	Accepted botanical names with publications	Synonyms* with publications	Homotypic synonyms* *with publications	Variety with publications	Distribution		Life-form/ Remark
					Major regions of the world	Major regions of India	
			#Basionym/ Replaced Synonym***				
4.	*Jatropha gossypiifolia* L., Sp. Pl.: 1006 (1753).	-	-	*Jatropha gossypiifolia* var. *elegans* (Pohl) Müell-Arg. in A.P.de Candolle, Prodr. 15(2): 1087 (1866). *Jatropha gossypiifolia* var. *gossypiifolia*.	Mexico to Trop. America, India	Naturalised in many parts of India.	Nanophanerophytes/ Cultivated as an ornamental and green dye plant.
5.	*Jatropha heynei* N.P.Balakr., Bull. Bot. Surv. India 3: 40 (1962).	*Jatropha heterophylla* B.Heyne ex Hook.f., Fl. Brit. India 5: 382 (1887).	-	-	S. India	S. India	Wild
6.	*Jatropha integerrima* Jacq., Enum. Syst. Pl.: 32 (1763).	*Jatropha hastata* Jacq., Select. Stirp. Amer. Hist.: 256 (1763).	*Adenoropium hastatum* (Jacq.) Britton & P.Wilson, Sci. Surv. Porto Rico & Virgin Islands 5: 485 (1924). *Jatropha integerrima* var. *hastata* (Jacq.) Fosberg, Rhodora 78(813): 102 (1976).	-	Cuba, Hispaniola, India	India	Nanophanerophytes or Phanerophytes/ Wild

Table 2: Contd.

S.No.	Accepted botanical names with publications	Synonyms* with publications	Homotypic synonyms* *with publications	Variety with publications	Distribution		Life-form/ Remark
					Major regions of the world	Major regions of India	
		Jatropha pandurifolia Andrews, Bot. Repos. 4: t. 267 (1802).	*Adenoropium pandurifolium* (Andrews) Pohl, Pl. Bras. Icon. Descr. 1: 14 (1826). *Jatropha diversifolia* var. *pandurifolia* (Andrews) M.Gómez, Anales Hist. Nat. 23: 51 (1894).	-	-	-	-
7.	*Jatropha maheshwarii* Subram. & Nayar, Bull. Bot. Surv. India 6: 331 (1965).	-	-	-	S. India	S. India	Wild
8.	*Jatropha multifida* L., Sp. Pl.: 1006 (1753).	*Adenoropium multifidum* (L.) Pohl, Pl. Bras. Icon. Descr. 1: 16 (1826).	#*Jatropha multifida* L., Sp. Pl.: 1006 (1753). *Manihot multifida* (L.) Crantz, Inst. Rei Herb. 1: 167 (1766). #Basionym/ Replaced Synonym***	-	S.W. Texas, Mexico, Caribbean, Venezuela, S. Brazil, Paraguay, India	India	Nanophanerophytes or Phanerophytes/ Cultivated in gardens.

Table 2: Contd.

S.No.	Accepted botanical names with publications	Synonyms* with publications	Homotypic synonyms* with publications	Variety with publications	Distribution		Life-form/ Remark
					Major regions of the world	Major regions of India	
9.	*Jatropha nana* Dalzell & Gibson, Bombay Fl.: 229 (1861).	-	-	-	India (W. Madhya Pradesh)	W. Madhya Pradesh	Hemicryptophyte/ Wild
10.	*Jatropha podagrica* Hook., Bot. Mag. 74: t. 4376 (1848).	-	-	-	Guatemala, NW. Honduras, Nicaragua, India	S. India	Succulent Nanophanerophytes/Wild
11.	*Jatropha tanjorensis* J.L. Ellis [illegible] J. [illegible] Soc. 58: 834 (1962).	-	-	-	India	S. India	Wild
12.	*Jatropha villosa* Wight, Icon. Pl. Ind. Orient. 3: t. 1159 (1846).	*Jatropha peltata* Wight, Icon. Pl. Ind. Orient. 4: t. 1169 (1848).	*Jatropha wightiana* Müell-Arg. in A.P. de Candolle, Prodr. 15(2): 1080 (1866).	-	S. India	S. India	Wild

Jatropha is a cross pollinated genus. It has more chances to form new hybrids or in other words the *Jatropha* species have good ability to interbreed. Sahu *et al*., (2009) based upon RAPD analysis of *J. curcas* accessions have inferred that genetic diversity is very low in the accessions being cultivated in Uttar Pradesh.

However ongoing studies in India on *J. curcas* have revealed considerable phenotypic variation among provenances and individuals for oil and seed morphology (Singh, *et al*. 2008).

CONCLUSION

Monoculture, the lack of biodiversity, has been a contributing factor to several agricultural disasters in history. So there is an urge to study the diversity of *Jatropha* while establishing large scale plantation in India. Global survey of *Jatropha* genetic diversity is important. For this purpose international collaboration is necessary. Twelve species excluding synonyms of *Jatropha* have been recognised to occur in India. Large phenotypic diversity occurs among species of this genus. Variation has been noticed among various species of *Jatropha* using DNA markers too. Genetic studies have shown low within-species variation in germplasm occurring naturally in a small area. Within-species phenotypic variation in *Jatropha curcas* has also been reported in India but it may not be wide enough to support its large-scale plantation. The available phenotypic diversity can be used for the genetic improvement. Since *Jatropha* is a highly cross-pollinated crop, each plant is genetically different. This offers great potential for selecting superior plants in a breeding programme. There is a lot of scope to produce natural interspecific hybrids in this genus. One natural hybrid of *Jatropha*, namely *Jatropha tanjorensis (J. curcas* x *J. integerrima)* is already reported in India. Extensive wasteland available across the country can be utilized for planting *Jatropha curcas* and superior material of diverse genetic make up should be planted in such plantations.

ACKNOWLEDGEMENTS

Dr R.K. Gupta, Reader, Pt. Lalit Mohan Sharma Government (P.G.) College, Rishikesh, is acknowledged for support, useful suggestions and encouragement. We thank the Library Incharge, Forest Research Institute, and Botanical Survey of India, Dehradun for help.

REFERENCES

Blatter, E (1923). Records of the Botanical Survey India, vol. 8 no. 4, p. 435

Dalzell, NA & Gibson, A (1861). The Bombay Flora. Bombay, p. 229

Dasmann, RF (1968). A Different Kind of Country. MacMillan Company, New York

Gaur, RD (1999). Flora of the District Garhwal – North West Himalaya, Transmedia, Srinagar (Garhwal)

Hooker, JD (1887). Flora of British India, L. Reeve and Co., London, **5:** 382-383

Eillis, JL & Saroja, TL (1962). Journal of the Bombay Natural History Society Published: **58:** 834-836

Kumar, D, Singh S, Chandra, H, Malhotra, K & Hussain, T (2009). Damage from frost: a major challenge to planting of *Jatropha curcas* in north India. In: National Conference on Biofuels: Potential and Challenges, Jabalpur, 25-26 February, 2009

Kureel RS (2006). Prospects and Potential of *Jatropha curcas* for Biodiesel Production. In: Biodiesel Conference Towards Energy Independence – Focus on *Jatropha, Hyderabad, 6-10 June, 2006*

Maheshwari, JK (1962). Studies on naturalized flora of India. In Proceedings of Summer School of Botany (Eds Maheshwari, P, Johri, BM & Vasil, IK), Sree Saraswati Press, Calcutta

NIIR Board of Consultants and Engineers. The Complete Book on *Jatropha* (Bio-Diesel) with Ashwagandha, Stevia, Brahmi and Jatamansi Herbs (Cultivation, Processing and Uses) 2008. Publisher: Asia Pacific Business Press Inc

Prabakaran, AJ & Sujatha, M (1999). *Jatropha tanjorensis* Ellis and Saroja, a natural interspecific hybrid occurring in Tamil Nadu, India. Genetic Resources and Crop Evolution 46: pp.213-218

Ram SG, Parthiban, KT, Kumar, RS, Thiruvengadam, V & Paramathma M (2007). Genetic diversity among *Jatropha* species as revealed by RAPD markers. Genet Resour Crop Evolution 55: pp. 803–809

Sahu, M, Akram, M, Kendurkar, PS & Kumar, M (2009). Genetic diversity among Jatropha curcas accessions revealed by RAPD markers in National conference on Biofuels: Potential and Challenges, Jabalpur, 25-26 Februrary, 2009

Sujatha, M (2005). Genetic improvement of *Jatropha curcas* (L.) possibilities and prospects. Indian Journal of Agroforestry, vol. 8 no. 2: pp.58-65

Sujatha, M & Prabakaran, AJ (2003). New ornamental *Jatropha* through interspecific hybridization. Genetic Resources and Crop Evolution 50: pp. 75-82

Singh, VRR, Srivastava, RK, Kumar, D, Kumar, V Sharma, R, Chauhan, B, Singh, S & Malhotra, K (2008). Variation in germination and seedling survival among Uttarakhand accessions of Jatropha curcas. *In* Envis Forestry bulletin, vol. 8 no.1, pp. 91-95

Yan H (2008). Jatropha International Congress, December 17-18, 2008. Singapore

Website search

http://*www.indiastat.com/* accessed on 11th March, 2009

http://www.kew.org/ last accessed on 10th October, 2010

19

Medico-Religious Plant Diversity of Uttar Pradesh

MUKESH KUMAR AND DEEPAK KUMAR[1]

Department of Botany, Sahu Jain P.G. College, Najibabad–246 763, Bijnor, (U.P.)
[1]Department of Botany, Dhampur Degree College, Dhampur–246 761, Bijnor, (U.P.)

ABSTRACT

Plants are of paramount importance to human life since the time immoral. Human beings depend upon them for food, shelter, clothing, medicines, ornamentals and tools for spiritual needs. In India documented knowledge about uses of plants can be traced back to vedic period (1000-5000 BC). In vedic compendia, i.e., Regveda, Yajurveda, Atharvaveda and Ayurveda there are ample evidences regarding the use of natural resources at many places. It is found that people use them in their daily life. Ayurveda is closely connected with the vedic culture which is believed to have developed after the arrival of Aryans in India. Man is using plants as a source of medicine since ancient times. In the beginning there was the main source of the folk or Ethnomedicine. As per Indian mythology the members of solar family (including Sun and Moon) are called the plates, i.e., GRAHA. Moon completes its journey in twenty seven days in its orbit, i.e., the starting location is achieved again after 27 days. Indian astrologers divided this span of journey into twenty-seven parts and each part consists a group of stars which are recognized as "NAKSHATRAS", each one of them is symbolized by one tree. These twenty seven nakshatras are connected with twenty-seven trees evolved on earth in different classes or species. These trees are believed to be consisted of divine effects of each nakshatra in them and therefore, the people worship these nakshatras. These trees are called as NAKSHATRA TREES and they have several medicinal uses also. The leaves of *Ocimum sanctum* contains a chemical named as ocimine. Basil, plant grows under wild conditions in forest. It is of important pharmaceutical uses. It is symbolically worshipped in Kartik month of Hindi calendar. The fruit of *Embilica officinalis* is used for curing gastric troubles. It is a rich source of gallic acid, vitamin A & vitamin C. It is also a sacred plant of Prabodhani Ekadasi. The symbol of Bharani Nakshatra *Asparagus racemosus* (xerophytic plant) generally known as Brahman Ganga. It is a rich source of carbohydrate and sucrose. Extract of cladodesvs used in leucorrhoea, seminal debility and reduces blood pressure, useful in acidity and ulcer patients.

The present chapter provides an enumeration of 53 species of phenarogames found in U.P., which are sacred to the local people and are therefore, worshipped by them. These species are also used in the treatment of various ailments of the human beings.

Keywords: Diversity, medico-religious plants, nakshatra plants.

INTRODUCTION

The plants represent an enormous pool of natural resources that can produce various products and chemicals for the advantage of all other life-forms. Uttar Pradesh is an old part of Gondwanaland. It lies within 23°52'-31°278' North latitude, 77°3'-84°39' East longitude. It is bordered by Delhi and eight other states. Internationally it is bordered by Nepal in its northern side. In the present investigation, 53 plant species medico-religious have been reported, there are used in the worship of Gods during the festivals and at social ceremonies. They belong to 36 genera and 53 species. The leaves/flowers/fruits/seeds of different plants are used for the worship of Gods in the state. Some species like *Aegle marmelos, Butea monosperma, Cannabis sativa, Ocimum tenuifolium, Ziziphus maurtiana*, etc. are very common to the people. Some plants, viz. *Aegle marmelos, Cannabis sativus, Datura inoxia, Nelumbo nucifera, Saccharum officinarum, Trapa natanus, Ziziphus maurtiana,* are sold in the vicinity of the temples. Based on the mythology, 27 Nakashtras are connected with 27 trees evolved on earth in different classes or species. These trees offer divine effects of each nakshatra in them and the worship of these trees is the worship the respective Nakshatras. These trees are called NAKSHATRA TREES OR SACRED PLANTS OR MEDICO-RELIGIOUS PLANTS and have several medicinal uses also. People also use the parts of sacred plants growing in their local area for various purposes.

In the present paper an effort has been made to present the identity, availability and utility of medico-religious plants occurring in the State of Uttar Pradesh. Present work is based on scrutining of literature (Srivastava 2006, Sharma 2004 and Nadkarni 1954) as well as interview with the local people, especially the elderly members of the community.

METHODOLOGY

In order to collect the medico-religious plants of U.P. regular field-trips were organized and interviews held with the people of different localities who used plants as medicine as well as socio-religious works. In the enumeration, the plants have been arranged in alphabetical order and presented in a tabular form. The family names are also given for each plant followed by names of the parts, used as medicine and remarks. The plants were preserved following the standard methods (Jain and Rao 1997) and herbarium specimens are prepared to keep a record. (Table 1).

RESULTS AND DISCUSSION

A wide range of medico-religious plants are traditionally used by the local people. During the present studies the medico-religious plant diversity of Uttar Pradesh state of India has been recorded. It includes 53 taxa belonging to 48 genera and 37 families of higher plants. Of these 42 taxa belong to dicotyledons, 8 from monocotyledonous and 3 to the gymnospermous families. As per the mythology of India the members of solar family, viz. sun and moon are called the planets, i.e., GRAH. Moon is related to complete its orbital journey in 27 days. The Indian astronomists have divided this span into 27 units and each of which consists of a group of stars,

Table 1: Medico-religious plants of Uttar Pradesh

Sl. No.	Botanical Names	Local Names	Families	Uses	Remarks
1.	*Acacia catechu* (L.) Willd.	Khair	Mimosaceae	Leaves, bark and gum used in asthma, diarrhoea, disease of urino-genital system.	Margsees N.P.
2.	*Achyranths aspera* L.	Chirchiri	Amaranthaceae	Whole plant used in dropsy, gonorrhoea, laxative, purgative anthelmintic, insect bite.	
3.	*Adansonia digitata* L.	Burhawa Baba	Bombacaceae	Antiperiodic in malaria, pneumonia, hydrophobic.	
4.	*Aegle marmelos* (L.) Correa	Bel	Rutaceae	Leaves and fruits are used in urinary troubles, laxative, febrifuge, ophthalmic, dysentery.	Chitra N.P.
5.	*Amaranthus caudatus* L.	Ramdana	Amaranthaceae	Cooling, pneumonia, hydrophoebia fever.	
6.	*Anthocephalous chinensis* (Roxb.) Miq.	Kadamb	Rubiaceae	Nerve tonic, memory promoter and epidepsy	Satbhisht N.P.
7.	*Artocarpous heterophyllus* Lamk.	Kathal	Moraceae	Inflammation, bronchitis and biliousness	Uttra Ashad N.P.
8.	*Asparagous recemosus* Willd.	Satavar	Aspargacaeae	Aphrodisiac, laxative, expectorant, tuberculosis.	
9.	*Azadirachta indica* A. Juss.	Neem	Meliacea	Whole plant is used in blood and skin diseases	Uttara Bhadrapada N.P.
10.	*Butea monosperma* (Lam.) Taubert	Palas	Fabaceae	Night blindedness, elephantiasis, aphrodisiac, piles	Purva Phalguni N.P.
11.	*Calotropis gigantea* (L.) R.Br.	Ark Madar	Asclepideaceae	Toothache, stomachache, appetizer, piles, asthma, tonic	Srwan N.P.
12.	*C. procera* (Aiton) R.Br.	Madar	Asclepideaceae	Toothache, stomachache, appetizer, piles, asthma, tonic	Srwan N.P.
13.	*Cannabis sativa* L.	Bhang	Cannabiaceae	Mental retardedness	
14.	*Crinum asiaticum* L.	Sudarshan	Amaryllidaceae	Skin diseases	
15.	*Cucumis sativus* L.	Khira	Cucurbitaceae	Indigestion, fever, purgative, emetic.	
16.	*Curcuma longa* L.	Haldi	Zingeberaceae	Asthma, fever, antibiotic.	
17.	*Cynodon dactylon* (L.) Pers.	Doob	Poaceae	Astringent, diuretic, genito-urinary wound.	
18.	*Dalbergia sissoo* Roxb.	Shishha	Fabaceae	Indigestion, purgative and emetic	Adra N.P.
19.	*Datura metel* L.	Dhatura	Solanaceae	Narcotive, sedative, intoxicant, ulcer disease.	
20.	*Dendrocalamus strictus* L.	Bans	Poaceae	Gonorrhoea, menstrual cycle	Punarvasu N.P.
21.	*Desmostachya bipinnate* (L.) Stapf.	Kush	Poeaceae	Astringent, laxative, biliousness, urinary disease.	
22.	*Embilica officinalis* Gaertn.	Anola	Euphorbiaceae	Gastric, ulcers and fever.	
23.	*Ficus benghalensis* L.	Bargad	Moraceae	Liver troubles, diabetes, cooling tonic	Magha N.P.
24.	*F. racemosa* L.	Gular	Moraceae	Biliousness, ulcer, vaginal complaints	Kritika N.P.
25.	*F. religiosa* L.	Pipal	Moraceae	Leucorrhoea, biliousness, ulcer, disease of vagina and uterus, alexipharmic	Pukhya N.P.

Table 1: Contd.

Sl. No.	Botanical Names	Local Names	Families	Uses	Remarks
26.	*F. virens* Aiton	Pakar	Moraceae	Spasmodic, diuretic, diaphoretic.	Uttar phalguni N.P.
27.	*Flacourtia indica* (Burm. f.) Merr.	Katia	Flacourtiaceae	Inflammatic, swelling, demulcent, diuretic	Visakha N.P.
28.	*Hibicus rosa-sinensis* L.	Deviphul	Malvaceae	Gonorrhoea, fever, antioxidant	
29.	*Lawsonia inermis* L.	Mehndi	Lythraceae	Digestive, skin disease, ulcers	
30.	*Madhuca indica* F. Ham. ex Gmel.	Mahua	Sapotaceae	Astringent, laxative sporadic	Ravati N.P.
31.	*Mangifera indica* L.	Aam	Anacardiaceae	Stomachic, duretic, laxative	Purva N.P. Bhadrapada
32.	*Mesua ferrea* L.	Nagkeshar	Clusiaceae	Weakness, tuberculosis	Ashlesa N.P.
33.	*Mimusops elengi* L.	Maulsari	Sapotaceae	Stimulant, weakness	
34.	*Mimusops hexacandara* Roxb.	Khinni	Sapotaceae	Astringent, laxative, sporadic	Anuradha N.P.
35.	*Nelumbo nucifera* L.	Kamal	Nympheaceae	Dysentery, diarrhoea	
36.	*Nerium indicum* Mill	Kaner	Apocynacea	Skin disease.	
37.	*Nytanthes arbor-tristis* L.	Harshingar	Nyetanthaceae	Aphrodisiac piles, anthelmintic	
38.	*Ocimum tenuiflorum* L.	Tulsi	Lamiaceae	Fever, parasitical skin disease.	
39.	*Oryza ruipogon* Griff.	Pasia	Poaceae	Dysentery, gonorrhoea, women's troubles	
40.	*Pinus rouxburghii* Sar.	Chir	Pinaceae	Mental retardness	Jyashtha N.P.
41.	*Prosopis cineraria* (L.) Druce	Sami	Fabaceae	Anthelmintic, alexiteric, leprosy	Ghanistha N.P.
42.	*Saccharum officinarum* L.	Ganna	Poaceae	Dysentery, indigestion.	
43.	*Salix tetrasperma* Roxb.		Salicaceae	Aphodisiac, althernative, diuretic	Purva Ashad N.P.
44.	*Sapindus mukorrossi* Gaertn.	Ritha	Sapindaceae	Mental retardedness	Hasta N.P.
45.	*Saraca asoca* (Roxb.) de Wilde	Ashok	Fabaceae	Menstrual cycle regulation	
46.	*Sesamum indicum* L.	Til	Pedaliaceae	Spleen troubles, piles bleeding, menorrhoea	
47.	*Shorea robusta* Gaertn. f.	Sal	Lipterocarpacea	Emetic, visceral obstruction	Mula N.P.
48.	*Strychnos nix-vomica* L.	Kuchala	Loganinacea	Fever, headache, mental reatrdness	Aswani N.P.
49.	*Syzygium cumini* (L.) Skeels	Jamun	Myrtaceae	Indigestion, haemorrhiza in kidney and uterus	Rohani N.P.
50.	*Terminallia alata* Heyne ex Roth.	Asna	Combretacea	Skin diseases	
51.	*Terminallia arjuna* Roxb.	Arjun	Cobretaceae	Heart diseases	Swati N.P.
52.	*Trapa bispinosa* Roxb.	Singhara	Trapaceae	Dysentery, diarrhoea	
53.	*Ziziphus mauritiana* Lamk.	Ber	Rhmanaceae	Cough, cold, skin disease, ulcerated gums.	

* N.P.= Nakshatra Plant.

which are recognized as "NAKSHATRAS". Each nakshatra has been symbolized by a specific plant species. All the nakshatra plants possess significant medicinal and economic values.

Shukla (2008) has reported that the stems and roots of *Achyranthus aspera* are being used by the people of India as herbal toothpaste. It is useful in toothache and also to remove the bad odour of the mouth. Srivastava (2006) has reported that *Mimusops elengi* has a powerful germicidal property. Its paste is applied for treatment of hydrocele and piles. The juice of *Ziziphus mauritiana* is used as a diuretic and ophthalmic medicine. Its leaves are used for curing night-blindedness, muscular pains, eczema and stomach tumours.

REFERENCES

Jain, SK & Rao, RR (1997). *A Handbook of Field Herbarium Methods.* Today and Tomarrow Printers and Publishers, New Delhi

Nadkarni, AK (1954). *Indian Materia Medica, Bombay*

Shukla, CP (2008). Floristic investigation on Ethnomedicinal flora of Dewghat Forest Koraon Range (U.P.), Proc. National Seminar on 'Natural Resources' held on August 30-31, 2008, Sahu Jain P.G. College, Najibabad (U.P.)

Srivastava, RC (2006). *Plant Resources in India*, New Delhi

Sharma, NK (2004). *Ethno-Medico-religious plants of Hadoti Plateau* (S.E. Rajasthan), pp. 136-150. Ethno-Medicinal Plants, Pointer Publishers, Jaipur

20

Floristic Diversity of Sikkim Himalayas with Notes on Ethnoveterinary Applications

KUMAR AVINASH BHARATI AND BL SHARMA

Department of Botany, N.R.E.C. College, Khurja–203 131, Bulandshahr, Uttar Pradesh, India
E-mail: kumaravinashbharati@rediffmail.com

ABSTRACT

The present paper deals with the survey and studies of plants in relation to treatment of livestock by major ethnic groups (Lepcha, Bhutia and Nepalese) of Sikkim Himalayas. The data have been collected between March 2006 and June 2009 through open conversation with local people and traditional healers. It has resulted in documentation of medicinal plants representing 30 families. The largest representation is of the family Rosaceae with five members followed by Ericaceae with three members, Caryophyllaceae, Saxifragaceae, Combretaceae with two members each and rest of the families have been represented by a single member. It indicates the rich biodiversity of Sikkim Himalayas. The indigenous people use these 40 species to cure 12 groups of ailments. Out of 40 species, 8 have potential to cure fever (maximum number of species), followed by diarrhoea and dysentery (6 species), injuries and wound (5 species), etc. It has been observed that 22 species are subjected to destructive harvesting for veterinary medicines using whole plant, root/rhizome, which may be related to their possible vulnerability towards endangerment. The richness of plant diversity decreases with increase in altitude but the dependence on herbal medicine shows inverse relationship, perhaps due to the preference given to herbal remedies and a combination of paucity of having no alternatives and poverty. The study, thus, underlines the floristic diversity and potentials of the ethnomedicines and need for documentation of traditional knowledge pertaining to the medicinal plant utilization for the greater benefit of mankind.

Keywords: Livestock, ethnoveterinary, plant diversity, Sikkim Himalayas, tribe.

INTRODUCTION

Plants have been the basis of several traditional medicine systems throughout the world for thousands of years and continue to provide mankind with new remedies. The potential for finding new compounds is still enormous since till date merely about 1% of species have been studied for their pharmaceutical potentials (Jachak and Saklani, 2007). Further, the World Health Organisation's (WHO) Traditional Medical Programme has provided the evidence that ethnomedical information can lead to valuable drug discovery. Several active compounds have been discovered from plants on the basis of ethnobotanical information, and used directly as patented drugs like, "Maprouneacin" isolated from *Maprounnea africana* is used as an anti-diabetic agent, "Taxol" obtained from *Taxus breviflora* is used as an anti-tumour drug, "Artemisinin", discovered from *Artemisia annua*, is used as a potent antimalarial compound against *Plasmodium* strains resistant to all known anti-malarials (Ajibesin *et al.*, 2007).

The ethnobotanical survey has been found to be one of the reliable approaches to drug discovery. The major advantage of choosing plants as the starting point in drug development through ethnobotanical survey is that active constituents of such plants which have undergone long-term use by man are likely to be safer than the active compounds isolated from plants with no history of ethnomedicinal use (Fabricant and Farnsworth, 2001).

Documentation of traditional knowledge is the urgent need of the hour since it transfers from generation to generation through oral communication. While transferring from one generation to another, a few medicines are lost in the process due to lack of proper records. Further, indigenous data bank will serve the important function of establishing community knowledge firmly in the public domain. This will provide the technical basis for rejecting patent claims that derive from indigenous knowledge. Patents like the one that was earlier granted on the wound healing properties of Haldi (*Curcuma domestica* Valeton) or the American patent on a liver drug derived from Amalaki (*Phyllanthus amarus* Schumach) used traditionally to cure liver disorders, should be challenged on the grounds of theft of indigenous knowledge (Sahai, 2003).

The survey has been conducted to document the ethnoveterinary wealth of Sikkim Himalayas. The state of Sikkim is situated on the flanks of Eastern Himalayas between 27°10' – 28°5' N latitude and 88°30' – 89° E longitude. It offers a number of features for ethnobotanical studies, including unique ethnic composition, apart from the three major ethnics—Bhutia, Lepchas, and Nepalese, a conglomerate of over 20 ethnic tribes and a still more number of sub-tribes are inhabit here (Rai and Sharma, 1994). All the major tribes have their unique culture, language and ethnobotany (Pradhan and Badola, 2008). Second, animal husbandry plays a significant role in economy; large number of livestock are maintained still in traditional methods (Balaraman and Golay, 1991). Another, due to the lack of modern infrastructure, at most of the localities, indigenous knowledge is still preserved but is depleting gradually because of the interaction with modern medicines, media and people.

The ethnobotany of Sikkim Himalayas had been described first of all by Atkinson in his premier publication "Economic Botany of the Himalayan Regions" (Atkinson, 1882). He was followed by Biswas (1956), Puri and Pandey (1980), Hajra and Chakraborty (1981), Bennet (1983), Krishna and Singh (1987), Srivastav *et al.* (1987), Srivastava (1990), Rai and Sharma (1994), Mayer (1995), Bhasin (1997), Jana and Chauhan (2000), Burman (2003), Maity *et al.* (2003), Dash *et al.* (2003), Dutta and Dutta (2005), Hussain and Hore (2006), Panda (2007), Brahma (2007), Sahoo and Ansari (2007), Panigrahi (2007), Pradhan and Badola, (2008), but the region has not yet been explored for livestock medicines. Keeping this in view, an attempt has been made to explore and document the traditional knowledge used in veterinary practices.

METHODOLOGY

The information had been gathered on datasheets which carried details such as local names, plant part(s) used, diseases treated, methods of preparation, methods of administration, doses and duration of treatment, gathered through personal interaction with community elders, forest dwellers and traditional healers. The field study has been carried out in nine villages besides a number of hamlets. The exploration route has been furnished in Fig. 1.

The data presented are based on the interviews conducted (March, 2006 to June 2009) with 42 informants (65% men and 35% women) who are natives of the studied locality. The methods adopted for extracting information from the locals have been those followed by Jain (1995), Martin (1995) and Pujadas *et al.* (2004). The conversation was held in Nepali, the local language of Sikkim. The prevalent animal diseases, their diagnostic details, medicinal plant species and other raw materials used in the treatments have been documented. These data have been verified with other informants among their communities. The specimen have been dried and preserved in the form of herbarium (Singh and Subramanium, 2008). Plants have been identified with the help of relevant flora like Hajara and Verma (1996), Herbaria of BSI at Gangtok (BSHC), and Dehradun Herbarium at Dehradun (DD), India. The herbarium specimens have been deposited at the herbaria of the Department of Botany, N.R.E.C. College, Khurja-203131, Bulandshahr, Uttar Pradesh, India.

RESULTS

The plants are being enumerated alphabetically with respect to their botanical names, families, English names, local names and ethnoveterinay uses in Sikkim.

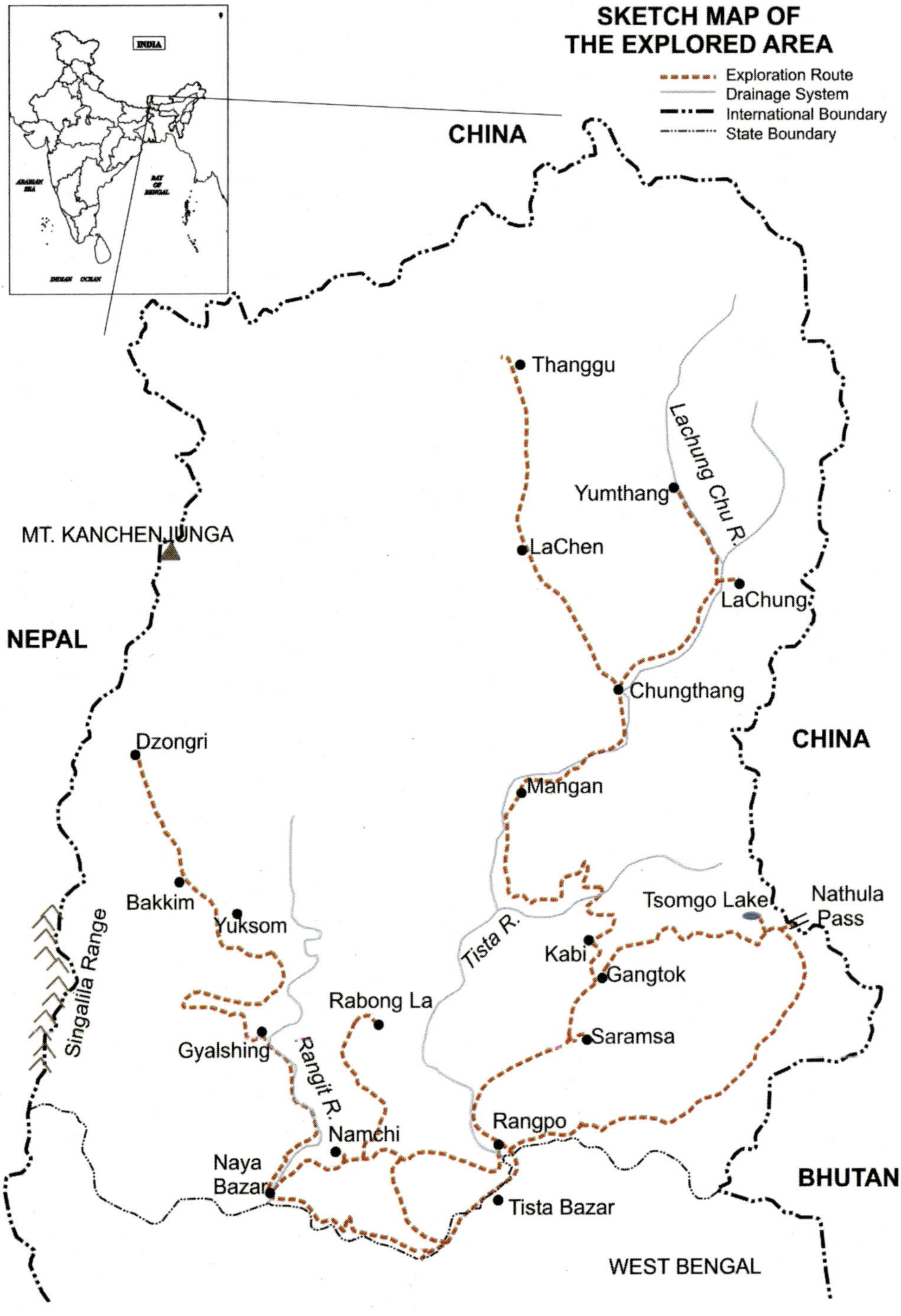

Fig. 1. Sketch map of the explored area.

Table 1: Enumeration of medicinal plants with ethnoveterinary notes

S.No.	Botanical Names and Families (in parenthesis)	English names	Local names	Ethnoveterinary medicinal uses
1.	*Abies densa* Griff. (Pinaceae)	The West Himalayan High-Level Fir	Ailey	When cattle lose hunger, feel uneasy, and exhibit signs of fever then juice of fresh leaves are administered orally.
2.	*Acorus calamus* Linn. (Acoraceae)	Calamus, Sweet flag, Sweet root	Bojo, Bojho	The paste of root and rhizome is orally administered twice a day to cure indigestion and paste of rhizome is applied on the wound for quick healing. Aromatic root powder is externally applied on the skin for protection against insects.
3.	*Amomum subulatum* Roxb. (Zingiberaceae)	Greater cardamom	Alaichi	The paste of seeds is used as antidote against scorpion-sting and other insect bites. Oil extracted from dry seeds is applied on eye-lids to allay inflammation of the eye.
4.	*Artemisia vulgaris* Linn. (Asteraceae)	Mugwort, Old Woman	Tetaypati, Teil	Leaves are crushed and extract is applied externally on skin to cure itching. Fresh leaves are cleaned, ground and the sap is extracted. It is used as a nasal-drop to stop nose bleeding.
5.	*Arundinaria maling* Gamble (Poaceae)	Hainan bamboo Cane	Malingo	Ash of the root is mixed with a few drops of mustard oil and the paste is applied externally on the ringworm. About 20 gm leaves are mixed with four to six black pepper (*Piper nigrum* L.) and fed with little salt to cattle to control diarrhoea.
6.	*Asparagus racemosus* Willd. (Asparagaceae)	Wild Asparagus	Satamuli	About 100 gm of root is cut into small pieces and mixed with hey or grain to feed the cattle for a fortnight to increase their milk-yield.
7.	*Astilbe rivularis* Buch.-Ham. (Saxifragaceae)	Astilbe	Buroo Okhatii, Padum	Internal part of the root is made into powder or paste and administered orally to cattle after delivery, to control blood secretion.
8.	*Bergenia ciliata* (Haw.) Sternb. (Saxifragaceae)		Pakhanbhed, Pakhin bet	Root is crushed and extract (≈ 40 ml) is given twice a day to cattle to cure diarrhoea and weakness.
9.	*Cannabis sativa* Linn. (Cannabaceae)	Hemp	Ganja, Bhang	Stem is cut into small pieces and fed to livestock to cure inflammation. The same formula has been used for good health as tonic during weakness.

Table 1: Contd.

S.No.	Botanical Names and Families (in parenthesis)	English names	Local names	Ethnoveterinary medicinal uses
10.	*Centella asiatica* (L.) Urban (Apiaceae)	Indian Pennywort	Ghod Tapre, Bhram Jhar	Plant is crushed and administered thrice a day to cure urinary disorders like little and/or frequent urine pass.
11.	*Costus speciosus* Sm. (Costaceae)	Crepe Ginger	Bet lauree	Rhizome powder (≈ 20 gm) is given twice a day to the animals to cure fever and inflammation.
12.	*Datura metel* Linn. (Solanaceae)	Thorn-apple	Kalo Dhaturo	Dried leaves are crushed and mixed with a few drops of mustard (*Brassica juncea*) oil. The paste is applied as an ointment to cure inflammation of udder of lactating animals.
13.	*Drymaria cordata* Willd. ex Schult. (Caryophyllaceae)	Chickweed, West Indian Chickweed	Abhijal	Paste of plant is applied externally on fractured bone and bandaged with the help of cotton cloth. After 20-25 days bandage is removed.
14.	*Equisetum debile* Roxb. ex Vaucher (Equisetaceae)	Horsetail, Scouring rush	Kurilo, Kurkure Ghans	Paste of stem shoot is applied externally on injury to stop bleeding. The plant is crushed and orally administered twice a day with *Jaggery* to cure passing of blood in the urine and blood secretion from vagina.
15.	*Fragaria nubicola* Lindl. ex Lacaita (Rosaceae)	Indian Strawberry	Bhui Ainselu	Leaves and fruits are administered with lukewarm water to cattle as diuretic.
16.	*Fraxinus floribunda* Wall. (Oleaceae)	Himalayan Manna Ash	Kangsu, Tuhasi	Thick paste of stem bark is applied externally on fractured or dislocated bone and bandaged with cotton cloth. After 20-25 days the bandage is removed.
17.	*Gypsophila cerastioides* D. Don (Caryophyllaceae)	Himalayan Baby's Breath	Lal Dana	Leaves and fruits are crushed and given orally during problems of alimentary canal. Such problems are diagnosed by behaviour of the animal.
18.	*Lycopodium clavatum* Linn. (Lycopodiaceae)	Lycopodium	Nagebeli	About 100 gm plant is crushed and given twice a day with little salt to cure muscle pull of the cattle.
19.	*Mallotus philippensis* Muell.-Arg. (Euphorbiaceae)	Monkey Face Tree	Numboongkor, Purva, Tukla, Sinduri	Seeds are administered to pigs along with food/grain to kill tapeworm.
20.	*Mentha viridis* Linn. (Lamiaceae)	Brown Mint	Mentha, Babri	The leaves are crushed with water and decocted. The decoction (≈ 100 ml) is given thrice a day to the animal to cure mild fever.

Table 1: Contd.

S.No.	Botanical Names and Families (in parenthesis)	English names	Local names	Ethnoveterinary medicinal uses
21.	*Oroxylum indicum* Vent. (Bignoniaceae)	Indian Trumpet Flower, Broken Bones Tree	Totilla, Tatelo, Shivnak, Ripmu	Paste of seed is mixed with little mustard oil and applied externally on crack-nipple of mulching animal for quick healing.
22.	*Picrorhiza kurroa* Royle ex Benth. (Scrophulariaceae)	Hellebore	Kutki	The root of plant is crushed and decocted. About 100 ml of decoction is given thrice a day to livestock to cure cold and fever.
23.	*Pieris formosa* D. Don (Ericaceae)	Re-Leaf Pearl Bush	Baloo, Aphin	The root is cut into small pieces and decocted. The preparation is used for dressing of poisonous wound. **Note:** This is a highly poisonous plant. Tender shoots are said to be more poisonous than older parts. When goats, sheep and cows eat the tender shoots they die. The local cattle realise the danger and do not eat tender foliage, but when cattle from other areas browse these shoots they die (Chopra *et al.,* 1965).
24.	*Plantago major* Linn. (Plantaginaceae)	Englishman's foot, Greater Plantain, Ripple Grass	Nasha Jhar	Paste of the root is used as ointment for quick healing of cut and wound.
25.	*Podophyllum hexandrum* Royle (Berberidaceae)	Duck's foot, Indian May Apple	Papari, Panchpatey	Paste of the root and rhizome is externally applied on hoof to cure hoof infection of cattle.
26.	*Potentilla fruticosa* Linn. (Rosaceae)	Shrubby Cinguefoil	Chinya phal, Bhairung	Plant is crushed and decocted. About 50 ml of decoction is given daily for fifteen days to regulate fertility of cattle.
27.	*Prunus cerasoides* D. Don (Rosaceae)	Himalayan cherry	Kongki, Paiyung, Painyu	Paste of the stem bark is applied externally on the sprain for quick healing.
28.	*Rauwolfia serpentina* Benth. ex Kurz. (Apocynaceae)	Serpentina	Sarpgandha	About 20 gm of root powder is orally administered to cattle as antidote to snake-bite.
29.	*Rhododendron arboreum* Sm. (Ericaceae)	Tree Rhododendron	Etok, Ghonas, Gurangs, Guras, Lal-Gurans, Lalli-Gurans	About 100 gm of flowers are crushed with water and given twice a day to cattle to cure diarrhoea and dysentery.
30.	*R. campanulatum* D. Don (Ericaceae)	Bell Rhododendron	Bargi, Cheriala, Teotosa, Gurans, Nilo Chimal	Leaves/twigs of the plant are mixed with leaves of tobacco (*Nicotiana tabacum* L.) in a ratio of 2:1 and

Table 1: Contd.

S.No.	Botanical Names and Families (in parenthesis)	English names	Local names	Ethnoveterinary medicinal uses
				crushed together. About 10 gm paste is administered to cattle to cure cold and chronic fever. **Note:** However, it is reported that young buds and leaves of the plant are poisonous to cattle (Anonymous, 2004).
31.	*Rubia cordifolia* Linn. (Rubiaceae)	Indian Madder	Manjito, Vhyem, Vhyeni, Soth	About 200 gm root is crushed on stone and decocted. It is given twice a day to cattle after delivery for quick recovery.
32.	*Rubus calycinus* Wall. (Rosaceae)	Creeping Raspberry	Ghordabri	About 20 gm of root-powder and leaves are given twice a day to cattle to cure little and frequent urine pass.
33.	*R. ellipticus* Sm. (Rosaceae)	Yellow Himalayan Raspberry, Asian Wild Raspberry	Ainselu, Asayloo, Kashyem	About 100 gm of new tender leaves and/or young stem-shoots are pounded and given thrice a day to the animal to cure cold and fever.
34.	*Rumex nepalensis* Spreng. (Polygonaceae)	Nepal Dock	Halhaly, Palu, Isomo	About 200 gm of root is crushed and decocted. It is administered thrice a day with little salt to the cattle to control diarrhoea and dysentery.
35.	*Schima wallichii* Choisy (Theaceae)	Chinese Guger Tree, Needle Wood	Chilone, Chilauni, Sambrong, Samching	Paste of stem bark (≈ 50 gm) is administered to cattle with little common salt to kill intestinal worms.
36.	*Swertia chirata* C.B. Clarke (Gentianaceae)	Chiretta	Chirowto	Fresh plant (≈ 100 gm) is cut into small pieces and decoction is prepared. This is later filter through cotton cloth and administered twice a day to cattle to cure fever and mouth ulcer.
37.	*Taxus wallichiana* Zucc. (Taxaceae)	Himalayan Yew	Cheongbu, Dhengre salla, Chharey salla	Thick paste of stem bark is applied externally on fractured bone and bandaged with cotton cloth for fortnight. **Note:** *Taxus wallichiana* (Himalayan yew) is also a source of the anticancer drug, paclitaxel (http://en.wikipedia.org/wiki/Taxus_wallichiana).
38.	*Terminalia bellirica* (Gaertn.) Roxb. (Combretaceae)	Belleric Myrobalan	Barra	About 100 gm of fruits are crushed with water and decocted. It is administered thrice a day to livestock for treatment of diarrhoea and dysentery.

Table 1: Contd.

S.No.	Botanical Names and Families in parenthesis	English names	Local names	Ethnoveterinary medicinal uses
39.	*Terminalia chebula* Retz. (Combretaceae)	Chebulic Myrobalan	Harra, Selim Pot	About 100 gm of fruit and/or bark are crushed and decocted. It is administered thrice a day with little rock salt to livestock to cure diarrhoea and dysentery.
40.	*Zingiber officinale* Roscoe (Zingiberaceae)	Ginger	Aduwa	Rhizome is ground and the extract is administered in the morning and evening with warm rice wash to cure cough.

Abies densa Griff.

Acorus calamus Linn.

Arundinaria maling Gamble

Bergenia ciliata (Haw.) Sternb.

Centella asiatica (L.) Urban

Datura metel Linn.

Fragaria nubicola Lindl. ex Lacaita

Mentha viridis Linn.

Plantago major Linn.

Podophyllum hexandrum Royle

Pieris formosa D. Don

Rhododendron campanulatum D. Don

Rhododendron arboreum Sm.

Rauwolfia serpentina Benth. ex Kurz.

Rubus ellipticus Sm.

Table 2: Comparison on the use of some plant species by different ethnic groups in veterinary and human medicine

Species	Tribes	Part(s) used	Used as veterinary medicine (present study)	Tribes	Part(s) used	Used as human medicine (Pradhan and Badola, 2008)
Acorus calamus Linn.	Nepali	Roots/rhizome	Wound healing and insects repellent	Lepcha	Roots/ rhizome	Cough and fever
Cannabis sativa Linn.	Nepali	Stem	Tonic, inflammation	Lepcha Bhutia	Seeds Seed and Leaves	Bodyache Burn, muscular pain and stomach pain
Costus speciosus Sm.	Nepali	Rhizome	Fever, inflammation	Lepcha	Roots/ rhizome	Urinary tract infection
Drymaria cordata Willd. ex Schult.	Nepali	Whole plant	Mouth ulcer, bone fracture	Lepcha	Whole plant	Sinusitis and nose blockage
Picrorhiza kurroa Royle ex Benth.	Nepali	Root	Appetizer, cold and mild fever	Lepcha Bhutia	Roots/ rhizome Roots/ rhizome	Cough and fever Jaundice, stomach ache and dysentery
Rhododendron campanulatum D. Don	Bhutia and Nepali	Leaf	Chronic fever, muscle weakness and cold	Lepcha	Leaves	Cough
Zingiber officinale Roscoe	Nepali	Rhizome	Cold & cough	Lepcha	Rhizome	Cough, fever and throat pain

Table 3: Flowering months of species in Sikkim Himalayas

Species	**Months of the year**											
	J	**F**	**M**	**A**	**M**	**J**	**J**	**A**	**S**	**O**	**N**	**D**
Abies densa Griff.				■	■							
Acorus calamus Linn.								■	■	■		
Amomum subulatum Roxb.				■	■							
Artemisia vulgaris Linn.							■	■	■	■		
Asparagus racemosus Willd.				■	■	■	■	■				
Astilbe rivularis Buch.-Ham.							■	■	■			
Bergenia ciliata (Haw.) Sternb.				■	■							
Cannabis sativa Linn.						■	■	■	■			
Centella asiatica (L.) Urban				■	■	■						
Costus speciosus Sm.						■	■	■				
Datura metel Linn.			■	■	■	■	■	■	■	■		
Drymaria cordata Willd. ex Schult.				■	■	■						
Fragaria nubicola Lindl. ex Lacaita				■	■	■						
Fraxinus floribunda Wall. in Roxb.				■	■							
Gypsophila cerasioides D. Don					■	■	■					
Mallotus philippinensis Muell.-Arg.									■	■	■	
Mentha viridis Linn.								■	■			
Oroxylum indicum Vent.							■	■				
Picrorhiza kurroa Royle ex Benth.						■	■	■				
Pieris formosa D. Don			■	■	■							
Plantago major Linn.					■	■	■					
Podophyllum hexandrum Royle					■	■	■	■				
Potentilla fruticosa Linn.						■	■	■	■			
Prunus cerasoides D. Don										■	■	
Rauwolfia serpentina Benth. ex Kurz.		■	■	■	■							
Rhododendron arboreum Sm.		■	■	■	■							
R. campanulatum D. Don				■	■	■						
Rubia cordifolia Linn.						■	■	■	■	■	■	
Rubus calycinus Wall.			■	■	■							
R. ellipticus Sm.		■	■	■								
Rumex nepalensis Spreng.				■	■	■	■					
Schima wallichii Choisy					■	■						
Swertia chirata C.B. Clarke								■	■	■		
Taxus wallichiana Zucc.				■	■							
Terminalia bellirica (Gaertn.) Roxb.			■	■	■							
Terminalia chebula Retz.				■	■							
Zingiber officinale Roscoe								■	■	■		

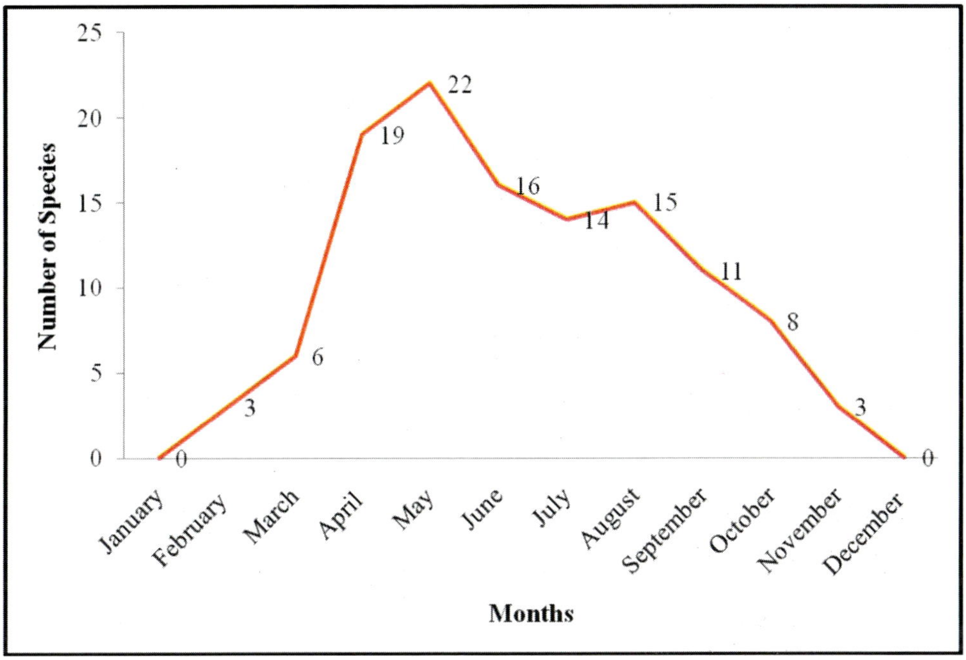

Fig. 2. Flowering periods of plant species.

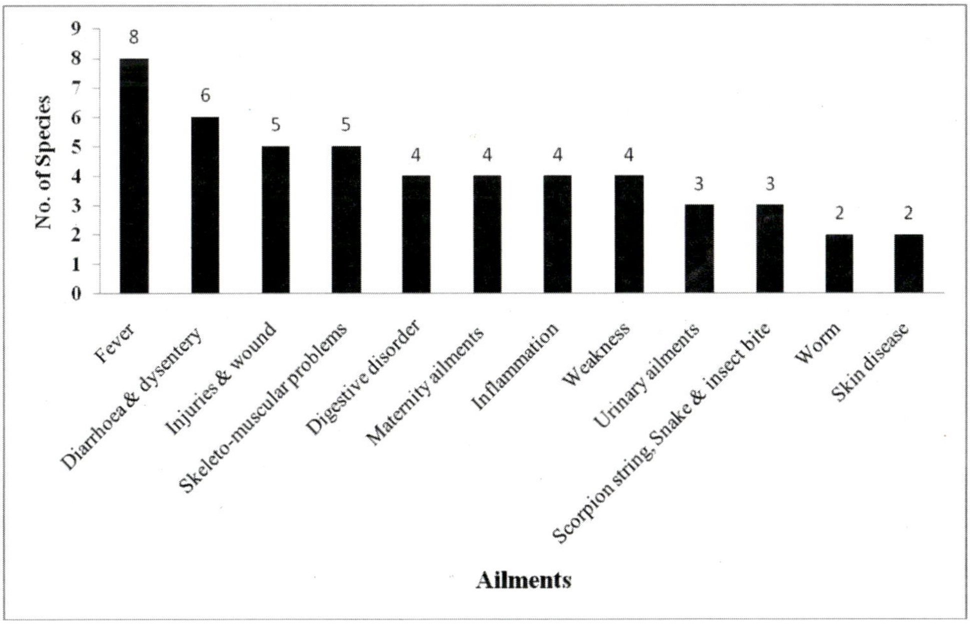

Fig. 3. Major ailment categories and use frequency (number of species) of plants.

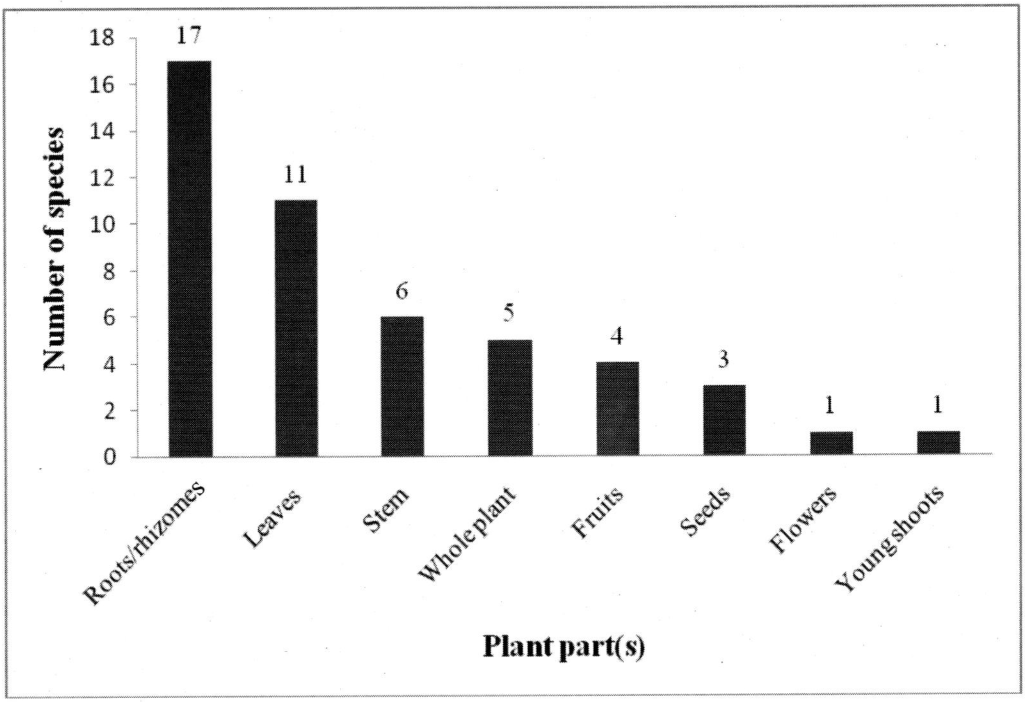

Fig. 4. Use frequency (number of species) of different plant parts in ethnoveterinary medicine preparation.

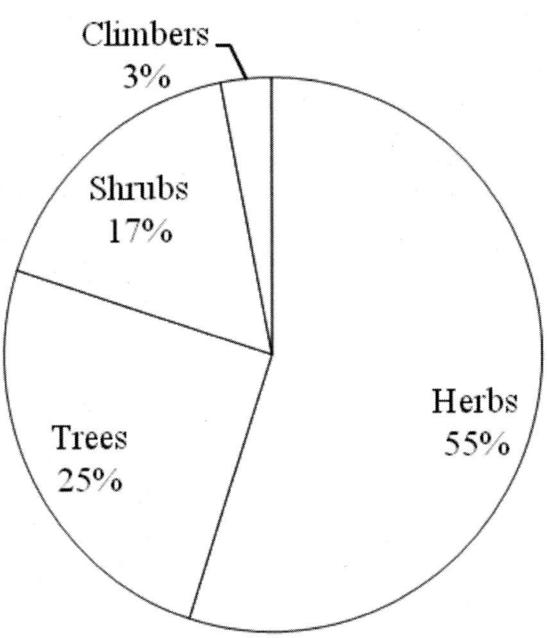

Fig. 5. Percentage distribution of medicinal plant species according to life-forms.

Table 4: Representation of species in various families

Families	No. of species
Rosaceae	05
Ericaceae	03
Caryophyllaceae	02
Saxifragaceae	02
Combretaceae	02
Zingiberaceae	02

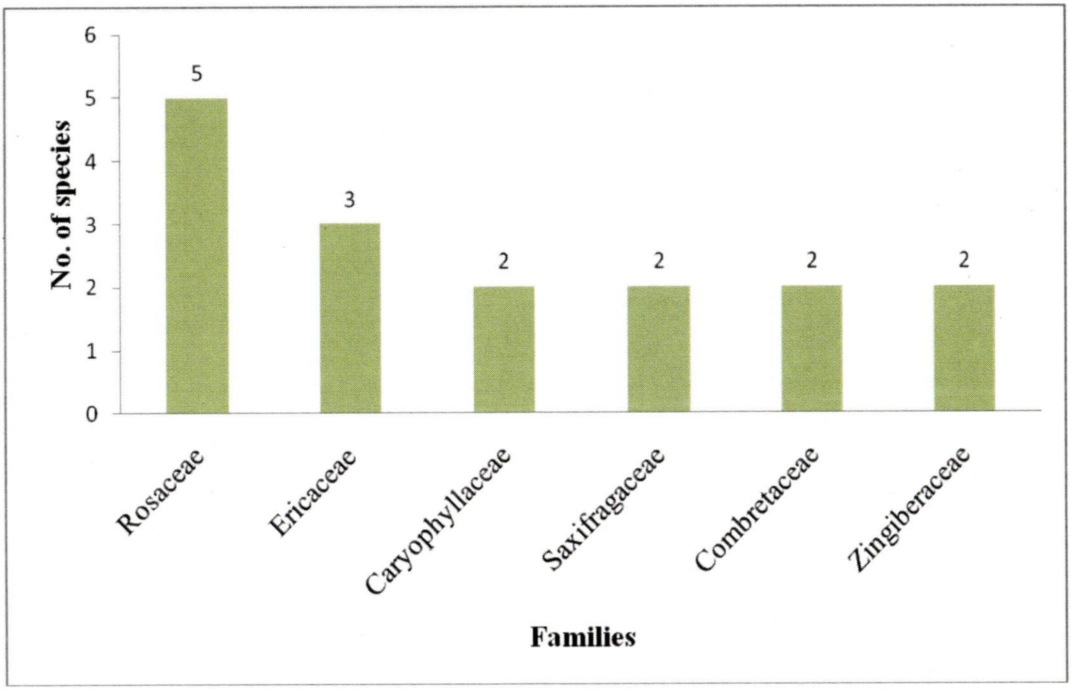

Fig. 6. Representation of species in various families.

Table 5: Distribution of ethnoveterinary plants

Altitude (m)	No. of species
Up to 900	16
900-1500	16
1500-2700	28
2700-3600	13
Above 3600	05

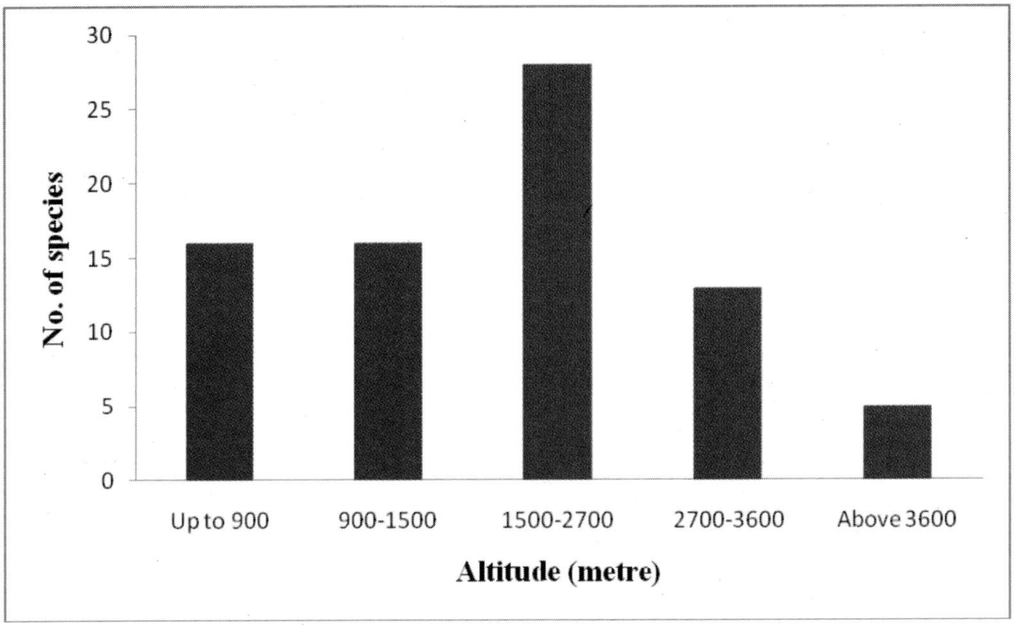

Fig. 7. Distribution of ethnoveterinary plants.

DISCUSSION

In the present study, a total of 40 species of plants have been recorded from Sikkim Himalayas which are being used in the treatment of about 50 ailments of livestock. Out of 40 species, eight species have been used to cure fever, followed by six species figured to treat diarrhoea and dysentery, five species for injuries and wound, etc. (Fig. 3).

It has been observed that the maximum number of species are harvested for roots and rhizomes (17 veterinary remedies), followed by leaves (11 veterinary remedies), whole plant (05 veterinary remedies), fruits (09 veterinary remedies), seeds (03 veterinary remedies), etc. (Fig. 4). The reason behind harvesting the highest number of species for roots is the presence of high concentration of bioactive compounds in them (Robinson, 1974). A total of 22 species are subjected to destructive harvesting for veterinary medicines using whole plant and root/rhizome which may be related to their possible vulnerability towards endangerment. The cases of *Acorus calamus, Bergenia ciliata, Podophyllum hexandrum, Picrorhiza kurroa, Swertia chirata*, etc. fall in this category (Fig. 4).

The herbs are the primary source of medicinal plants in terms of number of species (Fig. 5). This is perhaps because these are abundant and/or weedy species that are frequently found in the forest and it is believed that the more abundant a plant is, the more medicinal virtues it may possess (Coe and Anderson, 1996). The ease with which they can be collected, stored and transported and the ease with which bioactive compounds can be extracted are also factors that contribute to the preference of herbs. All the 40 species used in veterinary health-care by local people belong to 30 different families, the largest representation being made by Rosaceae family with five members followed by Ericaceae with three members, Caryophyllaceae, Saxifragaceae,

Combretaceae with two members each and rest of the families have been represented by a single member (Table 4; Fig. 6). It indicates the rich biodiversity of Sikkim Himalayas. In Sikkim, spring is the reproductive season for most of the plants, when they produce flowers during this short period (Fig. 2) because climatic condition becomes favourable for growth after winter and pollinators are easily available to them. At high altitudes, to avoid the chilled winter and to escape scarcity of water plants have adopted an unique evolutionary strategies to remain dormant by not engaging in growth, flowering and fruiting (Table 3; Fig. 2). Again, just after rains a few species flower and the rate of flowering decreases with the advent of winter, December and January record zero flowering (Table 3; Fig. 2).

Some plant species used in veterinary medicine by one tribe have but other tribes use it to treat human-being in entirely different disease. For example, rhizome of *Acorus calamus* has been used by Nepalese in veterinary medicine to cure wound but Lepchas use it to treat cough and fever in human-being, the stem of *Cannabis sativa* is used by Nepalese as veterinary tonic but Lepchas and Bhutias utilized the seeds of this species to cure bodyache and stomach ace of human-beings, respectively. Species like *Costus speciosus, Drymaria cordata, Picrorhiza kurroa Rhododendron campanulatum, Zingiber officinale* also fall under this category (Table 2).

There are a few species which are used in ethnoveterinary practices not only in Sikkim but also at some other parts of India too, as for example, *Asparagus racemosus* roots have been used in Sikkim to increase lactation in cow, the root is used for the same purpose for buffalo at some other place, *Datura metel* leaves have been used as ointment during inflammation of the mammary gland of lactating animals but at some other places leaf extract is used against snake-bite and scorpion sting, *Holarrhena antidysenterica* is used for stomach ailment but at some other places it is used to control coccidiosis in broiler chicks, *Swertia chirata* is given to cure fever but in other localities it is used as an antiseptic and antifungal ointment (Anonymous, 2004).

During the survey of remote areas of Sikkim Himalayas, it has been found that animal rearing is an integral part of subsistence economy in the State. Dependence of ethnic groups of this location is directly proportional to the increase of elevation. Therefore, economy is mainly dependent on it (Sundriyal, 1995). However, dependence on herbal medicine increases with altitude and occurrence of medicinal plants do not show such a correlation. Highest numbers of ethnoveterinary plants are occurring in warm or wet temperate (1500-2700 m) climatic conditions, whereas lowest numbers of medicinal plants have been found at higher altitude (Table 5; Fig. 7). Similar relationships have been found by studies (Kunwar *et al.*, 2006; Kunwar and Bussmann, 2008). The reason for this could be the lack of modern facilities and services, poverty and less development compel the high altitude population to use existing plant resources for their immediate needs. However, some tribal groups of comparatively lower altitude and lowland Terai, with poor access to the modern medicinal facilities due to poverty and illiteracy, are still utilizing local plant resources for livestock health management. Further, at high altitude, Buddhist monasteries and monks, traditional healers play a significant role in use and preservation of traditional knowledge. Most of the people of the remote areas, particularly elderly people, farmers, and mid-wives, are well acquainted to ethnoveterinary herbs and their usage.

It is fully borne out that the study and promotion of ethnoveterinary medicines is bound to help the communities to document and conserve information, and should integrate selected, tested practices into animal healthcare services in the rural areas. In trapping the potential of traditional veterinary medicines, two points are worth underlining: first, the local knowledge can be converted into valuable drugs or other commercial products. Local people and the originators of this knowledge must be recognized and appropriately compensated. Second, over-exploitation of medicinal plants is bound to put the survival of these at a risk and scientific measures are be needed to conserve them. The situation is made more urgent since in many regions from where both traditional knowledge and plant diversity are disappearing rapidly (Mathias, 2007).

ACKNOWLEDGEMENTS

The authors are highly indebted to Forest Research Institute, Dehradun and Botanical Survey of India, Gangtok, for their valuable help during identification of plants and references. We express our gratefulness to Prof G S Paliwal, New Delhi for his valuable help during writing of paper. Bharati K Avinash is thankful to Council for Scientific and Industrial Research (CSIR), India for providing financial assistance in the form of a Senior Research Fellowship (SRF-NET).

REFERENCES

Ajibesin, KK, Benjamin, AE, Danladi, NB, Etienne, EE & Saburi, AA (2008). Ethnobotanical survey of Akwa Ibom State of Nigeria. *J. Ethnopharmacol* **115:** 387-408

Anonymous, (2004). *Wealth of India.* Raw Material Series Vols 1-5. NISCAIR, CSIR, New Delhi

Atkinson, ET (1882). *The Himalayan Gazeteer* Vols 1, 2 Reprint 1973. Cosmo Publication, Delhi

Balaraman, N & Golay, MM (1991). *Livestock Production in Sikkim.* Sikkim Science Society, Gangtok

Bennet, SSR (1983). Ethno-botanical Studies in Sikkim. *Indian For.* **109:** 477-481

Biswas, K (1956). *Common Medicinal plants of Darjeeling and Sikkim Himalayas.* Kolkata Supdt Govt. Press, Kolkata

Brahma, KK (2007). *Status of Local Health Traditions of Sikkim Himalaya.* National Training workshop on folk medicine and Local health Tradition of Sikkim. pp. 16-21. National Research Institute Ayurveda. Gangtok, Sikkim

Chopra, RN, Badhwar, RL & Ghosh, S (1965). *Poisonous plants of India.* ICAR, New Delhi

Coe, FG & Anderson, GJ (1996). Ethnobotany of the Garifuna of Esatern Nicaragua. *Econ. Bot.* **50:** 71-107

Dash, SS, Maiti, A & Rai, SK (2003). Traditional uses of Plants among urban population of Gangtok in *Ethnobotany and Medicinal Plants of India and Nepal.* Pp. 313-324. Vol 1. In: *Ethnobotany and Medicinal Plants of India and Nepal.* Vol. 1. (Eds.) Singh, V & Jain, AP. Scientific Publishers, Jodhpur, India

Dutta, BK & Dutta, PK (2005). Potential of ethnobotanical studies in North-East India: an overview. *I. J. Trad. Knowl.* **4:** 7-14

Fabricant, DS & Farnsworth, NR (2001). The value of plants used in traditional medicine for drug discovery *Environmental Health Perspectives.* Supplement. **109:** 69-75

Hajra, PK & Chakraborty, P (1981). A survey of wild plants sold in the Lal market of Gangtok. *Indian J. For.* **4:** 217-220

Hajra, PK & Verma, DM (1996). *Flora of Sikkim.* Vol. I. BSI. Kolkata

Hussain, S & Hore, DK (2007). Collection and conservation of major medicinal plants of Darjeeling and Sikkim Himalayas. *Indian J. Trad. Know.* **6:** 352-357

Jachak, SM & Saklani, A (2007). Challenges and opportunities in drug discovery from plants. *Curr. Sci.* **92:** 1251-1257

Jain, SK (1995). *A Manual of Ethnobotany.* 2nd ed. Scientific Publishers, Jodhpur

Jana, SK & Chauhan, AS (2000). Ethnobiological studies on Lepchas of Dzongu North Sikkim, India. *Ann. For.* **8:** 131-144

Jana, SK & Chauhan, AS (1999). Studies on Lepcha medicobotany of Dzongu Sikkim Himalaya. *Himalayan Paryavaran* **6:** 121-126

Krishna, B & Singh, S (1987). Ethnobotanical observations in Sikkim. *J. Econ. Tax. Bot.* **9:** 1-7

Kunwar, RM & Bussmann, RW (2008). Ethnobotany in Nepal Himalaya. *J. Ethnobol. Ethnomed.* **4:** 24. [http://www.ethnobiomed.com/content/4/1/24]

Kunwar, RM, Nepal, BK, Kshhetri, HB, Rai, KS & Bussmann, EW (2006). Ethnomedicine in Himalaya: a case study from Dolpa Humla Jumla and Mustang districts of Nepal. *J. Ethnobol. Ethnomed.* **7:** 27 . [http://www.ethnobiomed.com/content/2/1/27]

Maiti, DC, Chauhan, AS & Maiti, G (2003). Ethnobotanical notes on some unexploited Plants used by Lepchas and Nepales Communities of North Sikkim. pp. 325-332. In: *Ethnobotany and Medicinal Plants of India and Nepal.* Vol. 1. (Eds.) Singh, V & Jain, AP Scientific Publishers, Jodhpur, India

Martin, JG (1995). *Ethnobotany.* Methods Manual. Chapman and Hall, London

Mathias, E (2007). Ethnoveterinary medicine in the era of evidence-based medicine: Mumbo-jumbo or a valuable resource? *The Vet. J.* **173:** 241-242

Panda, AK (2007). *Status of Local Health Traditions of Sikkim Himalaya.* National Training workshop on folk medicine and Local health Tradition of Sikkim. pp. 7-15. National Research Institute Ayurveda, Gangtok, Sikkim

Panigrahi, S (2007). *Status of Local Health Traditions of Sikkim Himalaya.* National Training workshop on folk medicine and Local health Tradition of Sikkim. pp. 22-24. National Research Institute Ayurveda. Gangtok, Sikkim

Pradhan, BK & Badola, HK (2008). Ethnomedicinal plant use by Lepcha tribe of Dzongu valley bordering Khangchendzonga Biosphere Reserve in North Sikkim India. *J. Ethnobiol. Ethnomed.* 4**:** 22. [http://www.ethnobiomed.com/content/4/1/22]

Pujadas, JJ, Coma, D & Roca, J (2004). *Etnografia.* Universitat Oberta De Catalunya, Barcelona

Puri, HS & Pandey, G (1980). Glimpses in to the crude drugs of Sikkim. *Bull Medico-Ethno-Bot. Res.* **1:** 55-71

Rai, LK & Sharma, E (1994). *Medicinal Plants of Sikkim Himalaya Status, uses and potentials.* G.B. Pant Institute of Himalayan Environment and Development, Gangtok

Robinson, T (1974). Metabolism and function of alkaloids in plants. *Sci.* **84:** 430-435

Sahai, S (2003). Importance of indigenous knowledge. *Indian J. Trad. Know.* **2:** 11-14

Sahoo, AK & Ansari, AA (2007). *Status of Local Health Traditions of Sikkim Himalaya.* National Training workshop on folk medicine and Local health Tradition of Sikkim. pp. 29-35. National Research Institute. Ayurveda. Gangtok, Sikkim

Singh, HB & Subramanium, B (2008). *Field Manual on Herbarium Techniques.* NISCAIR, New Delhi

Srivastava, RC (1990). A note on the arrow-poison used by Lepchas of Dzongu. Sikkim and its antidote. *Nat. Acad. Sci. Letters* **13:** 399-400

Srivastava, TK, Kapahi, BN & Atal, OK (1987). Ethno-medico botanical investigation in Sikkim. *J. Econ. Taxon. Bot.* **11:** 413-421

Sundriyal, RC (1995). Grassland forage production and management in the Himalaya. A review. *J. Hill Res.* **8:** 135-150

21

Coal-Fired Industries and the Anthropecology of Plants – A Review

Anil K Dwivedi[1*], Shashi[2], BD Tripathi[3] and A Tripathi[4]

[1]*Pollution and Environmental Assay Research Lab. (PEARL), Department of Botany, D.D.U. Gorakhpur University, Gorakhpur–273 009, U.P., India*
E-mail: anil.k.dwiwedi@gmail.com

[2]*Laboratory of Environmental Chemistry & Complexes, Department of Chemistry, Udai Pratap Autonomous P.G. College, Varanasi–221 002, U.P., India*

[3]*Coordinator, Centre of Environmental Science and Technology, Banaras Hindu University, Varanasi–221 005, U.P., India*

[4]*Pollution Ecology Research Laboratory, Department of Botany, Hindu College, Moradabad–244 001, U.P., India*

ABSTRACT

In recent age of science and technology the pollution problems are becoming more and more severe due to rapid and unplanned growth of industries. Though, the number of brick industries are less in comparison to other industries but, due to use of IIIrd and IVth grade of coal, huge amount of pollutants are released into the surrounding atmosphere. According to an estimate there are about one lakh brick industries in India where 20 million tons of coal is consumed per year. Thus, brick industries are considered as one of the most polluting industry among the coal fired industries. From literature survey, it was noted that there is a lack of information on the impact of brick industries on surrounding plants, its control and management. Therefore, present work was initiated to monitor the effects of pollutants on phytosociological behaviour of surrounding plant species and the change in the biodiversity. Present investigation was carried out at Varanasi, around five research sites i.e. five brick industries. Ambient air analysis and phytomonitoring was performed at all the sub-sites including the control site.

INTRODUCTION

Air, which has not been seen by any body on the Earth, is the real supporter of life. A person can live for five weeks without food, for five days without water; but his existence will cease within a five minutes without air (Raju, 1997). Food and water are required intermittently by a living organism but air is required continuously for the survival of life. Anthropecology is a new term coined in the year 2009, to explain the ecology of the system which is largely affected, rather directed by the multidimensional anthropogenic activities (Dwivedi *et al.*, 2009). Air is elastic, invisible and tasteless mixture of gases that surround the Earth, and the thick cover of air mantle that envelops the Earth is called atmosphere. Atmosphere sustains life on Earth and saves it from unfriendly environment of outer space in various ways.

There is no boundary or demarcation between the atmosphere and the void of outer space, yet the air remains condensed near the Earth surface. 75% of the Earth's atmosphere lies within 10 miles of the surface and 99% of the atmosphere lies below an altitude of 19 miles (Murray *et al.*, 1985). The density of atmosphere shows a sharp decrease with increasing altitude. The total mass of atmosphere is approximately 5×10^{15} tonnes, which is roughly one millionth of the Earth's total mass.

Troposphere, the innermost layer of atmosphere which is close to Earth surface extends till 8 kms (at poles) to 18 km (at equator). Rotation of the Earth keeps this layer in constant motion. Due to lapse rate, temperature falls at the rate of 6°C per km in altitude. The gases present in this layer are nitrogen, oxygen, argon and carbon dioxide as the major constituent. However traces of He, Ne, Kr, H_2, N_2O, CO, O_3, NH_3, NO_2, SO_2 etc. are also present in this layer. It is a turbulent, dusty zone containing water vapours. This region is also the locus of most of the weather phenomenon which affect different ecosystems of the biosphere (Fritts and Warne, 2000; Loutre, 2001).

The composition of air also varies from place to place and time to time, and it is the composition of constituents in the particular microenvironment which affects growth and development of the biota present in it (Edwards *et al.*, 1998; Jonnalagadda and Baloji, 1991). CO_2, SO_2 and NO_2 may be present in higher proportion in the place where fuel combustion is taking place (Li *et al.*, 1999); proportion of sulphur containing gases are reported to be relatively high in the paddy growing areas also (Yang *et al.*, 1998); H_2S and related gases are present in considerable amount in decaying and reducing environment (Gignac *et al.*, 1998; Tripathi and Dwivedi, 2003); Soil-air present in the lattices of farms shows CO_2 gas to be produced in higher amount due to microbial respiration (Atkin *et al.*, 2000; Lok, 2001).

MAJOR SOURCES OF AIR POLLUTION

In order to study the nature, intensity and dispersion behaviour of pollutants, acquiring information about pollution sources becomes a imperative for investigation. Broadly, the air polluting sources can be divided into two groups i.e., natural and anthropogenic.

Natural Sources

Some activities of nature which pollute the environment are, the volcanic eruption which emits particulates, SO_2, H_2S, CH_4, NO_x, etc. in the atmosphere; forest fire which releases CO, HC, NO_x, SO_x, smoke and ash in considerable amount; wind and storms uplift particulate matter and dust in air (Goudie, 1978 and Dwivedi *et al.* 2009); ocean which occupies the major portion of the planet continuously emits aerosols and salt particles. These alongwith pollen grains and spores also contribute to air pollution naturally.

In addition to the above, the cattle feed and their dung release in the form of ammonia and methane, respectively, thus dairy also contributes to air pollution (Kaharabata *et al.*, 2000). Farming practices are also a major contributor to ecological damage (Lok, 2001). Production of sulphur gases from the paddy fields in China has been reported by Yang *et al.* (1998). Cropland soil produces NO_2 due to microbial activity (Roelle *et al.*, 2000). Decomposing and decaying materials which produce foul odour due to hydrogen sulphide and ammonia also contribute to air pollution (Stanier *et al.*, 1987).

Anthropogenic Sources

The world's population is increasing relentlessly and so are people's needs and clearly, this is putting all sectors of the global economy, power generation, transportation, agriculture production etc. under increasing pressure (Verma, 2000, Dwivedi *et al.* 2006). All the above processes are the mother of pollution. Thus, anthropogenic sources that contribute to air pollution can be divided mainly into two categories, i.e. mobile sources and stationary sources. Certain activities of human such as deforestation (Potter, 1999), dead body cremation (Tripathi *et al.*, 1984; Mishra and Tripathi, 1999) also contribute to environmental pollution considerably.

Mobile sources mainly include the heavy (Luria *et al.*, 1984; Thompson *et al.*, 1984; Tyagi, 1988-89) or light vehicles (Oluwande, 1979) and traffic congestion (Newman and Kenworthy, 1987) which mainly pollutes the urban atmosphere (Mansfield and Freer-Smith, 1981). Composition of vehicular pollutants discharged has been studied by Pundir and Das (1985) and Joumard (1988).

Rapid industrialisation has given rise to various stationary sources of pollution (Scott *et al.*, 1999). Thermal power plants (Agrawal and Agrawal, 1989), brick industries (Tripathi *et al.*, 1999), cement industries (Oliveira and Jardim, 1999), pulp mill (Jarvensivu *et al.*, 2000), petroleum refinery (Sharma *et al.*, 1986) and various other industrial units (Jacobson, 2000) are prominent stationary sources of air pollution. Inspite of realizing that industries are the major source of pollution, the governing bodies fail to impose the law to check the pollution; further some time their policies become the cause of pollution (Dwivedi, 2008; Shashi and Dwivedi, 2008).

Alongwith this various sources of indoor air pollution have also been reported by Bruce *et al.* (2000) and Ormstad (2000). Room carpets have been reported to be the source of volatile organic compounds in the indoor environment by Won *et al.* (2000). While spirit cooking has been reported to be the source of CO (carbon monoxide) in kitchen (Rupp *et al.*, 2000, Dwivedi *et al.*, 2007).

Atmospheric NO_2 and SO_2 is released from various sectors of life such as by burning of fossil fuel (Cullis and Hirschler, 1990) and stack emission of industries (Novak *et al.*, 2001). Release of NO_2 has been noticed from agricultural sector also (Yu'e and Erda, 2000).

Various heavy metals are released in the atmosphere through a variety of agencies, the prominent among them being, lead by gasoline combustion (Snakin and Prisyazhnaya, 2000). Coal combustion released cadmium and zinc (Jaworowski and Grzybowska, 1980) as well as mercury (Wang *et al.*, 2000), nickel and copper by industrial source (Kashulin *et al.*, 2001) and many more. The presence of atmospheric halides i.e., fluorides has been reported in Tibet by Cao *et al.* (2000).

The fine particles of atmosphere which include particulates of diameter less than 100 μm, such as carbon particles aerosols, dust, tars, metallic dust etc. are mostly released through various industrial processes (Shu *et al.*, 2000), automobiles and domestic activities (Tong and Lam, 2000).

Evaporation and exhaust emission of unburnt gasoline accounts for the major part of hydrocarbon pollutants, which are chiefly benzene, toluene and benzopyrene. These are also discharged from automobiles and industries (Glenn *et al.*, 1993).

Chlorofluorocarbons such as Freon-11, Freon-12, Freon-114 and halons escape from industries, refrigeration, air conditioning, sprays, etc. (Montague and Perrine, 1990).

Above all, the rapidly increasing population is the main cause of pollution which has resulted surplus demand for food, cloth and shelter (Pudasaini, 1997). This has also enhanced the load on urban load alongwith requirement for energy (Nyamwange, 1995) (Foley, 1985, Dwivedi *et al.*, 2008). Highly populated land is also highly polluted (Vitousek *et al.*, 1997).

BRICK INDUSTRIES AND PROBLEMS INDUCED

Bricks are baked clay blocks of uniform size which are of prime importance in the construction of houses and buildings along with the cement. Bricks occupy a good rank among the highly demanded tools for the rapid urbanization and modernization of a developing nation like India. Evidences have proved the use of bricks eversince the days of Indus Valley civilization, i.e., even about 5 thousand years back (Maithel *et al.*, 2000).

In India, there are about 1 lakh small or large brick producing units which produce about 10,000 crore bricks per annum and about 50 lakh workers are employed in this industry (Maithel *et al.*, 2000). Annual turnover of this industry is above 8,000 crore rupees (Gupta and Gupta, 2000).

Baking of 1000 bricks require about 180 kg of coal. As an estimate, an average kiln consumes around 800 tons of coal in a season amounting roughly to rupees 30-35 lakhs (Jain and Singh, 2000). Indian brick industry with an estimated coal consumption of about 20 million tons per year is the third largest consumer of coal in the country after power plants and steel industries (Maithel *et al.*, 1999).

Brick kilns are provided with the most inferior quality of coal (Grade III/IV) which has very high ash and volatile matter content i.e., about 45 to 70% (Aslam, 1993). The coal shows large variation in calorific value, ash content, volatiles, sulphur content, moisture and fixed carbon content. Few kilns use wooden chips or saw dust for raising the initial ignition temperature which results in incomplete and improper combustion, causing wastage of fuel and enormous dust emission (Jain and Singh, 2000). Major air pollutants released from the brick industries include particulate matters, SO_2, NO_2, CO_2, CO etc. (Maithel, 1999).

The dried green bricks require baking between 600 to 1100°C to get the final fired bricks. The basic difference between the various types of brick kilns lies in the manner they heat the bricks. In view of the above following are some of the types of brick industries.

- Moving Chimney Bulls Trench Kiln
- Fixed Chimney Bulls Trench Kiln
- High Draught Kiln (zig-zag kiln)
- Vertical Shaft Brick Kiln

(1) Moving chimney bulls trench kiln

The oval or circular kiln consists of five zones- loading zone, brick pre-heating, combustion zone, the brick cooling zone and brick unloading area. The coal is fed from the top through the fire hole. In this type the chimney moves along with the fire so the chimney is shifted (replaced) daily. As this chimney requires regular shifting so it is of lower height ranging between 1.66 to 18.33 m. This smaller height is the greatest drawback of this type of kiln, as it releases pollutants at a lower height which results in their concentration at ground level, thus showing symptoms of their accumulation in the nearby vegetation affecting it drastically and reducing their yield.

Specific energy consumption is higher (1.25-1.75 MJ/kg of fired brick) and also specific fuel consumption is higher (25-30 tons of coal per lakh brick) in the moving chimney brick trench kiln, these have poor thermal efficiency i.e. only 18 to 20%.

(2) Fixed chimney bulls trench kiln

This also consists of the same five zones as in the moving brick trench kiln; the difference from the above lying in the chimney being stationary in this case. The height of the chimney is usually around 33.3 m and the flue gas passes through the flue channel to reach the base of the chimney. As the flue gas covers longer distance before exit, the particulate matters also settle down to some extent. Greater height of the chimney provides dilution of the pollutants and the high concentration of pollutants at the ground level is avoided. Specific energy consumption is 1.0 to 1.5 kg of fired brick and specific fuel consumption falls to 16 to 24 tons of coal per lakh of bricks.

After the announcement of emission standards by the government, two new designs: Priya brick (which was also adopted by Punjab State Council of Science and Technology) and the design developed by Central Building Research Institute (CBRI), Roorkee have been developed (Gupta and Gupta, 2000). These new designs are equipped by Gravity Settling Chambers also.

(3) High draught kiln (zig-zag kiln)

This kiln had been introduced by CBRI during 1970s. It is a rectangular gallery divided into 24 to 32 chambers by partition wall of green bricks, which run along the width of the gallery and are arranged so as to allow zig zag path of the hot air. At one end a space of about 60 cm is left to facilitate communication between the two chambers. For proper combustion of fuel, draught is created by an induced draught fan, with a 15 horse-power motor. The design is quiet similar to that of the Habla kiln, which has been in practice for centuries.

Coal in the range of 13 to 16 tonnes and 375 units of electric power is required for the burning of one lakh bricks. Its requirement of electricity is considered as a drawback of this type of kiln, since electricity is not easily available at such sites or the supply is not regular.

(4) Vertical shaft brick kiln

Vertical Shaft Brick Kiln technology is of Chinese origin and its testing in India has been conducted at a few sites by means of pilot plants. As the name itself suggests, the kiln is vertical, contrary to the other traditional kilns which are horizontal. Vertically, the kiln is divided into five zones, and movement of bricks takes place in this kiln, contrary to that of fire which takes place in other kilns. The green bricks are loaded on top, which move down the shaft through the trolley.

The specific energy consumption in VSBK pilot plants was found to vary between 0.72 to 1.06 MJ/kg of fired brick. Green bricks are loaded at the top most platform along with coal, this is followed by brick pre-heating zone, lower to which lies the firing zone where real baking of the bricks take place. Here fresh air enters the kiln through spy hole, having a brick cooling zone underneath which rests on support bars and beams. Last is the unloading zone, which consists of a trolley devised by single screw lifting mechanism.

According to Sengupta (1993), Vertical Shaft Brick Kiln has superiority over other industries in following ways:

1. Surface heat loss is less
2. Firing zone is stationery hence kiln mass attains and maintains the steady state temperature, thus conserving energy.
3. It requires less area.
4. Particulate matter released is much below the Government of India norms.

ENVIRONMENTAL APPROACH

Production of 3 main construction materials steel, cement and bricks requires burning of about 70 million tons of coal annually. Brick industry with an estimated coal consumption of about 20 million tons per year is the third largest consumer of coal in the country after power plants and the steel industry.

Various rules and regulations have been formulated from time to time to control the pollution (Khitoliya and Khitoliya, 1996). Central Pollution Control Board (CPCB) has prescribed the National Ambient Air Quality Standards for industrial, residential and the sensitive areas (Table 1). The Government of India has fixed National Emission Standards for brick kiln by a gazette notification number 141 dated 3rd April 1996. The gazette notification had also banned the use of movable steel chimney effective from December 31, 1997 (GOI, 1996) (Table 2). But the liberal government has granted a release to such industries from the order for time being.

Table 1: National Ambient Air Quality Standards

Pollutants	Average Time-spam	Concentration in ambient air		
		Industrial Area	Residential, Rural and Other Areas	Sensitive Areas
Sulphur dioxide (SO_2)	Annual*	80 μgm^{-3}	60 μgm^{-3}	15 μgm^{-3}
	24 hours**	120 μgm^{-3}	80 μgm^{-3}	30 μgm^{-3}
Nitrogen dioxide (NO_2)	Annual*	80 μgm^{-3}	60 μgm^{-3}	15 μgm^{-3}
	24 hours**	120 μgm^{-3}	80 μgm^{-3}	30 μgm^{-3}
Suspended Particulate Matter (SPM)	Annual*	360 μgm^{-3}	140 μgm^{-3}	70 μgm^{-3}
	24 hours**	500 μgm^{-3}	200 μgm^{-3}	100 μgm^{-3}
Respirable Particulate Matter (RPM) (size less than 10 nm)	Annual*	120 μgm^{-3}	60 μgm^{-3}	50 μgm^{-3}
	24 hours**	150 μgm^{-3}	100 μgm^{-3}	75 μgm^{-3}
Lead (Pb)	Annual*	1.0 μgm^{-3}	0.75 μgm^{-3}	0.50 μgm^{-3}
	24 hours**	1.5 μgm^{-3}	1.0 μgm^{-3}	0.75 μgm^{-3}
Carbon Monoxide (Co)	8 hours*	5.0 mgm^{-3}	2.0 mgm^{-3}	1.0 mgm^{-3}
	1 hour**	10.0 mgm^{-3}	4.0 mgm^{-3}	2.0 mgm^{-3}

* Annual Arithmetic mean of minimum 104 measurements in an year taken twice a week 24 hourly at a uniform interval.

** 24 hourly/8 hourly values should meet 98% at the time in an year. It may exceed by 2% but not on two consecutive days.

Table 2: GIO norms for brick kilns

Emission standards and stack height regulation

Size	Kiln Capacity	Maximum Limit for the Concentration of Particulate Matter (mg μm^{-3})	Stack height
Small	Less than 15,000 bricks per day (<5 m trench width)	1000	Minimum stack height of 22 m or induced draught fan operating with minimum draught of 50 mm water Gauze of 12 m stack height.
Medium	15,000-30,000 bricks per day (5-6.3 m trench width)	750	Minimum stack height of 27 m with gravitational settling chamber or induced draught fan operating with minimum draught of 50 mm water Gauze with 15 m stack height.
Large	More than 30,000 bricks per day (more than 6.3 m trench width)	750	Minimum stack height of 30 m with gravitational settling chamber or induced draught fan operating with minimum draught of 50 mm water Gauze with 17 m stack height.

Gazette notification No. 141 dated 3rd April 1996.

THE PROBLEM

From a general survey of air pollution sources at Varanasi, it was revealed that there are about 1588 small scale industries located in and around the city of Varanasi where coal and crude oils are used as a fuel material for power generation. Among these industries, though numerical strength of brick industry is very low in Varanasi i.e. 3.34%; but the brick industries are treated as most polluting industries because the brick kilns are provided with the most inferior quality of coal (Grade-III/IV) which has very high ash and volatile matter content i.e. 45-70% (Aslam, 1999). Higher quantity of sulphur dioxide is also released from burning of III/IV grade coal. Burning of 1 ton of grade-III coal releases 0.4536 kg of sulphur dioxide in the air (Pandey, 1997). At national level more than 1 lakh brick industries are working, which consume 20 million ton of coal per year. According to an estimate about 9.072 million kg (0.4536 × 20 million kg) SO_2 is released in the atmosphere every year.

In India, two reputed organisations are engaged in the development of brick industries such as Central Building Research Institute (CBRI), Roorkee and the Tata Energy Research Institute (TERI), New Delhi. These organizations are mostly working either for energy conservation in brick industries (Maithel *et al.*, 1999 and Maithel *et al.*, 2000) or development of techniques to reduce the pollution, especially particulate pollutants (Lakshmikantan *et al.*, 1999). Some work has also been oriented towards development of ecofriendly technique in brick making (Annon, 1999; Damle, 2000). However, scientific study still lacks essential information about the actual impact of the brick industries on the surrounding plants and its management. Therefore, the present study has been conducted at Varanasi which is situated in the eastern part of Uttar Pradesh, a province of India located in core of the Indian sub-continent (Fig. 1). On the globe, Varanasi finds its location at 25°18' North, 83°1' East and 76.19 m. above the sea level. For

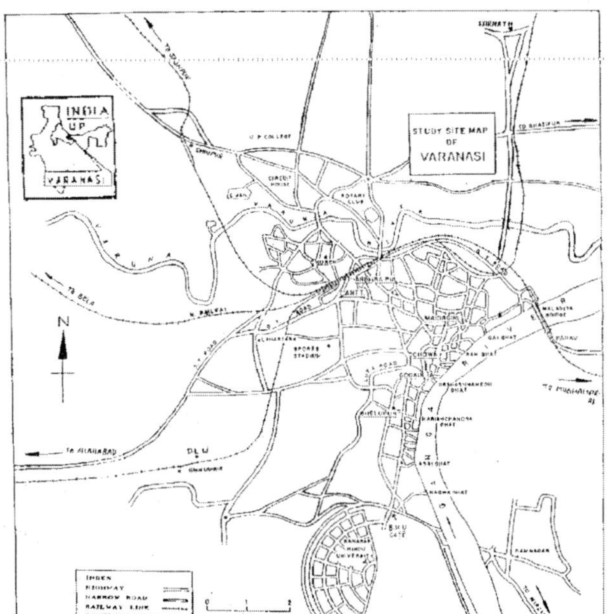

Fig. 1. Map of Varanasi.

detailed investigation, 4 sites i.e. brick kilns at Sandaha, Kotwan, Salarpur and Chamaraha (Table 3), located in different areas of out skirt of the city were selected and were coded as sites A, B, C and D respectively. In order to compare the results, one control site has also been selected near Sarnath, which was free from any kind of air pollution source. The control site was coded as site E.

Table 3: Details of the research sites

S.No.	Characters	*Site Code*				
		A	B	C	D	E
(1)	Location of the brick industry	Sandaha	Kotwan	Salarpur	Chamaraha	Sarnath
(2)	Type of the brick industry	Fixed Chimney	Fixed Chimney	Movable Chimney (Single)	Movable Chimney (Double)	Nil
(3)	Coal consumption (Tonnes per day)	7	6	3	5	Nil
(4)	Kiln capacity (Bricks per day)	32,000	24,000	12,000	16,000	Nil
(5)	Pollution control Device	With settling chambers	Unequipped	Unequipped	Unequipped	Nil
(6)	Total emission (m^3hr^{-1})	23,400	12,600	3240	2486 x 2	Nil
(7)	Stack height (m.)	29.4	28.8	12.6	10	Nil
(8)	Stack diameter (m.)					
	(a) Top	1.5	1.2	0.6	0.6	Nil
	(b) Bottom	4.2	3.0	0.9 x 1.8 = 1.27	0.75 x 1.8 = 1.16 (each)	Nil
(9)	Cross sectional area of the stack (m^2)					
	(a) Top	1.76	1.14	0.28	0.28	Nil
	(b) Bottom	13.86	7.06	1.62	1.35	Nil
(10)	Specific fuel consumption (Tonne coal per lakh bricks)	21.87	25.00	25.00	31.25	Nil
(11)	Gaussian most polluted point (m away from stack)	260	250	120	100	Nil

METEOROLOGY OF VARANASI

Atmosphere acts as a medium through which the air pollutants are transported and dispersed away from the source (Biscaye *et al.*, 2000; Placio *et al.*, 2000). Their transportation is affected by various climatic (Power, 2001) and meteorological factors (Lyons and Scott, 1990; Sanchez *et al.*, 1990).

The level of pollution varies from place to place due to difference in the weather condition and local meteorological factors. The location of receptors relative to source, as well as atmospheric influences affects pollutant concentration and the sensitivity of receptors to these concentrations determines the effect (Holzworth, 1972).

Thus, in light of all the above information it is evident that the prior knowledge of various meteorological factors is essential for the study of air pollution. The important meteorological factors interfering in dispersal and transportation of the air pollutants are: precipitation, relative humidity, temperature, wind velocity and the wind direction (Bowne, 1974; Turner, 1964), which ultimately reflects its impact through the vegetation cover (Auer, 1978).

City Varanasi, located in heart of the Indian subcontinent is a tropical city with marked monsoonal affect. Climatologically, the city surpasses three seasons distinctly in an year, viz. summer season (March to June), rainy season (July to October) and winter season (November to February). Monthly meteorological conditions of Varanasi for two years have been expressed in Fig. 2 (1998-1999) and Fig. 3 (1999-2000). All the meteorological data has been collected from Meteorological Unit, BHU, Varanasi.

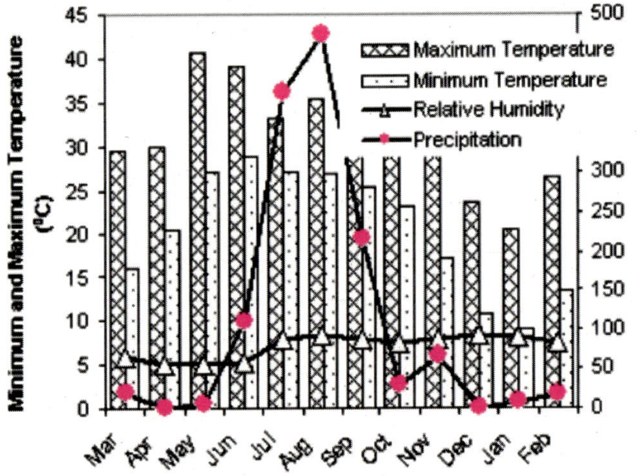

Fig. 2. Meteorological features of Varanasi. (1998-99)

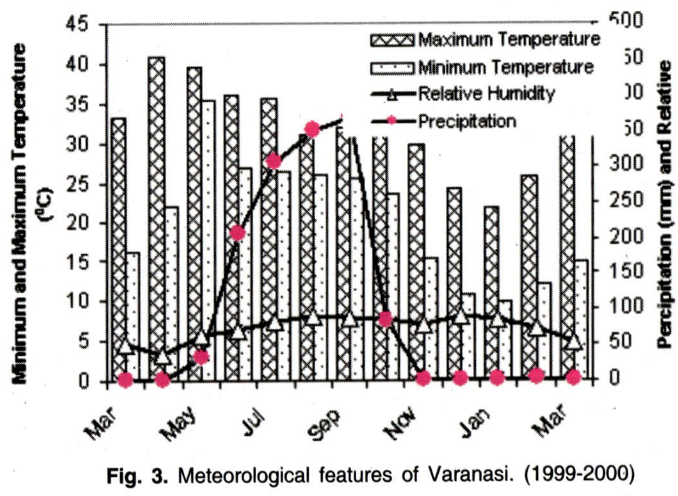

Fig. 3. Meteorological features of Varanasi. (1999-2000)

Precipitation

Precipitation or rainfall is the main natural scrubber of the atmosphere; it cleans both soluble and insoluble form of the pollutants present in the atmosphere. It rapidly reduces the particulate contents of atmosphere (Sanchez *et al.*, 1990) and also removes gases such as SO_2 and NO_2 present in atmosphere, in the form of acid rain (Gillett *et al.*, 2000) though the acid rain affects the biotic community adversely (Pal and Kumar, 2000; Tripathi and Dwivedi, 2002).

During 1998-1999, the highest rainfall (473.1 mm) has been recorded in the month of August, while it was zero in December. Total annual rainfall during March 1998 to March 1999 has been recorded to be 1347.8 mm. On the other hand during 1999-2000, precipitation showed an increasing trend from May to September and a reverse trend in October. Little winter rainfall has been noticed during December to March. Highest rainfall was recorded during September; whereas the total rainfall during March 1999 to March 2000 was 1346.3 mm.

Annual mean of monthly rainfall during 1999-2000 was found to be 103.5615 (SD ± 142.9798).

Relative Humidity

Relative humidity denotes the degree to which the air is saturated with moisture and is expressed in the terms of percentage. The range of relative humidity lies between 0%, in the case of absolutely dry air to 100%, in the case of fully saturated air. Relative humidity shows inverse relationship with temperature. Relative humidity of atmosphere is responsible for various secondary phenomena of environment, such as acid mist, fog etc. Relative humidity also affects the pollution uptake by the plants (Mclaughlin and Taylor, 1981).

Maximum relative humidity has been recorded during August (95.1%) and December (90.2%) in year 1998 and 1999 respectively. During 1998-1999, the relative humidity was lowest in April (56.4%) lagging by May (56.6%). On the other hand, the value of relative humidity was minimum in the month of April (36.2%) during 1999-2000. As an average, the relative humidity was highest during rainy months followed by winter and summer.

Temperature

Earth's surface absorbs solar radiation and a part of it radiates into the atmosphere, so air near the Earths surface becomes warm, and it expands and exerts pressure upon the cool air. A constant fall of 6.5°C temperature is seen per kilometer height in the troposphere which is termed as the lapse rate (Rao and Rao, 1989). When the reverse i.e., the negative lapse rate occurs the vertical mixing of the pollutants ceases and the pollutants remain concentrated beneath the inversion level. Highly irregular motion of wind in response to the temperature difference is called as Turbulence. Turbulence plays a critical role in the pollutant dispersal and plant growth (Yoshida *et al.*, 1998).

Highest monthly average temperature has been recorded during June (39.03°C) and May (39.61°C) during years 1998 and 1999 respectively. On the other hand the minimum temperature observed was 8.97°C and 9.85°C during January of 1999 and 2000 respectively. During both the study years highest temperature has been recorded in summer followed by rainy and winter seasons. Year 1999-2000 was observed to be hotter than year 1998-1999.

STACK MONITORING

At site A temperature of the stack emission has been recorded between 111 to 113°C. At site C it ranged from 198 to 200°C whereas at site D it was noted to be from 233 to 235°C. Velocity of the flue gas in the stack ranged between 0.62 to 0.67 $msec^{-1}$ at site A, 0.70 to 0.79 $msec^{-1}$ at site B, 0.82 to 0.92 $msec^{-1}$ in the brick industry at site C and 1.02 to 1.10 $msec^{-1}$ at site D. Particulate matters in the flue gas have been found between 549 to 568 mgm^{-3}, 1312 to 1340 mgm^{-3}, 1881 to 1897 mgm^{-3} and 2091 to 2121 mgm^{-3} at the brick industries at site A, B, C and D, respectively. Sulphur dioxide concentration in the stack was found between 205.1 to 207.36 μgm^{-3} at site A, 229.06 to 236.10 μgm^{-3} at site B, 238.10 to 243.91 at site C and between 237.30 to 240.65 μgm^{-3} at the site D brick industry. Concentration of nitrogen dioxide in the stack emission was between 48.03 to 49.13 μgm^{-3}, 47.61 to 48.31 μgm^{-3}, 43.10 to 45.74 μgm^{-3} and 41.00 to 43.11 μgm^{-3} in the brick industries at site A, site B, site C and site D, respectively.

AMBIENT AIR QUALITY

Average monthly Dust Fall Rate at the site A ranged between 0.51 to 3.46 gm^{-2} day^{-1} It was between 0.52 to 4.72 gm^{-2} day^{-1} at site B, 0.60 to 3.19 gm^{-2} day^{-1} at site C and between 0.47 to 4.91 gm^{-2} day^{-1} at site D. However, at the control site it ranged from 0.47 to 1.05 gm^{-2} day^{-1} only. Average monthly Suspended Particular Matters were found to be between 77.00 to 745.00 μg m^{-3}, 96.00 to 1213.00 μg m^{-3}, 78.00 to 1145.00 μg m^{-3}, 88.00 to 1154.00 μg m^{-3} and 63.00 to 137.00 μg m^{-3} at the sites A, B, C, D and E respectively. Average monthly sulphur dioxide in the ambient air was found to be between 47.10 and 198.00 μg m^{-3} at site A, 48.00 and 195.60 μg m^{-3} at site B, 50.80 and 178.80 μg m^{-3} at site C, 44.90 and 191.70 μg m^{-3} at site D and between 13.40 to 16.6 at site-E. Average monthly nitrogen dioxide in the ambient air was found to be between 20.90 to 41.30 μg m^{-3} at site A, 20.70 to 42.60 μg m^{-3} at site B, 15.90 to 32.80 μg m^{-3} at site C, 18.20 to 39.80 μg m^{-3} at site D and between 11.40 to 14.50 μg m^{-3} at the control site, i.e., site-E.

Thus, site D was found to be most polluted, followed by sites C and B. At the same time the least polluted site after the control site was the study site A.

AIR POLLUTION TOLERANCE INDEX

Biotic community is very sensitive to pollution and has been used by many workers as a monitor to study the extent of pollution (Lee, 1977; Page and Organiscak, 2000).

Plants have proved to be a better indicator of environmental change than the animals. Microscopic algae have been used to show the toxicity in air by Kabirov *et al.* (2000). Bryophytes have also been used to study the extent of air pollution by Ganeva (1998). Mandes(1998) showed fungi to be good environmental indicators.

Epiphytes have been used as a monitor of air pollution by Pisut and Lisicka (2000) and Gracia *et al.* (2000). Henderson (1999) and Von *et al.* (1990) used lichen as an indicator of environmental pollution. Montane meadows have been used as indicators of environmental changes by Debinski *et al.* (2000) and many more.

Various parameters used by earlier workers to define sensitivity or resistance of plants towards air pollution are - ascorbic acid content (Keller and Schwagar, 1977), relative water content (Rao, 1979), chlorophyll content (Bell and Mudd, 1976), membrane permeability (Farooq and Beg, 1980), leaf extract pH (Chaudhary and Rao, 1977) and peroxidase activity (Eckert and Houston, 1982).

However, categorization of plants into different groups as sensitive or tolerant can not be done on the basis of only one parameter since plant response to pollutants varies from species to species and within varieties of a given species. Similarly, plants also do not necessarily show similar susceptibility to different pollutants. One plant may be sensitive to one pollutant but resistant to the other.

Therefore, during present investigation ascorbic acid content, leaf extract pH, total chlorophyll content and relative water content was taken together in the form of mathematical expression to obtain an empirical value, signifying the Air Pollution Tolerance Index (APTI) (Singh *et al.*, 1991 and Raza *et al.*, 1985).

$$APTI = \frac{A(T+P)+R}{10}$$

Where,

A = Ascorbic acid content (mg g^{-1} fresh wt.)
P = Leaf extract pH
T = Total chlorophyll (mg g^{-1} dry wt.) and
R = Percent relative water content of leaf.

The plants recorded during survey of the brick industries and the control site were analysed for their APTI (Table 4). *Carica papaya* L. and *Ficus religiosa* L. were the two plants selected for detailed investigation, on the basis of their APTI value, their medicinal and economic importance (Nakamura *et al.*, 2000; Li and Kuo, 2000; Mandal *et al.*, 2000) and on the basis of their evergreen and deciduous nature respectively.

Plants have been the best tools for the study of various climatic and environmental changes. Phytomonitoring has been in practice for more than hundred years (Hooker, 1859; Smith and Robert, 1900). Plants are also considered as a mirror of the environment, because of difference in the range of sensitivity and resistance in the plant for specific changes (Tiwari *et al.*, 1993).

Table 4: Air Pollution Tolerance Index of plants at various study sites

Sl. No.	Plants	Family	*A*	P	T	R	APTI
1.	*Carica papaya* L.	Caricaceae	32.60	7.50	15.20	71.0	81.10
2.	*Ricinus communis* L.	Euphorbiaceae	25.10	6.20	17.20	70.0	65.73
3.	*Cynodon dactylon* (L.) Pers.	Poaceae	20.60	8.00	11.38	88.0	48.72
4.	*Cyperus rotundus* L.	Cyperaceae	20.40	7.80	10.96	85.0	46.77
5.	*Rumex dentatus* L.	Polygonaceae	15.22	6.90	10.56	68.0	33.37
6.	*Achyranthes aspera* L.	Amaranthaceae	17.90	6.00	7.20	82.0	31.83
7.	*Moringa oleifera* Lam.	Moringaceae	15.40	5.90	8.77	88.0	31.39
8.	*Alternanthera sessilis* (L.) DC	Amaranthaceae	12.80	6.10	11.70	74.0	30.18
9.	*Xanthium strumarium* L.	Asteraceae	20.35	7.40	4.50	54.0	29.62

Contd.

Table 4: Contd.

Sl. No.	Plants	Family	*A*	P	T	R	APTI
10.	*Parthenium hysterophorus* L.	Asteraceae	10.70	8.00	10.50	81.0	27.90
11.	*Digitaria adscendens* Henr.	Poaceae	9.60	7.70	10.81	86.0	26.37
12.	*Dactyloctenium aegyptium* (L.) Will.	Poaceae	9.50	7.90	10.88	85.0	26.34
13.	*Desmostachya bipinnata* (L.) Stapf.	Poaceae	9.60	7.80	10.86	84.0	26.31
14.	*Ziziphus mauritiana* Lamk.	Rhamnaceae	10.50	8.00	10.25	70.0	26.16
15.	*Capparis zeylanica* L.	Capparidaceae	10.40	8.00	9.96	70.0	25.68
16.	*Dolichos lablab* L.	Fabaceae	8.76	7.50	11.20	76.0	23.98
17.	*Saccharum munja* Roxb.	Poaceae	8.31	8.00	11.21	75.0	23.46
18.	*Asclepias curassavica* L.	Asclepiadaceae	12.30	5.10	7.70	76.0	23.34
19.	*Croton bonplandianus* Baill	Euphorbiaceae	8.78	6.00	11.40	79.0	23.18
20.	*Eragrostis tenella* (L.) Baeuv	Poaceae	8.61	7.20	9.41	80.0	22.30
21.	*Azadirachta indica* A. Juss.	Meliaceae	10.20	6.30	7.50	79.0	21.98
22.	*Coccinia indica* Wt. and Arn.	Cucurbitaceae	7.14	7.00	12.30	79.0	21.68
23.	*Eucalyptus citriodora* Hook.	Myrtaceae	14.50	5.00	4.25	80.0	21.41
24.	*Ipomoea fistulosa* Mart.	Convolvulaceae	11.00	6.50	6.20	74.0	21.37
25.	*Hibiscus rosa-sinensis* L.	Malvaceae	9.70	6.50	7.36	79.0	21.34
26.	*Phalaris minor* Retz.	Poaceae	7.50	7.00	9.98	85.0	21.24
27.	*Madhuca indica* J.F. Gmel.	Sapotaceae	12.74	5.80	5.26	64.0	20.49
28.	*Ficus religiosa* L.	Moraceae	7.69	7.10	11.40	55.0	19.73
29.	*Calotropis procera* Ait. Hort.	Apocynaceae	8.50	6.20	5.27	96.0	19.35
30.	*Ficus benghalensis* L.	Moraceae	7.49	8.00	6.95	79.0	19.10
31.	*Evolvulus alsinoides* L.	Convolvulaceae	7.16	7.00	8.13	79.0	18.73
32.	*Calotropis gigantea* (L.) R.Br.	Apocynaceae	8.50	6.20	5.27	88.0	18.55
33.	*Psidium guajava* L.	Myrtaceae	7.87	6.30	7.20	73.0	17.92
34.	*Eclipta alba* (L.) Hassk.	Asteraceae	8.50	6.00	5.29	79.0	17.50
35.	*Spilanthes indicus* Wall.	Asteraceae	5.11	7.00	11.25	81.0	17.43
36.	*Tamarindus indica* L.	Caesalpinaceae	7.90	6.10	4.85	78.0	16.45
37.	*Cleome viscosa* L.	Capparaceae	5.05	6.90	9.51	78.0	16.09
38.	*Nicotiana plumbaginifolia* Viv.	Solanaceae	8.60	6.40	2.50	84.0	16.05
39.	*Solanum nigrum* L.	Solanaceae	4.31	6.70	10.50	86.0	16.01
40.	*Polygonum plebeium* R. Br.	Polygonaceae	5.31	6.30	11.00	68.0	15.99
41.	*Aegle marmelos* (L.) Corr.serr.	Rutaceae	3.57	7.00	17.30	69.0	15.58
42.	*Anisomeles ovata* R. Br.	Lamiaceae	8.70	6.00	2.30	83.0	15.52
43.	*Argemone mexicana* L.	Papaveraceae	6.25	6.10	6.61	75.0	15.44
44.	*Amaranthus spinosus* L.	Amaranthaceae	6.25	6.00	6.67	74.0	15.32
45.	*Cuscuta reflexa* Roxb.	Cuscutaceae	6.32	6.00	7.11	69.0	15.19
46.	*Gnaphalium indicum* Pers.	Asteraceae	3.57	7.50	13.42	76.0	15.07
47.	*Chenopodium album* L.	Chenopodiaceae	4.20	6.00	11.60	75.0	14.89
48.	*Amaranthus viridis* L.	Amaranthaceae	6.20	6.00	6.67	70.0	14.86
49.	*Rungia pectinata* (L.) Nees	Acanthaceae	6.05	6.80	4.61	77.0	14.60
50.	*Emblica officinalis* Gaertn.	Euphorbiaceae	7.14	3.00	6.51	78.0	14.59
51.	*Datura metel* L.	Solanaceae	5.40	6.30	4.58	87.0	14.58
52.	*Malvastrum tricuspidatum* L.	Malvaceae	4.77	7.00	6.52	81.0	14.55
53.	*Acacia nilotica* (L.) Del. subsp. *indica* (Benth) Brenan	Mimosaceae	5.98	6.50	4.54	79.0	14.50
54.	*Cassia fistula* L.	Caesalpinaceae	5.94	6.50	4.56	79.0	14.47
55.	*C. alata* L.	Caesalpinaceae	5.96	6.50	4.54	78.0	14.38
56.	*Phyla nodiflora* (L.) Greene	Verbenaceae	4.27	6.00	10.00	75.0	14.33
57.	*Euphorbia hirta* L.	Euphorbiaceae	4.00	6.00	11.20	72.0	14.08
58.	*Vernonia cinerea* Less.	Asteraceae	4.00	6.00	11.20	72.0	14.08

Contd.

Table 4: Contd.

Sl. No.	Plants	Family	*A*	P	T	R	APTI
59.	*Trifolium alexandrium* L.	Fabaceae	4.61	7.50	7.50	70.0	13.92
60.	*Euphorbia thymifolia* (L.) Mill sp.	Euphorbiaceae	4.00	6.00	10.60	71.0	13.74
61.	*Oxalis corniculata* L.	Oxalidaceae	8.30	4.10	3.75	70.0	13.52
62.	*Lantana indica* Roxb.	Verbenaceae	4.63	7.60	7.51	65.0	13.50
63.	*Ageratum conyzoides* L.	Asteraceae	6.35	4.00	7.20	62.0	13.31
64.	*Ageratum houstonianum* Mill.	Asteraceae	6.16	4.00	7.00	63.0	13.08
65.	*Heliotropium indicum* L.	Boraginaceae	2.49	7.50	11.36	79.0	12.60
66.	*Mangifera indica* L.	Anacardiaceae	3.78	5.40	4.28	87.0	12.36
67.	*Ocimum basilicum* L.	Lamiaceae	3.41	6.30	5.66	82.0	12.28
68.	*O. tenuifolium* L.	Lamiaceae	3.43	6.20	5.64	81.0	12.16
69.	*Portulaca oleracea* L.	Portulacaceae	4.76	5.50	4.50	74.0	12.16
70.	*Punica granatum* L.	Punicaceae	3.56	6.50	8.81	67.0	12.15
71.	*Vitis vinifera* L.	Vitaceae	4.75	5.50	4.50	74.0	12.15
72.	*Antigonon leptopus* Hook. and Arn.	Polygonaceae	4.50	8.40	6.36	55.0	12.14
73.	*Vicia hirsuta* (L.) Gray	Fabaceae	4.75	5.50	4.50	70.0	11.75
74.	*Morus alba* L.	Moraceae	6.42	5.40	3.45	60.0	11.68
75.	*Lawsonia inermis* L.	Lythraceae	3.21	5.50	5.37	81.0	11.59
76.	*Lathyrus aphaca* L.	Fabaceae	3.98	7.00	4.69	68.0	11.45
77.	*Nepeta hindostana* (Roth.) Hains	Lamiaceae	3.23	6.20	5.56	76.0	11.40
78.	*Medicago lupulina* L.	Fabaceae	3.40	6.60	4.70	70.0	10.84
79.	*Thevetia peruviana* (Pers.) K.	Apocynaceae	2.51	6.50	5.86	76.0	10.70
80.	*Acalypha indica* L.	Euphorbiaceae	3.57	6.50	2.76	73.0	10.61
81.	*Mimosa pudica* L.	Mimosaceae	2.90	7.00	5.22	70.0	10.54
82.	*Syzygium cumini* (L.) Skeels	Myrtaceae	2.90	4.60	6.27	72.0	10.35
83.	*Melilotus indica* (L.) All	Fabaceae	2.60	6.40	6.27	68.0	10.09
84.	*Clerodendrum viscosum* vent.	Verbenaceae	1.00	6.00	10.00	82.0	9.80
85.	*Tridax procumbens* L.	Asteraceae	1.26	6.50	6.11	82.0	9.79
86.	*Commelina benghalensis* L	Commelinaceae	1.00	7.50	4.69	84.0	9.62
87.	*Abutilon indicum* (L.) Sweet	Malvaceae	1.00	6.50	21.70	65.0	9.32
88.	*Sida cordifolia* L.	Malvaceae	1.00	7.50	10.39	74.0	9.19
89.	*Centella asiatica* (L.) Urban	Apiaceae	2.10	6.40	4.20	69.0	9.13
90.	*Aerva lanata* (L.) Juss.	Amaranthaceae	1.00	7.00	10.30	71.0	8.83
91.	*Blepharis maderaspatensis* (L.) Roth.	Acantheceae	2.00	6.40	6.27	53.0	7.83
92.	*Justicia adhatoda* L.	Acanthaceae	1.00	7.00	5.60	65.0	7.76
93.	*Cannabis sativa* L.	Cannabinaceae	0.35	7.00	4.20	68.0	7.19
94.	*Blumea membranacea* DC.	Asteraceae	1.90	6.40	4.20	50.0	7.01
95.	*Boerhaavia diffusa* L.	Nyctaginaceae	1.90	6.40	6.27	45.0	6.91
96.	*Rauvolfia serpentina* (L.) Benth.	Apocynaceae	0.35	7.40	4.50	54.0	5.82
97.	*Vitex negundo* L.	Verbenaceae	0.35	7.40	4.50	54.0	5.82
98.	*Anthocephalus chinensis* (Lam.) A. Rich.	Rubiaceae	3.40	5.20	8.34	8.5	5.45
99.	*Lepidium sativum* L.	Brassicaceae	1.00	6.70	6.00	40.0	5.27

A great range of parameters have been used in different plants by scientists. Plant leaves have been used as a monitoring device by a number of workers (Rao, 1979; Rao *et al.*, 1990; Rao and Dubey, 1988). The concentration of antioxidants mainly ascorbic acid has been used to estimate the sensitivity of plants for certain pollutants (Arrigoni and De Tillio, 2000; Pierre and Queiroz, 1981; Potter *et al.*, 2000). A change in abscisic acid level in the leaves, in response to pollution has been studied by Gupta *et al.* (1991).

All the plants respond differently, as is evident by their APTI test (Rayappa and Singaracharya, 1993). Thus, the phytodiversity of a community is affected in response to various pollutants (Ambasht and Ambasht,). In other words, it can be said that the industrial air pollutants affect the phytodiversity of an area. Rosenberg (1979) has done one such study in response to a coal burning power plant, Darley (1966) in response to a cement industry; Sharma *et al.*(1986) in response to Mathura Refinery; Tripathi *et al.* (1999) and Gupta *et al.* (2001) around a brick industry; Narayan *et al.* (1994) around an aluminium factory and many more.

Every community is characterised by its species diversity (Kreb's, 1985) and it can be studied in detail by considering a number of parameters such as frequency, density and dominance. The relative study of these characters with respect to different plants along with the Importance Value Index (IVI) reflects the complete image of the community (Odum, 1971 and 1983; Whittekar, 1975).

Phytosociological study of all five study sites including Sarnath, the control site was performed. Relative frequency, relative density, relative dominance and I.V.I of the plants present at all the five sites are expressed in Tables 5.1 to 5.5. The result has been summarised in Table 10.

PHYTOSOCIOLOGICAL STUDY AT SITE A

Phytosociological study at Site A shows that I.V.I. of *Emblica officinalis* Gaertn. is highest (36.966) because of cultivated species in the nearby area followed by *Eucalyptus citriodora* Hook. (36.616), *Cynodon dactylon* (L.) Pers. (16.465), *Cyperus rotundus* L. (11.796) etc. Other plants with higher value of I.V.I. are *Xanthium strumarium* L. (8.825), *Polygonum plebeium* R. Br. (7.658), *Tridax procumbens* L. (6.466), *Calotropis procera* Ait. Hort. (6.296), *Rungia pectinata* (L.) Nees (5.539) etc. On the other hand, *Antigonon leptopus* Hook. & Arn., *Vicia hirsuta* L. and *Rauvolfia serpentina* (L.) Benth. had very low value of I.V.I. (Table 5). Total number of the plants recorded at this site was eighty-three.

Table 5: Phytosociological study of study site 'A' (around brick industry at Sandaha)

Sl. No.	Plants	Relative frequency	Relative density	Relative dominance	I.V.I.
1.	*Abutilon indicum* (L.) Sweet	0.35	0.317	.0203	0.6873
2.	*Acalypha indica* L.	0.70	0.23	0.0148	0.9448
3.	*Achyranthes aspera* L.	1.05	0.144	0.0092	1.2032
4.	*Adhatoda vasica* Nees.	1.05	0.835	0.2149	2.099
5.	*Aerva lanata* (L.) Juss.	0.70	.259	0.0106	0.9696
6.	*Ageratum conyzoides* L.	1.40	0.23	0.0148	1.6448
7.	*Ageratum houstonianum* Mill.	2.10	1.987	0.1278	4.2148
8.	*Alternanthera sessilis* (L.) DC	0.70	0.864	0.0088	1.5728
9.	*Amaranthus spinosus* L.	1.40	0.778	0.018	2.196
10.	*Anisomeles ovata* R. Br.	0.35	0.346	0.032	0.728
11.	*Anthocephalus chinensis* (Lam.) A. Rich	0.37	0.0288	1.482	2.12
12.	*Antigonon leptopus* Hook. and Arn.	0.35	0.317	0.1834	0.5804
13.	*Argemone mexicana* L.	2.45	0.634	0.3668	3.4508
14.	*Azadirachta indica* A. Juss.	0.35	0.001	0.2371	0.5881

Contd.

Table 5: Contd.

Sl. No.	Plants	Relative frequency	Relative density	Relative dominance	I.V.I.
15.	*Blepharis maderaspatensis* (L.) Roth.	0.35	0.403	0.0093	0.7623
16.	*Blumea membreneacea* DC.	1.05	0.605	0.0389	1.6939
17.	*Boerhaavia diffusa* L.	1.05	1.094	0.0113	2.1553
18.	*Calotropis gigantea* (L.) R.Br.	0.70	0.605	2.489	3.7949
19.	*Calotropis procera* Ait. Hort.	0.70	0.749	4.8167	6.2957
20.	*Carica papaya* L.	0.70	0.026	0.3267	1.0527
21.	*Cassia alata* L.	1.40	0.115	0.741	2.256
22.	*Cassia fistula* L.	0.70	0.346	0.5928	1.6388
23.	*Chenopodium album* L.	2.10	0.922	0.0379	3.0599
24.	*Cleome viscosa* L.	1.40	3.17	0.0203	1.7373
25.	*Clerodendrum viscosum* vent.	1.40	0.922	0.0592	2.3812
26.	*Coccinia indica* Wt. and Arn.	0.70	0.72	0.0166	1.4366
27.	*Commelina benghalensis* L.	1.05	0.806	0.0083	1.8643
28.	*Croton bonplandianus* Baill.	2.1	.835	.0085	2.9435
29.	*Cynodon dactylon* (L.) Pers.	3.50	12.96	0.0083	16.4683
30.	*Cyperus rotundus* L.	3.15	8.64	0.0055	11.7955
31.	*Dactyloctenium aegyptium* Beauv	1.05	1.0944	0.0007	2.1451
32.	*Datura metel* L.	0.70	1.325	1.3635	3.3885
33.	*Desmostachya bipinnata* (L.) Stapf.	1.05	0.432	0.0003	1.4823
34.	*Digitaria adscendens* Henr.	0.70	0.346	0.0002	1.0462
35.	*Eclipta alba* (L.) Hassk.	1.05	1.152	0.741	2.2761
36.	*Emblica afficinalis* Gaertn.	1.05	0.346	35.5699	36.9659
37.	*Eragrostis tenella* (L.) Baeuv	2.10	5.846	0.0037	7.9497
38.	*Eucalyptus citriodora* Hook.	0.70	0.346	35.5699	36.6159
39.	*Euphorbia hirta* L.	1.05	1.008	0.002	2.06
40.	*Euphorbia thymifolia* (L.) Mill. sp.	2.45	1.786	0.0046	4.2406
41.	*Evolvulus alsinoidis* L.	2.45	2.938	0.0075	5.3955
42.	*Ficus benghalensis* L.	0.35	0.0001	0.3001	0.6502
43.	*Gnaphalium indicum* Pers.	1.05	0.806	0.0518	1.9078
44.	*Heliotropium indicum* L.	0.70	0.058	0.00004	0.75804
45.	*Hibiscus rosa-sinensis* L.	0.35	0.002	0.0046	0.3566
46.	*Centella asiatica* (L.) Urban	1.05	0.115	0.0029	1.1679
47.	*Ipomoea fistulosa* Mart.	0.70	0.374	0.0963	1.1703
48.	*Lantana indica* Roxb.	1.40	0.922	2.1342	4.4562
49.	*Lathyrus aphaca* L.	1.40	0.979	0.0056	2.3846
50.	*Lawsonia inermis* L.	0.70	0.23	0.0593	0.9893
51.	*Lepidium sativum* L.	1.75	0.622	0.0017	2.4137
52.	*Malvastrum tricuspidatum* L.	2.80	0.806	0.0083	3.6143
53.	*Mangifera indica* L.	0.70	0.001	0.9263	1.6273
54.	*Medicago lupulina* L.	1.40	0.806	0.0005	2.2065
55.	*Melilotus indica* (L.) All.	1.75	1.094	0.0028	2.8468
56.	*Mimosa pudica* L.	1.05	0.115	0.0003	1.1653
57.	*Moringa oleifera* Lam.	0.35	0.017	0.4446	0.8116
58.	*Nepeta hindostana* (Roth.) Haines	0.70	0.490	0.001	1.191
59.	*Nicotiana plumbaginifolia* Viv.	1.05	0.288	0.0066	1.3446
60.	*Ocimum basilicum* L.	0.35	0.259	0.0166	0.6256
61.	*O. tenuifolium* L.	0.35	0.288	0.0185	0.6565
62.	*Oxalis corniculata* L.	2.45	2.938	0.0302	5.4182
63.	*Parthenium hysterophorus* L.	2.80	2.765	0.7117	6.2767
64.	*Phalaris minor* Retz.	1.05	1.872	0.0012	2.9232
65.	*Phyla nodiflora* (L.) Greene	1.40	0.052	0.00004	1.45204
66.	*Polygonum plebeium* R. Br.	1.75	5.904	0.0038	7.6578

Contd.

Table 5: Contd.

Sl. No.	Plants	Relative frequency	Relative density	Relative dominance	I.V.I.
67.	*Portulaca oleracea* L.	2.10	1.958	0.0012	4.0592
68.	*Psidium guajava* L.	0.70	0.002	0.0160	0.718
69.	*Rauvolfia serpentina* (L.) Benth.	0.35	0.023	0.0015	0.3745
70.	*Rungia pectinata* (L.) Nees	2.45	3.081	0.0079	5.5389
71.	*Rumex dentatus* L.	1.75	0.662	0.0426	2.4546
72.	*Saccharum munja* Roxb.	1.05	0.691	0.0004	1.7414
73.	*Sida cordifolia* L.	1.75	0.950	0.0611	2.7611
74.	*Solanum nigrum* L.	1.40	1.181	0.0759	2.6569
75.	*Spilanthes indicus* Wall.	0.35	0.922	0.0213	1.2933
76.	*Tamarindus indica* L.	0.35	0.0002	0.4268	0.777
77.	*Thevetia peruviana* (Pers.) K.	0.70	0.009	0.2223	0.9313
78.	*Tridax procumbens* L.	2.80	3.629	0.0373	6.4663
79.	*Trifolium alexandrium* L.	1.40	0.662	0.0068	2.0688
80.	*Vernonia cinerea* Less.	1.40	1.613	0.0166	3.0296
81.	*Vicia hirsuta* (L.) Gray	0.35	0.027	0.0003	0.3773
82.	*Vitex negundo* L.	0.70	0.001	0.0145	0.7155
83.	*Xanthium strumarium* L.	2.45	5.990	0.3853	8.8253

PHYTOSOCIOLOGICAL STUDY AT SITE B

Table 6 shows the result of phytosociological study around the brick industry at site B. At this site, eighty-two plants have been recorded with *Capparis zeylanica* L. showing the highest value of I.V.I. (29.676), followed by *Cynodon dactylon* (L.) Pers. (16.301), *Calotropis gigantea* (L.) W.T. Aiton (13.842), *Lantana indica* Roxb. (12.913) etc. Other important plants with high I.V.I. value are *Xanthium strumarium* L. (12.275), *Parthenium hysterophorus* L. (11.679), *Datura metel* L. (9.413) etc. The plants with low IVI value at this site are *Anisomeles ovata* R.Br. (0.874), *Mimosa pudica* L. (0.870), *Nepeta hindostana* (Roth.). Haines. (0.814), *Rauvolfia serpentina* (L.) Benth. (0.789) with *Punica granatum* L. and *Aegle marmelos* (L.) Corr. Serr., the least (0.380).

Table 6: Phytosociological study of study site 'B' (around brick industry at Kotwan)

Sl. No.	Plants	Relative frequency	Relative density	Relative dominance	I.V.I.
1.	*Abutilon indicum* (L.) Sweet	0.38	.3252	.1175	0.8227
2.	*Acacia nilotica* (L.) Del. Subsp. *indica* (Benth) Brenan	1.52	.00813	7.3476	8.87573
3.	*Acalypha indica* L.	.76	.1897	.0745	1.0242
4.	*Achyranthes aspera* L.	1.52	.1626	.1488	1.8314
5.	*Adhatoda vasica* Nees.	1.52	.7859	1.1364	3.4423
6.	*Aegle marmelos* (L.) Corr. Serr.	.38	.000271	.00006	0.380331
7.	*Ageratum conyzoides* L.	1.52	1.2737	.4603	3.254
8.	*Ageratum houstonianum* Mill.	1.89	1.8157	.6563	4.362
9.	*Alternanthera sessilis* (L.) DC	.38	1.0569	.0610	1.4979
10.	*Amaranthus spinosus* L.	1.14	.7046	.0917	1.9363
11.	*Amaranthus viridis* L.	1.52	.8401	.1093	2.4694
12.	*Anisomeles ovata* R. Br.	.38	.3252	.1692	0.8744
13.	*Antigonon leptopus* Hook. and Arn.	.38	.3252	1.0580	1.7632
14.	*Argemone mexicana* L.	2.27	.7046	2.2924	5.267

Contd.

Table 6: Contd.

Sl. No.	Plants	Relative frequency	Relative density	Relative dominance	I.V.I.
15.	*Asclepias curassavica* L.	.76	.0542	.0783	0.8925
16.	*Blumea membreneacea* DC.	.76	.5962	.2151	1.5713
17.	*Boerhaavia diffusa* L.	1.52	.9485	.0548	2.5233
18.	*Calotropis gigantea* (L.) W.T. Aiton	.76	.542	12.5399	13.8419
19.	*Cannabis sativa* L.	.38	.4336	.6269	1.4405
20.	*Capparis zeylanica* L.	.52	6.6233	22.5326	29.6759
21.	*Carica papaya* L.	.38	.02168	1.5361	1.93778
22.	*Cassia alata* L.	1.14	.1084	3.9187	5.1671
23.	*Chenopodium album* L.	1.98	.9214	.2131	3.1145
24.	*Cleome viscosa* L.	1.52	.3794	.1371	2.0365
25.	*Clerodendrum viscosum* vent.	1.14	.8401	.3037	2.2838
26.	*Commelina benghalensis* L.	.76	.7317	.0423	1.534
27.	*Croton bonplandianus* Baill	1.89	.7859	.0454	2.7213
28.	*Cuscuta reflexa* Roxb.	.38	.3794	.0013	0.7607
29.	*Cynodon dactylon* (L.) Pers.	3.79	12.466	.0450	16.301
30.	*Cyperus rotundus* L.	3.41	8.401	.0303	11.8413
31.	*Dactyloctenium aegyptium* (L.) Will.	1.52	1.0027	.0036	2.5263
32.	*Datura metel* L.	1.14	1.2195	7.0536	9.4131
33.	*Desmostachya bipinnata* (L.) Stapf.	1.52	.4065	.1470	2.0735
34.	*Digitaria adscendens* Henr.	1.14	.3523	.0012	1.4935
35.	*Eclipta alba* (L.) Hassk.	1.14	1.0027	.3625	2.5052
36.	*Eragrostis tenella* (L.) Baeuv	1.89	5.5826	.0202	7.4928
37.	*Euphorbia hirta* L.	1.14	.8672	.0011	2.0083
38.	*Euphorbia thymifolia* Linn.	2.27	1.5176	.0020	3.7896
39.	*Evolvulus alsinoides* L.	3.03	2.8184	.0408	5.8892
40.	*Ficus benghalensis* L.	.38	.0000813	.9522	1.332281
41.	*Ficus religiosa* L.	.38	.0000542	.5015	0.881554
42.	*Gnaphalium indicum* Pers.	1.52	.7588	.2743	2.5531
43.	*Heliotropium indicum* L.	1.14	.0813	.0294	1.2507
44.	*Centella asiatica* (L.) Urban	.76	.2168	.0031	0.9799
45.	*Lantana indica* Roxb.	1.14	.8401	10.9332	12.9133
46.	*Lathyrus aphaca* L.	1.52	.9485	3.0860	5.5545
47.	*Lawsonia inermis* L.	.76	.1897	.2743	1.224
48.	*Lepidium sativum* L.	1.52	.5962	.0086	2.1248
49.	*Madhuca indica* J.F. Gmel.	.38	.4336	.1920	1.0056
50.	*Malvastrum tricuspidatum* L.	2.27	.7317	.0423	3.044
51.	*Mangifera indica* L.	.38	.001355	4.8984	5.279755
52.	*Medicago lupulina* L.	1.52	.7317	.0026	2.2543
53.	*Melilotus indica* (L.) All.	1.52	.9756	.0141	2.5097
54.	*Mimosa pudica* L.	.76	.1084	.0016	0.87
55.	*Moringa oleifera* Lam.	.76	.01626	2.3512	3.12746
56.	*Nepeta hindostana* (Roth.) Haines	.38	.4336	.0006	0.8142
57.	*Nicotiana plumbaginifolia* Viv.	1.14	.271	.0004	1.4114
58.	*Ocimum basilicum* L.	.76	.271	.0979	1.1289
59.	*Oxalis corniculata* L.	3.03	2.8997	.1677	6.0974
60.	*Parthenium hysterophorus* L.	2.65	5.2845	3.7443	11.6788
61.	*Phalaris minor* Retz.	1.52	1.7344	.0062	3.2606
62.	*Phyla nodiflora* (L.) Greene	1.14	.0542	.0002	1.1944
63.	*Polygonum plebeium* R. Br.	1.52	5.5013	.0198	7.0411
64.	*Portulaca oleracea* L.	1.89	1.8428	.0066	3.7394
65.	*Punica granatum* L.	.38	.00000271	.0002	0.380203
66.	*Rauvolfia serpentina* (L.) Benth.	.76	.02168	.0078	0.78948

Contd.

Table 6: Contd.

Sl. No.	Plants	Relative frequency	Relative density	Relative dominance	I.V.I.
67.	*Rungia pectinata* (L.) Nees	2.27	2.8726	.0415	5.1841
68.	*Ricinus communis* L.	.38	.3252	1.8809	2.5861
69.	*Rumex dentatus* L.	1.89	.6775	.2449	2.8124
70.	*Saccharum munja* Roxb.	.76	.6233	.0022	1.3855
71.	*Sida cordifolia* L.	2.27	.8972	.3135	3.4807
72.	*Solanum nigrum* L.	1.14	1.1382	.4114	2.6896
73.	*Spilanthes indicus* Wall.	.76	.8672	.1128	1.74
74.	*Syzygium cumini* (L.) Skeels K.	.38	.0001355	.1763	0.556436
75.	*Thevetia peruviana* (Pers.)	1.14	.00813	1.1756	2.32373
76.	*Tridax procumbens* L.	2.65	3.4688	.0439	6.1627
77.	*Trifolium alexandrium* L.	1.52	.6504	.0375	2.2079
78.	*Vernonia cinerea* L. Less.	1.52	1.4905	.0862	3.0967
79.	*Vicia hirsuta* L.	.38	.025745	.0015	0.407245
80.	*Vitex negundo* L.	.38	.001084	.0767	0.457784
81.	*Xanthium strumarium* L.	2.65	5.5826	4.0427	12.2753
82.	*Ziziphus mauritiana* Lamk.	.38	2.3848	.0196	2.7844

PHYTOSOCIOLOGICAL STUDY AT SITE C

The result of phytosociological analysis at site C has been presented in Table 7. The plant with highest IVI value at this site is *Eucalyptus citriodora* Hook. (26.203) followed by *Calotropis procera* Ait. Hort. (20.775) and *Cynodon dactylon* (L.) Pers. (19.347). Other important plants with higher I.V.I. value are *Capparis zeylanica* L. (19.101), *Cyperus rotundus* L. (15.930), *Lantana indica* Roxb. (11.410), *Xanthium strumarium* L. (10.680) etc. The plants with low IVI value at this site are *Mimosa pudica* L. (0.979), *Acalypha indica* L. (0.969), *Ocimum tenuifolium* L. (0.908), *Anisomeles ovata* R.Br. (0.906) etc. however the lowest IVI is of *Punica granatum* L. (0.431). As many as sixty-nine tree species have been recorded at this site.

Table 7: Phytosociological study of study site 'C' (around brick industry at Salarpur)

Sl. No.	Plants	Relative frequency	Relative density	Relative dominance	I.V.I.
1.	*Acalypha indica* L.	.85	.096	.0233	0.9693
2.	*Achyranthes aspera* L.	1.28	0.256	.0624	1.5984
3.	*Adhatoda vasica* Nees.	1.71	.864	.8427	3.4167
4.	*Ageratum conyzoides* L.	2.56	.224	.0545	2.8385
5.	*Alternanthera sessilis* (L.) DC	.85	1.184	.0462	2.0802
6.	*Amaranthus viridis* L.	1.71	.96	.0843	2.7543
7.	*Anisomeles ovata* R. Br.	.43	0.352	.1235	0.9055
8.	*Antigonon leptopus* Hook. and Arn.	.85	.288	.6320	1.77
9.	*Argemone mexicana* L.	2.99	.832	1.8258	5.6478
10.	*Asclepias curassavica* L.	.85	.128	.1248	1.1028
11.	*Blepharis maderaspatensis* (L.) Roth.	.43	.416	.0364	0.8824
12.	*Blumea membreneacea* DC.	.85	0.64	.1560	1.646
13.	*Boerhaavia diffusa* L.	1.71	1.024	.0474	2.7814
14.	*Calotropis procera* Ait. Hort.	1.28	.768	18.7269	20.7749
15.	*Capparis zeylanica* L.	1.28	.704	17.1166	19.1006
16.	*Carica papaya* L.	.85	.032	1.5293	2.4113

Contd.

Table 7: Contd.

Sl. No.	Plants	Relative frequency	Relative density	Relative dominance	I.V.I.
17.	*Cassia alata* L.	1.28	.096	2.3408	3.7168
18.	*Cassia fistula* L.	.43	.0096	3.7454	4.185
19.	*Chenopodium album* L.	2.56	1.056	.1648	3.7808
20.	*Cleome viscosa* L.	1.28	.352	.0858	1.7178
21.	*Clerodendrum viscosum* vent.	1.71	1.024	.2497	2.9837
22.	*Coccinia indica* Wt. and Arn.	.85	.832	.0730	1.755
23.	*Croton bonplandianus* Baill	2.56	.896	.0349	3.4909
24.	*Cynodon dactylon* (L.) Pers.	4.27	15.04	.0365	19.3465
25.	*Cyperus rotundus* L.	3.42	12.48	.0304	15.9304
26.	*Dactyloctenium aegyptium* (L.) Will.	1.28	1.152	.0028	2.4348
27.	*Desmostachya bipinnata* (L.) Stapf.	1.28	.448	.0011	1.7291
28.	*Digitaria adscendens* Henr.	.85	0.32	.0008	1.1708
29.	*Dolichos lablab* L.	.43	.288	.0702	0.7882
30.	*Emblica afficinalis* Gaertn.	0.85	.0064	2.4969	3.3533
31.	*Eragrostis tenella* (L.) Baeuv	2.56	6.144	0.0150	8.719
32.	*Eucalyptus citriodora* Hook.	0.85	.384	24.9693	26.2033
33.	*Euphorbia hirta* L.	0.85	1.696	0.0015	2.5475
34.	*Euphorbia thymifolia* L.	2.99	1.824	0.0178	4.8318
35.	*Evolvulus alsinoides* L.	2.99	3.392	0.0331	6.4151
36.	*Ficus religiosa* L.	0.85	.000064	0.3995	1.249564
37.	*Gnaphalium indicum* Pers.	1.28	.896	0.2185	2.3945
38.	*Heliotropium indicum* L.	0.85	.096	0.0233	0.9693
39.	*Hibiscus rosa-sinensis* L.	0.43	.00224	0.0122	0.44444
40.	*Ipomoea fistulosa* Mart.	0.85	0.384	0.3745	1.6085
41.	*Lantana indica* Roxb.	1.71	0.992	8.7080	11.41
42.	*Lawsonia inermis* L.	0.85	0.256	.2497	1.3557
43.	*Malvastrum tricuspidatum* L.	2.14	0.832	0.0324	3.0044
44.	*Mangifera indica* L.	.85	.0016	3.9014	4.753
45.	*Melilotus indica* (L.) All.	1.71	1.12	0.0109	2.8409
46.	*Mimosa pudica* L.	.85	.128	.0012	0.9792
47.	*Morus alba* L.	.43	.00032	.4994	0.92972
48.	*Nicotiana plumbaginifolia* Viv.	.85	.576	.0501	1.4761
49.	*Ocimum basilicum* L.	.85	.32	.0780	1.248
50.	*Ocimum tenuiflorum* L.	.43	.384	.0936	0.9076
51.	*Parthenium hysterophorus* L.	3.85	2.944	1.407	8.201
52.	*Phalaris minor* Retz.	1.28	1.024	.0025	2.3065
53.	*Phyla nodiflora* (L.) Greene	1.71	.064	.00009	1.77409
54.	*Polygonum plebeium* R. Br.	1.71	6.432	.0157	8.1577
55.	*Portulaca oleracea* L.	2.14	2.08	.005	4.225
56.	*Psidium guajava* L.	1.28	.00064	.00224	1.28288
57.	*Punica granatum* L.	.43	.0000096	.0005	0.43051
58.	*Ricinus communis* L.	.85	.096	.3742	1.3202
59.	*Rumex dentatus* L.	2.56	.8	.195	3.555
60.	*Sida cordifolia* L.	2.14	1.12	.273	3.533
61.	*Solanum nigrum* L.	1.71	1.312	.3198	3.3418
62.	*Spilanthes indicus* Wall.	.85	.96	.0843	1.8943
63.	*Tamarindus indica* L.	.43	.00064	3.995	4.42564
64.	*Thevetia peruviana* (Pers.) K.	.85	.0096	.9363	1.7959
65.	*Tridax procumbens* L.	2.99	4.032	.1572	7.1792
66.	*Trifolium alexandrium* L.	1.71	.672	.0262	2.4082
67.	*Vernonia cinerea* L. Less.	1.71	1.662	.0649	3.4369
68.	*Vitex negundo* L.	.43	.00128	.0611	0.49238
69.	*Xanthium strumarium* L.	2.56	6.528	1.5918	10.6798

PHYTOSOCIOLOGICAL STUDY AT SITE D

A total of seventy-six plants were recorded at site D. *Eucalyptus citridora* Hook. had the highest value of IVI (68.171) followed by *Cynodon dactylon* (L.) Pers (19.631) and *Cyperus rotundus* L. (14.060). The other plants with higher value of IVI include *Calotropis procera* Ait. Hort. (11.970), *Xanthium strumarium* L. (10.487), *Polygonum plebeium* R.Br. (8.957), *Eragrostis tenella* (L.) Baeuv. (8.888) etc. On the other hand, the plants with low IVI value are *Asclepias curassavica* L. (0.613), *Mimosa pudica* L. (0.563), *Syzygium cumini* (L.) Skeels. (0.470), *Vitex negundo* L. (0.449) with *Vitis venifera* L. as the least (0.423; Table 8).

Table 8: Phytosociological study of study site 'D' (around brick industry at Chamaraha)

Sl. No.	Species	Relative frequency	Relative density	Relative dominance	IVI
1.	*Achyranthes aspera* L.	1.25	.2058	.0209	1.4767
2.	*Adhatoda vasica* Nees.	1.25	.1715	.0174	1.4389
3.	*Aegle marmelos* (L.) Corr.	.42	.00343	1.2547	1.67813
4.	*Aerva lanata* (L.) Juss.	.84	.2744	.0178	1.1322
5.	*Ageratum houstonianum* Mill.	2.51	2.2295	.2265	4.966
6.	*Alternanthera sessilis* (L.) DC	.84	1.2348	.0200	2.0948
7.	*Amaranthus spinosus* L.	.84	.7889	.0288	1.6577
8.	*Amaranthus viridis* L.	1.25	1.0976	.0401	2.3877
9.	*Anisomeles ovata* R. Br.	.42	.3773	.0551	0.8524
10.	*Anthocephalus chinensis* (Lam.) A. Rich.	.42	.000343	.3485	0.768843
11.	*Argemone mexicana* L.	2.51	.8232	.7532	4.0864
12.	*Asclepias curassavica* L.	.42	.1372	.0557	0.6129
13.	*Azadirachta indica* A. Juss.	.42	.000343	.2230	0.643343
14.	*Blepharis maderaspatensis* (L.) Roth.	.84	.4116	.015	1.2666
15.	*Boerhaavia diffusa* L.	1.25	1.2005	.0194	2.4699
16.	*Calotropis gigantea* (L.) W.T. Aiton	.84	.7546	4.9074	6.502
17.	*C. procera* Ait. Hort.	1.25	.9604	9.7591	11.9695
18.	*Cannabis sativa* L.	.42	.6174	.2509	1.2883
19.	*Carica papaya* L.	1.25	.1372	2.7325	4.1197
20.	*Cassia alata* L.	1.25	.0686	.697	2.0156
21.	*C. fistula* L.	.84	.1029	1.6729	2.6158
22.	*Chenopodium album* L.	2.51	1.1662	.0758	3.752
23.	*Cleome viscosa* L.	1.67	.343	.0348	2.0478
24.	*Coccinia indica* Wt. and Arn.	.84	.8232	.0301	1.6933
25.	*Commelina benghalensis* L.	1.25	.884	.0144	2.1484
26.	*Croton bonplandianus* Baill	2.51	.884	.0144	3.4084
27.	*Cuscuta reflexa* Roxb.	.42	.4459	.0004	0.8663
28.	*Cynodon dactylon* (L.) Pers.	4.18	15.435	.0156	19.6306
29.	*Cyperus rotundus* L.	3.76	10.29	.0104	14.0604
30.	*Dactyloctenium aegyptium* (L.) Will.	1.25	1.029	.001	2.28
31.	*Desmostachya bipinnata* (L.) Stapf.	1.25	.4459	.0004	1.6963
32.	*Digitaria adscendens* Henr.	.84	.3773	.0003	1.2176
33.	*Dolichos lablab* L.	.42	.2744	.0278	0.7222
34.	*Eragrostis tenella* (L.) Baeuv	2.09	6.7914	.0068	8.8882
35.	*Eucalyptus citriodora* Hook.	.84	.4116	66.9196	68.1712
36.	*Euphorbia hirta* L.	1.25	1.7493	.0071	3.0064
37.	*E. thymifolia* L.	2.93	1.7493	.0071	4.6864
38.	*Evolvulus alsinoides* L.	2.93	3.4986	.0142	6.4428
39.	*Ficus bengalensis* L.	.42	.0000686	.2258	0.645869
40.	*F. religiosa* L.	.84	.0000343	.0891	0.929134
41.	*Gnaphalium indicum* Pers.	.84	.9947	.101	1.9357

Contd.

Table 8: Contd.

Sl. No.	Species	Relative frequency	Relative density	Relative dominance	I.V.I.
42.	*Heliotropium indicum* L.	.84	.0686	.0069	0.9155
43.	*Ipomoea fistulosa* Mart.	.84	.3773	.1533	1.3706
44.	*Lantana indica* Roxb.	1.67	1.029	3.7642	6.4632
45.	*Lawsonia inermis* L.	1.25	.2058	.0836	1.5394
46.	*Madhuca indica* J.F. Gmel.	.42	.5831	.0376	1.0407
47.	*Malvastrum tricuspidatum* L.	2.09	.9261	.015	3.0311
48.	*Mangifera indica* L.	2.51	.001715	1.7427	4.254415
49.	*Melilotus indica* (L.) All.	1.67	1.097	.0044	2.7714
50.	*Mimosa pudica* L.	.42	.1374	.006	0.5634
51.	*Moringa oleifera* Lam.	.42	.01029	.4182	0.84849
52.	*Morus alba* L.	.42	.000343	.1254	0.545743
53.	*Nicotiana plumbaginifolia* Viv.	1.25	.343	.0125	1.6055
54.	*Ocimum basilicum* L.	.42	.343	.0348	0.7978
55.	*Ocimum sanctum* L.	.42	.4116	.0418	0.8734
56.	*Parthenium hysterophorus* L.	3.35	3.087	.6148	7.0518
57.	*Phalaris minor* Retz.	1.67	2.1266	.0021	3.7987
58.	*Phyla nodiflora* (L.) Greene	1.67	.0686	.00008	1.73868
59.	*Polygonum plebeium* R. Br.	2.09	6.86	.0069	8.9569
60.	*Portulaca oleracea* L.	2.09	5.5909	.0056	7.6865
61.	*Psidium guajava* L.	1.25	.001029	.015	1.266029
62.	*Ricinus communis* L.	.42	.4802	.7791	1.6793
63.	*Rumex dentatus* L.	2.51	.9261	.0009	3.437
64.	*Saccharum munja* L.	.84	.7546	.0007	1.5953
65.	*Sida cordifolia* L.	1.67	1.1319	.115	2.9169
66.	*Solanum nigrum* L.	1.25	1.4063	.1429	2.7992
67.	*Spilanthes indicus* Wall.	.42	.9947	.0363	1.451
68.	*Syzygium cumini* (L.) Skeels	.42	.0001372	.0501	0.470237
69.	*Thevetia peruviana* (Pers.) K.	.84	.01029	.4182	1.26849
70.	*Tridax procumbens* L.	3.35	4.2875	.0697	7.7072
71.	*Trifolium alexandrium* L.	1.67	.8575	.0139	2.5414
72.	*Vernonia cinerea* L. Less.	1.25	1.7836	.0289	3.0625
73.	*Vitex negundo* L.	.42	.001372	.0273	0.448672
74.	*Vitis venifera* L.	.42	.000343	.0022	0.422543
75.	*Xanthium strumarium* L.	2.93	6.86	.697	10.487
76.	*Ziziphus mauritiana* Lamk.	.42	.000686	.007	0.427686

PHYTOSOCIOLOGICAL STUDY AT SITE E

For phytosociological study of Sarnath, (the control site) the barren and abandoned land area near the Radio station Sarnath has been selected, because this area has been subjected to limited anthropogenic interference which affects the phytosociology of a community (Ashmore and Ainshorth, 1995; Oliveira and Jardim, 1999).

At this site ninety-seven plants were recognized (Table 9). *Eucalyptus citriodora* Hook had the highest value of IVI (49.075) followed by *Cynodon dactylon* (L.) Pers. (16.612), *Cyperus rotundus* L. (13.621), *Mangifera indica* L. (12.889) etc. The other plants with higher IVI values are *Eragrostis tenella* (L.) Beauv. (7.421), *Calotropis procera* Ait. Hort. (7.368), *Xanthium strumarium* L. (7.276) and *Polygonum plebeium* R.Br. (7.214). However, the plants with lower IVI value are *Ipomoea fistulosa* Mart. (0.649), *Moringa oleifera* Lam. (0.615), *Blumea membraneacea* DC. (0.527), *Vicia hirsuta* L. (0.355) etc., but the least value was of *Vitis venifera* L. (0.332).

Table 9: Phytosociological study at study site 'E' (Control Site i.e. Sarnath)

Sl. No.	Species	Relative frequency	Relative density	Relative dominance	IVI
1.	*Abutilon indicum* (L.) Sweet	.66	.3192	.0277	1.0069
2.	*Acacia nilotica* (L.) Del. subsp. *indica* (Benth) Brenan	1.32	.00798	1.7332	3.06118
3.	*Acalypha indica* L.	.66	.1862	.0161	0.8623
4.	*Achyranthes aspera* L.	1.32	.2926	.0254	1.638
5.	*Adhatoda vasica* Nees.	1.32	.1596	.0554	1.535
6.	*Aegle marmelos* (L.) Corr.	.66	.00266	.0819	0.74456
7.	*Aerva lanata* (L.) Juss.	.66	.1862	.0103	0.8565
8.	*Ageratum conyzoides* L.	1.32	1.2502	.1086	2.6788
9.	*Ageratum houstonianum* Mill.	1.99	1.729	.1502	3.8692
10.	*Alternanthera sessilis* (L.) DC	.99	1.2502	.0339	2.2741
11.	*Amaranthus spinosus* L.	.99	.665	.0208	1.6758
12.	*Amaranthus viridis* L.	.99	.8246	.0257	1.8403
13.	*Anisomeles ovata* R. Br.	.33	.3192	.0399	0.6891
14.	*Anthocephalus chinensis* (Lam.) A. Rich	.33	.002128	1.8488	2.180928
15.	*Antigonon leptopus* Hook. and Arn.	.33	.3192	.0415	0.6907
16.	*Argemone mexicana* L.	.99	.7182	.0623	1.7705
17.	*Asclepias curassavica* L.	.33	.0266	.0092	0.3658
18.	*Azadirachta indica* A. Juss.	.99	.000798	.4437	1.434498
19.	*Blepharis maderaspatensis* (L.) Roth.	.66	.3192	.0099	0.9891
20.	*Blumea membranacea* DC.	.33	.14858	.0485	0.52708
21.	*Boerhaavia diffusa* L.	.99	.9842	.0136	1.9878
22.	*Calotropis gigantea* (L.) R.Br.	.33	.14858	3.1060	3.58458
23.	*C. procera* Ait. Hort.	.33	.7982	6.2398	7.368
24.	*Cannabis sativa* L.	.66	.4522	.1571	1.2693
25.	*Capparis zeylanica* L.	.66	.1486	4.8531	5.6617
26.	*Carica papaya* L.	.33	.1596	2.7177	3.2073
27.	*Cassia alata* L.	.99	.1064	.9244	2.0208
28.	*C. fistula* L.	.66	.01596	2.2185	2.89446
29.	*Chenopodium album* L.	.66	.9842	.0547	1.6989
30.	*Cleome viscosa* L.	.66	.4522	.0393	1.1515
31.	*Clerodendrum viscosum* vent.	1.32	.8246	.0716	2.2162
32.	*Coccinia indica* Wt. and Arn.	.33	.665	.0208	1.0158
33.	*Commelina benghalensis* L.	.99	.6384	.0088	1.6372
34.	*Croton bonplandianus* Baill	1.99	.7448	.0103	2.7451
35.	*Cuscuta reflexa* Roxb.	.33	.266	.0002	0.5962
36.	*Cynodon dactylon* (L.) Pers.	3.31	13.3	.0018	16.6118
37.	*Cyperus rotundus* L.	2.98	10.64	.0014	13.6214
38.	*Dactyloctenium aegyptium* (L.) Will.	.99	1.0374	.0001	2.0275
39.	*Datura metel* L.	.66	1.2768	1.7748	3.7116
40.	*Desmostachya bipinnata* (L.) Stapf.	.99	.3724	.0003	1.3627
41.	*Digitaria adscendens* Henr.	.66	.3548	.0002	1.015
42.	*Eclipta alba* (L.) Hassk.	.99	1.064	.0924	2.1464
43.	*Emblica afficinalis* Gaertn.	.99	.0266	3.6976	4.7142
44.	*Eragrostis tenella* (L.) Baeuv	1.99	5.4264	.0047	7.4211
45.	*Eucalyptus citriodora* Hook.	.66	.3458	48.0696	49.0754
46.	*Euphorbia hirta* L.	.99	1.463	.0457	2.4987
47.	*Euphorbia thymifolia* (L.) Mill sp.	1.99	1.5694	.0054	3.5648
48.	*Evolvulus alsinoides* L.	2.32	2.8462	.0098	5.176
49.	*Ficus benghalensis* L.	.66	.000532	1.4975	2.158032
50.	*Ficus religiosa* L.	.66	.000266	.5916	1.251866
51.	*Gnaphalium indicum* Pers.	.99	.7714	.067	1.8284
52.	*Heliotropium indicum* L.	.66	.0532	.0046	0.7178
53.	*Hibiscus rosa-sinensis* L.	.66	.00532	.0166	0.68192
54.	*Centella asiatica* (L.) Urban	.99	.1862	.000005	1.176205

Contd.

Table 9: Contd.

Sl. No.	Species	Relative frequency	Relative density	Relative dominance	IVI
55.	*Ipomoea fistulosa* Mart.	.33	.3192	.00001	0.64921
56.	*Lantana indica* Roxb.	1.32	.8512	2.6623	4.8335
57.	*Lathyrus aphaca* L.	1.32	.9576	.0074	2.285
58.	*Lawsonia inermis* L.	.66	.1862	.0647	0.9109
59.	*Lepidium sativum* L.	1.65	.6384	.0022	2.2906
60.	*Malvastrum tricuspidatum* L.	1.32	.7182	.0099	2.0481
61.	*Mangifera indica* L.	1.32	.0133	11.5552	12.8885
62.	*Medicago lupulina* L.	1.32	.6916	.0005	2.0121
63.	*Melilotus indica* (L.) All.	1.65	.8512	.0029	2.5041
64.	*Mimosa pudica* L.	1.32	.7448	.0007	2.0655
65.	*Moringa oleifera* Lam.	.33	.00798	.2773	0.61528
66.	*Morus alba* L.	.33	.000266	.0831	0.413366
67.	*Nepeta hindostana* (Roth.) Hains	.66	.399	.0001	1.0591
68.	*Nicotiana plumbaginifolia* Viv.	.99	.266	.0083	1.2643
69.	*Ocimum basilicum* L.	.99	.266	.0231	1.2791
70.	*Ocimum tenuifolium* L.	.66	.3458	.0012	1.007
71.	*Oxalis corniculata* L.	1.99	2.9792	.0414	5.0106
72.	*Parthenium hysterophorus* L.	2.32	2.934	.4076	5.6616
73.	*Phalaris minor* Retz.	.99	1.6492	.0014	2.6406
74.	*Phyla nodiflora* (L.) Greene	1.32	.0532	.00005	1.37325
75.	*Polygonum plebeium* R. Br.	1.65	5.5594	.0048	7.2142
76.	*Portulaca oleracea* L.	1.99	4.4422	.0038	6.436
77.	*Psidium guajava* L.	.99	.000798	.0099	1.000698
78.	*Punica granatum* L.	.66	.000266	.0045	0.664766
79.	*Rauvolfia serpentina* (L.) Benth.	.66	.0532	.00005	0.71325
80.	*Rungia pectinata* (L.) Nees	1.99	2.793	.0096	4.7926
81.	*Ricinus communis* L.	.66	.3724	.5176	1.55
82.	*Rumex dentatus* L.	1.65	.6118	.0005	2.2623
83.	*Saccharum munja* Roxb.	.99	.7182	.0006	1.7088
84.	*Sida cordifolia* L.	1.65	.8512	.0007	2.5019
85.	*Solanum nigrum* L.	.99	1.0906	.0009	2.0815
86.	*Spilanthes indicus* Wall.	.33	.8778	.0274	1.2352
87.	*Syzygium cumini* (L.) Skeels	.33	.001064	.3327	0.663764
88.	*Tamarindus indica* L.	.66	.000532	1.1832	1.843732
89.	*Thevetia peruviana* (Pers.) K.	.66	.00798	.2773	0.94528
90.	*Tridax procumbens* L.	2.32	3.2984	.0458	5.6642
91.	*Trifolium alexandrium* L.	1.32	.665	.0092	1.9942
92.	*Vernonia cinerea* L. Less.	1.32	1.4896	.0206	2.8302
93.	*Vicia hirsuta* L.	.33	.025	.0003	0.3553
94.	*Vitex negundo* L.	.66	.0011	.0181	0.6792
95.	*Vitis vinifera* L.	.33	.000266	.0014	0.331666
96.	*Xanthium strumarium* L.	1.32	5.4796	.476	7.2756
97.	*Ziziphus mauritiana* Lamk.	.66	.00532	.0462	0.71152

It is clear from the study that the plants which are highly sensitive to air pollution are nearly absent at sites C and D. In other words only those plants are found at sites C and D which are less sensitive to air pollution. On the other hand, if the sensitive plants grow near the brick industries they show very low value of IVI. Thus to conserve the biodiversity, pollution control is essential (Jepson, 2001). As a general trend, the resistant plants had higher value of IVI near all the brick industries. At Sarnath, i.e., the control site, the value of IVI of different species did not show any correlation with their sensitivity. The plants were uniformly distributed without any external influence. By comparing the results of phytosociological characters of plants around different brick industries with that of the control site, it can be concluded that site D is most polluted followed by C sites and B, however site A is least polluted.

At nearly all the sites certain tree species have been recorded which influence the value of IVI. For inetance at site A, many plants of *Emblica officinalis* Gaertn. and *Eucalyptus citriodora* Hook. have been planted, consequently it had high density. *Anthocephalus chinensis* (Lam.) A. Rich., *Mangifera indica* L., *Tamarindus indica* L. etc. were also present but their density was very low. Similarly, the cultivated plants affected the IVI value at sites B, C and D also. The phytosociological data of a community was also affected by immigration or invasion of new species (Greene and Johnson, 2000; Mehrhoff, 2000).

At site A, 83 species have been recorded and among the wild plants *Cynodon dactylon* was the most dominant. At site B, 82 species were recorded with *Capparis zeylanica* as the most dominant flora. At C, D and E, 69, 76 and 97 plant species were recorded, respectively. *Calotropis procera* was the most dominant plant species of site C whereas *Cynodon dactylon* was the most dominant plant species of sites D and E. Various sensitive plant species have been recorded which were present at the less polluted and the control site but were not present at the highly polluted site such as sites C and D (Table 10).

Table 10: Performance of plant species at various sites

Sites	Total no. of plant species	Dominant wild plant species having higher IVI		Plants found at control site but not found at other sites
		Name of plant	IVI	
Site A	83	*Cynodon dactylon*	16.47	*Amaranthus spinosus*
		Cyperus rotundus	11.80	*Asclepias curassavica*
		Xanthium strumarium	8.82	*Cannabis sativa*
		Eragrostis tenella	7.95	*Capparis zeylanica*
		Polygonum plebajum	7.66	*Ricinus communis*
Site B	82	*Capparis zeylanica*	29.68	*Aerva lanata*
		Cynodon dactylon	16.30	*Blepharis maderaspatensis*
		Calotropis procera	13.84	*Coccinia indica*
		Lantana indica	12.27	*Ipomoea fistulosa*
		Xanthium strumarium	11.84	*Ocimum tenuifolium*
Site C	69	*Calotropis procera*	20.77	*Abutilon indicum*
		Cynodon dactylon	19.35	*Aerva lantana*
		Capparis zeylanica	19.10	*Centella asiatica*
		Cyperus rotundus	15.93	*Rauvolfia serpentina*
		Lantana indica	11.41	*Rungia pectinata*
				Vicia hirsuta
Site D	76	*Cynodon dactylon*	19.63	*Abutilon indicum*
		Cyperus rotundus	14.06	*Ageratum conyzoides*
		Calotropis procera	11.97	*Centella asiatica*
		Xanthium strumarium	10.49	*Lepidium sativum*
		Polygonum plebajum	8.96	*Rauvolfia serpentina*
				Rungia pectinata
				Vicia hirsuta
Site E	97	*Cynodon dactylon*	16.61	
		Cyperus rotundus	13.62	
		Eragrostis tenella	7.42	
		Calotropis procera	7.37	
		Xanthium strumarium	7.28	

DISCUSSION

Plants are the primary acceptor of pollutants because they are directly exposed to the field and do not posses any mechanism to escape or avoid the condition of high pollution concentration (Beckerson and Hofstra, 1979). As a result the pollutants are absorbed (Roger *et al.*, 1979, Rowland *et al.*, 1987) and accumulated in the plants (Hampp, 1992 and Hanson and Lindberg, 1991). These accumulated pollutants enter the chain of one or the other chemical reactions occurring in cells, resulting in malformation of the enzymes or any other change (Harmens *et al.*, 2000) which ultimately becomes detrimental to the sensitive plants (Badiani *et al.*, 1993), however the resistant ones remain unaffected (Robinson and Britz, 2000). Thus, the level of pollution decides the composition of a plant community.

Ricinus communis L. was present predominantly at all the sites except site A. *Parthenium hysterophorus* L. of family Asteraceae, one of the most problematic weeds of India had APTI value 27.90, and has been noted as a prominent resistant plant and the above data of this shows a clear correlation with the pollution level, corroborated by earlier studies (Backes, 2000).

The APTI value of *Carica papaya* L. was found to be 81.1 which shows it to be one of the most resistant plants. This plant was recorded at all the study sites including the most polluted site (site D). Though, it is a cultivated plant but its survival at all the sites is an indication of its resistant nature, parallel to its APTI value.

Phytosociological studies at all the five sites justify the above facts. *Abutilon indicum* (L.) sweet which is a sensitive plant with low APTI value 9.32 has been found only at site A and B which was less polluted and control site E. This plant was not recorded at highly polluted sites i.e. C and D. Similarly, *Aerva lantana* (L.) Juss. with low APTI value i.e. 8.83 was also noted as a sensitive plant species. This species has been recorded at the control site and at site A which was least polluted.

Similar type of distribution was recorded in case of several other plant species such as *Blepharis maderaspatensis* (L.) Roth. *Blumea membranacea* DC. and *Boerhaavia diffusa* L. were not evenly distributed on all the sites, and they had a very low APTI value. *Clerodendrum viscosum* Vent. of the family Verbenaceae had APTI value as far as 9.8 and this species was not found at site D, the most polluted site. *Centella asiatica* (L.) Gray Urban with APTI value 9.13 has been recorded at sites A, B, and E only. However, it has not been found at sites C and D which was due to its sensitivity for air pollutants.

Rauvolfia serpentina (L.) Benth. and *Lepidium sativum* L. which were highly sensitive plants with APTI value 5.82 and 5.27, respectively have been recorded only at the control site or at less polluted sites i.e. at site A and B. *Vicia hirsuta* (L.) Gray of family Fabaceae showed APTI value to be 11.75 but this plant was also absent from the sites which were more polluted i.e. sites C and D and on the other hand it was present at sites A, B and E.

On the other hand the resistant plants followed a different trend. Plants such as *Achyranthes aspera* L. (APTI value 31.83), *Alternenthera sessilis* (L.) DC. (APTI value 30.18), *Rumex dentatus* L. (APTI value 33.37), *Moringa oleifera* Lam. (APTI value 31.39) and *Xanthium strumarium* L. (APTI value 29.62) were evenly present at all the sites ranging from the site D (most polluted site) to site E (the control site) except the plant *Moringa oleifera* Lam. which has not been recorded at site C. The reason of distribution of *Moringa oleifera* Lam. is anthropogenically influenced, as it is a cultivated plant.

There were certain other wild plants which fall under highly resistant category, such as *Ricinus communis* L. (APTI value 65.73), *Cynodon dactylon* (L.) Pers. (APTI value 48.74), *Cyperus rotundus* L. (APTI value 46.77) etc. and showed wide distribution at most of the sites irrespective of the pollution level in the particular area (Arnone *et al.* 2000). *Cynodon dactylon* (L.) pers. was the most dominant flora at all the sites (as also reported by Whitmore and Mansfield, 1983) with a relative density of 12.96% at site A, 12.46% at site B, 15.04% at site C, 15.43% at site D and 13.30% at site E. Relative frequency of this plant was also higher at all the sites ranging from 3.5% at site A to 4.27% at site C. The next dominant species of all the sites studied was *Cyperus rotundus* L. with a relative frequency ranging between 2.98% at the control site and 3.76% at site D i.e. the most polluted site. Relative density of this plant was 8.64% at site A, 8.40% at site B, 12.48% at site C, 10.29% at site D and 10.64% at the control site i.e. site E. *Ricinus communis* L. was present predominantly at all the sites except site A.

Parthenium hysterophorus L. of the family Asteraceae, one of the most problematic weeds of India had APTI value 27.90. The value of relative frequency of this plant ranged between 2.32% at the control site and 3.85% at the site C. Relative density ranged from 2.76% at site A to 5.28% at site B. The IVI of this plant was 6.27 at site A, 11.67 at site B, 8.20 at site C, 7.05 at site D and 5.66 at site E. *Parthenium hysterophorus* L. was noted as the prominent resistant plant and the above data of this plant shows a clear correlation with the pollution level, it shows that the resistant plants grow independently irrespective of the pollution load (Backes, 2000).

Some tree species have also been recorded at all the sites but distribution of the trees at different sites was considerably low. Some of such species include *Aegle marmelos* (L.) Corr. *Anthocephalus chinensis* (L.) A. Rich, *Azadirachta indica* A. Juss., *Cassia fistulosa* L., *Emblica officinalis* Gaertn., *Ficus benghalensis* L., *F. religiosa* L., *Mangifera indica* L., *Syzygium cumini* (L.) Skeels, *Tamarindus indica* L. etc. These species showed random distribution. Presence of most of the above species at the control site shows that unpolluted atmosphere is best for the growth of the plants. The trees were much older as compared to the age of the brick kilns, so they cannot die in response to the emitted air pollutants but their normal morphology and physiology is affected. Similar reports were also published by Imsade and Touraine (1994) and Kainulainen *et al.* (1995). Reduction in foliar pigments (Singh and Rao, 1983 Wellburn *et al.*, 1972) and productivity (Sharma *et al.*, 1996 and Roose *et al.* 1982) has also been reported by previous workers. The previous studies also indicate that pollen germination and the pollen tube growth is affected in the presence of air pollutant such as SO_2 in the case of certain plants (Varshney and Varshney, 1981). If the pollination is affected, viable seed formation is also affected and the number of viable seeds produced determines the density of the plants, mostly wild plants. Ultimately, composition of the floral community is determined by pollution level.

CONCLUSIONS

It is clear that the plants which are highly sensitive to air pollution seem to be absent at sites B and C, in other words only those plants are found at sites B and C which are less sensitive to air pollution. The plants growing near brick industries with very low value of IVI seem to be sensitive. Thus to conserve biodiversity, pollution control is essential. As a general trend, the resistant plant species had higher value of IVI near both the brick industries. At the site A, the value of IVI of different plants did not show any correlation with their sensitivity, the plants were uniformly distributed without any external influence. The value of IVI is greatly influenced by certain tree species such as *Emblica officinalis* Gaertn. and *Eucalyptus citriodora* Hook. *Anthocephalus chinensis* (L.) A. Rich., *Mangifera indica* L., *Tamarindus indica* L. etc. which were artificially planted at the sites. The phytosociological data of a community is also affected by immigration or invasion of new species. At sites A, B and C, 97, 69 and 76 plant species were recorded respectively, which shows reduction in biodiversity of plants in the area by increase in pollutants in the ambient air. *Calotropis procera* was the most dominant wild plant species of site C while *Cynodon dactylon* was the most dominant plant species of site A and C both. Sensitive plant species such as *Abutilon indicum, Rauwolfia serpentina, Rungia pectinata, Vicia hirsuta* etc. have been recorded to be present at the site A i.e., control site but were not present at the polluted sites such as site B and C.

Conclusively, it was found that the plants with low APTI value were found only at the less polluted sites i.e. at site A and B. The plants with higher APTI value were found to be resistant and were present at most of the sites. Resistant plants also act as a sink for air pollutants as also reported by Dwivedi *et al.* in 2000 and Tripathi and Upadhyay 2002. As an outcome of this analysis, it is suggestible that the resistant plants marked in the investigation can be employed in abatement and control of air pollution. Thus, they can be grown in and around the coal fired industries to reduce the pollutants in air. Secondary benefits can be derived from them, as some are medicinally important plants example *Achyranthes aspera* L., *Adhatoda vasica* Nees., *Evolvulus alsinoides* L., *Nicotiana plumbaginifolia* Viv. and *Ocimum basilicum* L.; and few of the remaining are fruit-yielding, example *Carica papaya* L. and *Mangifera indica* L.

Thus, the phytodiversity of an area is highly affected by the pollution of an area, more particularly the air pollution. Quality as well as the quantity of air pollutants affects the phytodiversity, thereby regulating the biodiversity of plants.

REFERENCES

Agrawal, M & Agrawal, SB (1989). Phytomonitoring around a Thermal Power Plant. *Atmos Environ.* **23(4):** 763-769

Ambasht, RS & Ambasht, NK (1999). A Textbook of Plant Ecology XIIIth Ed. CBS Publishers and Distributors, New Delhi

Anon, (1999). CBRI develops eco-friendly technology for brick making. The Hindu, Sunday January 3

Arnone, JA III, Zaller, JG, Spehn, EM, Nikulaus, PA, Wells, CE & Korner, C (2000). Dynamics of root systems in native grasslands: Effects of elevated atmospheric CO_2. *New Phytologist* **147:** 73-85

Arrigoni, O & DeTillio, MC (2000). Role of ascorbic acid in cell metabolism: between gene-directed function and unpredictable chemical reaction. *J. Plant Physiology* **157:** 481

Ashmore, MR & Ainsworth, N (1995). The effects of cutting on the species composition of artificial grassland communities. *Funcional Ecology* **9:** 708-712

Aslam, M (1993). Environmental Concerns in Brick Industry. *Bricks and Tiles News* 33-39

Aslam, M (1999). Pollution prevention in brick kilns. *Bricks and Tiles News* 16-20

Atkin, OK, Edwards, EJ & Loveys, BR (2000). Response of root respiration to changes in temperature and its relevance global warming. *New Phytologist* **147:** 141-154

Auer, AH (1978). Correlation of land use and cover with meteorological anomalies. *J. Applied Meteorology* **17:** 636-643

Backes, A (2000). The ecology of the latifoliate forest of Morro do Coco in Viamao State of Rio Grande do Sul Brazil-III. Climate and Microclimate. Perquisas *Botanica* **50:** 119-136

Badiani MG, Scheucne AK, Paolacci AR & Fumagalli I (1993). Daily fluctuations of antioxidants in Bean (*Phaseotus vulgaris* L.) leaves as affected by the presence of ambient air pollutants. *Plant Cell Physiology* **34:** 92-112

Beckerson, DW & Hofstra, G (1979). Response of leaf diffusive resistance of radish cucumber and soyabean to O_3 and SO_2 singly or in combination. *Atmos Environ.* **13:** 1263-1268

Bell, JMB & Mudd, CH (1976). Sulphur dioxide resistance in plants: A case study of *Lolium perenne*, pp. 87-103. *In*: Mansifield, TA (Ed.) Effects of Air Pollution on Plants. Cambridge University Press, Cambridge

Biscaye, PE, Grousset, FE, Svensson, AM & Bory, A (2000). Eurasia air pollution reaches eastern North America. *Science* **290:** 2258

Bowne, NE (1974). Diffusion rates. *J. Air Pollution Control Association* **24:** 832-835

Bruce, N, Perezpadilla, R & Albalak, R (2000). Indoor air pollution in developing countries: a major environmental and public health challenge. *Bulleting of the World Health Organization* **78:** 1078

Cao, J, Yan, Z, Jianwei, L, Ruodeng, X & Sangbu, D (2000). Environmental fluoride content in Tibet. *Environmental Research* **83:** 333-337

Chaudhary, CS & Rao, DN (1977). A study of some factors in plants controlling their susceptibility to SO_2 pollution. *Proceedings of Indian National Science Academy* **43:** 236-241

Cullis, CF & Hirschler, MM (1990). A simple model for estimating emissions of carbon monoxide and hydrocarbons from the combustion of coal. *Atmos. Environ.* **24:** 1153-1160

Damle, A (2000). Utilization of waste materials in brick making. *Reg. Sem. on Methods of Pollut Contr and Energy Consv in Brick kilns.* (Presented)

Darley, EF (1966). Studies on the effects of cement-kiln dust on vegetation. *J. Air Pollution Control Association* **16:** 145-150

Debinski, DM, Jakubauskas, MJ & Kindscher, K (2000). Montane meadows as indicators of environmental change. *Environmental Monitoring and Assessment* **64:** 213-225

Dwivedi, AK & Shashi, (2006). Dawn to the darkness of misconception in science scholars. *Indian Science Cruisier* **20:** 9-12

Dwivedi, AK & Shashi, (2007). Species erosion the shrinking biodiversity-An approach. *Indian Science Cruiser* **20:** 8-14

Dwivedi, AK & Tripathi, SC (2010). Environment and Biodiversity. Publ. Lambert Academic Publishing Co., Germany

Dwivedi, S & Dwivedi, AK (2010). River Pollution and Solution. Publ. New Central Book Agency, Calcutta

Dwivedi, AK & Singh, MP (2003). Effect of coal fired industry on higher plant diversity and phytosociological behaviour. *J. Ecobiology* **15:** 345-353

Dwivedi, AK (1999). Effect of sulphur dioxide on growth behaviour of *Abelmoscus esculentus*, L. *Proc. Nat. Conf. for Better Environ.* in 21st Century Oct. 8-10 **1999:** 19-23

Dwivedi, AK (2005). Merit must be promoted. *Curr. Sci.* **89:** 1305

Dwivedi, AK (2008). An Express Issue. *Down To Earth* **16:** 4-6

Dwivedi, AK (2009). Ecological Backlashes of Ganga Express Way *Proc. 1st Rastriya Yuva Vaigyanik Sammelan* Nov. 28-30 2008 pp. 207-209

Dwivedi, AK, Hasan, H & Shashi, (2009). Anthropecology of Water Bodies. *In*: Trivedi, P.C. (Ed.) Plant Physiology in Agriculture and Forestry pp. 248-254. Aavishkar Publishers and Distributors, Jaipur, India

Dwivedi, AK, Pandey, S & Shashi, (2008). Hospital Waste *Indian Science Cruiser* **22:** 10-14

Dwivedi, AK, Pandey, S & Shashi, (2009). Fate of Hospital Waste in India *Biology and Medicine* **1(3):** 25-32

Dwivedi, AK, Pandey, S & Shashi, (2009) Hospital Waste: At a Glance pp. 114-119. *In*: Trivedi, P.C. (Ed.) Microbes Applications and Effect. Aavishkar Publishers and Distributors, Jaipur, India

Dwivedi, AK, Prajapati, UB & Shashi, (2007). Waste Water and its Management. *Indian Science Cruiser* **21:** 36-40

Dwivedi, AK, Prajapati, UB, Hasan, H, Singh, R, Shashi, & Singh, J (2008). Study of pre and post monsoon water quality in a tropical river Ami. *Modern J. Life Sciences* **¾:** 161-166

Dwivedi, AK, Shashi, & Singh, J (2006). Vermicomposting biotechnology – Demand of the date. *Proc. Nat. Sem. on Global Pollution: A Serious Threat to Indian Biodiversity* 31st Jan. and 1st Feb. 2006

Dwivedi, AK, Shashi, & Singh, J (2006). Water pollution and groundwater recharging. *Curr. Sci.* **91:** 407-408

Dwivedi, AK, Singh, R & Shashi, (2008). Groundwater Pollution: Causes and Impact *Indian Science Cruiser* **22:** 29-33

Dwivedi, AK, Tripathi, BD & Shashi, (2008). Effect of ambient air sulphur dioxide on Sulphate accumulation in plants. *J. Environ. Biology* **29:** 377-379

Dwivedi, AK, Upadhyay, AR, Pandey, T, Pandey, AK & Dwivedi, UP (2000). Variation in roadside vegetation with reference to vehicular pollution. *Bangladesh J. Forest Sciences* **29:** 53-60

Eckert, RT & Huston, DV (1982). Foliar peroxidase and acid phosphatase activity response to low level SO_2 exposure in eastern while pine clones. *Forestry Science* **28:** 661-664

Edwards, RD, Yurkow, EJ & Lioy, PJ (1998). Seasonal deposition of house dusts on household surfaces. *Sci. Total Environ.* **224: 11:** 69-80

Farooq, M & Beg, MU (1980). Effects of aquous sulphur dioxide on the membrane permeability of common Indian tree leaves. *New Botanist* **7:** 213-217

Foley, G (1985). Wood fuel and conventional fuel demands in the developing world. *Ambio.* **14(5):** 253-258

Fritts, DC & Werne, JA (2000). Turbulence dynamics and mixing due to gravity waves in the lower and middle atmosphere. In: Atmospheric Science across the stratopause Vol.123 Pub. by AMER Geophysical Union Florida AVE NW Washington DC 20009: 143

Ganeva, A (1998). Air borne pollution in parangalitza biosphere reserve (Rila mountain) estimated by means of bryophytes. *Herzogia* **13:** 113-118

Gignac, LD, Nicholson, BJ & Bayley, JE (1998). Utilization of bryophytes in bioclimatic modeling: Present distribution of peatlands in the Mackenzie river basin Canada. *Bryologist* **101 Winter:** 560-571

Gillett, RW, Ayers, GP, Selleck, PW, Mhw, T & Harjanto, H (2000). Concentration of nitrogen and sulphur species in gas and rain water from six sites in Indonesia. *Water Air and Soil Pollution* **120:** 205-215

Glenn, MA, Gomez-Bolea, Ropux, C & Lobella, R (1993). Lichen health in relation to vehicular traffic in two Mediterranean forest preserves. *Am. J. Botany* **80:** 8

GOI, (1996). The Gazette of India. Part II– Section 3- Sub-section (i). New Delhi: Government of India No.141 April 3

Goudie, AS (1978). Dust storms and their geomorphological implications. *J. Arid Environments* **1:** 291-310

Gracia, AZ, Nash, TH III & Hrvera Campos, MA (2000). Lichen decline in Desierto de los leones (Mexico City). *Bryologist* **103:** 428-441

Greene, DF & Johnson, EA (2000). Tree recruitment from burn edges. *Canadian J. Forest Research* **30(8):** 1264-1274

Gupta, AK & Dwivedi, AK (2002). Impact of air pollutants on biochemical characteristics of *Parthenium hysterophorus. Plant Archieves* **2:** 19-22

Gupta, AK & Gupta, RG (2000). Reduction in pollution and saving of energy through advanced design of brick kilns. Reg. Sem. on Methods of Pollut. Contr. and Energy Consv. in Brick kilns. (Presented)

Gupta, AK, Tripahi, BD & Dwivedi, AK (2001). Productive behaviour of a tropical deciduous plant in response to our pollution by brick industries. *International J. Mendel* **18:** 91-92

Gupta, GR, Sandhu, AK & Mulchi, C (1991). Changes in foliar concentration of abscissic acid in soyabean in response to SO_2 fumigation. *J. Environ. Qual.* **20:** 151-152

Hampp, R (1992). Comparative evaluation of the effects of gaseous pollutants acid deposition and mineral deficiencies on the carbohydrate metabolism of trees. *Agriculture Ecology and Environment* **42:** 333-364

Hanson, PJ & Lindberg, SE (1991). Dry deposition reactive nitrogen compounds: a review of leaf canopy and non-foliar measurements. *Atmos. Environ.* **25A:** 1615-1634

Harmens, H, Stirling, CM, Marshall, C & Farrar, JF (2000). Is partitioning of dry weight and leaf area within *Dactylis glomerata* affected by N and CO_2 enrichment? *Ann. Bot.* **86:** 833-839

Henderson, A (1999). Literature and air pollution and Lichens XLVIII. *Lichenologist* (London) **31:** 111-119

Holzworth, GC (1972). Mixing heights wind speeds and potential for urban air pollution throughout the contiguous United States. Office of Air Program. Pub. No. AP-101 Research Triangle Park N.C.: U.S. EPA

Hooker, JD (1859). On the flora of Australia its origin, affinities and distribution. *Bot. Antarctic Exp.* Pt-3 Fl. Tasm. **1:** 1-137

Imsade, JB & Touraine, A (1994). N demand and the regulation of nitrate uptake. *Plant Physio.* **105:** 3-7

Jacobson, GL (2000). Post-glacial changes in vegetation and climate in northern New England. *Rhodora* **102:** 246-247

Jain, SK & Singh, P (2000). Emission monitoring and air pollution control in brick kilns – A Case Study Reg. Sem. On Methods of Pollut. Contr. and Energy Consv. in Brick kilns. (Presented)

Jarvensivu, MT, Maenpaa, & Jounela, SLJ (2000). Field survey of reduced sulphur emission from a modern finish pulp mill. *Environmental Progress* **19:** 147-156

Jaworowski, Z & Grzybowska, D (1980). Natural radionuclides and heavy metals in soils - management of environment (Ed. Patel, B) Wiley Eastern Ltd

Jonnalagadda, NS & Baloji, R (1991). Studies on the level of SO_2 NO_2 NH_3 and HCl in ambient air of Harare Zimbabwe. *Environ Inter.* **17:** 461-467

Joumard, R (1988). Measurement of particle and gaseous pollution of the atmosphere due to buses. *Sci. Total Environ.* **76:** 55-62

Kabirov, RR, Sukhanova, NV & Khaibullina, LS (2000). Estimation of atmospheric air toxicity using microscopic algae. *Ekologiya* (Moscow) **3:** 231-232

Kaharabata, SK, Schuepp, PH & Desjardins, RL (2000). Estimating methane emission from dairy cattle housed in a barn and feed lot using atmospheric tracer. *Environ. Sci. Tech.* **34:** 3296-3302

Kainulainen, P, Holopainen, JK & Oksanen, JK (1995). Effects of SO_2 on the concentrations of carbohydrates and secondary compounds in scots pine (*Pinus sylvestris* L.) and Norway spruce (*Picea abies* (L.) Karst) seedlings. *New Phytologist* **130:** 231-238

Kashulin, NA, Ratkin, NE, Dauvalter, VA & Lukin, AA (2001). Air-borne pollution on the drainage area of subarcic lakes and fish. *Chemosphere* **42:** 51-59

Keller, T & Schwager, H (1977). Air pollution and ascorbic acid. *European J. Forestry Pathology* **7:** 338-350

Khitoliya, RK & Khitoliya, J (1996). Environmental protection and legislation. *Employment News* **21:** 1-2

Krebs, CJ (1985). Ecology 3rd ed. Harper and Row, New York

Lakshmikantan, KR, Kumar, EBV & Garg, N (1999). Vertical Shaft Brick Kiln Technology in India. *Bricks and Tiles News* 32-36

Lee, SD (1977). Biochemical Effects of Environmental Pollution. Ann Arbor Science Publishers Inc Ann Arbor Mich

Li, YC & Kuo, YH (2000). Four new compounds ficusal ficasesquilignan. A B and ficusolied diacetate from the heart wood of *Ficus microcarpa. Chemical and Pharmaceutical Bulletin* **48:** 1862

Lok, C (2001). Farming accused of eco-damage. *Nature* **409:** 969

Loutre, MF (2001). Earth history sediments to planetary motion. *Nature* **409:** 991

Luria, MT, Vinig, & Peleg, M (1984). The contribution of city buses to urban air pollution in Jerusalem. Israel. *J. Air Pollut. Control Ass.* **34:** 8828-8831

Lyons, TJ & Scott, WD (1990). Principles of Air Pollution Meteorology. Belhaven Press, London

Maithel, S (1999). Emission standards for brick kilns: an opportunity of technology upgradation. *Bricks and Tiles News* 37-47

Maithel, S, Kumar, A, Uma, R & Vasudevan, N (2000). Pollution control and energy conservation in brick kilns. Reg. Sem. On Methods of Pollut. Contr. and Energy Consv. in Brick kilns. (Presented)

Maithel, S, Uma, R, Kumar, A & Vasudevan, N (1999). Energy Conservation and Pollution Control in brick kilns. Proc Wk Sp on Prospects and Prospectives of Brick Industries in the Next Millennium 109-132

Mandal, SC, Maity, TK, Das, J, Saba, BP & Pal, M (2000). Anti-inflammatory evaluation of *Ficus racemosa* linn. Leaf extracts. *J. Ethno-Pharmocology* **72:** 87

Mansfield, TA & Freer-Smith, PH (1981). Effects of urban air pollution on plant growth. *Biol. Rev.* **56:** 343-368

McLaughlin, SB & Taylor, GE (1981). Relative humidity: important modifier of pollutant uptake by plants. *Science* **211:** 167-169

Mehrhoff, LJ (2000). Immigration and expansion of the New England flora. *Rhodora* **102:** 280-289

Mendes, B, Urbano, P, Alres, C, Morais, J, Lapa, N & Oliveira, JS (1998). Fungi as environmental microbiological indicators. *Science Water and Technology* **38:** 155-162

Misra, BP & Tripathi, BD (1999). Energetics of dead body cremation on the bank of river Ganga in Varanasi: An environmental and economic approach. *Pollution Research* **18:** 95-99

Montague, DC & Perrine, RL (1990). Preliminary assessment of the greenhouse warming implications of halocarbon substitutes for CFC-11 and CFC-12. *Atmos Environ.* **24A:** 1331-1339

Murray, F (1985). Changes in growth and quality characteristics of lucerne (*Medicago sativa* L.) in response to sulphur dioxide exposure under field conditions. *J. Exp. Bot.* **36:** 449-457

Nakamura, Y, Morimitur, Y, Uzu, T, Ohigashi, H, Murkami, A, Naito, Y, Nakgawa, Y, Osawa, T & Uchida, K (2000). A glutathione S-transferase inducer from Papaya rapid screening identification and structure activity relationship of isothiocynates. *Cancer Letters* **157:** 193

Narayan, D, Agrawal, M, Pandey, J & Singh, J (1994). Changes in vegetation characteristics downwind of an aluminium factory in India. *Ann. Bot.* **73:** 557-565

Newman, PWG & Kenworthy, JR (1987). The use and abuse of driving cycle research: Clarifying the relationship between traffic congestion energy and emissions. *Transpn Q.* **38:** 615-635

Novak, M, Jackova, I & Prechova, E (2001). Temporal trends in the isotope signature of airborne Sulphur in central Europe. *Environ. Sci. Tech.* **35:** 255

Nyamwange, M (1995). Population growth and development: The Kenyan experience. Scandianavian. *J. Development Alternatives* **14:** 149-160

Odum, EP (1971). Fundamentals of Ecology 3rd ed. WB Saunders and Co. Philadelphia

Oliveira, FPM & Jardim, MAG (1999). The floristic composition of a secondary forest in the municipality of Igarape-Acu State of para Brazil. *Botanica* **14:** 127-144

Oluwande, PA (1979). Automobile exhaust problems in Nigeria. *Ambio.* **8:** 26-29

Ormstad, H (2000). Suspended particulate matter in indoor air: adjuvants and allergen carriers. *Toxicology* **152:** 53

Page, SJ & Organiscak, JA (2000). Suggestion of a cause-and-effect relationship among coal rank air borne dust and incidence of workers pneumoconiosis. (AIHAJ) *J. Amer. Indust. Hygenic. Assoc.* **61:** 785

Pal, S & Kumar, N (2000). Effect of simulated acidic rain on yield and carbohydrate content of green pepper (*Capsicum annuum* L.). *Advances in Plant Sciences* **13:** 85-88

Pandey, GN (1997). Air Pollution and Its Control in Environmental Management. Vikas Publishing House, New Delhi

Pierre, M & Queiroz, O (1981). Enzymic and metabolic changes in bean leaves in continuous pollution by subnecrotic level of SO_2. *Environ. Pollut.* **25:** 41-51

Pisut, I & Lisicka, E (2000). Monitoring of epiphytes on permanent plot in the vicinity of Bratislava (Slovakia, S.W.). *Biologia* **55:** 369-373

Potter, CS (1999). Terrestrial biomass and the effect of deforestation on the global carbon cycle. *Bioscience* **49:** 769-778

Potter, G, Horemans, N, Caubergs, RJ & Asard, H (2000). Ascorbate and dehydroascorbate influence cell cycle progression in a tobacco cell suspension. *Plant Physiology* **124:** 17

Power, HC (2001). Estimating atmospheric turbidity from climate data. *Atmospheric Environment* **35:** 125

Pudasaini, SP (1997). Population environment and sustainable development in a global perspective. In: *Demographic Transition: The Third World Scenario:* 307-321

Pundir, BP & Das, S (1985). State of Report on vehicle emissions for Department of Environment Govt. of India

Raju, BSN (1997). Fundamentals of Air Pollution. Oxford and IBH Publishing Co. Pvt. Ltd., New Delhi

Rao, DN (1979). Plant leaf as pollution monitoring device. *Fertilizer News* 25-28

Rao, MN & Rao, HVN (1989). Air Pollution. Tata McGraw-Hill Publishing Company Limited New Delhi

Rao, MV & Dubey, PS (1988). Plant responses against SO_2 in field condition. *Asian Environ.* **3:** 1-9

Rao, MV & Dubey, PS (1990). Biochemical aspect (antioxidants) for development of tolerance in plants growing at different low levels of ambient air pollutants. *Environmental Pollution* **64:** 55-66

Rao, MV & Dubey, PS (1993). Response of hydrogen peroxide scavenging system in two soyabean cultivars exposed to SO_2: Experimental evidence for the detoxification of SO_2 by enhanced H_2O_2 scavenging components. *Environmental Pollution* **82:** 99-105

Rayappa, M & Singaracharya, MA (1993). Pollution tolerance index in some common plants around some major industries in Warangal city. Andhra Pradesh. *Pollution Research.* **12:** 57-59

Raza, SH, Kumari, NV, Murthy, MSR & Adeel, A (1985). Air Pollution Tolerance Index of Certain Plants of Hyderabad. In: Biological Monitoring of the State of the Environment (Bio-indications) Indian National Science Academy New Delhi 243-245

Robinson, JM & Britz, SJ (2000). Tolerance of a field growing soyabean cultivar to elevated ozone level concurrent with higher leaflet ascorbic acid level higher ascorbate-dehydroascorbate redox status and long term photosynthetic productivity. *Photosynthesis Research* **64:** 77

Roelle, PA, Aneja, VP, Gay, B, Geron, C & Piece, T (2001). Biogenic nitric oxide emissions from cropland soils. *Atmospheric Environment* **35:** 115

Rogers, HH, Jeffries, HE & Witherspoon, AM (1979). Measuring air pollutant uptake by plants: nitrogen dioxide. *J. Environ. Qual.* **8:** 551-557

Roose, ML, Bradshaw, AD & Roberts, TM (1982). Evolution to resistance to gaseous air pollutants. In: Effects of gaseous air pollutants in agriculture and horticulture (Eds. Unsworth, MH & Ormrod, DP). Butterworth London 379-409

Rosenberg, CR, Hutnik, RJ & Davis, DD (1979). Forest composition at varying distances from a coal-burning power plant. *Environ. Pollut.* **58:** 307-317

Rowland, AJ, Drew, MC & Wellburn, AR (1987). Foliar entry and incorporation of atmospheric nitrogen dioxide into barley plants of different nitrogen status. *New Phytol.* **107:** 357-371

Rupp, W, Nadjem, H & Thoma, R (2000). Spirit cooking stove as source of CO poisoning in campers. *Archiv. Fuer Kriminologie* **206:** 8-13

Sanchez, ML, Pascual, D, Ramos, C & Perez, I (1990). Forecasting particulate pollutant concentration in a city from meteorological variables and regional weather patterns. *Atmos. Environ.* **24A:** 1509-1519

Scott, E, Blake, DR & Rowland, FS (1999). New Directions: Rapid industrialization in developing countries: The challenge to Earth System research for the new Millennium. *Atmospheric Environment* **33:** 683-684

Sengupta, J (1993). Vertical Shaft kilns and innovative approach to brick burning *Bricks and Tiles News* 29-39

Sharma, HC, Thakre, RA & Aggrawal, AL (1986). Impact of refinery emission on vegetation around Mathura region. NEERI Report NEERI Nagpur

Sharma, YK, Leon, J, Raskin, I & Davis, KR (1996). Ozone-induced responses in *Arabidoposis thaliana* – the role of salicylic acid in the accumulation of defence related transcripts and induced resistance. *Proc. Nat. Acad. Sci. USA* **93:** 5099-5104

Shashi, & Dwivedi, AK (2004). Study of variation in pH of a river water as affected by industrial effluents. *Proc. Nat. Sem. on Strategies for Environment Quality_Management in Urban Area* 28th Oct. **2004:** 76-82

Shashi, & Dwivedi, A.K. (2008). Ganga Express Way – A path of wetland destruction *Curr. Sci.* **94:** 840-841

Shashi, Singh, J & Dwivedi, AK (2009). Numerical interdependence of pH acidity and alkalinity in a polluted tropical river *J. Environ. Bio.* **30:** 773-775

Shu, Jiong. J, Dearing, A, Morse, AP, Yu, L & Li, C (2000). Magnetic properties of daily sampled total suspended particulate in Shanghai *Environ. Sci. Techno.* **34:** 2393-2400

Singh, J, Shashi, & Dwivedi, AK (2003). Thermal behaviour of a lotic ecosystem in response to pollution load. *Proc. Nat. Conf. on Global Air Pollution and Water as Sustainer of Life* 9th-10th Feb. **2003:** 15-21

Singh, SK & Rao, DN (1983). Evaluation of plants for their tolerance to air pollution. In proceedings of the Symposium on Air Pollution Control. November 218-224

Singh, SK, Rao, DN, Agrawal, M, Pandey, J & Narayan, D (1991). Air pollution tolerance index of plants. *J. Environ. Manag.* **32:** 45-55

Smith, WG & Robert, (1900). Botanical Surgery of Scotland Edinburgh district colored vegetation map. *Scot. Geogr. Mag.* 1900

Snakin, VV & Prisyazhnaya, AA (2000). Lead contamination of the environment in Russia. *Sci. Total Environ.* **256:** 95-101

Stanier, RY, Ingraham, JL, Wheelis, ML & Painter, PR (1987). General Microbiology 5th Ed. Macmillan Education Ltd. London

Tiwari, S, Bansal, S & Rai, S (1993). Expected performance indices of some planted trees of Bhopal. *Indian J. Env. Health* **35:** 282-286

Tong, STY & Lam, KC (2000). Home sweet home? A case study of household dust contamination in Hong-Kong. *Sci. Total Environ.* **256:** 115-123

Tripathi, Anamika, Tewari, PB, Mahima & Singh, D (2009). Assessment of Air Pollution Tolerance Index of some Trees in Moradabad City, India. *J. Environ. Biology* **30:** 545-550

Tripathi, Anamika, Tripathi DS & Prakash V (1999). Phytomonitoring and NO_2 Pollution around silver refineries. *Environ. International* **25:** 403-410

Tripathi, A, Dwivedi, AK & Mahima (2010). Airborne Cu and Zn at some urban site in Pital Nagari, Moradabad, India. *J. Environ. Sci. Engg.* **52:** 53-56

Tripathi, BD & Dwivedi, AK (2002). Atmospheric pollution and its outcome – An analysis. *The Botanica* **52:** 88-92

Tripathi, BD & Dwivedi, AK (2003). Pollution control in brick industries and recent environmental legislation. *In*: (Srivastva, BK, Kumar, R & Jamal, A (Eds.). Environmental Management in Mines. Tara Book Agency, Varanasi, India pp. 509-514

Tripathi, BD & Dwivedi, AK (2007). Pollution tolerance and distribution pattern of plants in surrounding areas of coal fired industries. *J. Environmental Biology* **28:** 257-263

Tripathi, BD, Ambasht, RS & Sikandar, M (1984). Ecological investigation of energy consumption for burning of dead bodies at Varanasi Ganga ghats. *In*: Ambasht, R.S. and Tripathi, B.D. (Eds.). River Ecology and Human Health. NECA, Varanasi

Tripathi, BD, Dwivedi, AK, Pandey, T, Singh, MK, Upadhyay, AR, Pandey, AK & Dwivedi, UP (1999). Air foliage and Soil sulphur as a measure of coal generated pollution. *Advances in Biological Research* **17:** 1-5

Tripathi, BD, Pandey, T, Dwivedi, AK, Singh, MK, Upadhyay, AR, Pandey, AK & Dwivedi, UP (1999). Responses of *Mangifera indica* and *Ocimum sanctum* with respect to automobile exhaust. *Advances in Biological Research* **17:** 14-17

Tyagi, PC (1988-89). Assessment of vehicular pollution in metropolitan cities - Part X – Lucknow CUPS/26/1988-89 (CPCB New Delhi)

Varshney, SRK & Varshney, CK (1981). Effect of SO_2 on pollen germination and pollen tube growth. *Environ. Poll. Ser.* **A24:** 87-92

Verma, SS (2000). Environmental Pollution. *Employment News* **25:** 1-3

Vitousek, PM, Mooney, HA, Lubchenco, I & Melillo, JM (1997.) Human domination of Earth's ecosystems. *Science* **227:** 494-499

Von Arb, C, Mueller, C, Ahmann, K & Brunold, C (1990). Lichen physiology and air pollution. II. *New Phytologist* **115:** 431-437

Wang, Q, Shen, W & Ma, Z (2000). Estimation of mercury emission from coal combustion in China. *Environ. Sci. Tech.* **34:** 2711-2713

Wellburn, AR, Majernik, O & Wellburn, FAM (1972). Effects of SO_2 and NO_2 polluted air upon the ultrastructure of chloroplasts. *Environ. Pollut.* **3:** 37-49

Whittekar, RH (1975). Communities and Ecosystems. 2nd Ed. Macmillan New York

Won, D, Corsi, RL & Rynes, M (2000). New indoor carpet as an adsorptive reservoir for volatile organic compounds. *Environ. Sci. Tech.* **34:** 4193-4198

Yang, Z, Kong, L, Zhang, J, Wang, L & Li, S (1998). Emission of bioferic sulphur gases from Chinese ice paddies. *Sci. Total Environ.* **224:** 1-8

Yoshida, S, Kitano, M & Eguchi, H (1998). Growth of cucumber plants (*Cucumis sativus* L.) under diurnal control of air temperature. *Biotronics* **27:** 97-102

Yu'e, Li. & Erda, L (2000). Emission of NO_2, NH_3 and NO_2 from fuel combustion industrial processes and the agricultural sectors in China. *Nutrient Cycling in Agroecosystems* **57:** 99-106

22

Biodiversity of Sedges in Wetlands of Gyanpur

KA SIDDIQUI

Department of Botany, K.N. Govt. P.G. College, Gyanpur–222 001, Sant Ravidas Nagar (U.P.)

ABSTRACT

Sedges are the grass-like herbaceous plants with typical solid, triangular stems and leaf-sheath without a slit. These can be distinguished from grasses and rushes, though these are all often sympatric and occupy the same habitat. The taxa constitute a natural family called Cyperaceae, with 4,000 species spread over 70 genera, treated either in Poales or independently as Cyperales. The members of family are regarded as obnoxious weeds in agricultural fields. However, they essentially occupy wetland and aquatic ecosystems. The members serve as excellent ecological indicator of the habitats. Present paper deals with 15 species of sedges (Cyperaceae) associated with the flora of wetlands of Gyanpur Tract. It also includes essential features like field notes, growth-form, status, distribution, flowering and fruiting.

Keywords: Biodiversity of sedges, Cyperaceae, ecological indicators, sedges.

INTRODUCTION

Wetlands have been identified as one of the most useful natural resource systems. They play a vital role in controlling the water cycles and cleaning the environment. They have emerged as essential life support systems. Increased attention has been diverted for the conservation and management of such systems (Walker, 1992).

The members of the monocotyledonous family Cyperaceae are commonly known as sedges. These are cosmopolitan in distribution and are mostly found in arctic and temperate regions of the northern and southern hemispheres, constituting about 10% flora of arctic regions. The taxa usually occur in marshy places and grow well in humus soils. Sedges are usually perennial herbs and bush-like in appearance. They perennate by means of creeping sympodial rhizomes, either solitary or in clusters.

Gyanpur Wetlands are situated in the upper Gangetic Plains 72 km apart in Western belt of Varanasi district, located between 25°30' N longitude and, 82°28' N longitude above 75.12 msl. It was felt necessary to undertake a systematic work on Cyperaceae, the sedges, since the taxa are a major constituent of the sedges-meadow stage in hydrosere.

METHODOLOGY

The specimens have been collected during 2005-2007 from the wetlands spread over different blocks of Gyanpur. The specimens were identified by standard taxonomic methods (Dwivedi and Singh, 1985). Some of the specimens were processed for herbarium preservation. The taxonomic account, thus prepared, includes key to the genera and of the species. In case of the genera with more than one species, citiation, basionym, familiar synonyms and information about flowering and fruiting reasons of the year, field notes, status their distribution and specimens have been examined. The species under each genus have been arranged alphabetically.

RESULTS AND DISCUSSION

Cyperus is the most dominant genus with 12 species, it is followed by *Fimbristylis* having 4 species. These two are characteristic amongst the four major genera of Cyperaceae of the world (Judd *et al.*, 2002). *Eleocharis* and *Kyllinga* occur in the scenario of wetlands in Gyanpur with two species each while *Bulbostylis, Mariscus* and *Schoienoplectus* have single species each. From the field notes it is clear that most of the species occur in the moist fringe or peripheral part of the wetlands. However, some of them like *Cyperus compressus. C. exaltatus, C. distans, C. iria, C. platistylis,* and *Eleocharis atropuropurea* stretch themselves to somewhat shallow water regimes. The distribution of Cyperaceae in wetland scenario of the Gyanpur indicates a later stage (sedge-meadow) of ecological succession (hydrosere), which proceeds from the periphery towards centre.

CYPERACEAE JUSS.

Key to the Genera

1. Glumes spirally arranged; hypogynous bristles present:
2. Style base dilated and constricted or articulated above the nut:
3. Leaves well developed:
4. Style base caducous....... *Fimbristylis*
4. Style base persistent....... *Bulbostylis*
3. Leaves capillary................ *Eleocharis*
2. Style base not dilate, continuous with the nut.............. *Schoenoplectus*
1. Glumes distichous; hypogynous bristles absent:
5. Nuts trigonous:

6. Rachilla of spikelets articulated at base............................. *Mariscus*
6. Rachilla of spikelets not articulated.............................. *Cyperus*
5. Nuts lenticular.. *Kyllinga*

1. *Bulbostylis Kunth nom. cons.*

1. **Bulbostylis barbata** (Rottb.) Kunthex Clarke in Hook f., FI. Brit. India **6:** 651. 1893. *Scirpus barbatus* Rottb. Programme 27, 1772.

   FI. and **Fr.:** June to October.

   Field notes: Grows in wet soils and all along the wetlands.

   Status: Common

   Distributions Throughout India; all warmer countries of the world; native of Australia.

   Specimen examined: Siddiqui, K.A. 540

2. *Cyperus L.*

Key to the species

1. Glumes spirally arranged; hypogynous bristles present:
1. Stems with bladed leaves:
3. Plants annual:
4. Glumes nearly as wide as long, obtuse at the apex, hardly imbricating at maturity; nuts almost as long as the glumes....... *C. iria*
4. Glumes much longer than wide, acute at the apex, imbrication at maturity; nuts much shorter than glumes .... *C. compressus*
3. Plants perennial with woody rhizomes;
5. Plants with slender stolons:
6. Glumes mostly 4 mm or more long; stolons usually ending in tubers .... *C. rotundus*
6. Glumes less than 4 mm long; stolons with or without tubers.... *C. pilosus*
5. Plants without stolons:
7. Spikelets more than 1 mm, wide glumes close together on a straight rachilla........................... *C. exaltatus.*
7. Spikelets not more than 1 mm wide glumes rather remote on a zigzag rachilla....................... *C. distans.*
2. Stems without bladed leaves.................................. C. *articulatus.*
1. Spikelets in digitately or stelately spreading clusters or in spherical heads:
8. Plants perennial with a woody rhizome. .... *C. platystylis*
8. Plants annual:
9. Glumes with long recurved mucros:
10. Spikelets at anthesis 1.7-3 mm wide, nuts-obovoid................... *C. cuspidatus.*

11. Spikelets at anthesis not more than 1.5 mm wide, nuts cylindrical...... *C. castaneus.*
9. Glumes either without mucros or with short and straight mucros:
Spikelets in almost spherical cluster: about as long as the subtending part................................ *C. difformis*

Cyperus difformis L. in Torner, Cent. Pl. 2:6. 1756.

Vernacular name: Behua ghas.

Fl. and Fr.: September to December.

Field notes: Gregariously grows on the margins of the pools, ponds, ditches, and marshes.

Status: Very common.

Distributions: Almost throughout India; Tropical and warmer temperate regions of the Old World and Australia.

Specimens examined: Siddiqui, K.A. 580.

Cyperus distans L., f., Suppl. 103. 1781.

Vernacular name: Pani malanga.

Fl. and Fr.: May to October.

Field notes: In wet places, edges of pools, ditches and other waterbodies, and often in shallow water.

Status: Very common.

Distribution: Throughout India; Pantropic.

Specimen examined: Siddiqui, K.A. 753

Cyperus exaltatus **Retz.,** Obs. fasc. 5:11.1789.

Fl. and Fr.: Throughout the year, mainly on September to November.

Field notes: In the marshy places, ponds etc. and it grows gregariously.

Status: Common.

Distribution: Throughout India, Sri Lanka, Indo China, Tropical Africa, Southeast Asia and Australia.

Specimen examined: Siddiqui, K.A. 641.

Cyperus iria **L.** Sp. Pl. 45. 1753.

Vernacular name: Bara chuncha.

Fl. and Fr.: Almost throughout the year, but mainly in April to September.

Field notes: Along the margins of the water bodies, often in shallow water.

Status: Very common.

Distribution: Throughout India; Old World Tropics.

Specimens examined: Siddiqui, K.A. 581.

***Cyperus pilosus* Vahl,** Enum. Pl. 2:354. 1806.

Fl. and Fr.: April to September.

Field notes: In seasonally flooded wetlands, edges of ponds.

Status: Not very common.

Distributions: Throughout India; Old world tropics.

Specimen examined: Siddiqui, K.A. 643

***Cyperus platystylis* R. Br.,** Prodr. 214. 1810.

Fl. and Fr.: April to September.

Field notes: Grows in shallow water or in floating mats of vegetation; usually gregarious.

Status: Common.

Distributions: Throughout India Tropical Asia and Australia.

Specimen examined: Siddiqui, K.A. 710

***Cyperus rotundus* L.,** Sp. Pl. 45. 1753.

Vernacular name: Mutha ghash.

Fl. and Fr.: June to October.

Field notes: In embankments of ponds, beels and other-waterbodies.

Status: Very Common.

Distributions: Major parts of India; warmer parts of the world.

Specimen examined: Siddiqui, K.A. 754

3. Eleocharis R. Brown

Key to the species

1. Stem usually slender; glumes membranous .... *E. artopurpurea.*
1. Stem usually stout; glumes sub-rigid......*E. dulcis*
   Eleocharis artopurpurea (Retz.) Kunth, Enum. 2: 151. 1837

Fl. and Fr.: September to December.

Field notes: Grows in marshes and drying margins of water bodies; gregarious in shallow water and in seasonally flooded places.

Status: Common.

Distributions: Throughout India; a cosmopolitan species.

Specimen examined: Siddiqui, K.A. 582

*Eleoharis dulcis (***Burm. f.***)*

Vernacular name: Chenchka

Fl. and Fr.: July to October.

Field notes: Everywhere in the ditches, embankments of ponds and other wetlands.

Status: Common.

Distribution: Throughout India; Sri Lanka, Malaya, China, Japan, Madagascar, Formosa, Tropical west Africa and Tropical Australia..

Specimen examined: Siddiqui, K.A. 547.

4. *Fimbristylis* Vahl. nom. cons.

Key to the species

1. Styles bifid:
2. Nuts conspicuously longitudinally striate:
3. Spikelets angular from the acute keels of the glume ....*F. bisumbellata.*
3. Spikelets terete................*F. dichotoma.*
2. Nuts Smooth........................*F. aestivalis.*

***Fimbristylis aestivalis* (Retz.) Vahl**. Enum. 2:288. 1806.
Scirpus aestivalis Retz. Obs. 4:12:1786.

Fl. and Fr.: March to July.

Field notes: Gregariously grows in the embankments of waterbodies.

Status: Common.

Distributions: Throughout India, Southeast Asia, Malaysia and Tropical Australia.

Specimen examined: Siddiqui, K.A. 648.

***Fimbristylis bisumbellata* (Forssk.)** Bubani, Dodec. 30. 1850. Scirpus bisumbellatus Forssk., Fl. Aegy. Arab. 1:15:1775.

Fl. and Fr.: Round the year, notably in summer.

Field notes: Grows on the wetland embankments.

Status: Common.

Distributions: Throughout India; warmer parts of the world.

Specimen examined: Siddiqui, K.A. 760.

***Fimbristylis dichotoma* (L.)** Vahl. Enum. 2:287.1806.
Scirpus dichotomus L. Sp. Pl. 50. 1753.

Vernacular name: Bara nirbishi.

Fl. and Fr.: June to November.

Field notes: Gregariously grows on the embankments of waterbodies.

Status: Common.

Distribution: Throughout India; Sri Lanka.

Specimen examined: Siddiqui, K.A. 761.

Mariscus Vahl.

***Mariscus aristatus.* (Rottb.)** Tang and Wang in Fl. Reipubl. Pop. Sin 11: 178. t. 59. f. 12-13. 1961. *Cyperus aristatus* Rottb., Oeaser. et. Icon. 23.t.6.f.1. 1773.

Fl. and Fr.: Almost throughout the year.

Field notes: In moist places near the edges of ponds and weels.

Status: Common.

Distributions: Throughout India, most of the Tropical countries.

Specimen examined: Siddiqui, K.A. 552.

Schoenoplectus Palla.

***Schoenoplectus articulatus* (L.)** Palla in Bot. Jahurb. 10: 229. 1889.

Venacular name: Chatpati.

Fl. and Fr.: Throughout the year, sepecially in December to March.

Field notes: Gregariously grows in the shallow wetlands as an emergent hydrophyte.

Status: Common.

Distributions: Throughout India, Old World tropics.

Specimen examined: Siddiqui, K.A. 563.

REFERENCES

Dwivedi, JN & Singh, RB (1985) Essentials of Plant Techniques. Scientific Publishers, Jodhpur, (India)

Judd, SW, Campbells, C, Kellogg, A, Elizabeth, SF, Peter & Donoghue, JM (2002). Plant Systematics-Phylogenetic approach, pp 284-286. (2nd Ed.). Sinauer Associates, Inc Publ Sunderland; Massachusetts, USA

Walker, D (1992). Energy, Plants and Man. Pa Kard Publication, Montague, Brighton

23

Phytodiversity Conservation in India

Iqbal Habib and UK Chaturvedi[*]

Department of Botany, Govt. Degree College, Badaun, Uttar Pradesh–243 601, India
**Department of Botany, Bareilly College, Bareilly–243 001, Uttar Pradesh*

ABSTRACT

Plants have been exploited continuously for their medicinal, ornamental and horticultural purposes and this has led to their copious depletion on the earth. With the increase of human population, pressure on Indian forests has reached to a level where their productive potential is narrowing down. Therefore, major efforts are needed to restore the forest cover to meet the demand of several forest products. With more and more alternative methods available for *in-situ* preservation of individuals, populations or their germplasm, it is not too optimistic to hope that we should be able to contain or considerably slow down the current rate of extinction or depletion of our endangered flora.

Keywords: Diversity, endangered flora, germplasm, phytoconservation, population, protection.

INTRODUCTION

The growing human population and its increasing demands for food and materials have already taken a heavy toll of the earth's natural resources. The growth of large urban areas, construction activities such as dams, buildings and roads, encroachment on vast areas of forest lands for agriculture and mining operations are examples of direct onslaughts on Nature which have steadily degraded and depleted our natural resources. The pollution caused by large scale urbanization, industrialization, excessive use of chemical fertilizers and pesticides have also contributed to gradual degradation of our soil, water and air. As a result, the very survival of wildlife is in danger. Besides this, human greed has led to uncontrolled hunting, trapping and poaching of wild animals so much so that some species are already reaching the state of their extinction.

India with its mainland area of about 3,267,500 km^2. extends from South to North for 3200 km. between 804' and 370.6'N and from West to East for 3000 km. between 6807' and 97025'E.

It is situated in the Southern Peninsula of the Asian Continent with its triangular outline extending into the Indian Ocean and has sometimes been called as a subcontinent. The world's loftiest and longest mountain ranges, viz. the Himalayas extending for a distance of about 2,500 km from East to West along its northern borders and the 5,700 km long coast line of the peninsula, the great plains of the Ganga-Brahmaputra basin, the extremely arid Thar desert in Rajasthan, the uplands in the Deccan plateu surrounded by the hill ranges of the Vindhya-Satpura system in the north, the western ghats along the western coastline and the eastern ghats along the coast converging with the Nilgiri mountains in the South and the extensive riverine systems form the salient physiographic features. Besides the mainland area, the Lakshadweep (Laccadeve and Minicoy Islands) in the Arabian sea and the Andaman and Nicobar groups of islands in the Bay of Bengal are also within the territory of India.

It is estimated that there are 5-10 million species of plants and animals on earth. India has about 45,000 species of plants comprising 15,000 vascular plants, 600 pteridophytes, 2,700 bryophytes, 5,000, algae, 20,000, fungi and 1,600 species of lichens. The earliest organised efforts of listing endangered plant species of India were made around 1970, when the Botanical Survey of India and the Forest Research Institute at Dehradun jointly drafted a list of about one hundred rare and endangered species. The rate of extinction of species has been particularly fast during the last 100 years. Unfortunately, it was not until the latter half of the present century that measures were initiated for the conservation and protection of wildlife. Even then, the rapid disappearance of some beautiful animals boosted, efforts to save wild fauna, and flora was given hardly any attention. It was only in 1968, when UNESCO sponsored conference on the problem of conservation of flora caught the attention of conservationsists and several recommendations were made urging the International Biological Programme (IBP), the International Union for Conservation of Nature (IUCN) and various International and National organizations to initiate studies and to look into the problems of protection and preservation of wild fauna and flora in their natural habitats.

LOSS OF BIODIVERSITY

Nature continuously adjust and maintain a balance between the biotic and abiotic components of the ecosystem as well as biodiversity. A clear and rapid decline of biodiversity may occur due to natural calamities such as floods, droughts, earthquakes and forest fires, etc. Further, as the man began to use his superior knowledge to understand nature, and his greedy nature, started disrespecting and destroying the nature. Owing to the anthropogenic activities along with the explosion of human population, the loss of biodiversity has set in. Being the second most populous country of the world, India faces severe strains on her ecosystem. Population pressure creates the need for more land for agriculture, haphazard growth of rural and urban areas, migration of people and animals, deforestation and degradation of forest cover, pollution and depletion of natural resources. Besides these, the mushroom like growth of industries along with the construction of roads, dams, canals, mining or quarrying activities threaten to disrupt the delicate ecological balance. The clearing of the natural forests to allow agricultural activities and commercial plantations, cause great damage to the wildlife.

EFFECT OF POLLUTION

The main cause of loss of biodiversity or the cause that affect the biodiversity, is pollution.

In fact, in the countries where there have been the greatest technological advances, the maximum pollution occurs. In the most industrialized and technologically advanced countries, the disposal of certain kinds of waste products and by-products, which when injected into the biosphere in large quantities, affect the normal functioning of ecosystems and thus adversely affect the plants, animals and man. Exposure of plants to low pollution of SO_2 and NO_2 can cause temporary or permanent injury to them. Similarly, when plants are exposed to acidic aerosols during periods of foggy weather, light, rains, relative humidity and moderate temperatures, which are injurious to them and adversely affect their productivity. For example, it has been observed that Pines are more susceptible than broadleaf trees and react by partial defoliation and reduced growth.

Most of the Indian rivers and freshwater streams are greately polluted by industrial wastes. For example, chlorine which is added in water to control the growth of algae and bacteria in the cooling system of power stations, may remain present in streams and rivers which cause mortality to planktons and fish fauna. In one observation, it was reported that heavy fish mortality in river Sone in Bihar was caused by free chlorine content of the chemical wastes discharged by factories near Mirzapur in U.P. Toxic chemicals are agents of water pollution which originate in industrial operations, acid mine drainage, washing of herbicides and insesticides, radioactive fall out from atomic explosion and oil spills.

Pesticides and herbicides used in agriculture to protect the crops from pests and wild growth, have created serious problems which were not previously anticipated. Many of their chemicals present in their constituents accumulate in the environment and contaminate numerous animals and plants not intended to be the targets. Similarly, herbicides have also been reported to cause serious ecological effects through destruction of forest areas, mangrove forests, rubber plantations, killing of insectivorous birds, increase in mosquito populations and arugmented soil erosion.

THE ENDANGERED SPECIES

We consider a species as endangered when its natural regeneration is not able to keep pace. While the rare species are the failing species whose individuals are becoming genetically weaker and unable to withstand species competition. Recent studies indicate that about 35% of vascular plants in India are endemic and about 900 taxa of plants are rare or endangered. 'Red Data Book' is the term applied to the books dealing with threatened plants or animal species of any region. Several countries have prepared their own 'Red Data Books'. IUCN has published two volumes covering a few hundred species. There is another book popularly called as the 'Green Book' which lists rare plants growing in protected areas such as botanical gardens. In India also, a similar 'Green Book' has been published which deals with about one hundred rare plants growing in gardens under Botanical Survey of India. Rare and threatened species can be found in remote areas which are rarely touched by human-beings. For example, *Nepenthes khasiana* which is a very rare and endangered pitcher plant, is located in Meghalaya while *Coptis teeta* occurs in rather remote hilly regions of north-east India. Another species *Peucedanum dehradunensis* has so far been found growing only inside the compound of a bungalow in Dehradun.

Forests in the country are under increasing stress due to over-exploitation and habitat disturbances. There has been a dramatic decline in plant and animal diversity in recent years and reduced availability of forest resources has greatly affected on the livelihood of the poor people. Over exploitation of the forests is one of the main causes of disappearance of not only economic plants but also biological curiosities like the insectivorous and primitive plants such as *Gnetum* and *Psilotum* etc. In the over-exploited category, some medicinal plants such as *Podophyllum, Coptis, Aconitum* etc. and horticultural plants like Orchids and Rhododendrons come. Medicinal plants such as *Rauwolfia* and *Coptis* yield drugs from their roots have been over exploited are particularly threatened and are rarely found. Absence of collections at regular intervals or decline in plant populations and failure to locate any particular species under research are a few signals of their rarity. A species is called extinct if all of its members and generations are dead or vanished from the environment for ever. For example, *Dichanthium plampedicellatum* was discovered in Manipur. It was later searched in that locality but could not be found and it has probably become extinct.

PROTECTION AND CONSERVATION OF BIODIVERSITY

The methods of protection and conservation of endangered species and genetic resources include:

1. Management of protected areas and habitat restoration.
2. Identification and long term conservation of ecosystems.
3. Domestication of wild economic biodata
4. Biosphere observation network (BON)
5. Knowledge of local technology to be applied
6. Rehabilitation of endangered and threatened species
7. Captive breeding programmes
8. Wildlife education and interpretation of the data
9. Research and monitoring
10. Control on over-exploitation
11. Establishment of National Parks and Biosphere Reserves
12. Establishment of gene sanctuaries for selected species in their natural habitat
13. Cultivation of rare species in more areas in forests and Botanical Gardens.
14. Collection and preservation of 'Germplam' in the form of seed, pollen, tissue etc.
15. Legislation for conservation.

To restore endangered plants to their natural habitats, the conventional methods in use might not be successful. Therefore, to meet the commercial requirement and to protect the endangered plant species, biotechnology offers several advantages and that too in a much shorter time than that of the conventional methods and techniques. For controlling over exploitation, main approach is to impose control on the trade of threatened species. India is a signatory to the convention of International Trade in Endangered Species of Wild Fauna and Flora. Some restrictions on the collection of plants and their products from wild populations, and their export have been in force in India even before the ratification of the convention. Sanctuaries are protected places and genes are the unique segments of chromosomes which constitute the specific genetic make-up of an organism. Gene sanctuary, therefore, is a protected place where plants of specific genetic material and characters grow. Genetic engineering techniques can bring about the transfer of useful genes from one species to another. Plant cell and tissue culture techniques play a useful role in understanding the complicated mechanisms which are involved in gene transfer. India now has 44 National Parks and 210 sanctuaries covering some 88,000 Sq. Km. These constitute about 11.7% of the total forest area of India.

Further, plant tissue culture and other biotechnological methods and techniques can play a vital role in conserving forest flora and fauna. *In-vitro* culture techniques have now been used in several countries of the world for conservation of plant species. Plant cell, tissue and organ culture methods are the basic parts of biotechnology and they can revolutionize the forestry, agriculture and horticulture. The main idea of plant tissue culture is based on the large-scale regeneration of plantlets and then into full plants. As such millions of plants can be produced with their improved progeny and vigour in shortest periods using 'micropropagation' technique. This includes the regeneration of somatic embryos from 'callus tissue' or from 'cell mass'. This technique has an enormous potential to produce plants for genetic transformation experiments. This method of plant propagation will help in propagating endangered and rare plant species also.

Some government and semi-government institutions are trying to take up steps for protection and conservation of threatened plant species by establishing new national parks, sanctuaries and Biosphere Reserves. Some strong public movements like 'Chipko' and 'Save Silent Valley Society' organised by some committed individuals and institutions have created general awareness for conservation activities in India. Indian cultures, traditions and classical literatures are full of references of 'Ashrams' and 'Forest Reserves' where all kinds of wildlife was protected.

CONCLUSION

In India, forest resources have been under mounting pressure owing to increase in human population. Forest resources are being denuded rapidly which has become a cause of serious concern for the government and the people of this country. Therefore, remedial measures are needed urgently to restore our depleting forest flora in a sustainable manner. It is alarming to note that several hundred plant species are becoming extinct everyday due to habitat disturbances and population stress. Therefore, major efforts are required to restore our forest flora so that a continuous supply of forest products and goods may be maintained. Here, the role of taxonomists, ecologists, horticulturists, agriculturists, biotechnologists, plant breeders, geneticists and to a larger extent our government, becomes all the more important to rescue our forest resources from their total depletion.

REFERENCES

Allen, T & Rana, SVS (2000). Biodiversity conservation in National and International Perspectives pp. 25-40 Agarwal, VP & Gupta, SK (Eds.) Conservation of Biodiversity

Agarwal, A, Narain, S & Sharma, A (1999). Green Politics. Global Environmental Negotiations. pp. 123-161. Centre for Science and Environment, New Delhi

Anonymous, (1977). Convention on International Trade in Endangered Species of Wild Founa and Flora. Govt. of India, New Delhi

Chatterjee, D (1940). Studies on the endemic flora of India and Burma. *J. Asiat. Soc. Bengal* **5:** 19-67

Cook, CDK (1980). The status of some Indian endemic plants. *Threatened Plants Committee News Letter* **6:** 17-18

Das, P (1997). Forest Conservation. If there is a will. *The Hindu Survey of Environment* 189-194

Gadgil, M (2000). Defining biodiversity resources. *Current Science,* **79:** 280-282

Ganguly, JK (1978). Conservation of forest wild life and genetic resources. *J. Sci. Club.* **32:** 23-27

Henry, ANK, Vivekanathan, & Nair, NC (1978). Rare and threatened flowering plants of South India. *J. Bombay Nat. Hist. Soc.* **75:** 684-697

Jain, SK (1976). Rare and endangered plants, the problem and its solution. *Indian Sci. Ann.* **2:** 20-23

Jain, SK (1978). Conservation of endangered plant species and genetic resources. *J. Sci. Club.* **32:** 19-22

Jain, SK (1980). Orchids - Gems of mountain flora in India. *Botanical Survey of India*, Howrah

Jain, SK and Hajra, (1976). Orchids in some protected habitats in Assam in Eastern India. *American Orchid Soc. Bull.* **45:** 1103-1109

Jain, SK & Mehra, KL (1983). Conservation of Tropical Plant Resources. *Bot. Survey of India* pp. 1-253

Jain, SK & Sastry, ARK (1980). Threatened plants of India-A state of the Art Report, *Botanical Survey of India*, Howrah

Jain, SK & Sastry, ARK (1985). Threatened Plants and Habitats-A Review of work in India. *Environment and Natural Resources* pp. 105-117

Kataki, SK, Jain, SK & Sastry, ARK (1984). Threatened and Endemic Orchids of Sikkim and North-Eastern India. pp. 1-95. *Bot. Survey of India*, Howrah

Khoshoo, TN (1986). Environmental priorities in India and sustainable development. Presidential address, 73, Session Indian Science Congress Association, Calcutta

Khshoo, TN (1996). India needs a Natural Biodiversity Conservation Board. *Curr. Science,* **71:** 506-513

Kothari, A (1996). Conserving Biodiversity: Demand without people. *The Hindu Survey of the Environment* 129-136

Maheshwari, JK (1977). Conservation of rare plants-Indian scene vis-a-vis world scene. *Bull. Bot. Surv. India* **19:** 167-173

Mehrotra, A (1979). A few rare Indian grasses. *Bull. Bot. Surv. India* **21:** 237-238

Mukherjee, Rajat, (1999). Environmental Pollution and Loss of Biodiversity. *Science and Culture,* **65:** 356-358

Mallaiah, B, Mustafa, M & Caroline, VJE (2008). Micropropagation of an endangered medicinally important cucurbit *Colosynthes vulgaris* L. through multiple short production. *Proc. Nat. Acad. Sci.* Indian, Sect. B, Vol. **78 pt III:** 256-267

Melville, R (1970). Endangered Plants and Conservation in the Islands of the Indian Ocean. *IUCN Publ.* **18:** 103-107

Nayar, MP (1977). Changing patterns of Indian flora. *Bull. Bot. Surv. India* **19:** 145-155

Nayar, MP (1980). Endemism and patterns of distribution of endemic genera (Angiosperms) in India. *J. Econ. Taxo. Bot.* **1:** 99-110

Qureshi, IM & Kaul, ON (1970). Some Endangered Plants and Threatened Habitats in South-East Asia. *IUCN Publ.* **18:** 115-126

Rao, AS (1974). The vegetation and phytogeography of Assam-Burma in ecology and biogeography of India. (Ed. Mani, M.S.) pp. 204-246

Rao, CK (1972). Angiosperm genera endemic to the Indian floristic region and its neighbouring areas. *Indian Fore.* **98:** 560-566

Sahni, KC 1970. Protection of rare and endangered plants in the Indian Flora. *IUCN Publ.* **18:** 95-102

Santapau, H (1970). Endangered species and their habitats. *IUCN Publ.* **18:** 83-88

Singh, NP (1979). On a recollection of a few rare plants from Karnataka. *Bull. Bot. Surv. India* **21:** 231-234

Singh, JS (1986). Biodiversity and ecosystem function in conserving biodiversity for sustainable development, (Ramakrishnan, PS, Das, AK & Saxena, KG) (Eds.) pp. 117-130. *Indian National Science Academy*, New Delhi

Subramanyam, K & Jain, SK (1972). Conservation of rare plants-A timely awakening. *Souv. National Fair, Calcutta* pp. 73-79

Subramanyam, K & Sreemadhavan, CP (1969). Nature conservation–A losing battle. *Indian Forester* **95:** 719-723

Subramanyam, K & Sreemadhavan, CP (1969). Endangered plant species and their habitats. A review of the Indian situation. *IUCN Publ.* **18:** 108-114

24

Ethnobotanical Flora of Dewghat Forest, Uttar Pradesh: Need for Conservation

CP Shukla and Mukesh Kumar*

Department of Botany, T.D.P.G. College, Purvanchal University, Jaunpur–222 002 (U.P.)
**Department of Botany, Sahu Jain P.G. College, Najibabad–246 763, Bijnor (U.P.)*
E-mail: mukesh.najibabad@gmail.com

ABSTRACT

Biodiversity is the source of natural biological wealth on which human life and well-being is dependent. The biodiversity of forest forms a very fragile ecosystem. The present communication throws light on 118 species of flowering medicinal plants (Angiosperms), occurring in Dewghat Forest, which are used by the tribal community and some other forest dwelling people. The botanical names, local/trade/vernacular names, habit, parts used, and medicinal uses etc. are also being provided in this Chapter.

Keywords: Biodiversity, biological wealth, ethnobotanical flora.

INTRODUCTION

Historically, plant products have been a rich source of medicinally active compounds and play an important role in drug discovery. Plants have always been an exemplary source of drugs and several of the currently available drugs have been derived directly or indirectly from them. Traditional medicines are used by about 60 per cent of the world's population. These are not only used for primary health care in rural areas in developing countries, but also in developed countries where modern medicines are predominantly used.

In the western world also, the use of herbal medicines is steadily growing, with approximately 40% of the population. Public, academic and government interest in traditional medicines is growing exponentially due to the increased incidence of the adverse drug reaction and the enormous financial burden of the modern system of medicine.

Plants have long supplied active ingredients for drugs. However, the researches on natural products are changing over the years. In the last 10 years NP (Natural Product) researchers have cast a sharper eye on herbal products to get medicinally important compounds. Recent studies have revealed that increasing number of patients have now begun to rely on herbal medicines for overall health or treatment. Hence, with the growing use of herbal remedies, the medicinally important plants are considered to be a rich source of potential drugs defined as Phytomedicines or phytopharmaceuticals. There is considerable interest in the investigation of potential plant products that may have an impact on blood sugar and cholesterol lowering activity when used as nutraceuticals.

Medicinal plants have the advantage of having no or only few side effects. Some of them are being used in traditional systems of medicine since hundreds of years in several countries of the world. Information on medicinal plants in India has been systematically organized. India is the largest producer of medicinal herbs and is appropriately called the 'Botanical Garden of the world'.

Plant folklore can be used as a tool for predicting sources of new chemo-preventive agents with antidiabetic activity. Nature has provided several effective antidiabetic agents in current use. Since herbal drugs or phytomedicines have gained importance in recent years because of their safety, efficacy and cost-effectiveness, therefore, plants can be considered as a potential source of drugs.

Public, academic and government interest in traditional medicines is growing exponentially due to the increased incidence of the adverse drug reactions and economic burden of the modern system of medicines. Thus the popularity of plant based drugs in the modern system of medicines is on increase day by day because they are claimed to be safe, economical, efficacious and easily available.

The major hindrance in the amalgamation of herbal medicines into modern medical practices is the lack of scientific and clinical data, and better understanding of efficacy and safety of the herbal products. To ensure the quality and safety of its products and practices, standardization is of vital importance. Most of the herbal products do not have drug regulatory approval to demonstrate their safety and efficacy. A systematic approach through experimental and clinical validation of efficacy is required for a plant identified for traditional medicine, as is done in modern medicines.

The use of terrestrial plants in the treatment of human ailments is as old as mankind itself. Nature has been a source of medicinal treatment for thousands of years, and plant-based system continues to play an essential role in the primary health care of 80% of the world population. In fact the traditional medicines from different parts of the world are largely based on the plants. The traditional herbal remedies have long been used to treat chronic disease in China and Japan (Kampo), in India (Ayurveda) and in Greece and Middle East (Unani) among others.

Medicinal plants have been used as a major source of therapeutic agents by man for thousands of years. In India, out of 15,000 species of flowering plants about 17% are considered to be of medicinal value. Most of these plants are collected from nature indiscriminately for commercial exploitation, resulting in depletion of natural population and some to the extent of becoming rare and endangered. With the patronage of herbal medicines and their increasing products, there is an urgent need to conserve the endemic diversity in the medicinal plants before it is moved out from nature. The factors contributing to the erosion of diversity are: bringing more land under cultivation due to biotic pressures, commercialization of agriculture for increased production, introduction of selected improved cultivars and thrust on socio-economic development. Therefore, considering these factors, an inventory of medicinal plant species, which evolved and adapted over a long period, assumes great significant.

The diversity of plant life is an essential part of most terrestrial ecosystems in India. India is rich in medicinal wealth, having about 2,500 plant species of known medicinal value. The country is thus endowed with one of the world's richest biodiversity and cultural traditions in respect of medicinal plants. This area is well-known for its forest luxuriant vegetation and medicinal plants. In the light of this fact both traditional medicinal knowledge as well as the medicinal plants are important for the human health. The present studies provide an additional knowledge about Dewghat Forest which has a large number of plants of economic importance and medicinal use. The human population subsisting on this unique natural resource exerts a pressure on the locality which lacks sustainability and necessitates conservation of biodiversity. The importance of retrieving information on medicinal flora of the State of U.P. is to initiate the proper scientific management and exploitation for economic development of the area. A review of trends in medicinal plant research and development, and scope of medicines has been given by Kumar (2003) and Tiwari (2000). Conservation and sustainable use of biological resources based on local knowledge system and practices is ingrained in the Indian Ethos.

Study Area

Dewghat forest is located between geographical limits of 25°15'-35°30' N Latitude, 81°45'-82°15' E longitude. The forest has specific geographical situation, comprising the treasure of medicinal plants used by all sections of the society. The climate is hot and humid is summer and dry-cool in winters. The annual rain fall ranges from 100-160 cm, mainly received in June to September months of the year. Average, maximum and minimum temperatures range from 30-35°C in summer and 20-30°C in winter, respectively.

METHODOLOGY

The present study is based on the survey of Dewghat Forest. Periodic extensive survey has been made from August, 2005 to September, 2006. In order to collect the data on medicinal plants, regular field trips were organized and interviews held with the tribal community of Dewghat Forest, who use plants as medicine. In the enumeration, plants are arranged in a tabular form. Table 1 indicates the indcxing of medicinal plants on the basis of their botanical names, trade/vernacular names, families, habit, parts used and medicinal uses. The present communication describes 118 plant species which are used as medicine by the rural population. The plants have been preserved following the standard method of Jain (1998) and herbarium specimens have also been prepared.

RESULTS AND DISCUSSION

In all 118 plants species belonging to 95 genera and 54 families have been documented (Tables 1, 2). Fabaceae with 10 genera (28 species) was highly represented, followed by Amaranthaceae with 5 and Cucurbitaceae with 4 genera. It has been found that most of the families represent plants with medicinal values. Of the 30 families, majority are represented by a single genus and species. Table-2 shows habit-wise distribution of medicinal plants. As indicated, most of them are herbs (64) followed by shrubs (28) and trees (21). Generally, under such a harsh climate, herb and shrub type of vegetation is prominent. There is only one perennial sedge, two climbers and two perennial grasses. The latter (both belonging to the family Poaceae) have been reported to have medicinal potential.

Table 1: Ethnobotanical plant species of Dewghat Forest, Uttar Pradesh

S. No.	Botanical Names	Local/Trade/ Vernacular Names	Families	Medicinal Uses
1	2	3	4	5
1.	*Abrus precatorius* L. **(S)**	Ghumchi	Fabaceae	Sore throat, cough, chronic inflammation of mucus membrane of eyeball and eyelid.
2.	*Abutilon indicum* L. Sweet **(S)**	Kanghi	Malvaceae	Astringent, anthelmintic, diuretic, alexiteric, toothache, demulcent tonic, rheumatism, piles, laxative, gonorrhoea.
3.	*Abutilon Indicum* (Lam.) Sweet **(S)**	Kanghl	Malvaceae	Tonic, gonorrhea, bladder stone.
4.	*Acacia catechu* (L.) Willd. **(T)**	Khadira	Mimosaceae	Useful in diarrhoea, sore throat, stomatitis and ulcerations.
5.	*Acacia farnesiana* (L.) Willd. **(T)**	Vilayati Babul	Mimosaceae	Bark and pods astringent. Gum demulcent, nutrient and emollient.
6.	*Acacia nilotica* (L.) Delile **(T)**	Babool	Mimosaceae	Eye sores in children, sexual impotency, diseases of urino-genital system, Astrigent, demulcent, asthma, diarrhea.
7.	*Acacia senegal* (L.) Willd. **(T)**	Kumatia	Mimosaceae	Emollient, demulcent, burns, sore nipples, haemorrhage, intestinal mucous, food for diabetics.
8.	*Achyranthes aspera* L. **(H)**	Apamarg	Amaranthaceae	Asthma, cough, cholera, renal dropsy, gonorrhea, eye disorders, piles, laxative, purgative, anthelmintic, anti-periodic in malaria, insect bit, pneumonia, hydrophobia.
9.	*Adhatoda zeylanica* Medik. **(S)**	Adusa	Acanthaceae	Diuretic bronchitis, gonorrhoea, jaundice, anti-spasmodic, ophthalmic, heart trouble, loss of memory, leucoderma.
10.	*Aegle marmelos* (L.)' Correa **(T)**	Bel	Rutaceae	Abdominal pain, heart palpitation, urinary troubles, laxative, febrifuge, ophthalmic, deafness, dysentery.
11.	*Aerva javanica* (Burm.f.) Juss.ex schultes **(T)**	Bui	Amaranthaceae	Inflammation, swelling demulcent, diuretic.
12.	*Ageratum conyzoides* L. **(H)**	Uchanti	Asteraceae	Styptic, sores cuts; Root: antilithic.

Contd.

Table 1: Contd.

S. No.	Botanical Names	Local/Trade/ Vernacular Names	Families	Medicinal Uses
1	2	3	4	5
13.	*Alangium salvifolium* (L.f.) Wangerin **(T)**	Akola	Apiaceae	Bark and root used in jaundice.
14.	*Albizia lebbek* (L.) Benth. **(T)**	Siris	Mimosaceae	Bark decoction used in Leprosy. Leaf and seeds used for eye troubles.
15.	*Alstonia scholaris* (L.) R.Br. **(T)**	Chhitwan	Apocynaceae	Bark decoction used in malarial fever and dysentry. Leaves used in beri-beri dropsy etc.
16.	*Amaranthus tricolor* L. **(H)**	Marsa	Amaranthaceae	Plants are astringent and emollient, used in diarrhoea, dysentery and haemorrhage.
17.	*Amaranthus viridis* L. **(H)**	Jangli Chauli,	Amaranthaceae	Emollient, laxative.
18.	*Anagallis arvensis* L. **(H)**	Neel	Primulaceae	Inflammation, diseases of kidney, epilepsy, ophthalmia, cerebral problems.
19.	*Anogeissus latiofolia* (DC.) Wall. ex Bedd. **(T)**	Dhaura	Combretaceae	Leaves used in dysentry
20.	*Argemone mexicana* L. **(H)**	Satyanashi	Papaveraceae	Eczema, skin disorders, blood purifier, laxative, emetic, expectorant, demulcent, jaundice, asthma, diuretic, leucoderma, liniment of head, aphrodisiac, whooping cough.
21.	*Asparagus racemosus* Willd. **(S)**	Satawari	Liliaceae	Aphrodisiac, laxative, expectorant, galactagogue, tuberculosis, leprosy, epilepsy, night blindness, diseases of kidney, liver, blood, eye and throat.
22.	*Asphodelus tenunifolius* Cav. **(H)**	Banpiyazi	Liliaceae	Seeds diurectic
23.	*Azadirachta indica* A. Juss. **(T)**	Neem	Meliaceae	Anthelmintic, diuretic, blood and skin diseases, leprosy, insecticidal, ophthalmia, biliousness.
24.	*Barleria prionitis* L. **(S)**	Bajra-danti, Kala-bans	Acanthaceae	Alexiteric, bronchitis, blood complaints, leucoderma, laceration, cracking of feet, fever, catarrhal affections, cough, diarrhea, toothache, boils, glandular swellings, dropsy.
25.	*Bauhinia variegata* L. **(T)**	Kachnar	Caesalpiniaceae	Bark powder used in worms, piles, diarrhoea and skin diseases etc.
26.	*Blepharis madaraspatensis* (L.) Heque ex Rath. **(H)**	Bhangri	Acanthaceae	Earache, tonic, increase milk production.
27.	*Boerhaavia diffusa* L. **(H)**	Punarnava	Nyctaginaceae	Astringent, biliousness, anaemia, night blindness, leucoderma, inflammation, blood purifier, diuretic, laxative, expectorant, intestinal inflammation, fever.
28.	*Bombax ceiba* L. **(T)**	Semal	Bombacacea	Fruits stimulant, expectorant and diuretic used in calculus affections and ulceration of bladder and kidneys.
29.	*Bambusa arundinacea* **(H)**	Bans	Poaceae	Used in thread worm and bronchial troubles, fever.
30.	*Borrerla articularis* (L.f.) Williams **(H)**	Guthari	Rubiaceae	Leaf used in haemorrhoids and gall stone. Seeds used in diarrhoea and dysentery.

Contd.

Table 1: Contd.

S. No.	Botanical Names	Local/Trade/ Vernacular Names	Families	Medicinal Uses
1	2	3	4	5
31.	*Boswellia serrata* Roxb. ex Colebr. **(T)**	Salai	Burseraceae	Biliousness, dysentery skin diseases, ulcer, blood purifier, leucoderma, piles, antipyretic, astringent, emmenagogue, rheumatism, vaginal discharge, diabetes, bronchitis.
32.	*Butea monosperma* (Lam.) Taubert. **(T)**	Palas, Dhak	Fabaceae	Night blindness, elephantiasis, aphrodisiac, laxative, dysentery, stomach worms, piles, astringent, tonic eye disease, diuretic, chronic diarrhoea, round worms.
33.	*Cadaba fruticosa* **(S)**	Kodhab	Capparaceae	Anthelmintic, emmenagogue, antiseptic, purgative, urinary obstruction.
34.	*Calotropis procera* (Aiton) R.Br. **(S)**	Madar	Asclepiadaceae	Toothache, stomachic, appetizer, piles, asthma, tonic.
35.	*Capparis zeylanica* L. **(S)**	Ker	Capparedceae	Carminative, aphrodisiac, appetizer, emmenagogue, alexipharmic, lumbago, rheumatism, hiccup, analgesic, diaphoretic, laxative, anthelmintic, ulcer, cough, asthma, piles, cardiac troubles, inflammation.
36.	*Caralluma edulis* **(H)** Benth. ex Hook. f.	Pimpa	Asclepiadaceae	Anthelmintic, leprosy, blood diseases.
37.	*Cassia auriculata* L. **(S)**	Tarawar	Caesalpiniaceae	Skin diseases, astringent, anthelmintic, diabetes, urinary disorders, conjunctivitis.
38.	*C. fistula* L. **(T)**	Amaltas	Caesalpiniaceae	Fruit pulp in dengu fever, gastric complains, in constipation.
39.	*C. occidentalis* L. **(C)**	Anwal	Caesalpiniaceae	Skin diseases, astringent, anthelminitc.
40.	*C. siamea* Lam. **(T)**	Sami	Caesalpiniaceae	Leaves used as tonic.
41.	*C. sophora* **(C)**	Kasundi	Caesalpiniaceae	Bark, leaves and seeds are cathartic, bark infusion or seeds powder given in diabetes.
42.	*C. tora* L. **(H)**	Chakwar	Caesalpiniaceae	Useful in skin diseases (ringworms, foul, ulcers)
43.	*Celosia argentea* L. **(H)**	Surli Garke	Amaranthaceae	Antipyretic, aphrodisiac, liver tonic, gonorrhoea.
44.	*Chenopodium album* L. **(H)**	Bathua	Chenopodiaceae	Appetizer, anthelmintic, diuretic, laxative, aphrodisiac, abdominal pain, eye disease, piles, tonic, disease of blood, heart and spleen.
45.	*Chrozophora rottleri* Klotz **(H)**	Sahadevi	Euphorbiaceae	Emetic, corrosive.
46.	*Citrullus colocynthis* (L.) Schrader **(H)**	Indrayan	Cucurbitaceae	Purgative, jaundice, rheumatism, urinary disease
47.	*Cleome gynandra* L. **(S)**	Karalia	Cleomaceae	Anthelmintic cough sores, rubefacient, intermittent fever, muscular pain, rheumatism, headache, intestinal wounds.
48.	*C. ornithopodoides* L. **(H)**	Khiramar	Cleomaceae	Carminative, anthelmintic, rubefacient, vesicant, piles, round worms leucoderma, skin diseases, earache, fever, dysentery.

Contd.

Table 1: Contd.

S. No.	Botanical Names	Local/Trade/ Vernacular Names	Families	Medicinal Uses
1	2	3	4	5
49.	*C. viscosa* L. **(H)**	Handi-bagro, pili-hulhul	Cleomaceae	Laxative, anthelmintic, diuretic, ulcer, leprosy, malaria, piles, lumbago.
50.	*Clitoria ternatia* L. **(S)**	Gokarni, Aparajita	Fabaceae	Laxative, diuretic, alexiteric, anthelmintic, brain tonic, corneal ulcer, elephantiasis, leucoderma.
51.	*Cocculus hirstutus* (L.) Diels **(M)**	Pilawan	Menispermaceae	Skin disease
52.	*Commelina benghalensis* L. **(H)**	Bukana	Commelinaceae	Emollient, leprosy.
53.	*Commelina paludosa Blume* **(H)**	Kanjuna	Boraginaceae	Vertigo, laxative, biliousness, fever.
54.	*Convolvulus microphyllus* Sieb. ex. Spreng. **(H)**	Santari	Convolvulaceae	Laxative, brain tonic.
55.	*Cucumis colocynthis* L.	Khat-Kachrio	Cucurbitaceae	Indigestion, fever; purgative, emetic.
56.	*Cuscuta reflexa* Roxb. **(S)**	Amar bel	Cuscutaceae	Purgative, itch, protracted fevers, sores, chest pain.
57.	*Cynodon dactylon* (L.) Pers. **(P/G)**	Doob	Poaceae	Astringent, diuretic, dropsy, cut, wound, genital-urinary disorder.
58.	*Cyperus rotundus* L. **(P/S)**	Motha, Mothee	Cyperaceae	Diuretic, emmenagogue, anthelmintic, diaphoretic, astringent, stimulant, stomach disorder, bowel irritation.
59.	*Datura alba* Nees **(S)**	Dhatura	Solanaceae	Toothache, insanity, catarrhal, cerebral complaints, skin diseases, lumbago, fever asthma, hydrophobia, malarial fever.
60.	*D. metel* L. **(H)**	Dhatura, Kalo-dhaturo	Solanaceae	Toothache, insanity, catarrhal, cerebral complaints, skin diseases, lumbago, diarrhoea, asthma, hydrophobia, malarial fever.
61.	*D. stramonium* L. **(H)**	Dhatura	Solanaceae	Antispasmodic, narcotic, anodyne, sedative, intoxicant, carbuncles.
62.	*Disoma tomentosa* **(S)**	Vajradanti	Asteraceae	Pyorrhoea, febrifuge, febrile attacks.
63.	*Digera muricata* (L.) Mart. **(H)**	Khanjuru	Amaranthaceae	Astringent, laxative, biliousness, urinary discharge.
64.	*Eclipta alba* (L.) Hassk. **(H)**	Bhangro	Asteraceae	Emetic, purgative, antiseptic, tonic, hepatic and spleen enlargement, jaundice, catarrh, skin diseases, hair dye.
65.	*Ehretia laevis* Roxb. **(S)**	Tombolan	Ehretiaceae	Venereal diseases.
66.	*Enicostema hyssopifolium* (Willd.) I.C. **(H)**	Chhota-chirayata	Gentianaceae	Blood purifier, dropsy, rheumatism, abdominal ulcer, hernia, swellings, itches, malaria.
67.	*Euphorbia hirta* L. **(H)**	Dubhali	Euphorbiaceae	Dysentery, worms, colic, bowel complaints, cough, asthma, vermifuge, diseases of urino-genital tract, diarrhoea, leucorrhoea, menorrhagia.
68.	*Evolvulus alsinoides* L. **(H)**	Shankhpushpi	Convolvulaceae	Bronchitis, asthma, fever, diarrhoea, tonic, vermiguge, dysentery.
69.	*Feronia limonia* (L.) Swingle **(T)**	Kaitha	Rutaceae	Cough, dysentery; alexiteric, heart diseases, aphrodisiac, leucorrhoea, opthalmia, biliousness.

Contd.

Table 1: Contd.

S. No.	Botanical Names	Local/Trade/ Vernacular Names	Families	Medicinal Uses
1	2	3	4	5
70.	*Ficus benghalensis* L. **(T)**	Bargad	Moraceae	Biliousness, ulcer, erysipelas, vaginal complaints, fevers, liver troubles, diabetes, cooling tonic.
71.	*F. religiosa* L. **(T)**	Peepa	Moraceae	Leucorrhoea, biliousness, ulcer, diseases of vagina, and uterus, alexipharmic.
72.	*Fumaria indica* (Haussk.) Pugsley **(H)**	Pittpapara	Fumariaceae	Diuretic, diaphoretic, aperients, anti-spasmodic, CNS depressant.
73.	*Glinus lotoides* L. **(H)**	Gandhi-butti	Moluginaceae	Diuretic, purgative, boils, wounds, indigestion, bilious attacks.
74.	*Grewia asiatica* L. **(S)**	Gangeran	Tiliaceae	Cough, pain, diarrhoea.
75.	*Haloxylon recurvum* Bunge ex Boobs. **(S)**	Khar	Chenopodiaceae	Ulcers, headache.
76.	*Heliotropium ellipticum* Ledeb. **(H)**	Kalibui	Boraginaceae	Ulcers, earache, emetic.
77.	*Indigofera tinctoria* L. **(H)**	Neel, Surmaineel	Fabaceae	Anthelmintic.
78.	*I. linifolia* (L.f.) Retz. **(H)**	Sidio bakario	Fabaceae	Febrile eruptions, amenorrhoea.
79.	*I. trita* L.f. **(S)**	Khuara	Fabaceae	Root: Appetizer, rheumatism; Whole plant: spleen and liver problems.
80.	*Indigofera tinctoria* L. **(S)**	Neel	Fabaceae	Laxative, expectorant, alexipharmic anthelminitc, leucoderma, abdominal complaints, heart disease.
81.	*Ipomoea aquatica* Forssk. **(H)**	Kirpan-beli	Convolvulaceae	Laxative, carminative, fever, abdominal and liver troubles, leucoderma.
82.	*Justicia procumbens* L. **(H)**	Kagner	Acanthaceae	Biliousness, intoxication, fever, diuretic, enriches blood, leprosy, mental and blood diseases.
83.	*Leucas aspera* (Willd.) Link **(H)**	Chota-halkusarq	Lamiaceae	Chronic rheumatism, psoriasis, skin diseases, swellings.
84.	*L. cephalotes* (Roth.) Spreng. **(H)**	Gomia	Lamiaceae	Diaphoretic, stimulant, laxative, bronchitis, jaundice, dyspepsia, paralysis, leucoderma, urinary, fever, scorpion sting.
85.	*Malva parviflora* L. **(H)**	Khumbasi	Malvaceae	Nerve tonic, profuse menstruation, wounds and swellings; Seed: demulcent in cough, bladder ulcer.
86.	*Mimosa pudica* L. **(H)**	Shiah-Kanta, Jinjanio	Mimosaceae	Stimulant, weakness.
87.	*Moringa oleifera* Lam. **(T)**	Sainjna	Moringaceae	Aphrodisiac, alexiteric, analgesic, anthelmintic, ulcers, heart troubles, ophthalmia, muscular and spleen diseases.
88.	*Ocimum basilicum* L. **(H)**	Kali tulsi	Lamiaceae	Plants useful in fever, cough, worms, stomach complaints and gout.
89.	*O. americanum* L. **(H)**	Tulsi	Lamiaceae	Leaf juice is dropped in the ear to relieve earache. Also used in skin diseases.

Contd.

Table 1: Contd.

S. No.	Botanical Names	Local/Trade/ Vernacular Names	Families	Medicinal Uses
1	2	3	4	5
90.	*Opuntia dillenii* auct.pl **(S)**	Nagphani	Cactaceae	Baked fruit used in whooping cough, syrupincreases the flow of bile and control spasmodic cough and expectoration.
91.	*Oxalis corniculata* L. **(H)**	Khatari, Khatti-buti	Oxalidaceae	Appetizer, dysentery, diarrhoea, refrigerant, stomachic, anti-scorbutic, piles, skin diseases.
92.	*Panicum antidotale* Retz **(G)**	Gramma, Garmano	Poaceae	Small pox, would.
93.	*Parkinsonia aculeata* L. **(S)**	Vilayati kikar	Caesalpiniaceae	Plant diaphoretic, antipyretic, leaves aborfi facient.
94.	*Pedalium murex* L. **(H)**	Baragokhru	Pedaliaceae	Anti-biliousness, gonorrhoea, dysuria, anti-spasmodic, aphrodisiac, diuretic, demulcent, emmenagogue.
95.	*Peganum harmala* L. **(H)**	Harmal, Isband, Gandhio	Zygophyllaceae	Emmenagogue, gonorrhoea, aphrodisiac, abortifacient, toothache, Seed: expectorant anthelmintic, lumbago, colic, urinary troubles, opthalmia, rheumatism bronchitis.
96.	*Periploca aphylla* Decne **(S)**	Barri	Periplaceace	Tumours, swellings, purgative.
97.	*Phaseolus trilobatus* (L.) Ant. **(H)**	Jangli-Moth	Fabaceae	Sedative, intermittent fever.
98.	*Phyllanthus amarus* Webster **(H)**	Gugaria, Bhui Amla	Euphorbiaceae	Diuretic, dropsical affections, gonorrhoea, genitor-urinary diseases, sores, stomachic, dysentery.
99.	*P. maderaspatensis* L. **(H)**	Hazarmani	Euphorbiaceae	Expectorant, diaphoretic, carminative, laxative, diuretic, bronchitis, earache, ophthalmia, liver tonic.
100.	*Plumbago zeylanica* L. **(S)**	Chitrak	Plumbaginaceae	Dysentery, leucoderma, piles, inflammation, rheumatism, bronchitis, anaemia, liver and intestinal complaints.
101.	*Portulaca oleracea* L. **(H)**	Lunkha	Portulaceaceae	Alexipharmic, laxative, diarrhoea, asthma, ulcer, blisters, boils, dysentery, leproxy, piles, kidney and spleen diseases, burning sensation.
102.	*P. quadrifida* auct. L. **(H)**	Lunki	Portulacaceae	Alternative, laxative, asthma, cough, urinary discharge, ulcers, eye and skin diseases.
103.	*Prosopis cineraria* (L.) Druce **(T)**	Khejri	Mimosaceae	Rheumatism, astringent.
104.	*Pterospermum acerifolium* (L.) Willd. **(S)**	Karnikara Kanakdhmpn	Sterculiaceae	Flowers used for ulcers, tumors and leprosy.
105.	*Salvadora oleoides* Decne **(T)**	Mitho-jal	Salvadoraceae	Vesicant, purgative, cough, aphrodisiac, laxative, carminative, bronchitis, spleen enlargement, piles.
106.	*S. persica* L. **(T)**	Kharo-jal	Salvadoraceae	Rheumatism, scurvy, astringent, anthelmintic, diuretic, analgesic, liver tonic, carminative, deobstruent, aphrodisiac.

Contd.

Table 1: Contd.

S. No.	Botanical Names	Local/Trade/ Vernacular Names	Families	Medicinal Uses
1	2	3	4	5
107.	*Salvia aegyptiaca* L. **(S)**	Tukham malanga	Labiatae	Diseases of eye, diarrhoea, gonorrhoea, haemorrhage.
108.	*Saraca asoca* (Roxb.) de Wilde **(T)**	Ashok	Caesalpiniaceae	Dried bark – astringent, used in excessive uterine bleeding, flower used in haemorr hagic dysentery. Seeds used in urinary discharges.
109.	*Sesamum indicum* L. **(H)**	Til	Pedaliaceae	Aphrodisiac, spleen troubles, piles bleeding, menorrhoea, diuretic diaphoretic, cooling, hari growth, diarrhoea, lungs diseases, small-pox,syphilis.
110.	*Sida acuta* Burn. f. **(S)**	Bal	Malvaceae	Aphrodisiac, demulcent in gonorrhoea, astringent, nervous and urinary trouble.
111.	*S. cordifolia* L. **(S)**	Bariar	Malvaceae	Astringent, cooling, tonic, emollient, aphrodisiac, urinary troubles, blood disease, bleeding piles.
112.	*S. retusa* L. **(S)**	Bal, Dhabi	Malvaceae	Lumbago, tonic, cooling effect.
113.	*S. rhombifolia* L. **(S)**	Bariara	Malvaceae	Aphrodisiac, tonic, febrifuge, nervous and urinary troubles, strangury, heart diseases, piles, rheumatism.
114.	*Solanum indicum* L. **(S)**	Nar-kanta	Solanaceae	Astringent, anthelmintic, digestive, cardiac troubles, bronchitis, leucoderma, fever.
115.	*S. nigrum* L. **(H)**	Makoi, Chirpoti	Solanaceae	Diseases of eye, ear, nose, ulcer on the neck, laxative, boils, headache, aphrodisiac, alternative, diuretic, inflammation, asthma, bronchitis, fever, diarrhoea, hydrophobia, leucoderma, cathartic, liver enlargement, piles, dysentery.
116.	*S. surattense* Burm. f. **(H)**	Ringni, Kateli, Bhoringi	Solanaceae	Expectorant, cough, asthma, chest pain, catarrhal fever, vomitting, rheumatism, cough, toothache, burning feet. Malaria.
117.	*Tamarindus indica* L. **(T)**	Imli	Fabaceace	Anodyne, aperient, astringent, diuretic and febrifuge.
118.	*Tamarix dioica* Roxb. **(T)**	Lal-jhau	Tamaricaceae	Astringent, aphrodisiac, eczema.

G = Grasses, H = Herbs, S = Shrubs, T = Trees

Table 2: Habit-wise distribution of medicinal plants of Dewghat Forest (U.P.)

S. No.	Habit of the Medicinal Plants	No. of Species
1.	Herbs	64
2.	Perennial Climbers	2
3.	Perennial Grasses	2
4.	Perennial Sedges	1
5.	Shrubs	28
6.	Trees	21
	Total	118

Dewghat Forest represents the biodiversity of a highly fragile ecosystem. The increasing demand of medicinal plants has resulted in the rapid dwindling of these natural resources. There is an urgent need of systematic conservation and sustainable production of medicinal plants involving local communities, university students and departmental field groups with stronger linkage of collaborative work to meet future demand in a sustained manner.

REFERENCES

Jain, SK & Rao, RR (1997). *A Handbook of Field Herbarium Methods.* Today and Tomorrows Printers and Publishers, New Delhi

Shushil, K, Sanjaya, K & Sharma, A (2003). Trends in Plant Medicinal Research and Development and Scope of Medicines–A Review. Journal, **46:** 101-117

Tiwari, DN (2000) *Report of the Task force on conservation and sustainable use of medicinal plants.* Govt. of India, Planning Commission, New Delhi. pp. 175

25

Muga Silkworm Biodiversity: A Profile

DP Paliwal, RK Khatri[1], CJ Prabhakar[2], AK Paliwal[3] and B Choudhury[2]

Research Extension Centre, Central Silk Board, Bageshwar–263 462, Uttarakhand.
[1]Zonal Silkworm Seed Organization, Central Silk Board, Dehradun–249 001, Uttarakhand.
[2]Muga Silkworm Seed Organization, Central Silk Board, Guwahati–781 001, Assam
[3]Department. of Botany, Govt. P.G. College, Bageshwar–263 462, Uttarakhand.

ABSTRACT

India has the pride to be the only producer country in the world of the fabulous golden silk "Muga". This special kind of silk comes through the saliva of the silkworm *Antherea assama* Helfer. Presently, muga silk is cultivated only in North-Eastern states of India. The muga culture occupies a distinct place in the cultural, social and religious life of the people of these states and more particularly Assam, which produces nearly 99% of the total muga raw silk.

Muga culture being a potential forest-based industry holds great promise for afforestation as a supplementary activity and apart from this, it permits gainful utilization of the vast natural wealth. The muga silkworm is reared on host plants *Persea bombycina* King and *Litsea monopetala* Juss. in the N-E states where these plants occupy an area of 4,000 hectares of land which can support the rearing of more than 1.50 crore of dfls (disease free laying). However, even rearing of worms at such a large scale is not sufficient to fill the gap between the production and supply of this precious silk in the world. The muga industry has rich natural sources and manpower and the challenge is to utilize these to bring about a balanced development without disturbing the forest ecology, traditional culture and way of life of the primary producers. Muga industry offers tremendous scope to develop the rural economy through sustainable income as well as employment generation. Therefore, this agro-based cottage industry has the promise to satisfy the livelihood needs for rural masses in Uttarakhand also. Particularly because this region has the advantage of having a favourable weather from May to October, for muga rearing.

Keywords: Biodiversity, muga silkworm, rearing.

INTRODUCTION

China is known as the original home of natural silk in the world as has been described in the volume "Notes of Ancient and Modern" compiled in the Han Dynasty in the 3rd century B.C. "Mulberry wild silkworms formed cocoons then the silkworm in the cocoon became a moth, then the moth laid eggs" and people began to use tussah silk in the Dong Mou mountains in the country, Dong Lai China. Later China started its production at a commercial scale. Though the actual cultivation of the natural silk started in China more than 2,000 years ago, the rearing of non-mulberry silk 'Tussah' has been initiated only 1200 years ago. The worth of the fiber made the Chinese to give equal importance to both silkworm rearing as well as rice growing as a part of their agricultural practice. Henceforth, on account of the foresight and wisdom of the Chinese people their country is exporting quality lustured and colourful, natural silk to the other parts of the world and ranks first with regard to the sericulture countries of the world.

The silk rearing has been practiced in India also from times immemorial. It is believed that sericulture which evolved in China, travelled to different parts of India and its adjoining countries through the famous trade route "silk route" which existed in the river valley. However, the present form of sericulture, which is perhaps the largest and oldest cottage industry in the Indian textiles, has become an integral part of country's economy and social culture, particularly because of the fact that silk is being used not only for making garments but also for decorating the houses, pandals and premises for all auspicious ceremonies. As such, its use has profitted both the rural as well as urban masses and sericulture has grown into an agri-based, small-scale cottage industry which is mostly confined to rural areas but its produce i.e silk is utilized by urban people at a large scale in various forms. India has the unique distinction of being the only country in the world which produces mulberry and all kinds of non-mulberry silk i.e. tropical and temperate tasar, eri and muga. Non-mulberry silks are produced by wild sericigenous insects such as *Antheraea assama* Ww., *A. mylitta* Drury, *A. proylei* Jolly, and *Philosamia ricini* Hutt., etc. Although Eri, Tasar and Muga are commercially exploited in different states of India, the commercial rearing is predominatly confined to Arunachal Pradesh, Assam, Bihar, Manipur, Meghalaya, Mizoram, Nagaland, Orissa, and Tripura. Tasar silk is commercially cultured extensively in Andhra Pradesh, Bihar, Madhya Pradesh, Maharashtra, Manipur, Mizoram, Orissa and Uttar Pradesh whereas the Muga culture is the exclusive prerogative of north-east India and is restricted to the Brahmaputra valley of Assam and its adjacent states due to favourable climatic conditions and a wide distribution of its host plants in the region.

Although, the origin of the muga silk industry is still obscure, the historical records suggest that the exquisite varieties of muga silk were sent by the King Bhaskara of Kamrup (Assam) as a gift to King Harsha Vardhana about 1,300 years ago. In Arthashastra, Koutilya states that weavers of ancient Assam had earned a great reputation in production of quality silk. The first official record of muga silk dates back to 1662 AD in connection with the Mughal's invasion of Mirjumla who remarked that quality muga silk is produced in the region which was insufficient even for the people of Assam where fabulous, golden yellow muga silk played an important role in enriching the cultural heritage of Assamese people. Both men and women wore the golden silk "muga pat" costumes at each and every auspicious occasion. The popular women-folk costume of muga pat is called "mekhla". Thus historical evidences clearly support the origin of muga silk to be from Assam.

The muga industry holds high promise as an employment intensive occupation especially for people in rural and semi-urban localities. Presently more than 27,000 families are engaged in muga culture in Assam and produce 99 MT of muga raw silk annually (Source Annual Report, CSB, 2000-2001).

The muga silkworm, *Antherea assama* Ww. (Syn. *A. assamensis* Ww.) belongs to the Phylum Arthropoda, Class Insecta, Order Lepidoptera, Superfamily Bombycidae and the Family Saturnidae. It is cultivated exclusively in Assam, the foothills of Meghalaya and Nagaland. The silkworm is a polyphagous insect and thrives on various endemic plants, mainly on representatives of the family Lauraceae and secondarily on a few representatives of families Celastraceae, Melastomaceae, Rhamnaceae, Rutaceae, Verbenaceae, etc. The family Lauraceae comprises about 6 genera and 2,000 species, a vast majority of which are trees or well formed shrubs (Table 1). Its members are mostly restricted to the tropical and subtropical forests. With regard to India, the family occupies an important economic position since some of its species like *Litsaea citrata, L. polyantha* (Fig. 1a) and *L. salicifolia* as well as *Persea bomycina* (Table 1) are cultivated in the north-eastern part i.e chiefly in Assam and Meghalaya for rearing the muga silkworms. Otherwise, these muga host plants are widely distributed in Burma, India, Indonesia, Malaysia and Nepal (Jolley *et al*., 1979). Isa & Thangavelu (1988) have reported the natural distribution of one of the muga host plant, *Litsea polyantha* in the Doon valley and its adjoining hilly areas up to an altitude of 800m. It is locally known as Katmara or Kankava whereas the occurrence of the other host plants *P. bombycina* (Fig. 1b) and *M. odoratisima* locally known as Kaul, Dhin or Kaula have been reported between the altitude of 800–2,700m in the hills of Uttarakhand State. It is to be pointed out that the muga silkworm host plants cover roughly 0.1 per cent of the total forest vegetation of the Uttarakhand hilly regions.

MUGA SILKWORM HOST PLANTS

Table 1: List of host plants of muga silkworm* (*Antheraea assama* Ww.)

S. No.	Botanical names	Common names	Families
	Primary Hosts		
1.	*Litsaea polyantha* Juss. (Syn.*Litsaea monopetala* Pers.)	Soalu	Lauraceae
2.	*Machilus bombycina* King.	Som	Lauraceae
	Secondary Hosts		
3.	*Anthaera angustifolia* B1.	Bagnala	Lauraceae
4.	*Actinodaphne obovata* B1.	Patihonda	Lauraceae
5.	*Celastrus monosperma* Roxb.	Bhumloti	Celastraceae
6.	*Cinnamomum glanduliferum* Meissn.	Gansarai	Lauraceae
7.	*C. obtusifolium* Nees.	Chhamejam	Lauraceae
8.	*Gmelina arborea* Roxb.	Gamari	Verbenaceae
9.	*Litsaea citrata* Blume.	Mejankari	Lauraceae
10.	*Litsaea nitida* Roxb.	Kathalua	Lauraceae
11.	*L. salicifolia* Roxb.	Digloti	Lauraceae
12.	*Magnolia sphenocarpa* Roxb.	Chapa/Panchopa	Magnoliaceae
13.	*Zanthoxylum rhesta* DC.	Bajramoni	Rutaceae
14.	*Ziziphus jujuba* Lamk.	Bogori/Ber	Rhamnaceae

*(Muga silk, Thangavelu, 1988)

Fig. 1:(a) *Litsaea polyantha* Juss. (Soalu) (b) *Persea bombycina* King. (Som)

LIFE CYCLE OF MUGA SILKWORM

Like other lepidopterans, muga silkworm is a holometabolous insect, passing through a complete cycle of metamorphosis from egg to adult stage through two intermediate stages of larva and pupa. The entire life-cycle lasts for about 50 days in summer and 150 days in winter (Table 2). The days involved in subsequent stages are as follows:

Table 2: Various stages of muga silkworm life cycle

S. No.	Stages	Summer (Minimum days)	Winter (Maximum days)
1	Egg	7	15
2	Larval	24	70
3	Spinning	3	7
4	Pupal	14	55
5	Moth	2	3
Total days		**50**	**150**

The complete life-cycle is termed as one brood. The muga silkworm is normally multivoltine and has four or five generations in a year. Moreover some chrysalids in winter are found in a hibernating state. The broods are locally known as Katiya, Jarua, Jethua and Bhodiya. The crops of muga silkworm and different stages are given below:

Eggs

The eggs are oval, dorsoventrally flattened and bilaterally symmetrical. Each egg is about 2.4 to 2.7mm in length and 2.5mm in diameter, weighing approx. instantly 0.0073mg. A female lays 100-170 eggs (Fig. 2) in one season.

Fig. 2. Eggs of Muga silkworm.

Larva

The larvae hatch in the morning hours by nibbling through the egg shell. Soon after hatching these eat the leaves of the host plants. The larvae undergo four moults and pass through five instar stage.

Ist Stage: The newly hatched larva is 0.7–1.2 cm in length and weighs approximately 0.0069-0.007 gm. The colour of the body is black with distinct yellow lines at the intersegmental regions. The body tubercles are yellow in colour, provided with setae. The head is small but distinct black in colour (Fig. 3).

IInd Stage: Such a larva is light yellow in colour, measuring about 1.4-1.8 cm in length and having a weight of around 0.083-0.0912 gm. The colour of the head changes from black to dark brown whereas the tubercles are blue.

Fig. 3. First Stage larvae.

Fig. 4. Second Stage larva.

IIIrd Stage: But now, the larva green in colour which measures about 1.8-2.5 cm just after second moult weighs approximately 0.442-0.632 gm. The tubercle is violet in colour. A faint yellow lateral line is visible in lateral region of the body. The clasper becomes triangular in outline and yellowish in colour (Fig. 5).

Fig. 5. Third Stage larva.

IVth Stage: Larva is green in colour which attains a length of about 2.5-3.5cm just after second moult and weighs approximately 2.00-3.5gm. The dorsal surface of the body is dark green, whereas the ventral surface is light green in colour. The tubercle is red in colour and provided with setae. A yellow lateral line is visible in lateral region of the body. The clasper becomes triangular in outline. The head is also triangular with a copper hue (Fig. 6).

Vth Stage: The newly transformed fifth instar larva is approximately 4-5.5cm in length and weighs about 4.121-5.213gm. Although the dorsal surface of the body is light green, the ventral surface is deep green in colour. On either side of the epicranial suture, eye spots are prominent. The body tubercles are brick red in colour and provided with setae. Each segment of the body bears a pair of dorsal tubercles, a pair of upper lateral tubercles and a pair of lower lateral tubercles. The lateral line is prominent, which extends from posterior part of the body to the anterior abdominal segment. The lower portion of the lateral line is yellow and upper portion is chocolate coloured. The fully mature larva attains 10-15gm weight. Generally, the female larvae are larger and heavier than the males. The male and the female larvae can be easily distinguished from sexual markings which develop distinctly on the eight and ninth abdominal segments on the ventral side of the fourth and fifth instar larva. In the female, the sexual markings appear as a pair of milky white spots in each of the eighth and ninth segments. In male, a small milky white body appears at the centre on the ventral side between eight and ninth segments (Fig. 7).

Fig. 6. Fourth Stage larva.

Fig. 7. Fifth Stage larva.

Pupa (Chrysalis)

The ripe worm ceases to feed and remains motionless for sometime. The size of the worms shrinks and the body colour is slightly altered. The change is a prelude to formation of cocoon. The worm then wanders aimlessly in search of a crevice or a suitable place for forming the cocoon. After making the cocoons, the larva transforms into a pupa through an intermediate stage, known as pre-pupal stage. During this transformation phase, the pre-pupa slowly changes colour from green to brown. Its shape and size also change alongwith the external transformation, the internal organs also undergo changes and the last larval skin known as 'cremaster' is shed inside the cocoon. The pupal stage is a preparatory one for larva to transform itself into a moth.

Moths

The moths now emerge from the cocoons. The male and female moths exhibit distinct sexual dimorphism (Fig. 8) and can be easily identified with the characters mentioned hereunder:

- **(i) Curvature of wings:** In the male moth, the tips of the forewings have a sharp curve which is easily distinguishable from that of the females.
- **(ii) Antenna:** The antennae of the male moths are broader than those of females.
- **(iii) Abdomen:** The abdomen of the male moths is narrow and small, whereas in female the abdomen is broad, larger and swollen.
- **(iv) Colouration:** The colouration of the wings and the body of the male moth is copper brown to dark brown whereas in the female it is yellowish light brown. Some wild muga moths exhibit colour variations like the larvae and these wild moths are black in colour.

Fig. 8a. Male moth of muga silkworm.

Fig. 8b. Female moth of muga silkworm.

The different aspects of life of muga silkworm are highly sensitive to the effect of photoperiodism (Table 3). The effect is evident in every phase of the worms activity as indicated below:

Table 3: Various life activities of muga silkworm related to the photoperiodism

Activities	Periods
Emergence of moth	Dusk to dawn
Pairing of moths	Dusk to dawn
Deposition of eggs	Evening to dawn
Hatching of worms	Morning
Moulting of larva	Any time of the day
Maturing of the worm	Dusk

MUGA SILKWORM REARING

Rearing of muga silkworm is carried out on the host plant in the field. In this process the hatched larvae are placed on the leaves of the plant in the open habitat till maturation of the worm and during rearing the rearers pay their attention to the supervision of the concerned activities like cleaning of plant base, handling of moulting larvae, transfer of larvae, handling of infected larvae, preparation in respect of suitable cocoon age, collection of worms, etc. On commencement of spinning the mature worms are collected from plants by the rearers for cocooning. The muga silkworm is multivoltine in nature and therefore, rearing can be conducted all through the year.

GRAINAGE

Quality seed is of prime importance in the sericulture industry. For producing quality disease-free layings means dfls. The grainage operation involves four steps:

(i) Selection and preservation of seed cocoons in the grainage hall.
(ii) Moth emergence and coupling.
(iii) Oviposition
(iv) Microscopic examination of mother moth and preparation of disease-free laying dfls.

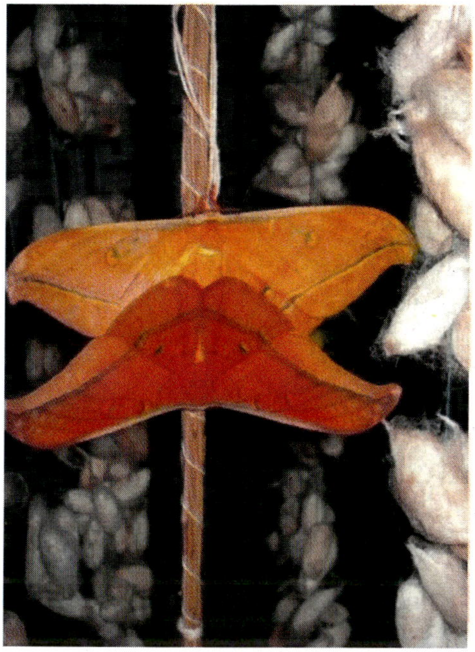

Fig. 9: (a) Coupled moths.

(b) Oviposition.

Apart from the above, the disinfection and incubation are also important aspects of grainage operation.

MUGA SILKWORM CROPS

Muga silkworm is polyphagous and multivoltine in nature, producing 4-6 generations in a year. According to the Assamese Calendar the season-wise crops have been put in Table 4.

Table 4: Seasonal chart for muga silkworm rearing under the weather conditions of Assam

S.No.	Crops	Seasons	Months/periods	Types of the crops
1.	Jarua	Winter	December-February	Seed
2.	Jethua	Spring	March-May	Commercial
3.	Aherua	Early Summer/ Early Monsoon	June-July	Seed
4.	Bhodia	Late Summer/ Late Monsoon	July-September	Seed
5.	Kotia	Autumn	September- October	Commercial

*(Muga Silk, Thangavelu 1988)

These generations often overlap each other. Further, two crops viz. Kotia and Jethua are used for the production of reeling cocoons whereas the rest are used for seed production or simply for stock maintenance for the seed purpose.

The climatic conditions of the areas of Garhwal and Kumaon hills are more or less similar to the North-Eastern region being subtropical which vary according to the altitude all over the hills. The climate of Doon valley is very much similar to North-East as it lies at the foot-hills of North-Western Himalaya at an altitude of 900m MSL. The average rainfall of this valley is 2,000 mm/year. Furthermore, the highest temperature is 38°C and the lowest 4°C with a range of relative humidity from 50% to 95%, favouring luxuriant growth of *L. polyantha.* In this congenial atmosphere the species show a highly luxuriant growth which exceeds even to that of Assam. Therefore, four crops of muga silkworm w.e.f. May-October could also be taken in the favourable weather conditions of Uttarakhand. The present status of the crop pattern (Fig. 10) of muga silkworm cultivation in Assam and Uttarakhand has been presented below in Tables 5 and 6.

Fig. 10. A view of crop pattern in Kumaon region of Uttarakhand.

Table 5: A comparision of the crop patterns of Assam and Uttarakhand

S.No.	Assam		Uttarakhand	
	Crops	Periods	Crops	Periods
1.	Chhotua	February –March	-	-
2.	Baisakhi	April-May	Jeshta	May-June
3.	Aherua	June-July	Ashadh	June-July
4.	Bhodia	July-August	Bhadra	August-September
5.	Kotia	September-October	Kartika	September-October
6.	Ahenua	November-December	-	-

Table 6: Present status of muga culture in north–eastern regions

S.No	Parameters	Estimated figures
1.	Number of families engaged	30000
2.	Area under plantation	6755 hectares 1559 ha in Govt. Sector 5196 ha in Private sector
3.	Average land-holdings of muga plantation	½ acre
4.	Rearing size/family	202-286 dfls
5.	Cocoon production per family	16500
6.	Average income per family through sale of cocoons (Rs.)	₹ 7425

(Source : Chakarvorty 2004)

CONSTRAINTS IN MUGA CULTURE

The slow progress in muga culture is attributed to several constraints which are given below:

- Lack of pure lines and narrow genetic base in muga silkworm.
- The seed is transacted in the form of cocoons and not in disease-free layings.
- The farmers are reluctant in respect to maintenance of plantation through input applications and cultural operations, which results in yield gap and poor cocoon harvest.
- The gap in seed requirement and supply is about 30% as a result of which plantations remain unutilized.
- Shortage of systematic bush plantation for management of seed crops.
- Unscientific multiplication of stock by the farmers.
- Absence of clear-cut demarcation of seed zone.
- Lack of co-ordination at different levels of multiplication.
- High-cost involvement in the establishment of commercial grainages due to the requirement of larger space, more man power and transportation constraints.
- Inadequate cold storage facilities for skipping rearing during adverse seasons.
- No systematic seed organization under Muga Sector and Licenced Seed Producers are inadequate.

There is a series of crop specific constraints in muga culture, which affect the cocoon productivity and seed quality. The crop specific constraints recorded during the cultivation of muga silkworm have been listed in Table 7.

Table 7: Constraints during different crops of muga silkworm

S.No.	Crops	Constraints
1.	Aherua (Pre–Seed)	High humidity and temperature-their fluctuations, high rainfall and inclement weather conditions, virosis, attack of ants and wasps.
2.	Bhodia (Seed)	High humidity and temperature-their fluctuations, high rainfall and inclement weather condition, virosis, bacteriosis and attack of wasps.
3.	Kotia (commercial)	Lack of adequate quantity of dfls due to low rate of multiplication.
4.	Aghenua Jarua (Pre-seed)	Prolonged larval duration, low temperature, muscardine/virosis, Uzi fly attack and poor quality of foliage.
5.	Chhotua (Seed)	Uzi fly, heavy wind, dry spell bacteriosis and virosis.
6.	Jethua (Commercial)	Lack of adequate seed, Uzi fly, heavy wind and hail- storm.

(*Source:* PK Das & AK Sahu 2003)

Other Limitations in culture of *Antheraea assamensis*

The muga silkworm has the following limitations:

- Sensitive to biotic and abiotic factors.
- Lack of technology to control silkworm diseases.
- Lack of commercial aptitude in muga growers.
- Uncertainty of silkworm crops due to climatic adversities.
- Gap in 'Research and Development' on seed technology and evolution of high-yielding breeds.

Quality Foliage

The silk is essentially a converted form of leaf protein. As such if the feed quality is superior the silk will be of good quality. There are different morphotypes of *Persea bombycina* Kost. and *Litsea monopetala* (Roxb.) Pers. but the seed crops should be raised on superior morphotyes confirmed by the study viz. S3 and S5 of *Persea bombycina* Kost. Repeated studies conducted indicate that the silkworm fed on *Litsea monopetala* lays more eggs as compared to the other host plants. Hence, these characteristics of host plants should be utilized while raising the seed crops. The plantations used for raising seed crops should be maintained through agronomical practices and culture operations so that quality leaf can be obtained.

Quality Basic Seed

The basic seed is a very crucial material for raising commercial crops. In muga industry the silkworm seed requirement always remains unfulfilled due to poor quality of basic seed. The basic seeds should be prepared with the cocoons of good quality. The potent moths should only be selected for egg laying and seeds should be provided for raising the commercial crops through standard grainage technique and production protocol. The Fuziwara Technique for detection of pebrine should be practiced during the preparation of basic seed material.

Rearing of Wild Race of Muga Silkworm

The wild race which thrives exclusively on *Litsea monopetala* and exhibits diapausing characters during winter and completes four generation, in a year should be collected and basic seeds be prepared and multiplied in order to solve the seed constraints faced in this production.

Disinfection

Disinfection is the backbone of sericulture industry. Though silkworm rearing is done outdoors but disinfection and sanitation can be maintained very well by base cleaning, sprinkling of bleaching powder and ash around the surface area of the trees so that disease occurrence could be minimized and seed crop can be saved.

Disease Monitoring

In muga culture disease-monitoring aspect has not been practiced at a large scale but disease monitoring by microscopic examination at each stage from egg-shell to pupa during the rearing span should be undertaken through Fuziwara Technique so that the remotest chance of pebrine can be avoided.

Cellular Rearing During Chawkie

Cellular rearing for seed crops should be conducted alongwith microscopic examination at each stage to ensure that the seeds are disease free, only the mass rearing should be conducted. In this way, good seed cocoons can be generated for commercial seed production.

Maintenance of Hygiene in Rearing

In seed crop rearing hygiene should be maintained properly. Hands and chalnis should be cleaned and washed with formalin water while transferring the worms, so that disease causing pathogens could be destroyed. During rearing, personal hygiene aspects should also be taken care of.

Use of Botanicals to Control Pests and Predators

Quite a heavy loss due to pest and predators during rearing is observed. Though the use of chemicals is in vogue but the botanicals like leaf extracts of *Ocimum tenuifolium* and *Phlogapcanthus* spp. can be used to control the ants and bugs, which do not have toxic, deterent effect on the silkworm.

Selection of Seed Cocoons for Seed Preparation

In muga industry the seed preparation is done on the basis of randomly selected cocoons without following any metric method. Therefore a great need has emerged for the selection of seed cocoons on the basis of compactness and cocoon weight. The sexing in the grainage is also not done which can also be practiced for preparation of the seeds.

Use of Disinfectants During Rearing

The practice of using disinfectants on the body of the silkworm has also not been tried as yet. The disinfectants, which prevent the silkworm diseases, should be used after screening of the disinfectants for controlling bacteriosis, virosis and muscardine disease since the occurrence of the disease is more severe during the cultivation of Bhodia and Aghanua crop and muscardine during winter crops.

Proper Cocooning and Harvesting

The cocooning aspects in muga culture have also not been addressed as yet. It is suggested that the mature worms should be put for cocooning in their specific halls where humidity and temperature conditions are controlled. The harvesting of cocoons is also not carried out in proper time as sometimes it is done even before the formation of pupae which directly affects the moth emergence. The proper time for harvesting of cocoon is only after assessment of pupa formation.

Synchronization of Moths

The heavy loss of female moth is attributed to the shortage of male at the later part of the grainage operation. To avoid loss of female moth, the synchronization of cocoons should be done properly. Though putting the cocoons in natural conditions during night is being practiced, it needs a more scientific approach to minimize the egg loss, which will certainly contribute to solve the problem of seed dearth.

Use of Nylon Net

Use of nylon net provides protection to worms from pests and predators, minimizes their mortality and improves cocoon productivity. It also checks Uzi infestation.

Impact of Effective Seed-crop Management

The effective seed-crop management will bring an improvement in the cocoon productivity from the present 20-30 cocoons per gram seed to 40-50 cocoons per gram. The quality of basic seed will be improved which in turn will improve the cocoon productivity in commercial crops qualitatively and quantitatively. The farmers will get more price for their cocoons and will derive more benefits from the muga culture by rearing disease-free seeds.

Muga Silkworm Seed Requirement and Arrangements in our Country

The total seed requirement of Muga industry is around one crore commercial seeds. The Central Silk Board contributes 1 lakh commercial seeds, DOS, Assam 6.5 lakh, private graneurs 30 lakh and private rearers 62.5 lakh seeds each year. There is a need to scale up the seed production under government sector so that the quality seeds can be made available to the farmers. The private rearers produce a large quantity of seeds without following scientific methods and production protocol. Such seeds give poor cocoon-yield due to high incidence of disease. Hence, there is a need to produce good quality seeds by raising good seed crops. The details of seed requirement and supply arrangements of the country have been given below in Table 8.

Table 8: Muga silkworm seed requirement and availability

S.No.	Total Seed requirement of the country (lakh)	% of seed requirement met by Central Silk Board (lakh)	Contribution of the Central Silk Board		Left over seed requirement met by State grainage (lakh)		Private graineruers (lakh)		Private rearers (lakh)	
			Qty (lakh)	%	Qty (lakh)	%	Qty (lakh)	%	Qty (lakh)	%
1.	Pre Basic Seed	2.78	0.834	30	1.3	47	–	–	–	–
2.	Basic Seed	16.66	0.667	4	11.4	68	–	–	–	–
3.	Commercial Seed	100.00	1.00	1	6.5	7	30	30	62.5	62.5

In recent years the muga culture has also been spread in other parts of the country viz. West Bengal, Uttarakhand and Andhra Pradesh. The culture in these states is still as its initial stage and will gain momentum in the years to come and increase hopefully as seed pockets for muga culture.

The studies of various parameters like rearing and grainage operations have already been carried out under Doon valley by Khatri *et al.* (2000). The results have revealed that congenial weather of Uttarakhand can support muga silk cultivation in four crops i.e. Jethua, Aherua, Bhodia and Kotia. Presently, on the basis of the results of the studies undertaken, muga rearing and grainage operations are being conducted successfully in District Bageshwar of Kumaon region of Uttarakhand since April, 2010 and it has been found that there is tremendous potential for its culture in this region of the State.

Comparison of the Crop Pattern in Assam and Uttarakhand

The rearing performance suggests a four-crop calendar for muga culture in Uttarakhand from May-October period unlike Assam, where six-crop calendar is in operation. A comparative crop pattern of Assam and for Uttarakhand has been presented in Table 9.

Table 9: Crop patterns of muga silkworm in Assam and Uttarakhand

S.No.	Assam		Uttarakhand	
	Crops	Periods	Crops	Periods
1.	Chotua	February –March	-	-
2.	Baisakhi	April-May	Jeshta	May-June
3.	Aherua	June-July	Ashadh	June-July
4.	Bhodia	July-August	Bhadra	August-September
5.	Kotia	September-October	Kartika	September-October
6.	Ahenua	November-December	-	-

Constraints in Muga Silkworm Rearing at Bageshwar

The constraints in muga silkworm rearing in District Bageshwar, Uttarakhand may be summarized as follows:

1. Low/abnormal hatching in dfls received from Assam.
2. Loss of the worms due to predators i.e. birds, ants and wasp attacks.
3. Lack of man-made muga host plantation.
4. Availability of muga host plants at undesirable locations.
5. Lack of know-how about rearing among local masses.
6. Availability of natural grown muga host plant in far-flung hilly areas.
7. Lack of suitable technical staff.
8. Lack of facilities for the transportation at the rearing sites.
9. Lack of nylon nets to cover the host plant and to protect the worms from predators.
10. Lack of round the clock watch during rearing span.

Prospects of Muga Culture at Bageshwar

1. Availability of nature grown muga host plants.
2. Successful rearing and grainage performance.
3. Availability of congenial climate.
4. Unemployment among rural masses.

Strategies Suggested

1. Man-made muga host plantation may be raised in clusters.
2. Nature grown plantation may be maintained at a desired height for the purpose of rearing.
3. For watching the muga technology activities in adjoining areas of Uttarakhand, a permanent infrastructure must be established at a central place. The State Farm at Jyoli kot, District Nainital, Government of Uttarakhand is suitable for it, where around 4.5 acres of land is available, which may be utilized for raising of muga host plantation for muga silkworm seed generation.
4. A training programme for technical staff as well as farmers may be organized in Uttarakhand itself.
5. Nylon nets are to be used on the host plants which can protect the worms from predators attack during rearing.
6. Proper transformation facility is needed for the mobility of technical staff in far-flung hilly tracts at rearing sites and transportation of seeds, cocoons and rearing articles from time to time.

Prospectives of the Attempts

1. Enhancement of avenues of employment generation opportunities in the hilly rural areas.
2. Enhancement of the production of golden glittering fabulous muga silk in India.
3. Effective use of low productive land which is still not being utilized for agricultural activities.
4. Uttarakhand is likely to emerge as an alternative muga seed growing zone, of the country. Since the summer and late summer crops which are known as Aherua and Bhodia crops are the most successful crops in the State, as the temperature and humidity remains most congenial for muga silkworm rearing in the seasons. It is hoped that the trouble faced by muga culture may be solved to a great extent through muga seed production in this zone.
5. Due to pleasant weather in the State more than two commercial crops may be raised right from March to October. Rest of the period w.e.f. December–February rearings cannot be achieved due to severe winter and this leisure period would help as natural cold storage for seed cocoons preservation or induce diapauses in pupae for further preparation of seeds as Jarua seed for the March seed crop Chhotua.

CONCLUSION

The rearing results on existing plantations and suitable weather are favourable factors for fostering growth of muga culture in Uttarakhand. Hence, it is suggested that a suitable strategy may be chalked out to tie up seed production programme of the Uttarakhand State, with the Government of Assam so that the seed-raising problem can be solved by creating another resource for seed and Uttarakhand will be a new locality of production of muga seed for North-Eastern states.

REFERENCES

Das, PK, Bhatt MM & Chaudury, B (2004). Problems and prospects in Muga silkworm Seed Production. Proceedings of National Workshop on 'Potential and Strategies for Sustainable Development of Vanya Silks in Himalayan States', pp. 249-251

Isa, M & Thangavelu, K (1988). North-west sub-Himalayan Belt-A potential zone for muga culture. Indian Silk **27:** 7-11

Jolly, MS, Sinha SS & Razdan, JL (1971). Influence of temperature and photoperiod on termination of pupal diapause in tasar silkworm, *A. mylitta. J. Insect Physiol.* **19:** 753-760

Jolly, MS, Sen, SK, Sonwalker, TN & Prasad, GK (1979). Manuals on Sericulture (Non-mulberry Silks) & FAO Agriculture Service Bulletin, Tome 127-128

Khatri, RK, Singh N, Paliwal, DP, Babulal, Siddiqui AA & Sharma AK (2000). Correlation studies on environmental factors, rearing and grainage behaviour of *Antheraea assama* Ww. under Doon Valley conditions. Proceedings of National Workshop on 'Potential and Strategies for Sustainable Development of Vanya Silks in Himalaya States', pp. 160-166

Thangavelu, K, Chakarborty, AK, Bhagowati, AK & Isa, Mohammed (1988). Hand-book of Muga Culture. Central Silk Board, Bangalore

26

Medicinal Plants: Local Heritage with Global Importance

CP Shukla and Mukesh Kumar

Department of Botany, T.D. P.G. College, Jaunpur–222 002, (U.P.)
Department of Botany, Sahu Jain P.G. College, Najibabad–246 763, Bijnor, (U.P.)
E-mail: cp.shukla@yahoo.com

ABSTRACT

The drug plants have been a prime resource of forests for a number of necessary commodities for the tribals living in and around them. Traditional medicine is a comprehensive term used to refer to the traditional (Chinese, Ayurveda. Unani) and various forms of indigenous medicine systems. The study dealing with medicines derived from plants that are used in the treatment of various diseases and ailments in accordance with indigenous pharmacopoeia, folklore and herbal treatment is referred to as ethno-medicine. In the present Chapter, a whole gamut of issues related to the traditional and modern knowledge regarding medicinal, cosmetic and other uses of Indian plants have been discussed and future strategies for strengthening alternative therapy have been suggested, placing enough emphasis on the public sensitization on the utility of medicinal plants, IPR of plant products, *in-situ* and *ex-situ* conservation, especially of endangered plants, setting up of National and State Medicinal Plants Board, development of Research and Development infrastructure, market managements at Domestic and International levels, as an overview on this holistic health science having diversity, accessibility, affordability, safety and preventive as well as curative emphasis to meet potential challenges of curbing human population, treating chronic disease like diabetes, asthma, malaria, heart disorders, etc. distressing and detoxification packages of Ayurveda, Homoeopathy, Biochemic and Unani systems to solve psychosomatic problems.

Keywords: Ethnobotany, herbal therapy, plant biodiversity.

INTRODUCTION

Traditional Knowledge and Glimpses of Ethnobotany

Many traditional societies often referred to as indigenous or tribal people, have accumulated a whole lot of empirical knowledge on the basis of their experience whilst dealing with nature and natural resources. This traditional wisdom is based on the intrinsic realisation that man and nature form part of an indivisible whole and therefore should live in partnership with each other. This eco-centric view of traditional societies is widely reflected in their attitudes towards plants, animals, rivers and the earth. This whole body of knowledge centered around economic value of plant arid animal species is part of ethnobiology, and has potential value for the society at large.

India with rich biodiversity is one of the greatest storehouse of traditional knowledge and wisdom. It has the potential of becoming a major player in the global trade in herbal formulations, ethnomedicine and products. But the tribal wisdom and their indigenous knowledge, unfortunately lacks due recognition and adequate modern documentation and is fast diminishing.

Therefore, there is an urgent need to promote and protect these fragile knowledge systems and the tribal's wisdom. Further, there is need to revive, update and link this wisdom with modern science in order to get value-added products like medicines, nutraceuticals, cosmaceuticals etc. for human welfare. Ethnobotany is defined as the study of relationship which exists between people of a primitive society and their plants environment (*see also* Harshberger, 1885). It is the study of contextualized plant use. Jain (1987) elaborated it, as the total nature and traditional relationship and the interaction between man and his surrounding plant wealth.

Wisdom of Tribals on Medicinal and Economic Plants

As is well known, Wickens (1990) defines Ethnobotany as the study of useful plants prior to the commercial exploitation and eventual domestication. In fact Ethnobiology must have been the first knowledge on plants and animals, which the early man had acquired by sheer necessity, intuition, observation and experimentation in the forests. It is now almost universally taken as the total direct relationship between man and plants. It includes the use of plants by both tribals and non-tribal communities without any implication of primitive or developed societies. Further, Ethnobotany by nature is an interdisciplinary science and has relevance to sociology, anthropology, taxonomy, phytochemistry, archaeology, ecology, agriculture, medicine, linguistics, etc. Today, Ethnobotany has become an important and crucial area of research and development in resource management, sustainable utilization and conservation of biodiversity and for socioeconomic development. All over the world, there has been an increasing interest towards the scientific study of man-plant interaction in the natural environment among the botanists, social scientists, anthropologists, and practitioners of indigenous systems of medicine. The Indian tribes and their future have been the subject of vigorous discussion during the past several decades and great concern has been expressed for their protection, welfare, development and upliftment by the Government as well as voluntary organisation, Tribal Sub Plan (TSP) was implemented by the Government and the strategy has been in operation since the fifth Five Year Plan for improving the quality of life, of the tribal communities. There are 500 tribal communities of different ethnic groups inhabiting India, which constitute about 8% of the total population of the country. The tribal population in India has gone up from about 20 million (Census 1951) to 66.5 million (Census 1981). Among these the most important tribal groups are Gonds, Bhils and Santhals, inhabiting central and western zones of the country.

The State of Uttar Pradesh presents unique physical and ethnic diversity and is inhabited by large numbers of tribals, viz. Bhotiya, Raji, Jaunsari in the hill regions, Tharu and Bhoxa in the terai region of Uttaranchal and Gond, Kol, Baiga, Chero, Kharwar, Korwa, Ghasias, Agaria, Panika, Oraon, Mushhar, Sahariya, etc. in the plains of Uttar Pradesh. Among them there are food gatherers and hunters (Raji, Korwa, Kol), nomads (Bhotiyas) and cultivators (Tharus and Bhoxas). The tribes of Uttar Pradesh belong to two great stocks the Mongoloid and Dravidian. The ethnobotanical studies have been conducted amongs the tribes Agaria, Baiga, Bhoxa, Bhotia Kharwar, Kol, Korwa, Jaunsari, Tharu, Raji, Sahariya, Panika, Oraon, Gond and other hill communities, who live in the tribal tracts and forest areas and villages of Kheri, Sonbhadra, Allahabad, Varanasi, Banda, Garhwal, Jhansi, Lalitpur, Nainital, Dehradun, Gonda, Bahraich, Pithoragarh and Gorakhpur districts of Uttar Pradesh and Uttarakhand amongst others by Maheshwari *et al.* (1981), Singh and Maheshwari (1989), Singh and Maheshwari (1992). The tribal areas are botanically rich on account of its diverse forests, river, valleys and hillocks, which provide them their basic needs and livelihood. They practice primitive agriculture, observing traditional practices like fishing, hunting, gathering foods, religious ceremonies, taboos, believes and ethnoherbal therapy. They collect and utilize various herbs, leaves, roots, rhizomes, tubers, flowers, fruits and seeds of many wild as well as cultivated plant species grown in ambient vegetation for their day-to-day needs and healthcare.

What is Ethnobotany?

The knowledge on the use of bioresources hidden in the traditional culture is vital not only for ethnic societies but also for the urban cultures. India is fortunate in having more than 550 ethnic tribes with a vast indigenous knowledge coupled with an amzing diversity of flora and fauna.

Traditional societies who are the custodians of such knowledge systems are to be amply rewarded by sharing benefits arising from out of sale of drugs/products based on traditional knowledge. Confidence and co-operation of tribal communities at all stages is vital. National and International level conventions for exchange of information among all tribal resource persons essential. Such cross-cultural ethnic knowledge among different developing nations help in establishing the species of plants and animals that are used for a specific ailment among different ethnic communities. The number of species common to treat a particular ailment within and between communities, or where diverse species are used which ones are better? Collective and Co-operative efforts of all developing nations in the study of cross-cultural ethnobotany can provide solutions for many such issues and result in establishing safer, cheaper and locally acceptable drugs for many of the ailments rampant among the ethnic communities in the entire region. The vital role of taxonomist in such studies and the need for strengthening the taxonomic base in all the developing nations is emphasized.

With the modernity, primitive tribal societies are fast dwindling, hence, traditional knowledge possessed by them must come out for further conservation and development of food and herbal medicines so that it is not forgotten and lost in antiquity. Ethnobotany is holistic living and alternate system(s) of medicine emphasizing both physical and mental well being of an individual while, Vriksayurveda and Homeopathy believe in the harmony with the nature. The literature on the subject is far too numerous for the purpose of overview. Some important reviews and books include those by Jain (1981, 1989, 1995, 1996, 2001 a and b), Trivedi (2002), Baburaj (2003), Trivedi and Sharma (2004) and Bhattacharya and De, (2005); Shukla (2006), Mishra and Shukla (2006), and Shukla (2008)

History of Ethnobotany: An Indian Perspective

India is a country of varied topography, soils, agro-climates and its forests are good abodes of several fascinating ethnic communities, floristic regions and diverse emporia of natural plant vegetation germplasm. Knowledge about them therefore must generate a huge scope for patronage of the present and future generations. Indeed being one of the twelve biodiversity mega hotspots of the world, India has one of the oldest, richest and diverse folk tradition associated with the use of medicinal plants in its healthcare system to promote safeguard and relieve pain, heal wounds and refresh mind, muscles and nerves. One of the oldest records in traditional medicine is Vrikshayurveda known to be compiled by Parashara and Manohar. Glorious traditions in this context date back to Rigveda (4500-2500 BC), Atharveda (2500-2000 BC), Samhita era i.e. Brihad Samhita, Charak Samhita (2000 BC) and Sushruta Samhita (1000-800 BC). Later came the Ayurveda, a practice in Sanskrit. The vedic and post-vedic periods (pre-Budhist) had celebrated Indian herbalists and physicians like Bhikshu, Atrey, Mahabharata, Nagarjuna, Vaghbhatta Jr. (800 AD), Chakradatta, Sharangadhar, Bangasen and Dhanvantari. Vaghabhatta wrote Ashtang Hriday Samhita (Rawat and Uniyal, 2003). In Mughal Period, Hindu Materia, Raja Nighantu, Naradhan Pundia appeared. In British period, Madana Pala Nighantu by Madan Pala and Bhava Prakash by Bhava Mishra came into being. Since then, many authentic and excellent publications like Pharmacopoea of India and books of Homeopathy and Materia Medica have come out on the subject. Rawat and Uniyal (2003), Bhattacharya, (2004), Shukla (2006) and Shukla (2008) have fairly discussed the topic.

Ethnobotany at the Service of Mankind

During the past few years, a substantial wealth of information has accumulated using knowledge collected from Sadhus, Maharajas, Rishies, Giri, Puri and Nath, etc. in groves, forests and hills as well as from those involved in witchcrafts, magic folklores, herbal practitioners as doctors and villagers of ethnic communities, foresters, shepherd, migratory people and classical literature, travelogues, notebooks, gazetteers, herbarium etc. and pertinent previous literature which existed in various languages in the form of poems, exaggerated eulogy, religious taboos and beliefs, archaeological data, sacred groves, questionnaires, field work, grandma's prescriptions, voucher specimens, etc. Jain and Mudgal (1999), Mitra and Jain (1991), Jain, (2001). Parabia and Reddy, (2008) have also made use of ethnobotany. It has advanced from an agrobased tribal society to an industry and technology-based society and involves conservation of biodiversity, discoveries of new herbal drugs, characterization of drugs, updating and documentation of existing knowledge and marketing of herbs and their products, etc. Some basic issues of ethnobotany in general human welfare have been described in Table 1. Ethnobotanical knowledge in various Indian states has been summarized in Table 2. Major Indian contributions on medicinal plants have been reviewed in Table 3. The herbal cure of population growth and major diseases has been depicted in Table 4. An annoted list of some important plants used to treat a plethora of diseases has been given in Table 5. The systematic exploration of plant wealth in conjunction with work on ethnobotanical data is needed so as to assess the plant genetic resources of different areas (Singh *et al.*, 2004).

Table 1: Contributions of ethnobotany in general human welfare

S. No.	Economic Values	Author(s) and Years
1.	Making of plates *(Butea monosperma)*	G.S. Singh (2002)
2.	Timber and agricultural implements *(Albizia lebbek)*	Tosh (2004)
3.	Edible plants	Awasthi (1957)
4.	Food from forests	Soni (1995)
5.	As fodder *(Indigofera astragalina)*	Negi (1977), Balaraman (1981), Bhatt and Joshi (2004)
6.	For religious ceremonies	Majumdar and Boissya (1980)
7.	For making brooms *(Aristida adscensionis)*	Bhatt *et al.* (2002)
8.	Gums and resins *(Gardenia resiniferum, Acacia chundra and A. senegal)*	Singh (2002), Saini (2004)
9.	Tannins and dyes *(Acacia chundra, Cassia fistula)*	Sen *et al.* (1983)
10.	As detergents *(Madhuca longifolia)*	Prasad *et al.*, (2002)
11.	Beverages and narcotics	Tosh (2004)
12.	Plants as fertilizers	Kavitha (2004)
13.	Bioenergy crops (Petro-crops)	Elamathi *et al.*, (2005)
14.	As cosmetics *(Aloe barbadensis, Citrus limon)*	Sharma (1989)
15.	Ethnomagico-religious beliefs	Jain (1963) Hajrat (1987)
16.	For making ropes, mats, hats and basket	Thakur and Thakur (1994)
17.	Fibre-yielding *(Helicteres isora, Urena lobata)*	Singh and Singh (1982), Islam (1984)
18.	For musical instruments *(Ailanthes excelsa)*	Parmar (1978)
19.	As a source of fatty oil *(Madhuca longifolia)*	Mahapatra (1983)
20.	Spices and condiments *(Piper nigrum)*	Gopinatham (2004)
21.	Bio-diesel formation *(Jatropha curcas)*	Lal and Lata (1980),
22.	As antifertility drugs *(Abroma augusta, Laccardia* sp.)	Jain *et al.* (2005)
23.	For stuffing pillows and mattresses *(Bombax ceiba, Salmalia malabarica)*	Singh (2002)
24.	For making bidi wrappers *(Diospyros melanoxylon)*	- -
25.	Biofences *(Lawsonia inermis)*	- -
26.	Bows and Arrows *(Bambusa arundinacea)*	- -
27.	Tea preparation *(Anogeissus latifolia)*	- -
28.	Fish Poison (*Securinega virosa)*	- -
29.	Plant foliage as vegetables	Bhatt *et al.* (2003)
30.	Used for tooth-brushes *(Azadirachta indica, Salvadora persica)*	Bhatt *et al.* (2000)
31.	As modulators of radiation *(Ocimum tenuiflorum, Withania somnifera)*	Goyal (2004)

Table 2: Publications on ethnobotany in general human welfare

S. No.	Contents	Author(s) and Years
1.	Antivenomous plants of East Khandesh	Bagul and Yadav (2003)
2.	Ethnobotanical plants of Pachchham hills of Kutchh district	—
3.	Ethnobotanical studies in East Sikkim	Janmeda (2004), Verma (2005)
4.	Ethnobotanical studies of Karwa tribe in Madhya Pradesh	Maheshwari and Dwivedi (1990)
5.	Ethnobotanical studies of Madhya Pradesh Prajapati and Khare (2004)	Sahu (1982), Gupta (1999),
6.	Ethnobotanical studies of sacred groves at Ajwali, Ramanujam Pune district *et al.* (2002)	Sadhab, *et al.*, (1991),
7.	Ethnobotany and Medicinal plants of North-East India	Singh *et al.* (2004)
8.	Ethnobotany in Bargarh district in Orissa	Sen, *et al.*, (2001)
9.	Ethnobotany of Bastar	Sahu and Sahu (2002)
10.	Ethnobotany of Eastern Ghats	Prasad *et al.* (2002)
11.	Ethnobotany of Gandhmardan hills	Braham and Saxena (1990)
12.	Ethnobotany of Golapara, Assam	—
13.	Ethnobotany of Kadars, Malasara, Muthervars,	—
14.	Ethnobotany of Mahendragarh	—
15.	Ethnobotany of Saurashtra	Shah, *et al.*, (1981), Bhattacharya (1996), Jadeja, *et al.*, (2005)
16.	Ethnobotany of Western Maharashtra	Tosh (1996), Kulkarni and Umbhojkar (2002)
17.	Ethnogynaecological plants used by Irulars of Siruvani hills, Western Ghat	Kathikayani (2003)
18.	Ethnomedical plant lore from Mukundara ranges, Jhalawar distt.	Sharma, *et al.*, (1989)
19.	Ethnomedicinal plants of Shatrunjaya hills of Palitana	Bhatt and Mitaliya (1999)
20.	Ethnomedicinal plants of Victoria Reserve Park	Bhatt, Mehta and Mitaliya (1999) Forest, Bhavnagar
21.	Ethnomedicines of Amarkantak Plateau	Brijlal and Dubey (1966)
22.	Ethno-medico Religious Plants of Hadoti Plateau	Sharma (2002)
23.	Ethno-phyto-toxicological studies in Sidhi district.	Saini (2004)
24.	Ethnoveterinary Medicinal plants in North-West Udaipur	Deora, *et al.*, (2004) Sharma and Trivedi (2004)
25.	Folk Medicines of Mahendragarh district, Haryana	—
26.	Folklore medicine from Assam and Meghalaya	Majumdar *et al.* (1978)
27.	Herbal treatment of children among tribals of Shadol district	Khan, *et al.*, (2005)
28.	Herbal wealth of Mizoram	—
29.	Medicinal and Aromatic Plants of Chamba Forest Division	Gupta (1964)
30.	Medicinal Flora of Guru Nanak Dev University, Amritsar	Dhiman and Khanna (2001), Kaul (2004)
31.	Medicinal incense and edible Gymnosperms, Pteridophytes and Mushrooms of East Sikkim	Singh, (2004)
32.	Medicinal plants of Bhavnagar	Bhatt *et al.* (2001)
33.	Medico-botanical Profile of Western Maharashtra	—

Contd.

Table 2: (Contd)

S. No.	Place	Author(s) and Years
34.	Noteworthy medicinal plants of Vallendu block	Natarajan and Manickam (2003)
35.	Religious trees of Kutchh and their significance	Joshi *et al.* (2000)
36.	Sacred plants of Gujarat	Bhatt *et al.* (2001)
37.	Sacred plants used for fumigation purpose in Gujarat	Mitaliya *et al.* (2003)
38.	Some medicinal plants used by Rabha tribe in Golapara district, ssam	—
39.	Some wild edible Mushrooms of Siang Valley (Arunachal Pradesh)	Mishra (2003)
40.	Traditional medicines of Brahamputra Valley, Assam	Kalita and Bakshi (2004)
41.	Traditional phytotherapy for ophthalmic diseases of Saurashtra	Jadeja, *et al.*, (2005)
42.	Traditional plants of Karaikal ranges (Pondicherry)	Nadanakunjidam and Kamakshi (2003)
43.	Trees of Sikkim	—
44.	Useful Plants of Mayurbhanj State	Bal (1942)

Table 3: Parts of plants used for medicine

S.No.	Plant Parts	Percentage used
1.	Whole plant	16%
2.	Roots	30%
3.	Rhizome	4%
4.	Stem	6%
5.	Bark	14%
6.	Wood	3%
7.	Leaves	6%
8.	Flowers	10%
9.	Fruits	7%
10.	Seeds	4%

(*After* Jakhar *et al.,* 2004)

Table 4: Family-wise distribution of medicinal plants

S.No.	Families	Medicinal plants
1.	Asteraceae	419
2.	Euphorbiaceae	214
3.	Lamiaceae	214
4.	Fabaceae	214
5.	Rubiaceae	208
6.	Poaceae	168
7.	Acanthaceae	141
8.	Rosaceae	129
9.	Apiaceae	118

(*After* Rawat and Uniyal, 2003)

Table 5: Major Indian contributions on medicinal plants

S.No.	Publications	Author(s) and Years
1.	A Dictionary of Economic Products of India	Bhatt (1889-1893)
2.	A Handbook of Ethnobotany	Jain and Mudgal (1999)
3.	A Handbook of medicirial plants	Prajapati and Kumar (2003)
4.	A Manual of Medicinal trees	—
5.	A Textbook of Pharmacognosy	Shah and Qadry (1989)
6.	Adivasi Aushadh (Homeopathy) Parts 1-6	—
7.	Dictionary of Medicinal Plants	Prajapati and Kumar (2003)
8.	Anti-carcinogenic effect of some plant products	Aruna and Shivaramakrishnan (1992)
9.	Antistress activity of *Ocimum sanctum*	—
10.	Astanga Hridaya Samhita	—
11.	Astanga Sangrah	Chhangani (1945)
12.	Bhavprakash	—
13.	Bhavprakash Nighantu	—
14.	Bio-energy	Elamathi, *et al.*, (2005)
15.	Biovedic cultivation of medicinal plants in India	—
16.	Black Pepper	Gopinathan (2004)
17.	Botany of Homeopathic medicine	—
18.	Brihad Samhita	—
19.	Charak Samhita	—
20.	Chemistry of cultivated medicinal plants	—
21.	Commercial cultivation of medicinal crops	Upadhyaya, Kumar and Srivastava (2003)
22.	Commercially important medicinal plants of India	Chadha, Gupta and Nagarajan (1980)
23.	Compendium of Indian medicinal plants	Rastogi and Mehrotra (1993)
24.	Dictionary of Indian folk medicines and ethnobotany	Jain (1991)
25.	Drugs from plant resources: An overview	Rao and Gurjar (1990), Sivarajan and Balachandran (1994)
26.	Economic medicinal plants of Indian desert	Singh and Pandey (1998)
27.	Effect of Liv. 52 against radiation sickness	Saini *et al.* (1984)
28.	Endangered species of medicinal herbs in India	Jain (1987)
29.	Ethnobotany of India: A status report	Pushpangadan (1994)
30.	Ethnomedicinal plants	Trivedi and Sharma (2004)
31.	Ethnomedicinal studies of Mahendragin ranges of hills Eastern ghats	Senapati *et al.* (2001)
32.	Ethno-veterinary medicinal plants in Udaipur district	Deora *et al.*,
33.	Glimpses of Indian Ethnobotany	Jain (1991)
34.	Glossary of Indian medicinal plants	Chopra, (1956)
35.	Glossary of Indian medicinal plants with active principles	—
36.	Handbook of medicinal plants	Bhattacharya (2004)
37.	Homeopathic medicinal plants	Baburaj (2003)
38.	Homeopathic medicinal plants of India	Baburaj *et al.* (1995)
39.	Indian Materia Medica	Nadkarni and Nadkarni (1954) Nadkarni (1926)

Contd.

Table 5: (Contd.)

S.No.	Publications	Author(s) and Years
40.	Indian Medicinal Plants	Kirtikar and Basu (1933)
41.	Kalpriksha	Panarea (1996)
42.	Medicinal Plant Research in India	Mitra and Jain (1991)
43.	Medicinal Plants	Jain (1991, 1995)
44.	Medicinal Plants	Bhattacharya and De (2005)
45.	Medicinal Plants in India and their conservation	—
46.	Medicinal Plants of Arunachal Pradesh	Haridasan, Shukla and Beniwal (1995)
47.	Medicinal Plants of Chhakinal watershed in the north-western himalayas	Dobriyal *et al.* (1996)
48.	Medicinal Plants of India and Pakistan	Dastur (1952)
49.	Medicinal Pteridophytes: An overview	Dixit and Singh (2004)
50.	Modulators of Radiosensitivity	—
51.	Neem-A Boon	Vanathi and Rathika Das (2004)
52.	Neem for disease management	Venkatesan and Shannlingam (2004)
53.	New vistas in ethnobotany in South Asia	Maheshwari (1996)
54.	Phytopharmaceutical Industry	Aruna Mathur and Ramamurti (2003)
55.	Plants their medicinal parts, chemical ingredients and uses	Bhattacharya and De (2005)

Role of NCP for Plant Biodiversity

To evolve a proper and precise methodology for conserving the ecology, national biodiversity strategy and action plan (NBSAP) is required to cover areas like national ecosystems, wild species and their relatives, agriculture ecosystems and domestic species. Issues such as conservation, sustainable use, ethical practices and cultural, scientific and economic aspects of biodiversity assume importance herein involving villagers, NGOs, ethnic groups, and government officials to make Joint Forest Management programmes aimed at people's participation. The role of biodiversity need not be overstressed for the life on Earth still remains largely unexplored at the species and intraspecies level, diversity varies in different ecosystems, geographical areas and also among different taxonomic groups, human-beings have already obtained enormous economic benefits from diversity and Germplasm centres of biodiversity which are required for the upliftment of country's economy and betterment. Indeed with a mosaic of geoclimatic conditions India possesses a rich diversity of plants, animals and microbes, as also of the ecosystems, rated as one amongst the world's megadiversity centres.

Conservation Strategies

Since 95% of the medicinal plants are harvested from the wild, to revive these renewable natural resources both *in-situ* and *ex-situ* conservation strategies are required. The former aims at conserving plants in their natural habitats, for that we have 140 Botanical Gardens, Biosphere Reserves, Sanctuaries and National Parks whereas the latter does so through the Gene bank, Seed bank, Pollen bank, DNA libraries, Plant tissue culture repositories and Cryopreservation (Deb, 1975; Malik, 2004).

Following eight Government departments and 40 NAGS (National Active Germplasm Structures) deal with strategies to conserve PGR: DAC (Department of Agriculture and co-operation); DARE (Department of Agriculture Research and Education); DOEF (Department of Environment and Forestry); DOC (Department of Commerce); MHFW (Ministry of Health and Family Welfare); MTX (Ministry of Textiles); CSIR (Council of Scientific and Industrial Research); Department of Biotechnology and NGOs.

Sustainable Management and Applied Sciences

With the aim of sustainability, ICAR society in its 74th General Body Meeting claimed to have attained following measures towards making India a developed nation through sustainability efforts. Over 2.5 lakh germplasm stored in the National Gene Bank, for posterity, Nutrient deficiencies in the feed and fodder mapped for the whole of the country, Soil erosion maps in different studies printed, Soil map of India on 1:1 million scale prepared and published, Perfected resource conserving technologies adopted, Watershed management technology developed, Integrated plant nutrient system evolved, Shifting cultivation banned, Large capacity, evaporatively cooled storage structures developed for enhancing the shelf-life of fruits, vegetables and efforts are on for herbal products (10) Tractor mounted harvester for turmeric developed and about 6 lakh farmers, rural youths and extension functionaries trained by the KVKS.

Distributional Pattern of Medicinal Plants

From a perusal of Tables 1-6, it is amply clear that life-forms ranging from herbs, shrubs, climbers, lianas to trees have medicinal properties which are distributed to almost all the categories of the plants viz.Herbs (28%), Herbaceous climbers (7%), Lianas (3%), Shrubs (21%), Woody climbers (5%), and Trees (36%).

A macroanalysis of the distribution of medicinal plants of our country shows that these are distributed in diverse ecosystems and landscape elements—Western and Eastern Ghats, the Vindhyas, Chhota Nagpur Plateau, Aravallis and Himalayas (with dry and moist deciduous vegetation of north-eastern) are ideal hotspots for such plants in India. Of the 386 families and 2200 genera of medicinal plants recorded in the country, the familywise enumeration has been given in Table 3.

Since India has centres of origin and diversity of more than 20 major agrobotanical crops and over 8000 species are of ethnobotanical interest (Mukerjee, 2004), Bioprospecting, therefore, is of immense value of the total angiospermic plants, utilitywise distribution of plants is as Medicinal (7500), Edible (3900), Culturally important (700),Used for fibre (525), Fodder plants (400), Insecticidal and Pesticidal (300), Gums, Resins and Dye plants (300) and Incense and perfume plants (100).

Soldiers deployed at high altitudes generally suffer from *acute mountain sickness* (headache, lassitude, anorexia, drowsiness, weakness, insomnia, palpitation and muscular pain, etc.), High altitude *Pulmonary oedema* (breathlessness, cough, pink sputum and cyanosis), *Chronic mountain sickness* (clubbing of fingers and right ventricular hypertrophy), *High altitude cerebral oedema* (ataxia, irritability, hallucinations, coma, cranial nerves palsies and tremors) and *High altitude Retinopathy* (vasoconstriction of retinal vessels and head haemorrhage). Under the above conditions, high altitudes plants are used either as tonics for soldiers or to treat various categories of illness.

Table 6: Herbal cure of major diseases

S.No.	Attributes	Author(s) and Years
1.	Antifertility drugs and agents	Jain, *et al.*, (2005)
2.	Contraceptives	Sharma, Chopra, Bhushan and Munjal (2003)
3.	Ethnogynaecological plant	Kartikayani (2003)
4.	Healthcare during pregnancy and post-pregnancy	Nargas and Trivedi (2004)
5.	Jaundice	Bhatt, *et al.* (2001) Singh, *et al.* (2004)
6.	Preventive herbals against malaria	Bhatt, *et al.* (2001) Mukherjee (1991)
7.	Herbal treatment of diabetes	Bhatt, *et al.* (2000) Balachandran, *et al.* (2004)
8.	Dental protectors	Gupta and Mishra (2000)
9.	Phytotherapy of opthalmic diseases	Jadeja, *et al.* (2005)
10.	Insecticidal plants	Roy and Saraf (2005)
11.	Cervical spondylosis	—
12.	Anticancer	—
13.	Pedalediema (Leg swelling, bone fracture)	Karthikayani (2003)
14.	Abodminal ulcer, mouth and stomach ulcers	Karthikayani (2003)
15.	Leucorrhoea *(Gymnema sylvestris)*	Karthikayani (2003)
16.	Rheumatism *(Grewia hirsuta)*	Gopal (2004)
17.	Menstrual troubles *(Polygonum glabrus)*	Gopal (2004)
18.	Leucoderma	Biswas (1956), Misra (2004)
19.	Rheumatism	Hemadri (1981), Misra (2004)
20.	Infectious diseases	Saxena and Vyas (1981)
21.	Influenza	Menon (1919), Misra (2004)
22.	Asthma and respiratory diseases	Jha (2001), Tosh (2004)
23.	Venereal, gonorrhoea, syphilis and aphro-diseases	Misra (2004)
24.	Natural healers	Parabia and Reddy (2008)
25.	Dysurea	Sen *et al.* (2001)
26.	Piles	—
27.	Miscellaneous phytotherapy	Shukla *et al.* (2001), Bhatt and Mitaliya (2000)

Table 7: List of some plants used against diseases

S. No.	Diseases	Species
1.	Abscesses	*Sida rhombifolia; Cucumis trigonus*
2.	*Amoebiasis*	*Syzygium cuminii*
3.	Anemia	*Olax' scandens*
4.	Anorexia	*Crataeva nurvala* (bark), *Piper nigrum*
5.	Antibacterial against *Pseudomonas aeruginosa*	*Citrus aurantifolia, Syzygium cuminii, Musa pardisiaca, Emblica officinalis*
6.	Anticancer	*Taxus* sp., *Heliotropium indicum, Withania somnifera*
7.	Antifertilityin woman	*Hyptis suaveolens*
8.	Antifungal activity	*Argyria speciosa*
9.	Antiguinea worm	*Crotalaria juncea, Commiphora wightii*
10.	Antiseptic to Cattle worm	*Emblica officinalis, Ipomoea carnea*
11.	Aphrodisiacs (enhance sexual potential)	*Asparagus racemosus, Abrus precatorius, Aloe barbadensis, Bombax ceiba, Digera muricata,*
12.	Arthritis	*Withania somnifera*
13.	As Antidotes of Scorpion and Snake bite	*Aristolochia indica, Allium cepa, Berberis asiatica, Cyperus rotundus*
14.	Asthma	*Solanum surattense, Calotropis* procera, *Adhatoda vasica*
15.	Bachache	*Tylophora indica*
16.	Blood in urine	*Euphorbia thymifolia*
17.	Blood purifier	*Imperata cylindrica, Hibiscus mutabilis*
18.	Body Pain and Rheumatism	*Achyranthes aspera, Alpinia speciosa*
19.	Body swelling	*Arylosia* sp., *Plumbago zeylanicum*
20.	Bone Fracture	*Terminalia arjuna*
21.	Bronchitis	*Acacia nilotica*
22.	Cancer	*Heliotropium indicum, Withania somnifera, Hyptis suaveolens, Gloriosa superba*
23.	Cardiac stimulant	*Terminalia arjuna*
24.	Catarrh	*Hyptis suaveolens*
25.	Cervical spondylosis	*Actaea spicata*
26.	Chicken pox	*Ficus religiosa*
27.	Cholera	*Calotropis procera*
28.	Constipation	*Murraya koenigii, Piper betle, Cassia angustifolia*
29.	Contusions	*Schefflera venulosa*
30.	Cuts and Burns	*Tridax procumbens, Beta vulgaris, Ziziphus jujuba*
31.	Dental ailments	*Madhuca indica, Barleria pronitis, Salvadora oleoides*
32.	Diabetes	*Acacia nilotica, Adiantum caudatum, Dillenia indica*
33.	Diarrhoea	*Anthocephalus cadamba, Ficus benghalensis*
34.	Dropsy	*Mucuna pruriens, Baliospermum montanun*
35.	Dysentery	*Helicteres isora*

Contd.

Table 7: Contd.

S. No.	Diseases	Species
36.	Earache	*Cymbidium aloifolium*
37.	Eczema (bleeding)	*Solanum surattense*
38.	Eczema (dry)	*Smilax lanceifolia*
39.	Elephantiasis	*Hybanthus enneaspermus*
40.	Epilepsy	*Benincasa hispida*
41.	Eye problems	*Vitex nugendo, Albizzia lebbek, Desmodium gangeticum*
42.	Fever and gout	*Holarrhena antidysenterica*
43.	Filaria	*Caesalpinia decapetata*
44.	Nervous system	*Cannabis satwa, Datura metel, Celastrus paniculatus*
45.	Fractured bones	*Cylista scariosa, Woodfordia fruticosa*
46.	Gastric trouble and stomachache	*Chrysanthemum americanum*
47.	Giddiness	*Zingiber rubens, Viscum orientale*
48.	Goitre	*Butea monosperma*
49.	Gout	*Solanum stramonium, Piper nigrum*
50.	Hair growth	*Evolvulus alsinoides*
51.	Headache	*Artocarpus heterophyllus*
52.	Helminthes	*Chenopodium album*
53.	Hydrocele	*Dioscorea sp., Mimosa pudica*
54.	Impotency	*Anellema scapiflorum*
55.	In livestock diseases	*Acacia polycantha. Cassia tora, Cleome brachycarpa, Cassia fistula*
56.	Inflammation	*Breynia retusa*
57.	Insect repellant	*Pongamia pinnata*
58.	Insomnia	*Biophytum reinwardtii*
59.	Jaundice	*Argemone mexicana, Boerhaavia diffusa, Thespesia sp., Cuscuta reflexa*
60.	Leprosy	*Albizia odoratissima, Anagallis arvensis*
61.	Leucoderma	*Argemone mexicana*
62.	Lung and throat ailments	*Acalypha indica, Boerhaavia diffusa, Adiantum capillus-veneris*
63.	Malaria	*Cleome gynandra, Ricinus communis, Cinchona sp., Melia, azadirach*
64.	Mastitis	*Zingiber officinale* (rhizome), *Tragia involucrata*
65.	Migraine	*Ocimum canum, Clerodendrum sp., Withania somnifera*
66.	Mouth and teeth diseases	*Barleria prionitis, Dalbergia volubilis*
67.	Mumps	*Tragia involucrata*
68.	Muscular pain	*Clausena excavata.*
69.	Nausea	*Glochidion lanceolarium*
70.	Nose bleeding	*Phyllanthus acidus*
71.	Painful urination	*Cochlospermum religiosum*
72.	Paralysis	*Terminalia chebula, Butea monosperma*
73.	Peptic ulcer	*Achyranthes aspera*
74.	Piles	*Citrus limon, Capparis decidua, Ficus glomerata, Abutilon indicum*
75.	Pimples	*Bombax ceiba*

Contd.

Table 7: Contd.

S. No.	Diseases	Plants
76.	Pinworm	*Benincasa hispida*
77.	Pregnancy	*Butea superba*
78.	Prevention of conception	*Ficus religiosa*
79.	Pyorrhea	*Mimusops elengi Averrhoea* sp.
80.	Rheumatism	*Grewia* sp.
81.	Ringworm	*Ocimum canum*
82.	Scabies	*Argemone mexicana*
83.	Scurf fever	*Antidesma ghaesembilia*
84.	Skin Diseases	*Ageratum conyzoides, Cassia fistula, Adiantum caudatum*
85.	Snakebite	*Scilla hyacinthiana*
86.	Spleen inflanmation	*Woodfordia floribunda, Gloriosa superba*
87.	Sprains of cattle	*Erycibe paniculata*
88.	Sterility in women	*Cuscuta reflexa*
89.	Stomach problem	*Butea superba*
90.	Tapeworm	*Carica papaya*
91.	Threadworm, filaria, colic pain	*Andrographis paniculata*
92.	Tonic	*Bauhinia vahlii, Curcuma pseudomontana*
93.	Toothache	*Tamarindus indica*
94.	Tuberculosis	*Actiniopteris radiata, Barleria prionitis*
95.	Typhoid	*Portulaca castrim*
96.	Urinogenital diseases	*Bixa orellana, Cocculus hirsutum*
97.	Uterine infection menorrhagia	*Saraca asoca*
98.	Veneral diseases	*Acacia catechu (gonorrhoea), Abelmoschus esculentus (syphilis), Acacia nilotica*
99.	Vermicidal	*Aegle marmelos, Ficus benghalensis*
100.	Vomitting, diarrhoea and liver disorders of cattle	*Abutilon indicum, Abrus precatorius*
101.	Whooping cough	*Achyranthes aspera*
102.	Wound of cattle	*Erythrina ruberosa*

Medicinal, Dietary and Cosmetic Properties of Some Selected Plants

Manjishta: Rubia cordifolia

The roots of this plant are valued in the Ayurveda for the maintenance of general health and normal blood circulation.

The other important plants of this category are: *Cinchona pubescens* (used against malaria), *Acorus calamus* (brain tonic), *Andrographis paniculata* (hepatoprotectant, anthelmentic, blood purifier), *Oroxylum indicum* (fever), *Rauwolfia serpentina* (treatment of blood pressure), *Tinospora cordifolia* (vitaliser), *Coptis trifolia* (antipyretic, jaundice), *Koelpinia linearis* (for lymphatic diseases, gastrointestinal disorders, bone fracture, inflammation), *Juniperus macropoda* (kidney diseases, joint pains), *Codonopsis clematidea* (arthritis, anti-infectious), *Bistorta affinis* (lung disorders), *Swertia thomsonii* (bile disorder), *Taraxacum* (vomitting of blood, stomach ulcers), *Pedicularis longiflora* (alcohol problems) and *Sida cordifolia* (antibacterial and antifungal).

Ghrita-Kumari: (*Aloe barbadensis*)

This species is used as a dietary supplement and as a vital ingredient in beauty products. The gel of this plant is very potent with 200 components and 74 known nutrients including vitamin B-12, 21 amino acids, iron, manganese, calcium, zinc, minerals and enzymes.

Aloe barbadensis products are also used as health drinks and antacids. The medicinal properties of this gel are useful in arthritis, diabetes and have antibacterial, antifungal and antioxidant and other related properties.

The calcium isocitrate salts present in the species prevent hardening of arteries and improve blood circulation. It decreases the bad cholesterol and increases the good cholesterol and the presence of potassium in it maintains the heart rhythm and stabilizes the blood pressure.

Further the vitamins antioxidants like A, C and E present in this plant neutralize the free radicals and thereby reduce heart diseases. The prescribed dose is 60 ml every morning on an empty stomach.

The beauty industry swears by the cosmetic properties of *Aloe barbadensis*. It increases (1) Cell regeneration (2) Healing and boosting of the natural immune system (3) It is a very powerful moisturiser and cleanser. These days wound healing creams with it gel have flooded the national and international market.

Antidiabetic Plant: *Stevia rebaudiana*

Stevia is a shrub whose leaves are 20 times sweeter than sugarcane. It is free of carbohydrates and fat. It does not affect blood sugar level and also does not taste bitter. It is ideal for the patients of diabetes. Its market within the next 5 years is going to rise upto 1,000 crore rupees. Shri Rajaram Tripathi, President of Central Herbal Agro Marketing Federation of India and an organic farmer has developed MDS-13 and MDS-14 as new varieties which can tolerate temperatures between 4-15°C. "Grow More", biotech has also developed a few varieties of *Stevia*. It is available in the market as leaf paste, powder and herbal tea.

A Petro Plant: *Jatropha curcas*

Jatropha is a pollution free bio-diesel producing petrocrop with several medicinal and industrial uses. Plant extract is used in folk medicines for the remedies of cancer, alopaecia, burns, cough, dermatitis, purgative, diuretic, vermifuge, dysentery, dyspepsia, eczema, fever, gonorrhea, inflammation, jaundice, paralysis, scabies, rheumatis, sores, stomachache, syphilis, ulcers and uteroses. Besides, latex of the plant can be applied topically to bee and wasp stings. Decoction of root is used against dysentery. Leaves are antiparasitic, and applied to scabies. Juice of the leaves is used externally for piles. Fruits are anthelmintic and useful in anaemia, fistula and heart diseases. Oil of the plant is used for cleaning wounds and sores. Curcin is its active principle. Oil contains oleic and linoleic acid whereas the bark contains tannins, wax, resins, saponins and reducing sugars.

The other uses of *Jatropha* include as a substitute for oil in engines, preparation of soaps and candles. Latex of the plant inhibits watermelon mosaic virus. Bark is a fish poison. Seeds are used as a contraceptive and insecticide. Plant prevents soil erosion and can be grown on wastelands. Uttarakhand and Chhattisgarh Governments have taken the initiative to raise the crop commercially. The advantage of biodiesel obtaired from this plant is that it can be used as a substitute for oil engines without polluting the atmosphere and causing global warming (Somasundaram and Manjunath, 2004).

Guggul: Mahaushadhi of Ayurveda *(Commiphora wightii)*

The gum oleoresin extracted from *Commiphora wightii* is well-known for its anti-rheumatic, anticholesterolaemic and anti-inflammatory action. It is widely distributed in Gujarat, Rajasthan and parts of Madhya Pradesh, Maharashtra and Karnataka. Due to over-exploitation, it is facing the danger of becoming extract. Its propagation, is now being done, in Guggulu Herbal farm at Mangliawas (Ajmer district) and Central Council for Research in Ayurveda and Siddha (Department of ISM and H Government of India).

Versatile Neem (Azadirachta indica)

Almost every part of neem is bitter. The plant is useful as timber, fuelwood and fodder. Neem oil (Margosa) is used as a lubricant. It is also used in soaps and cosmetics. De-oiled neem cake has multiple uses in agriculture, poultry and animal husbandry. Neem can reclaim wasteland and can utilize industrial effluent water. Limnoids like Azadirachtin-A, Selanin, Nimbin and Melinatriol are effective against insects as biopesticides. Azadirachtin is a potent locust anti-feedant; Nimbin is effective against fungi like *Tinea rubra* and *Coccidiodes immites* whereas Selanin has been reported to deter feeding of striped cucumber, beetle, houseflies and Japanese beetle (Das, 2004). Neem products are effective against many diseases like Tungro virus; Sheath blight *(Acrocylindrium oryzae);* Blast *(Pyricularia oryzae);* Sheath blight *(Rhizoctonia solani)* of Paddy; Rust *(Puccinia arachidis)* and rot *(Sclerotium rolfsii)* of groundnut; Ganoderma wilt *(G. lucidum)* of coconut; Powdery mildew *(Erysiphae polyconi);* Root rot *(Macrophomina phaseolina)* of Black gram and wilt of betelvine *(Phytophthora parasitis).* Dixit *et al.* (1989) found that Neem is useful in ulcers, fertility regulation, diabetes, antherosclerosis and human fungal pathogens. Neem wood is resistant to termites (Vanathi and Rathika, 2004). The structural activity and relationship of the chemicals obtained from Neem and allied species like China Berry tree *Melia azadirach* have been suggested as potential pesticide (Nag *et al.*, 2004). Neem is, therefore, good enough for biodynamic farming.

Safed Musli (*Chlorophytum borivilianum*)

Safed Musli is an endangered species which is distributed in southern Rajasthan, western Madhya Pradesh and north Gujarat. The plant has high aphrodisiac values and acts on Central Nervous System. Its dry roots are known to remove impotency. Chemical composition of the plant includes steroids, resins, tannins, carbohydrates, calcium, magnesium and potassium. Tubers contain steroid sapoginin and are fat free. It is useful in Pitta and Vata. It is also effective for removing fatigue and purifying blood. Its beneficial effect is also observed in diseases like renal calculus, leucorrhoea and diabetes.

Turmeric: An Important Domestic Medicinal Herb (*Curcuma domestica*)

It has great application in traditional system of medicines, for treatment of stomache, dyspepsia, as a complexion promoter, eliminator of body waste products, treatment of poisoning, antiseptic, anti-ulcer, anti-parasitic, wound healing etc. The medicinal properties of turmeric as an antioxidant, free radical scavenging, anti-gastric ulcer, hepatoprotective, hepatoglycemic, anti-lithogenic, hypo-cholesterimic, hypolipidemic, anti-asthmatic, anti-allergic, anti-microbial including anti-HIV are well known. Besides cucuminoids also exhibit supression of nuclear factor kapp'a beta (NF-KB), this blockage of estrogen-mimicking chemicals that promote cancer growth, suppression of inflammation inherent to certain forms of cancer, destroying abnormal cells that can become cancerous, stopping cancer cells from multiplying, enhancing immunity and inhibiting angiogenesis. Curcumin of turmeric even protects body from cervical cancer caused by Human Papilloma virus by disallowing its expression (Kounteya Sinha, TNN. 25th October, 2005), According to the Director of the Institute of Cytology and Preventive Oncology (Das, 2004), every fifth women suffering from cervical cancer in the world is Indian.

Black pepper: perennial woody climber (Piper nigrum)

Gopinathan (2004) mentioned fourty five varieties of the plant which are cultivated in 1.8 lakh hectares. Kerala is the main pepper growing state of India. Pepper is known to be an effective cure for dyspepsia, malaria and hemorrhoids. It has also been used as an aphrodisiac and as a insect repellent. Dry pepper is effective against asthma, cough, heart diseases and pains. It is also advised in the treatment of diabetes and piles. The use of pepper in homeopathy is for treating malaria, nausea, vomiting and inflammation of penis with burning pain. Coconut oil with pepper soothes itching of the body. Pepper with milk is good for dandruff. Fruit of pepper is used in vertigo and coma. Volatile oils containing alkaloids and piperin are active principles.

Ashwagandha (Withania somnifera)

The plant is wild in western and north-west, India. Its bark is used to cure migraine and also to increase masculine power and seminal strength. It relieves asthma and cough. Fine powder of the stem with mustard oil is applied locally on genitals to increase potency and masculine strength (Kumar, 2004). It is anti-cancerous and is also used in arthritis and constipation. As a "Rasayana", the plant promotes health and longevity by increasing disease resistance, arresting ageing and revitalizing body. It improves memory and has significant anxiolytic and antidepressant effects. Roots are anti-inflammatory, anti-cancerous and antioxidant.

Gudbel (Tinospora cordifolia)

It is obtained from *Tinospora cordifolia,* a climber. It is used in constipation; diabetes, jaundice, urinary troubles, pimples and weakness (Kumar, 2004).

Peppermint (Mentha piperita)

Peppermint is aromatic, stimulant, carminative and used for allaying nausea, flatulence and vomiting (Wealth of India, 1962). It contains eugenol, caffeic acid, rosmarinic acid and a tocopherol. The plant has radio-protective properties.

Ghar ki devi : Tulsi (Ocimum tenuiflorum)

It is an important religious and medicinal plant belonging to family Lamiaceae. Its leaves are boiled with tea and also consumed orally for relief from cold and cough. It has anti-stress properties. Flavanoids, orientin and vicenin have been reported to have radio-protective effects. Extract of it serves as an antioxidant when taken with tea.

Amla (Emblica officinalis)

Decoction of amla is used for stomach disorders. It prevents hair-fall. It is an antioxidant, rich in Vitamin C.

Bhuiaonla (Phyllanthus amarus)

It is effective against hepatitis. Its grinded plants are used for different liver disorders.

Sarpagandha (Rauwolfia serpentina)

It grows in waste lands and in shady forests from Punjab to Sikkim, Bhutan, Nepal and Western Ghats. The roots of the plant are very useful. Root decoction is used for uterine contraction and for expulsion of foetus, to treat intestinal disorders, anti-helminthic, and induction of hypotension. It is also used in psychosis and schizophrenia (Trivedi and Nehra, 2004).

Rosemary (Rosmarinus officinalis)

It is an aromatic perennial herb mainly cultivated for its oil used in perfumes, antiseptic creams, cosmetics and medicine and also in aromatherapy. Leaves are added in tea for flavour (Prabhu, 2005). The plant grows well in loamy soil and can be propagated through cuttings.

Anti-Oxidant Plants

Free radicals are highly reactive chemical species which possess an unpaired electron in the outer shell of the molecules and can react with proteins, lipids, carbohydrates and DNA.They attack the nearest stable molecules, stealing its electron (Dubey *et al*., 2005) leading to many destructive diseases. Free radicals are produced in the body as by-products of normal cellular metabolic activities such as prostaglandin synthesis, mitochonodrial electron transport, endoplasmic reticulum enzyme activity, oxy-haemoglobin, auto-oxidation and phagocytosis, as well as exposure to pollutants and ionizing radiation. A balance is maintained between oxidative attack of free radicals and the anti-oxidants.

The anti-oxidative defence system present in human body may be of two types, viz., enzymatic, e.g. catalse, glutathione peroxidase and reductase, s-transferase, sequential removal of reactivo oxygen species (ROS) and non-enzymatics, e.g. metal binding proteins and low molecular weight non-enzymatic antioxidants like cytochrome-C, carotenoids, ascorbic acid, riboflavin, lipoic acid, selanin and zinc, which modulate radiation damage (Jain, 2001). Many plants with Vitamin A, E and C are rich in antioxidants also e.g. *Withania, Ocimum, Mentha, Emblica, Cassia* etc.

National Medicinal Plants Board

To coordinate, administer and give technical advice, NMPB was set up under a government resolution notified on 24th November, 2000. It included: 1. Registration of farmers, cultivators, traders and State Medicinal Plant Boards. 2. General and specialized survey for domesticated and global markets. 3. Simplification of Transit permits. 4. R and D studies in post-harvest management at CSIR, CIMAP, IEFRE, RRLs, DBT, Horticultural and Forest Departments. 5. Constitution of State Medicinal Plants Board. 6. Ensure participation of NGOs, manufacturers and Pharmaceutical industry. 7. Contractual farming to grow prioritized medicinal plants. 8. Proposing Herbal Gardens. 9. Overall development of the sector (Rawat and Uniyal, 2003).

The number of State medicinal plant board is now 25 in 21 States and 4 in Union Territories of India. The following issues have been identified by the NMPB;

1. Policy and legal issues for the development of Ayurveda and medicinal plants.
2. Development of Medicinal Plants Sector.
3. Ayurvedic drugs development and product standardization.
4. Globalization of Ayurvedic and Medicinal Plant Sector.

Market Scenario of Herbal Trade

It is daunting, vast, secretive and largely unregulated trade of medicinal plants, mainly from the wild.

Domestic Market

India has 7,000 plant species for traditional medicines. The annual production of medicinal and aromatic plant raw materials are worth 200 crores. Besides, about 300 natural products are used as raw materials in flavour and fragrance industry in the form of essential oils, extracts, oleoresin, concrete, absolutes, resinoids and tinctures. As per WHO estimate, medicinal plant related trade in India is estimated to be 550 crore per year. According to Mukherjee (2004), India's total turnover is of Rs. 2,300 crore for Ayurvedic and herbal products, major OTC products contribute around Rs. 1,200 crores. Other formulation fetch around 650 crore, and classical Ayurvedic formulations contribute remaining 450 crores. The market of medicinal plants used in the Indian systems in the country has been estimated as follows:

S. No.	Systems	Percentage	Market size (₹ Crores)
1.	Ayurveda	84	3,500
2.	Unani	02	100
3.	Siddha	14	5
4.	Homeopathy	0	600
5.	Tibetan Medicines	-	-
	Total		**4,205**

Global Market

According to WHO, EXIM bank of India, the value of medicinal plant related trade in India is of the order of 5.5 billion dollars and is growing rapidly. Export : India's share in International market is about 0.5%. The quantity (kg) of medicinal plants export in 2001 was 47,477,464 kg whose value in rupees was 3,15,77,40,878. The countries to which India exports crude drugs are USA, Germany, France, Switzerland, UK and Japan who share between them 80% of the total export of crude drugs from India. Principal drugs exported are *Aloe, Alropa, Acorus, Cinchona, Cassia tora, Dioscorea, Digitalis, Ephedra, Plantago, Senna* etc.

Constraints and Future Strategies

Following constraints are associated with the use, development and marketing strategies: Poor agricultural practices, Poor harvesting practices, Poor propagation of herbs, Poor quality control, Lack of cross cultural comparisons of herbs, Similarity and dissimilarity of uses, Lack of research on high yielding varieties, Paucity of pharmacognosity evaluation, Inefficient processing, High energy losses during processing, Standardization of raw and final products, Toxicological studies, Chemistry of raw materials, Lack of analytical procedures to detect residues of underground pollutants and heavy metals, Validation of plants by ethnic tribes, Testing the efficacy of active ingredients, Storage and shelf life of herbs and their products, Difficulties in marketing, Lack of local market, Lack of trained personnel and equipment, Lack of access to latest technology for market information, Safety, stability of the product, Quality control to check losses in impotency, Lack of knowledge of optimal stage of harvest, drying, grinding, processing, packing and transport of herbs and their products.

Intellectual Property Rights

IPR are protected in several ways: Trade secrets, patents, plant breeder's right, farmers right and copyright. Indian Patents Act was passed in 1970 (Sahai, 2003). In 1983, Paris convention established equal protection of Industrial IPR. India joined it in 1998. TRIPS is administered by WTO. Benefits of IPR are many like transfer, exchange, encourage and safe guard new ideas and to encourage investment in R and D. As to the Indian scenario, there are (1) Fast alterations in IPR (2) Indians believe in heritage, tradition and benefits to the society at the cost of individual interest (3) Continued and systematic efforts to survival and (3) Checking of biopiracy. TRR (Traditional Resource Rights) and TPR (Traditional Plant Resources Rights) and their database are equally important (Mashelkar, 2001).

Standards and Quality Control

Government of India has set up pharmacopoeal committees for Ayurveda, Siddha, Unani and Homeopathy systems. PLIM and HPL located at Ghaziabad are providing technical back up to these committees.

I. The Ayurvedic Formulatory of India

It contains 634 formulations involving 500 medicinal plants. Of these, standards for 80 single plant drugs are included in Part-I and 78 plants in IInd Part and 100 in 3rd part of Ayurvedic Pharmacopoea of India.

II. The Central Council of Unani Medicines

The council has published two volumes on single plant drug standardization covering 100 medicinal plants. The Unani Pharmacopoea Committee has published one National formulatory of 441 formulations of unani drugs. A word about. Al-Tib medicines shall not be out of place. According to Padmashree Hakim Saifuddin, an expert of unani medicine (June 3, 2005, Meerut Plus of Times of India) Al-Tib is based on humoral theory, viz., Dam (blood-hot and moist), Balgam (phlegm-cold and moist), Safra (yellow bile) and Sauda (black bile) in the human body unique for every individual. Unani medicines help the body to regain Quwate-Mudabbira to an optimum level and restore humoral balance which helps retain a healthy state of body. Al-Tib says about seven components of human body-Al-Umar Al-Tabiya (health maintenance), Al-Arkan (elements), Al-Mizaj (temperament), Al-Akhlat (human), Al Aza (organs), Al-Arwah (vital spirit), Al-Quwa (powers) and Al-Afal (functions) and six essential causes which directly or indirectly affect our health. These are: Al-Hawa Al-Muhit (atmospheric air), Al-Makul Wal Mashrub (food and drinks), Al-Harkat Walsukun-Albadniya (movements and body response), Al-Harkat Walsukun-Al Nafsanya (mental movements), Naum-Wa-Yaqzah (sleep and wakefulness) and Istifrag Wa Ihtibas (evacuation and retention). Other non-essential factors are profession, habits, sex, temperament and social factors. Al Tib's main emphasis is on diagnosis of a disease through nabz (pulse), baul (urine) and baraz (stool). The system consists of many therapeutics like Ilaj Bil Tadabeer (Regimental therapy), Bil Ghija (Diet therapy), Ilaj Bil (Pharmacotherapy) and rarely surgery. It believes in immunization against disease.

Documentation of Traditional Knowledge

Karnataka Peoples Biodiversity Registers (PBRs) and NGO initiatives like FRLHT (Foundation for the Revitalization of Local Health Traditions, Banglore) SRISTI (Society for Research and Initiatives for Sustainable Technologies, Ahmedabad), BBA (Beej Bachao Andolan, Tehri Garhwal) and BCPP (Biodiversity Conservation and Prioritization Programmes- 56 sites in 7 states) and TKDL (Digital Library of Traditional Knowledge) are some of the initiatives in this direction.

Training of Organic Farming

Since 400 major forest based products are available in Indian markets, forest cover is on the decrease, world wild life funds are available and we already have Indian Botanical Garden, Kolkata; National Botanical Research Institute, Lucknow; Llyod Botanical Garden, Darjeeling; Botanical Garden, FRI, Dehradun; Tropical Botanical Garden, Mysore and IIFM, Bhopal. Right now, 14 state universities conduct M.Sc. in Forestry. Some of the important places for the purpose are : (1) Indira Gandhi National Forest Academy, Dehradun (2) Forest Research Institute, Dehradun (3) Tropical Forest Research Institute, Jabalpur (4) Arid Forest Research Institute, Jodhpur (5) Himalayan Forest Research Institute, Shimla (6) Institute of Forest Productivity, Ranchi (7) Institute of Wood Science and Technology, Bangalore (8) Institute of Forest Genetics, Coimbatore (9) Rain Forest Research Institute, Jorhat (10) Centre for Social Forestry, Allahabad.

Herbal Drug Development in Uttar Pradesh

After the formation of Uttrakhand the forest cover of Uttar Pradesh has been reduced very much. Hence, first priority in both the states especially in UP is to bring forest cover up to a National requirement level of 33% because forests are the repository of wild flora and fauna and of wood and non-wood products especially medicinal and edible plants, which are required very much for the sustainable development.

Ethnobotanical studies conducted in UP, and Uttaranchal states have been presented in Table 7 but they are not commensurate with the size of UP.

Four spear heading units of UP working on Indian probing for healthier future are CDRI, CIMAP, NBRI and ITRC. All net stationed at Lucknow and work under Council of Scientific and Industrial Research. CSIR with its 38 labs are party to the prestigious Global Research Alliance with the objective of applying global knowledge pool for global good through. A brief account of its four unit's infrastructure and activities is given as under :

Central Drug Research Institute

The Institute has research centres at Bangalore, Hyderabad, Pantnagar and Purara. CDRI's strategic areas include Development of new plant drugs/products, identification of new biochemical and molecular targets for the development of target specific bioassays, collaboration with industry, to provide Science & Technology support to the Pharma industry and to offer trained personnel.

CDRI is multidisciplinary institution which inroads into pharmacology, toxicology, pharmaco-kinetics and drug delivery etc. Important landmark discoveries with commercial source from this institution are: Anti-materials e.g. Spermicidal cream (con sap), Memory enhancer *pro-mind* from *Bacopa monniera*, bronchodilator-l-ephedrme. Besides, CDR1 has identified and developed potential drug molecules under collaborative license with industry.

Central Institute for Medicinal and Aromatic Plants

CIMAP is continuously evolving towards excellence by steering research and technology development in medicinal and aromatic plants under the leadership of its Director. Using a number of equipments like Electron microscope, Atomic Absorption Spectrophotometer, Automated DNA sequeucer, Synthesizer, Micro-array system, PCR, Proteomics work lab, Flow cytometer, Bio-reactors and Safety cabinets. CIMAP is extracting phytoceuticals from plants and has about 300 patents for its products. The thrust areas being Agro-Biotech chemical sciences.

National Botanical Research Institute

It was established in 1953 and is maintaining Nature's bounty. The thrust area of the Institute aims at self reliance in food, nutrition and primary health care. Some of the products of the Institute which serve all and sundry comprise Nutraceuticals, Anti-ageing creams, Face packs, Anti-bacterial toothpicks, Anti-cig, After-shave lotions, Herbal gulal and enzyme based antioxidants. Nanobiotechnology has been used and the NBRI has, Traditional Knowledge Digital Library, Kits to check purity of milk, ghee, herbal products, and food adulterants, especially *Argemone* detection kit.

Healthcare awareness and research developments about preventive and curative herbs which are safe and affordable have come out in a big way from agro-based ethnobotany to industry-based Pharma industry both in concept and compass. Herbal medicines and products laying emphasis on mind-body relationship have been successfully and increasingly used to cure Stress-related, Anxiety-related, Psychosomatic, Post-traumatic stress related, Digestion-related, Radiation-related, Toxic-related, Disability-related and Infection-related disorders. Herbal gardens and Spas, the centres of alternative healing on picturesque locale using herbal oils (Abhyanga), fragrant essential oils (Aromatherapy), Seaweed wrap, Shiastu (Acupressure) and Swedish massages and circular bath soothe the nerves and muscles rejuvenating mind, body and soul relationships. 7500 medicinal plants, 3900 edible plants, 100 perfume plants have a domestic market of Indian systems of medicine of the tune of 4,205 crores and International market of Rs. 315,77,40,878. Seabuckthorn, *Aloe*, Mushrooms, Liv.-52, Brahm-rasayana etc. are excellent cosmetic and dietary supplements. Over and unscientific-exploitation of medicinal and other utility plants, denudation of the forest cover and degradation of the ecosystems are the harbingers of a disaster that is already upon us in the form of habitat losses and endangered and vulnerable biodiversity. Therefore, we should treat forests with respect, care, concern and consideration for sustainable progress.

Since to undo the damage caused to nature is difficult, to conserve natural fauna and flora with responsibility and sensitivity is easy for sustainable development and has been discussed therefore. CIS and tissue culture techniques, cryopreservation, forest management etc. should be tried, alongwith the development and strengthening of herbal gardens and medicinal, aromatic and spices boards based on existing constraints. Standardization and quality control of herbal drugs, management of herbal trade and documentation of Traditional Knowledge by establishing Digital Library. Future strategies in this context have been suggested. But for exports of our herbal medicines/preparations like Hamdard's Safi, Dabur's Shilajit, Himalaya Drugs Karek capsules, Zhandu Pharma's Maha Sudarshan Churna, etc. component labelling has been made compulsory from Jan. 2006 by Health Ministry, Government of India.

REFERENCES

Aruna, K & Sivaramkrishnan, V (1992). Anticarcinogenic effects of some plant products. *Food Chem. Toxicol.,* **30:** 935

Awasthi, DD (1957). On new lichens from the Himalaya. *Proc. Ind. Acad. Sci.,* **B45:** 119-139

Baburaj, S (2003). Homeopathic Medicinal Plants: An overview. *Agrobios News Letter* **1:** 43-44

Baburaj, S, John, D, Britto, S & Mathew, GK (1995). Exotic medicinal plants useful in Homeopathy found in Nilgiris District, Tamil Nadu. *CCRH Quart. Bull.* **14:** 5-10

Bagul, RM & Yadav, SS (2003). Antivenomous traditional medicines from Satpuda, East Khandesa. *Plant Archives* **3:** 319-320

Bal, SN (1942). Useful plants of Mayurbhanj state in Orissa. *Rec. bot. Surv. India* **6:** 1-112

Balachandran, DA, Rajsekhar, P & Mullaianathan, L (2004). Antidiabetic properties of Fenugreek and Bittergourd. Plant *Archives* **4:** 235-339

Balaraman, N (1981). Feeds & fodder resources for livestock in Sikkim. *Res. Bull.,* No. 13, ICAR Research Complex for NE Hill region, Shillong

Bhatt, DC & Mitaliya, KD (1999). Ethnomedicinal plants of Victoria Park (Reserve Forest) near Bhavnagar. *Ethnobotany* **11:** 81-84

Bhatt, DC, Mehta, SK & Parmar, RP (2002). Studies on some ethnomedicinal plants from Talaja Taluka of Bhavnagar district, Gujarat, pp. 295-310. *In:* Trivedi, PC (Ed.) *Ethnobotany.* Aavishkar Publishers, Jaipur

Bhatt, DC, Mitaliya, KD, Maheta, DP & Prajapati, MM (2001). Preventive herbal medicine against Malaria. *Ad. Plant Sci.* **14:** 7-10

Bhatt, DC, Mitaliya, KD, Parmar, RP, Lashkari, PI & Dodia, SK (2003). Observations on foliage employed as vegetable by tribals and rurals in Gujarat. *Ad. Plant Sci.* **16:** 413-416

Bhatt. DC, Nurani, MA, Mitaliya, KD & Baxi, US (2001). Ethnobotanical medicinal plants of Barda hills of Saurashtra, Gujrat, *Ad. Plant Sci.,* **14:** 427-431

Bhatt, DC, Mehta, SK & Mitaliya, KD (1999). Ethnomedicinal plants of Shatrunjaya Hill of Palitana. *Ethnobotany* **11:** 22-25

Bhatt, DC, Mitaliya, KD & Mehta, SK (2000). Plant twigs as tooth brushes. *Ad. Plant Sci.* **13(1):** 19-22

Bhattaeharya, SK & De, LC (2005). *Medicinal Herbs and Flowers.* Aavishkar Publishers, Jaipur

Bhattacharya, SK (1996). Medicoethnobotanical value of Saurashtra weeds. *J. Econ. Tax. Bot.* **12:** 166-168

Bhattacharya, SK (2004). *Handbook of Medicinal Plants.* Pointer Publishers, Jaipur

Biswas, K (1956). *Banausadhi* Vols. 1-6. Calcutta Univ. Publication (in Bengali)

Biswas, K (1956). Vegetable drugs in the treatment of Leucoderma. *J. Asiat. Soc.* **22:** 61-63

Brahm, M & Saxena, SD (1990). Ethnobotanical value of Gandhmardan hills, Saurashtra. *J. Econ. Tax. Bot.,* **12:** 166-168

Brij Lal & Dubey (1966). Ethnomedicines of Amarkantak plateau in central India. *Agric. Biol. Res.,* **8:** 29-37

Chadha, YR, Gupta, R & Nagarjan, S (1980). Scientific appraisal of some commercially important medicinal plants of *India. Indian Drugs* and *Pharma Industry* **15(2):** 7

Chhangani, GS (1945). *Astanga Sangrah.* Varanasi: Chowkhamba Sanskrit Series

Chopra, RN, Nayar, SL & Chopra, TC (1956). *Glossary of Indian Medicinal Plants.* CSIR, New Delhi

Das, A (2004). Versatile uses of Neem. *Agrobios* **3:** 19-20

Dastur, JF (1952). *Medicinal Plants of India and Pakistan.* D.B. Taraporewala & Co. Pvt. Ltd., Bombay

Deb, DB (1975). Economic Plants of Tripura state-Edible plants, food grains, vegetables and fruits. *Indian Forester* **101:** 399-405

Deora, GS, Jhala, SGP & Rathore, MS (2004). Ethnoveterinary medicinal plants in north-west part of Udaipur distt, pp. 74-81. *In:* Trivedi, PC & Sharma, NK (eds) *Ethnomedicinal Plants.* Pointer Publishers, Jaipur

Dhiman, AK & Khanna, DR (2001). Notes on Medicinal Flora of Guru Nanak Dev University Campus, Amritsar. *Environ. Conserv. J.* **2:** 45-52

Dixit, RD & Shweta, Singh, (2004). Medicinal Pteridophytes- An Overview, pp. 269-298. *In:* Trivedi, PC (Ed.). *Medicinal Plants.* Aavishkar Publishers, Jaipur,

Dobriyal, RM, Singh, GS, Rao, KS & Saxena, KG (1996). Medicinal plant resources in Chhakinal watershed in the north-western Himalayas. *J. Herbs, Spices and Med. Plants* **5:** 15-27

Dubey, PK, Edwin, E & Sheeja, E (2005). Oxidative Stress. *Plant Archives,* **5(1):** 1-8

Elamathi, S, Abraham, Thomas & Singh, SS (2005). Bioenergy - A new hope for the millennium. *Agrobios News letter,* **3(8):** 15-16

Gopinathan, KM (2004). Black Pepper: The King of spices. *Agriobios* **3:** 55-58

Goyal, PK (2004). Modulation of Radiosensitivity by certain plants and plant products, pp. 226-239 *In:* Trivedi, PC & Sharma, NK (eds) *Ethnomedicinal Plants.* Pointer Publishers, Jaipur

Gupta, AK & Mishra, SK (2000). Folk lore-dental protector plants of Chattisgarh, M.P. India. *Ad. Plant Sci.* **13:** 501-503

Gupta, AK, Mishra, SK & Khan, AA (1999). Ethnobotanical studies on medicinal plants of Amarkantak forest, Shahdol district, M.P. *Ad Plant. Sci.* **12:** 451-456

Gupta, R (1964). Survey record of medicinal and aromatic plants of Chamba Forest Division, Himachal Pradesh. *Indian For* **90:** 454-458

Haridasan, K, Shukla, GP & Beniwal, BS (1995). Medicinal plants of Arunachal Pradesh SFP: Itanagar, SFR1 Information Bulletin No. 6

Harshberger, JW (1885). The purpose of Ethnobotany. *Bot. Gaz.* **31:** 148-158

Islam, M (1984). Certain fibre-yielding plants of north-eastern region. *J. Econ. Taxon. Bot.* **5:** 767-783

Jadeja, BA, Bhatt, DC & Odedra, NK (2005). Plants used in Abscess in the Saurashtra region of Gujarat. *Plant Archives* **5:** 113-117

Jadeja, BA, Odedra, NK & Pethani, SJ (2005). Traditional Phytotherapy for Opthelmic diseases in Saurashtra. *Plant Archives* **5:** 41-48

Jain, SK & Mudgal, V (1999). *A Handbook of Ethnobotany.* Bishan Singh-Mahendra Pal Singh Publishers, Dehra Dun

Jain, SK (1963). Observations on ethnobotany of the tribals of M.P. *Vanyajati,* **11:** 177-183

Jain, SK (1965). Medicinal plant lore of the tribals of Bastar. *Econ. Bot.* **19:** 236-250

Jain, SK (1981). *Glimpses of Indian Ethnobotany.* Oxford and IBH Publishing Co. New Delhi

Jain, SK (1987). Endangered species of medicinal herbs in India. *Medicinal Herbs Indian Life* **16:** 44-53

Jain, SK (1989). *Methods and Approaches in Ethnobotany.* Soc. of Ethnobotanists. Lucknow

Jain, SK (1991a). *Dictionary of Indian Folk Medicine and Ethnobotany.* Deep Publications, New Delhi

Jain, SK (1995). *A Manual of Ethnobotany.* Scientific Publishers, Jodhpur

Jain, SK (1996). *Medicinal Plants.* National Book Trust, New Delhi

Jain, SK (2001). *Bibliography of Indian Ethnobotany,* Scientific Publishers, Jodhpur

Jain, SK (2001b). Ethnobotany in Modern India. *Phytomorphology,* Golden Jubilee Issue pp. 39-54

Jain, Vishal, Sharma, T, Dashone, K, Saraf Swamlata, & Saraf, S (2005). Phytosome: A novel approach for herbal drug delivery system. *Plant Archives,* **5:** 325-330

Janmeda, BS (2004). Ethnobotanical studies in East-Sikkim. Ph.D. Thesis, Choudhary Charan Singh University Meerut, UP, India

Jha, RK (2001). Ethnomedicinal plants used against asthma at Chhotanagpur, Jharkhand (India). *Ad. Plant Sci.* **14:** 95-97

Joshi, Hema & Pandey, PC (2002). Studies on peculiar traditional methods of medicine in Kumaon region of Uttaranchal, pp. 317-321. *In:* Trivedi, PC (Ed.) *Ethnobotany.* Aavishkar Publishers, Jaipur,

Kalita, D & Bakshi, D (2004). Traditional medicines used by the Sonowal Kacharis of Brahmputra valley, Assam. *Plant Archives* **4:** 77-80

Karthikayani, TP (2003). Studies on Ethnogyenaecological plants used by the Irulars of Siruvani hills, Western Ghats. *Plant Archives* **3:** 159-166

Kaul, MK (2004). *Medicinal Plants of Kashmir and Ladakh.* Indus Publishing Co. Tagore Garden, New Delhi

Kavitha, K (2004). *Azolla-A Nature Friend. Agrobios News Letter* **3:** 30-31

Khan, AA, Singh, Pragyan and Rajshree Pandey, (2005). Herbal treatment curing children disease among tribals of Shakdol district, M.P. **5:** 159-163

Kirtikar, KR & Basu, BD (1933). *Indian Medicinal Plants.* Vols **1-4.** Allahabad

Kulkarni, DK & Kumbhojkar, MS (2002). Status and Prospects of Ethnobotany in Western Ghats of Maharashtra, pp. 98-118. *In:* Trivedi, PC (Ed.) *Ethnobotany.* Aavishkar Publishers, Jaipur

Kumar, S (2004). The Economic Plants of North-East India. Scientific Publishers, Jodhpur

Lal, SD & Lata, K (1980). Plants used by the Bhat community for regulating fertility. *Econ. Bot.,* **34:** 273-275

Mahapatra, AK (1983). Essential oil yielding grasses in Thar desert. *J. Econ. Taxon. Bot.* **4(1):** 173-175

Maheshwari, JK & Dwivedi, RP (1990). Notes on the Ethnobotany of the Oraon and Karwa tribes of M.P. *J. Econ. Taxon. Bot.* **14:** 75-90

Maheshwari, JK (1996). New Vistas in Ethnobotany (editorial), pp. 1-11. *In:* Ethnobotany in South Asia. *J. Econ. Taxon. Bot.* Scientific Publishers, Jodhpur

Maheshwari, JK, Singh, KK & Saha, S (1981). *The Ethnobotany of the Tharus of Kheri district in Uttar Pradesh.* N.B.R.I. Lucknow

Majumdar, JK (1978). Some folk-lore medicines from Assam and Meghalaya. *J. Crude Drug Reg.* **16:** 4

Malik, CP (2004). Intellectual Property Rights, Growth and Competitiveness of Indian Pharmaceutical Industries, pp. 27-35. *In: Ethnomedicinal Plants. Aavishkar Publishers,* Jaipur

Mashelkar, RA (2001). Intellectual Property Rights and the Third World. *Curr. Sci.* **81:** 955-965

Mathur, Aruna & Ramamurti, BN (2003). The Growing Phyto-pharmaceutical Industry. *Agrobios News Letter* **1(8):** 25-26

Menon, CK (1919). *Embelia ribes* - A medicine for Influenza. *Indian For* **45:** 210

Mishra, R (2003). Some wild edible Mushrooms of Siang valley: Arunachal Pradesh. *Plant Archives* **3(1):** 81-84

Mishra, KN & Shukla CP (2006). Medico-Religious Plants of Uttar Pradesh, pp 145-148. National Seminar on Env. Deg. Con., Bihar Sharif, Nalanda

Misra, MK (2004). Ethnomedicobotany of Orissa-A Review, pp. 82-115 *In: Ethnomedicinal Plants,* Avishkar Publisher, Jaipur

Mitra, RG & Jain, SK (1991). Medicinal Plants Research in India: *A Review. Ethnobotany,* **3:** 65-77

Mitaliya, KD, Bhatt, DC, Maitreya, BB, Dodia, SK & Parmar, RP (2003). Note on sacred plants used for fumigation purpose in Gujrat. *Plant Archives* **3:** 311-313

Mukherjee, T (1991). Anti-malarial herbal drugs-A review. *Fitoterapia* **62:** 197-204

Mukherjee, T (2004). Medicinal Plants: Need for protection. *In: Medicinal Plants Aavishkar* Publishers, Jaipur pp. 391-404

Nadanakunjidam, S & Kamakshi, G (2003). Traditional Medicinal knowledge of a few plants of Karaikal region, Pondicherry. *Ad. Plant Sci.* **16:** 405-412

Nadkami, KM & Nadkami, AK (1954). *Indian Materia Medica.* Bombay. Dhootpapeshwar Prakashan Ltd. Vols **2**

Nadkarni, KM (1926). *Indian Materia Medica* I and II. Popular Prakashan, Bombay

Nag, Sibrik, Oria, BK & Mukesh, K Raikwar, (2004). Chemistry and structure activity relationship of chemicals from Neem and allied species as potential botanical pesticide. *Everyman's Science* **38(5):** 266-277

Natarajan, K & Manickam, VS (2003). Noteworthy medicinal plants of Vellandu block of Buck Sanctuary in Tamil Nadu. *Ad. Plant Sci.* **16:** 385-387

Negi, S (1977). Fodder tress of Himachal Pradesh. *Ind. For.* **103:** 616-622

Panara, BP (1996). *Kalpvriksha.* Limdo Gola Publishers, Ahmedabad.

Parabia, H & Reddy, MN (2008). Protocol for ethnomedicinal studies, pp. 381-393. *In:* Trivedi, PC (Ed.) *Ethnobotany.* Aavishkar Publishers, Jaipur

Parmar, S (1978). A note on folk musical instruments of Madhya Pradesh. *Folk lore,* pp. 26-27.

Prajapati, ND & Kumar, V (2003). Agros Dictionary of medicinal plants. *Agrobios,* Jodhpur, India.

Prajapati, PL & Khare, PK (2004). Medicinal Plants used by Tribals of Panna district, M.P. *Plant Archives* **4(1):** 113-115.

Prasad, V Krishna, Rajgopal, T & Badrinath, KVS (2002). Ethnobotany of Kondareddis of Rampur Agency, Eastern Ghats, Andhra Pradesh. In: *Ethnobotany* (Ed. Trivedi, PC). Aavishkar Publishers, Jaipur, pp. 205-220

Pushpangadan, P (1994). Ethnobotany of India: A status Report. New Delhi, IICRRP, Project, MOEF, Govt. of India

Ramanujam, MP, Kadamban, D, Kumaravelu, G & Praveen Kumar, K (2002). *Sacred Groves- An Overview,* pp. 13-53. *In: Ethnobotany.* Aavishkar Publishers, Jaipur

Rao, NR & Gurjar, MK (1990). Drugs from plant resources: An overview. *Pharma Times* **22(5):** 19-27

Rao, RR (1994). Biodiversity in India. *Bishen Singh Mahendra Pal Singh, Dehradun*

Rao, RR and Jamir, NS (1982a). Ethnobotanical studies in Nagaland 1. Medicinal Plants. *Econ. Bot.* **36:** 176-181

Rao, RS (1977). Medicobotany of Mysore. *J. Res. Indian Med.* Yoga Homeo. **12:** 53-58

Rastogi, RP & Mehrotra, BN (1993). *Compendium of Indian Medicinal Plants.* Vols. 1-4. CDRI, Lucknow and PID. New Delhi

Rawat, RBS & Uniyal, RC (2003). National Medicinal Plant Board. *Agrobios News Letter* **1(8):** 12-16.

Roy, A & Saraf, S (2005). Insecticidal Plants and Screening Methods: A Review. *Plant Archives* **5(1):** 9-21.

Sahai, S (2003). Indias Plant Variety Protection and Farmer's Right Act, 2001. *Curr. Set.* **84:** 407

Sadhab, A, Majumdar, AM & Pendse, GS (1991). Ethnobotanical studies of sacred groves at Ajwali, Pune distt. *J. Econ. Tax. Bot.* **14:** 257-262

Sahu, TR (1982). Ethnobotanical studies of M.P. *Ancient Sci. Life* **1:** 178-181

Sahu, TR & Sahu, P (2002). Ethnobotanical Scenario of Madhya Pradesh: Knowledge and Culture in Tribal belt of Bastar pp. 73-97. In: Ethnobotany. Aavishkar Publishers, Jaipur

Saini, DC (2004). Ethno-phyto-toxicological studies in Sidhi district of Madhya Pradesh. In: *Ethnomedicinal Plants,* Pointer Publishers, Jaipur, pp. 192

Saini, MR, Kumar, S, Jagetia, GC & Saini, N (1984). Effect of Liv. 52 against radiation sickness and mortality. *Ind. Pract.* **37:** 1133-1138

Saxena, AP & Vyas, KH (1981). Curing Obsterics and gynaecological disorders. Ethnobotanical records on infectious diseases from tribals of Banda distt. *J. Econ. Tax. Bot.* **2:** 191-194

Sen, R, Pal, DC & Guha, A (1983). *Kuchila: A Tribal Dye. Vanyajati* **31:** 3-4

Sen, SK, Pradhan, NB & Behera, LM (2001). Ethnobotany of Bargarh distt. in Orissa. Ethnomedicinal Plants used against disurea at Barhgarh district in Orissa. *Ad. Plant Sa.* **14(2):** 459-462

Shah, GL, Menon, AR & Gopal, BV (1981). An account of the ethnobotany of Saurashtra in Gujrat state. *J. Econ. Taxon. Bot.* **2:** 173-181

Shah, CS & Qadry, JS (1989). *A Textbook of Pharmacognosy.* Shah Prakashan, Ahmedabad

Sharma, SC, Sarabhai, R & Ahmad, SA (1989). Medicinal hydrophytes of Shahjahanpur (U.P.). *Vegetos,* **2:** 235-239

Sharma, HM & Devi, AR (2004). Ethnomedicinal uses of plants in the treatment of urinary tract diseases by the Meiteis of Manipur. *In: Ethnomedicinal Plants,* Pointer Publishers, Jaipur, pp. 151-159

Sharma, NK (2002). Ethno-medico-religious plants of Hadoti plateau (S.E. Rajasthan) *In: Ethnobotany.* Aavishkar Publishers, Jaipur, pp. 394-411

Sharma, Neelu & Trivedi, PC (2004). Medicoethnobotany of Dausa district, Rajasthan. In: *Ethnomedicinal Plants,* pp. 46-73

Sharma, RK, Chopra, G, Bhushan, G & Kanchan, Munjal (2003). Use of Indian plants in fertility regulation in mammals. *J. Exptt. Zool. India* **6:** 57-74

Shukla, KML, Khan, AA, Shabina, & Verma, AK (2001). Wild edible plants of Bilaspur, M.P. *Ad: Plant Sci.* **14:** 57-60

Shukla, CP (2006) Harnessing the Dewghat forest biodiversity for medicinal uses. *Proc. Nat. Acd. Sci.* **76B:** 182-193

Shukla, CP (2008). Floristic Investigation on Ethnomedicianal flora of Dewghat Forest Koraon Range (U.P.) Proc. in National Seminar on Natural Resources held on August 30-31, 2008, Sahu Jain P.G. College Najibabad, Bijnor (U.P.)

Singh, AK (2002). Management of Plant genetic resources: An Indian perspective. *Science Letters* **25:** 74-87

Singh, GS (2002). Minor forest products of Sariska National Park: An Ethnobotanical Profile. pp. 221-238. In: Trivedi, PC (Ed.) *Ethnobotany*. Aavishkar Publishers, Jaipur

Singh, GS, Kuniyal, JC & Vishvakarma, SCR (2004). Agrobiodiversity of cold desert of Lahaul valley: Present Scenario. *Everyman's Science* **38:** 331-334

Singh, KK & Maheshwari, JK (1989). Traditional herbal remedies among the Tharus of Bahraich district, U.P. *Ethnobotany* **1:** 51-56

Singh, KK & Maheshwari, JK (1992). Folk medicinal uses of some plants among the Tharus of Gorakhpur district, U.P. *Ethnobotany* **4:** 39-43

Singh, L, Vats, Preeti, Garg, Shivani, & Ranjana, (2004). Medicinal Incense and edible Gymnosperms, Pteridophytes and Mushrooms of East Sikkim. *Acta Botanica Indica* **32:** 75-80

Singh, Mrigendra Pratap, Khan, AA & Khan, Shabina, (2004). Medicinal Plants of Ethnobotanical importance curing Jaundice from Maikal hills (Amarkantak) Shahdol, M.P. *Ad. Plant Sci.* **17:** 41-44

Singh, V & Singh, P (1982). *Fibre yielding plants of Rajasthan.* CSIR Project Univ. Rajasthan

Singh, V & Pandey, RP (1998). *Ethnobotany of Rajasthan, India.* Scientific Publishers, Jodhpur

Sivarajan, VV & Balachandran, Indira, (1994). *Ayurvedic Drugs and Their Plant Sources.* Oxford and IBH Publishing Co. Pvt. Ltd. New Delhi

Somasundaram, E & Manjunath, Gondi, (2004). *Jatropha curcas*-A pollution-free power tree. *Agrobios* **3:** 43-44

Soni, PL (1995). Food from forest. *Ind. For.* **121:** 837-884

Thakur, D & Thakur, DS (eds; 1994). *Tribal Life and Forests.* Deep Publications, New Delhi

Tosh, J (1996). Ethnobotany of western Maharashtra. *J. Econ. Taxon. Bot.* **12:** 169-174

Tosh, J (2004). Ethnobotany - Green Gold branch in botanical sciences. In: *Ethnomedicinal Plants* (Eds. Trivedi, PC & Sharma, NK). Pointer Publishers, Jaipur, pp. 1-77

Trivedi, PC & Sharma, NK (2004). *Ethnomedicinal Plants.* Pointer Publishers, Jaipur

Trivedi, PC & Nehra, S (2004). Potential Medicinal Plants: Botany, Medicinal uses and chemical constituents. pp. 405-424. *In:* Trivedi PC (Ed.) *Medicinal Plants,* Aavishkar Publishers, Jaipur

Trivedi, PC (2002). Ethnomedicinal plants of Rajasthan state. In: *Ethnobotany.* Aavishkar Publishers, Jaipur, pp. 412-439

Vanamala, K Naidu (1966 a,b). Medicinal Plants of Tirupati Hills and Environs. I and II. *Nagarjun* **10:** 1-8; 54-83

Vanathi, D & Rathika, S (2004). Neem-A boon to Biodynamic Farming. *Agrobios* **3:** 40-41

Verma, V (2005). Ethnobotanical studies in East Sikkim. Masters Thesis, Sikkim Manipal University of Health, Medical and Technological Science, Manipal

Watt, G (1889-1893). The Dictionary of Economic Products of India, **Vols. 1-4**. Government Press, New Delhi

27

Indigenous Traditional Knowledge of Medicinal Plant Biodiversity amongst Tribals of Rajasthan

MANOJ KUMAR PALIWAL AND POONAM PALIWAL

Department of Botany, I.P. College, Bulandshahr (U.P.)
E-mail: manojpaliwal_512@yahoo.co.in, paliwalgk1@rediffmail.com

ABSTRACT

Beginning with the study of plants used by tribal people for food and shelter, the Traditional Knowledge now encompasses facets like conservation practices used by the tribal communities, ethnopharmacology, ethnopharmacognosy, ethnogynecology, ethnobotany and biodiversity etc. The technical knowledge on traditional medicines is considered to be one of the most vital parts of information for the modern drug hunters. This is because such a information is taken as valuable lead in developing the therapeutic aids. Quite a good number of examples can be cited where the plant products used by present civilized world have been developed through indigenous knowledge. In this paper it is presented as to how the tribals have been utilizing the plant wealth of forests of Rajasthan, in the vicinity of their villages, for their own welfare. This has really widened the horizon of knowledge since the village people have been using, either the complete plants or their specific organs for curing a series of 15 kinds of ailments like fever, cough, skin disease, cardiovascular problems, gastrointestinal troubles, and the problems of the male as well as female reproductive systems, etc. This is obviously on account of their close association with the forest biodiversity.

Keywords: Biodiversity, indigenous traditional knowledge, modern drug, hunters, tribals.

INTRODUCTION

India has a rich heritage of indigenous knowledge and practices of healing and care. People's narratives in stratified villages throw light on such knowledge and practices that had guided their lives earlier. The indigenous health culture-related knowledge and practices, which was socially knit into people's daily lives even till mid-20 Century, has been later threatened by aggression of 'development' and profit-making markets. Traditional knowledge on medicinal plants includes information, whether documented or non-documented, of the different kinds and functions of plants developed in or known since ancestral times but subjected to contemporary improvements and adaptations.

The utilization and management of traditional knowledge on medicinal plants follows customary laws enforced by a group of elders who exercise leadership in a community or tribe.

Women play a central role in keeping families and communities healthy at large through providing preventive and curative care. There was a time when they were the sole manager of childbirth. Important roles that women played in biological and social reproduction have been ignored and marginalized in recent times. They protected and transferred their knowledge and practices to the following generations systematically and skillfully integrating these into the processes of socialization and these were even passed on or inherited from one generation to the other.

Over the decades, the villages have grown and socio-economic organizations have changed. Thus there have been intense debates on whether health is a social service or product bred by the rapid commodification of the knowledge drug industry. At the same time, within the knowledge health medicine sector, there is an emerging trend towards the consumption of herbal health products as alternative treatments to modern medicines. Ayurveda is not an exception under this journey since ayurvedic drug companies in contemporary India are aggressively marketing various health products and laxatives known as herbal and natural remedy and have no side effects.

Ayurveda, whose history goes back to 5000 BC, is one the ancient health care systems. It was developed through daily life experiences with the mutual relationship between mankind and nature. The ancient text of Ayurveda reports more than 2,000 plant species for their therapeutic potentials. Besides Ayurveda, other traditional and folklore systems of health care were developed in the different time periods in the Indian subcontinent, where more than 7,500 plant species have been used for various ailments.

According to a WHO estimate, even now about 80% of the world population relies on traditional systems of medicines for primary health care, where plants form the dominant component over other natural resources.

Beginning with the study of plants used by tribal people for food and shelter, the gamut of ethnobotany now includes studies like biodiversity conservation practices used by the tribal communities, ethnopharmacology, ethnopharmacognosy, ethnogynecology, etc. Technical knowledge on traditional medicines is considered to be one of the most vital information for those who are still searching new drugs. Such information is taken as valuable leads in developing desirable therapeutic aids. Quite a good number of examples may be cited where plant products used by the present civilized world have been developed through indigenous knowledge. Several drugs viz. reserpine, quinine, digoxin, ephedrine, cocaine, emetin, khellin, colchicine, artemisinine, gugulipid etc. have been derived from the plants which play on important ethnomedical role in tribal societies. *Lobelia inflata* smoked earlier by Indian tribal people, is now used as a substitute for Tobacco. *Podophyllum peltatum*, formerly used by tribals to remove skin warts is now used in treating uterine warts in modern system of medicine (Singh and Pandey, 1998). Moreover, out of 120 active compounds currently isolated from higher plants and used in modern medicine, 74 per cent have positive correlation between the modern therapeutic use and the traditional use of the species from which they were derived or discovered. Several other tribal medicines have also been incorporated in the organized system of medicine and some more miraculous medicines known to the tribal communities are still secrets of certain families.

About the Tribals

Tribal, Adibasi, Girijan, Vanvasi, Janjati, Vanyajati, Adim jati, Aboriginal, etc. are the words synonymous to each other. Such distinct ethnic groups are mentioned as Scheduled Tribes, in Article 342 of Constitution of India.

India has been rightly called as a "Melting Point" of races and tribes. It is one of the richest countries from the ethnomedicinal and biodiversity point of view due to the presence of multiethnic groups of ancient lineage. The country presents a colorful mosaic of about 500 tribal communities representing 7.76 per cent of the total population, spread over 19 per cent of total area of the nation (Singh and Pandey, 1998).

Their entire style of the life revolves around the forest. Their knowledge about medicinal usage of plants has survived only by oral communications from generation to generation. The field studies among such societies for advancing our knowledge of medicinal plants are of direct value since these have been tested by them.

METHODOLOGY

A floristic survey of ethnomedicinal plants occurring in tribal areas of Rajasthan was conducted to asses the potentiality of the plant resources for modern treatment. The information on medicinal plant is based on the observations, participation with tribals and exhaustive interviews with local physicians practicing indigenous system of medicine, village headmen, priests or *bhopas*.

Generally two types of interviews were taken, firstly of individuals and secondly of group of individual. Interviews were taken at different site as and when situation demanded. A group of tribal people was taken to the forest and specimen of ethnomedicinally important plants were collected along with notes. The samples of such plants were identified using the local floristic work of Singh *et al.* (1983), Sharma (1984), etc.

OBSERVATIONS

Repeated visits to the tribal rich belts of three districts of the State, Banswara, Baran and Udaipur, proved to be highly rewarding information especially as to how they have been utilizing the plant wealth of the forest in the vicinity of their villages, for their own welfare. This has really widened the horizon of the knowledge since the village people have been using, either complete plants or their specific organs for curing a series of 15 kinds of diseases (Table 1) like fever, cough, skin diseases, cardiovascular problems, gastrointestinal troubles and problems of male and female reproductive system etc.

Alongwith this botanical names, common/vernacular names and families names have been furnished Table 2 reveals that different parts of the plants such as the roots and other underground parts, flowers, flower buds, fruits, seeds, gums, resins are variously utilized by the tribal people for curing the ailments.

Table 2 is segregated disease-wise into different sections. As many as 129 ways are listed in which tribes of this region are using approximately 77 plant species belonging to 45 families of angiosperms. Alongwith 40 families of dicotyledons species of 5 families of monocotyledons viz. Arecaceae, Cyperaceae, Liliaceae, Musaceae and Poaceae are used by the tribal people for treatment of the diseases. It is quite clear from the Table that approximately 20 plant species of both dicots and monocots are consumed by tribes for the treatment of more than one kind of disease. Species like *Allium cepa, A. sativa, Acacia nilotica, Albizia lebbek, Ficus benghalensis, F. religiosa,* and *Terminalia chebula* are commonly used for treating more than four types of ailment, hence valued by Bhopas or persons practicing Indigenous system of medicine in these Districts of Rajasthan.

It has also been observed during the screening of results that these tribal people are more aware about the diseases related with skin, male reproductive system, female reproductive system and poisoning and the nature as well as utility of plants growing in their vicinity, hence using an array of plant species for these diseases.

Table 1: Use of plants for medical purposes

S.No.	Ailments	Number of species being used
1.	Fever	05
2.	Cough	05
3.	Eye Ailment	05
4.	Ear troubles	03
5.	Dental troubles	05
6.	Tuberculosis	05
7.	Skin Diseases	14
8.	Brain related problems	08
9.	Stomach Diseases	09
10.	Piles	06
11.	Heart Diseases	03
12.	Male Reproductive system	19
13.	Female Reproductive System	19
14.	Remedies against Poison Consumption	16
15.	Diabetes cure	06

Table 2: Statement of the utilization of the various plant species for curing ailments

Sr. No.	Botanical Names	Families	Common/ vernacular Names	Parts Used
	Fever			
1	*Cordia dichotoma* var.wallichii (Cl.) Maheshwari	Ehretiaceae	Lasora	Bark
2	*Cynodon dactylon* (L.) Pers.	Poaceae	Dub	Entire plant
3	*Moringa oleifera* Lamk.	Moringaceae	Sahjan	Roots
4	*Nyctanthes arbor-tristis* L.	Oleaceae	Harsingar	Leaves
5	*Terminalia chebula* Retz.	Combretaceae	Harad	Fruit powder
	Cough			
6	*Adhatoda vasica* Nees	Acanthaceae	Vasak, Arusa	Leaves
7	*Allium sativum* L.	Liliaceae	Lahsun	Extract of bulb
8	*Cordia dichotoma* var. wallichii (Cl.) Maheshwari	Ehretiaceae	Lasoda	Fruit with seeds
9	*Ficus religiosa* L.	Moraceae	Peepal	Fruits
10	*Terminalia belirica* (Gaertn.) Roxb.	Combretaceae	Bahera	Fruit, seed-coat

Table 2: Contd.

Sr. No.	Botanical Names	Families	Common/ vernacular Names	Parts Used
	Eye Ailments			
11	*Acacia nilotica* (L.) Willd.	Mimosaceae	Babool	Leaves
12	*Albizia odoratissima* L. f.	Mimosaceae	Kala siris	Leaves
13	*Allium cepa* L.	Liliaceae	Pyaj	Extract of bulb
14	*Boerhaavia diffusa* L.	Nyctaginaceae	Punarnava	Root
15	*Ficus benghalensis* L.	Moraceae	Burh, vat	Latex
	Ear Troubles			
16	*Allium cepa* L.	Liliaceae	Pyaj	Extract
17	*A. sativum* L.	Liliaceae	Lahsun	Extract of bulb
18	*Acacia nilotica* (L.) Willd.	Mimosaceae	Babool	Leaves
	Dental Troubles			
19	*Albizia lebbek* (L.) Benth.	Mimosaceae	Kala siris	Leaves
20	*Euphorbia neriifolia auct*, pl. non. L.	Euphorbiaceae	Thuhar Chhota	Extract of entire plant
21	*Ficus benghalensis* L.	Moraceae	Bargad, Bad	Latex, Fruit
22	*Helianthus annuus* L.	Asteraceae	Surajmukhi	Seed oil
23	*Syzygium aromaticum* (L.) Merr. & Perry	Myrtaceae	Laung	Bud, Clove oil
	Tuberculosis			
24	*Adhatoda vasica* Nees	Acanthaceae	Vasak, Arusa	Leaves
25	*Allium sativum* L.	Liliaceae	Lahsun	Bulb
26	*Dendrocalamus strictus* (Roxb.) Nees	Poaceae	Laathibans	Alcoholic extract
27	*Elettaria cardamomum* Maton	Zingiberaceae	Chhoti elaichi	Alcoholic extract
28	*Tinospora cordifolia* (Willd.) Miers. ex Hook. F. & Thoms.	Menispermaceae	Giloe	Alcoholic extract
	Skin Diseases			
29	*Albizia lebbek* (L.) Benth.	Mimosaceae	Siris	Bark
30	*Allium cepa* L.	Liliaceae	Pyaj	Extract of Bulb
31	*A. sativum* L.	Liliaceae	Lahsun	Extract of bulb
32	*Annona squamosa* L.	Annonaceae	Sharipha	Leaves
33	*Azadirachta indica* A. Juss.	Meliaceae	Neem	Leaves, Bark
34	*Buchnania lanzan* Spreng.	Anacardiaceae	Chironji	Seeds
35	*Crotalaria juncea* L.	Fabaceace	Sanai	Leaves
36	*Cynodon dactylon* (L.) Pers	Poaceae	Doob	Entire plant
37	*Dalbergia sissoo* Roxb.	Fabaceae	Sheesham	Oil,Leaves
38	*Datura stramonium* L.	Solanaceae	Dhatura	Fruits
39	*Ficus religiosa* L.	Moraceae	Peepal	Bark
40	*Mangifera indica* L.	Anacardiaceae	Aam	Leaves
41	*Nicotiana tabacum* L.	Solanaceae	Tambaku	Leaves
42	*Terminalia chebula* Retz.	Combretaceae	Harad	Fruits, Leaves
	Brain Related Diseases			
43	*Acorus calamus* Linn.	Araceae	Bach	Powder
44	*Albizia lebbek* (L.) Benth.	Mimosaceae	Siris	Seeds
45	*Bacopa monnieri* (L.) Pennell	Scrophulariaceae	Nira-Brahmi	Plant extract
46	*Cynodon dactylon* (L.) Pers.	Poaceae	Doob	Entire plant
47	*Evolvulus alsinoides* L.	Convolvulaceae	Shankh-pushpi	Extract

Table 2: Contd.

Sr. No.	Botanical Names	Families	Common/ vernacular Names	Parts Used
48	*Moringa oleifera* Lamk.	Moringaceace	Sahjan	Flower buds
49	*Pongamia pinnata* (L.) Pierre	Fabaceae	Karanj	Seeds
50	*Terminalia chebula* Retz.	Combretaceae	Harad	Fruits, Seeds
	Stomach Diseases			
51	*Acacia nilotica* (L.) Willd.	Mimosaceae	Babool	Bark, Seeds
52	*Albizia lebbek* (L.) Benth.	Mimosaceae	Siris	Extract of leaves and twigs
53	*Boerhaavia diffusa* L.	Nyctaginaceae	Punarnava	Root
54	*Cinnamomum tamala* (Buch.-Ham.) Nees & Eberm.	Lauraceae	Tejpat	Bark
55	*Euphorbia tirucalli* L.	Euphorbiaceae	Sehund	Latex
56	*Ficus religiosa* L.	Moraceae	Peepal	Aqueous extract of the bark
57	*Operculina turpethum* (L.) Silva Manso	Convolvulaceae	Nisoth	Root, bark
58	*Terminalia belirica* (Gaertn.) Roxb.	Combretaceae	Baheda	Fruit, Seed
59	*T. chebula* Retz.	Combretaceae	Harad	Fruit, Seed
	Piles			
60	*Allium cepa* L.	Liliaceae	Pyaj	Bulb
61	*Cynodon dactylon* (L.) Pers.	Poaceae	Doob	Stem
62	*Ficus benghalensis* L.	Moraceae	Vat, Bargad	Latex, aerial roots
63	*Plumbago zeylanica* L.	Plumbaginaceae	Chitrak	Root, milky juice
64	*Pyrus pyrifolia* (Burm. F.) Nakai	Rosaceae	Nashpati	Fruit pulp
65	*Thespesia populnea* (L.) Soland. ex Corr.	Malvaceae	Paras Peepal	Seed, Young buds, flowers
	Heart Diseases			
66	*Bacopa monnieri* (L.) Pennel	Scrophulariaceae	Nira-Brahmi	Whole plant extract,leaves
67	*Terminalia arjuna* (Roxb.) W. & A.	Combretaceae	Arjun	Bark
68	*Tinospora cordifolia* (Willd.) Miers. ex Hook. F. & Thoms	Menispermaceae	Giloe	Stem extract
	Ailments of the Male Reproductive system			
69	*Acacia nilotica* (L.) Willd.	Mimosaceae	Babool	Bark, Thorn
70	*Acorus calamus* Linn.	Araceae	Bach	Rhizome
71	*Albizia lebbek* (L.) Benth.	Mimosaceae	Siris	Bark
72	*Allium cepa* L.	Liliaceae	Pyaj	Bulb
73	*Asparagus racemosus* Willd.	Liliaceae	Shatavar	Root Juice
74	*Butea monosperma* (Linn.) Taub.	Fabaceace	Palash, chhola	Gum
75	*Catharanthus roseus* (L.) G. Don	Apocynaceae	Sada Vahar	Roots
76	*Ceiba pentandra* L.	Bombacaceae	Semal	Roots
77	*Chlorophytum borivilianum* Sant.	Liliaceae	Safed musli	Rhizome
78	*Curculigo orchioides* Geartn.	Hypoxidaceae	Kali musli	Stem
79	*Ficus benghalensis* L.	Moraceae	Vat, Bargad	Bark,Latex
80	*F. religiosa* L.	Moraceae	Peepal	Bark,Latex
81	*Hygrophila auriculata* (K. Schum.) Heine	Acanthaceae		Leaf, Seed, Root
82	*Mucuna prurita* Hook.	Fabaceace	Kaunch, Kavanch	Seed
83	*Peganum harmala* L.	Zygophyllaceae	Gokhru	Seed
84	*Phoenix sylvestris* (L.) Roxb.	Arecaceae	Khajuri	Fruits

Table 2: Contd.

Sr. No.	Botanical Names	Families	Common/ vernacular Names	Parts Used
85	*Rhinacanthus nasuta* Kurz	Acanthaceae	Palak juhi	Leaves, roots, seeds
86	*Tamarindus indica* L.	Caesalpiniaceae	Imli	Pulp of the fruit
87	*Terminalia chebula* Retz.	Combretaceae	Harad	Fruit
	Ailments of the female Reproductive system			
88	*Acacia nilotica* (L.) Willd.	Mimosaceae	Babool	Bark powder, Seed powder
89	*Acorus calamus* Linn.	Araceae	Bach	Stem extract
90	*Anogeissus pendula* Edgew	Combretaceae	Dhaura, Dhav	Bark
91	*Azadirachta indica* A. Juss.	Meliaceae	Neem	Gum of the stem
92	*Butea monosperma* Taub.	Fabaceae	Palash, Chhola	Gum of the stem, Bark
93	*Crocus sativus* L.	Iridaceae	Kesar	Petal, steman
94	*Crotalaria juncea* L.	Fabaceae	Sanai	Seeds
95	*Cyperus scariosus* R. Br.	Cyperaceae	Nagarmotha	Rhizome
96	*Phyllanthus embilica* Linn.	Euphorbiaceae	Aonla, Amla	Fruit
97	*Ficus benghalensis* L.	Moraceae	Vat, Bargad	Arial root, young leaves
98	*Ficus glomerata* Roxb.	Moraceae	Gular	Fruit
99	*F. religiosa* L.	Moraceae	Peepal	Fruit
100	*F. rumphii* Blume	Moraceae	Pakhar	Fruit
101	*Musa paradisiaca* L.	Musaceae	Kela	Fruit
102	*Drypetes roxburghii* Wall.	Euphorbiaceae	Putranjiv	Bark
103	*Tectona grandis* L. f.	Verbenaceae	Sagwan	Secondary wood
104	*Terminialia belerica* (Gaertn.) Roxb.	Combretaceae	Behda	Fruit,Seed
105	*T. chebula* Retz.	Combretaceae	Harad	Fruit,Seed
106	*Tribulus terrestris* L.	Zygophyllaceae	Gokhru	Fruit, Leaves, root
	Remedies against Poison Consumption			
107	*Albizia odoratissima* Benth.	Mimosaceae	Kali siris	Gum of the stem
108	*Allium cepa* L.	Liliaceae	Pyaj	Bulb extract
109	*Brassica juncea* (L.) Gaertn. & Coss.	Brassicaceae	Rai	Seeds
110	*Butea monosperma* Taubert.	Fabaceae	Palash	Stem gum
111	*Cordia dichotoma* wallichii (C) Maheshwari	Ehretiaceae	Lasoda, Labheda	Fruit
112	*Gossypium hirsutum* L.	Malvaceae	Kapas	Seeds
113	*Heliotropium indicum* L.	Boraginaceae	Hatta Juri	
114	*Leptadenia pyrotechnia* (Forsk.) Dence	Asclepiadaceae	Khip	Stem
115	*Malus pumila* Mill.	Rosaceae	Seb	Fruit pulp
116	*Mangifera indica* L.	Anacardiaceae	Aam	Fruit pulp, Leaves
117	*Michelia champaca* L.	Magnoliaceae	Champaca	
118	*Piper nigrum* L.	Piperaceae	Kali mirch	Fruit
119	*Salvadora persica* L.	Salvadoraceae	Pilu	Leaf, Root, Fruit
120	*Sesamum indicum* L.	Pedaliaceae	Til	Seed, leaves
121	*Tephrosia purpurea* (L.) Pers.	Fabaceae	Sarphonka	Leaf, Leaf powder
122	*Tinospora cordifolia* (Willd.) Miers. ex Hook. F. & Thoms	Menispermaceae	Giloe	Aerial stem extract

Table 2: Contd.

Sr. No.	Botanical Names	Families	Common/ vernacular Names	Parts Used
	Cure for Diabetes			
123	*Aegle marmelos* (L.) Corr.	Rutaceae	Bael	Fruit pulp
124	*Azadirachta indica* A. Juss.	Meliaceae	Neem	Fruit, seed
125	*Ficus glomerata* Roxb.	Moraceae	Gular	Fruit
126	*Gossypium hirsutum* L.	Malvaceae	Kapas	Seed
127	*Pterocarpus marsupium* Roxb.	Fabaceae	Bijasal	Gum, Flower, Seeds
128	*Syzygium cumini* (L.) Skeels	Myrtaceae	Jamun	Fruit, seed

CONCLUSION

Any research on indigenous knowledge, therefore, must focus on people's assertions and aspirations for national development. Regulatory forest management strategies do not correspond any longer to the political reality. Social and political processes at the level of communities reflecting different interests in forests require more attention in policy analysis. National regulation can only be successful if they are meaningful to and accepted by indigenous people. At the global level, forests have become part of worldwide concern and subject to political efforts in order to develop a more consistent cooperation on their management.

It could be concluded with the above observations that the tribal population living in and around the forest, which is heavily dependent upon forest for its daily need, viz. fuelwood, food, fodder, etc., is able to maintain its health and problems related with diseases with their time-tested wealth of information on medicinal plants. Their entire style of the life revolves around the forests. Their knowledge about medicinal usage of plants has survived only by oral communications from generation to generation. The field observations among such societies for advancing our knowledge of medicinal plants are of great importance.

REFERENCES

Bhatt, DC, Mitaliya, KD & Mehta, SK (2000). Plant twigs as tooth brushes. *Ad. Plant Sci.* **13:** 19-22

Deora, GS, Jhala, SGP & Rathore, MS (2004). Ethnoveterinary medicinal plants in north-west part of Udaipur distt, pp. 74-81. *In:* Trivedi, PC & Sharma, NK (eds) *Ethnomedicinal Plants.* Pointer Publishers, Jaipur

Jain, SK (1963). Observations on ethnobotany of the tribals of M.P. *Vanyajati,* **11:** 177-183

Jain, SK (1995). *A Manual of Ethnobotany.* Scientific Publishers, Jodhpur

Sahu, TR & Sahu, P (2002). Ethnobotanical Scenario of Madhya Pradesh: Knowledge and Culture in Tribal belt of Bastar, pp. 73-97. *In:* Ethnobotany. Aavishkar Publishers, Jaipur

Sharma, NC (1984). Studies on the Flora of Banswara District of Rajasthan. Kalyani Publishers. New Delhi

Sharma, NK (2002). Ethno-medico-religious plants of Hadoti plateau (S.E. Rajasthan), pp. 394-411. *In: Ethnobotany.* Aavishkar Publishers, Jaipur

Sharma, Neelu & Trivedi, PC (2004). Medicoethnobotany of Dausa district, Rajasthan. *In: Ethnomedicinal Plants,* pp. 46-73

Sharma, RK, Chopra, G, Bhushan, G & Kanchan, Munjal (2003). Use of Indian plants in fertility regulation in mammals. *J. Exptt. Zool. India* **6:** 57-74

Singh, V & Pandey, RP (1998). Ethnobotany of Rajasthan, India. Scientific Publishers. Jodhpur

Singh, V & Singh, P (1982). *Fibre yielding plants of Rajasthan.* CSIR Project Univ. Rajasthan

Singh, U, Wadhwani, AM & Johri, BM (1983). Dictionary of Economic Plants in India. ICAR. New Delhi

28

Conservation of Phytodiversity Through *In-Vitro* Strategies

BN Pandey

Department of Botany, Government PG College, Lansdowne, Pauri–246 139, Uttarakhand
E-mail: drpandeybn@yahoo.co.in

ABSTRACT

Conservation *in-vitro* is wholly dependent upon the techniques of plant tissue culture. It is appropriate in situations where either conventional seed storage cannot be employed or it is felt that a greater conservation effort is required than would be provided by any attempts *in-situ*. It is especially employed for vegetatively propagated plants, species with recalcitrant seeds and material modified by techniques of molecular biology. The simplest method to conserve *in-vitro* for short to medium term is to maintain cultures by repeated subcultures with extended subculture duration by lowering temperature, applying osmotic stress or growth retardants. For long-term storage, cryopreservation of protoplasts, cells, meristem or somatic embryos has been employed, by storing them at ultralow temperature, through using liquid nitrogen.

Keywords: Conservation, cryopreservation, germplasm, *in-vitro*, phytodiversity.

INTRODUCTION

Plants are vital part of the world's biological diversity and an essential resource for the planet. In addition to the small number of crop plants used for basic food and fibres, several thousand of wild plants have great economic and cultural importance and potential, providing food, medicine, fuel, clothing and shelter for a vast number of people throughout the world. Plants also play a key role in maintaining the planet's basic environmental balance and ecosystem stability, and provide an important component to the habitats for the world's animal life.

It is estimated that the total number of plants may be in the order of 3,00,000 species. A good number of these species are in danger of extinction, threatened by habitat transformation, over-exploitation, alien invasive species, pollution and climate change. The disappearance of such vital and large amounts of biodiversity poses one of the greatest challenges for the world community: coupled with halting the destruction of the phytodiversity (plant diversity) that is very essential to meet the present and future needs of mankind.

APPROACHES TO PHYTODIVERSITY CONSERVATION

Conventional Approaches

Conservation of plant diversity can be exercised in a number of complementary ways. The two basic approaches to conserve the phytodiversity are: *in-situ* conservation (in the original or natural place or site) and *ex-situ* conservation (outside the original or natural place or site). In the former approach plants are conserved in their natural habitat whereas in *ex-situ* conservation the plants or their propagules are removed from their natural habitats and are stored as collections in seed genebanks or field genebanks.

In-situ Conservation

In *in-situ* conservation the species are allowed to remain in their ecosystems within a natural or properly managed ecological continuum. Species conserved in this matter have the potential for continued evolution of a particular trait within the species and are subject to natural selection. Wild species maintain their original characteristics best in the habitat to which they are adapted and are conserved most effectively *in-situ.* This method is also of significance to the wild relatives of crop plants and a number of tree crops and forest species where there are limitations on the effectiveness of *ex-situ* methods (Prescott-Allen, 1981). However, there are several problems to establishing this type of reserve, for example, cost, size and maintenance aspects, political and social issues and the danger of genetic wipe out as a result of natural disasters, fire, etc. Further, for species which are widely distributed the conservation of their total genetic diversity *in-situ* is difficult.

Ex-situ Conservation

With a view to conserve such a germplasm (the sum total of all genetic material in a species), which is difficult to be conserved *in-situ* or where additional protection is required, *ex-situ* conservation strategies are adopted. There are two basic criteria that must be met in any *ex-situ* method. These are: (a) minimal maintenance, and (b) genetic integrity of the stored material (Rajdan and Cocking, 1997). There are also two basic objectives of a germplasm conservation programme: (i) to provide safety against loss of genetic resources, and (ii) to make available these resources for crop improvement at present as well as in future. Conservation of germplasm, therefore, calls for long-term commitment and hence cost (establishment and running) is an important creation in the choice of the conservation method (Bhat, 1995).

Seed gene bank

Conventionally, plant germplasm is stored as seeds in seed genebanks. The advantages are:

(a) these occupy a completely small space; (b) can be stored at low temperatures for many years; and (c) can be dried and packed for transmission. The limitations of this method, however, are: (a) non-applicability for: clonal plants, plants producing no seeds (vegetatively propagated plants), and plants producing recalcitrant seeds (seeds that die when exposed to desiccating conditions or low temperatures); (b) loss of viability of seeds over the passage of time; and (c) susceptibility to insect or pathogen attack.

Field genebanks

The clonal plants, vegetatively propagated plants, and plants producing recalcitrant seeds are traditionally maintained by the use of the field genebanks. Here, the germplasms are maintained as living collections in a field or orchard or plantation. Barring few advantages (relating to the opportunities that are offered for evaluation of the germplasm), it has must been disadvantages (Withers and Engels, 1990) such as: (i) high labour costs, (ii) vulnerability to pests and pathogens, (iii) vulnerability to natural hazards, and (iv) requirement of large spaces. Moreover, the field genebanks do not represent the entire range of genetic variability within the respective crop genepool and most of these represent only a fraction of the variability which should be conserved (Withers and Williams, 1985).

In-vitro **Approaches**

To overcome the above mentioned deficiencies/problems in conservation methods, one has to find alternative methods. This has led to the consideration of *in-vitro* techniques for germplasm conservation (Withers and Alderson, 1986).

In-vitro conservation refers to maintenance of germplasm in a relatively stable form under more or less defined nutrient conditions in an artificial environment. It is suitable for vegetatively propagated material, species with recalcitrant seeds, and material modified by the techniques of molecular biology. The latter may have been costly to produce, be limited in quantity and have characteristics whose stability during repeated mitotic divisions, and following meiosis are unknown.

The potential advantages of using *in-vitro* methods of germplasm conservation are (Bhojwani and Rajdan, 1996; Rajdan and Cocking, 1997; Sarasan *et al.*, 2006):

1. Conservation of germplasm of endangered plants;
2. Conservation of germplasm of vegetatively propagated plants, and plants with recalcitrant seeds;
3. Conservation of rare germplasms arising through somatic hybridization, somaclonal and gametoclonal variations and genetic engineering;
4. Relatively little space and time is needed for the storage of large number of clonally multiplied plants;
5. Maintenance of the plants free from pest, pathogen, virus and natural hazards;
6. Production of a large number of plants rapidly, when required, from the preserved material, which servey as nucleus stock to propagate;
7. Being free from known pathogens and viruses the material has easy movement in international trade or germplasm exchange with minimum quarantine requirements.

It is expected that the development of the full potential of *in-vitro* culture storage and associated biochemical techniques could revolutionize the handling of germplasm (Chandel and Pandey, 1991; Sarasan *et al.*, 2006).

INFRASTRUCTURE REQUIREMENTS FOR *IN-VITRO* CONSERVATION

There are certain infrastructure requirements recommended for a model *in-vitro* conservation programme. These are: greenhouse facility, tissue culture facility, cryopreservation facilty, facility for disease indexing, facility for monitoring genetic stability, and computer facility (Mandal, 1995).

Greenhouse Facility

This is essential for pre-inoculation planting of plant material received in dormant stage or in a highly contaminated state. It is also required for preliminary transfer of tissue culture raised plantlets to *in-vivo* growth.

Tissue Culture Facility

The basic tissue culture facilities include: (i) preparation area, (ii) transfer area, and (iii) growing area (Bhojwani and Rajdan, 1996).

The preparation area is essentially a kitchen with three basic functions viz. cleaning of glassware, preparation and sterilization of media, and storage of glassware and supplies. This area is equipped with baths for glassware, sink, draining racks, hot air oven for drying glassware, cabinets/racks for storage of glassware, refrigerator for storage of chemicals/stocks/media, distilled water plant, hot plate, magnetic stirrer, pH meter, balance, burner, metal trolley, etc.

Transfer area is the place, which is free from any microorganisms; where explants are inoculated into culture media; and subcultures to fresh media take place. Transfer is most conveniently done in an open-sided laminar airflow hood where filtered air is passed from the rear of the hood outward on a positive pressure gradient. Less expensive alternatives such as an enclosed walk-in room or various enclosed transfer boxes can be used if these are carefully sterilized [with UV lamps] before use. The various requirements during aseptic inoculation are: forceps, scalpels, spirit lamp/glass bead sterilizer, etc.

Growing area is the place where the cultures are kept under controlled light, temperature and humidity in a separate room i.e. culture room. This area is equipped with shelves fitted with fluorescent tubes, connected to a timer for the control of photoperiod. The temperature is controlled by cooling and heating devices. For storage of germplasm, a separate room or incubator is available with all the facilities of a normal culture room but with the facility to reduce the temperature up to 1°C.

There are various types of containers available for growing cultures: test tubes, Erlenmeyer flasks, Petri-plates, and bottles of various shapes and sizes. These may be of borosilicate glass; or less expensive soda glass; or reusable or disposable plastic. Various kinds of closers for the containers are used e.g. nonabsorbent cotton plugs, metal covers or polypropylene caps. For germplasm conservation culture tubes (to reduce space requirement) with polypropylene caps (to reduce water evaporation) are preferred.

Cryopreservation Facility

The various equipment required for a cryopreservation laboratory are: (1) reliable source of liquid nitrogen (LN); (2) safety equipments (gloves, apron, face shield and goggles), pumps for dispensing LN from and to large storage tanks, and trolleys for transport of tanks; (3) LN tanks: small cryotanks (1-3 l); large cryotanks for routine storage (1000 l, etc.); and tanks for long term storage equipped with alarm system which is activated if the LN falls below a critical level; (4) cryovials, boxes, canes and racks; (5) programmable freezer; and (6) water bath for thawing at 35-50°C.

Facility for Disease Indexing

This facility may either be a part of the *in-vitro* programme or may be available in a separate unit. It includes infrastructure for the following tests/studies: grow out infectivity test, direct detection through TEM/SEM, serological test and detection through molecular techniques.

Facility for Monitoring Genetic Stability

Like disease indexing facility, this facility also may either be a part of the *in-vitro* programme or may be available in a separate unit. The following studies are carried out in this facility: (a) determination of cytological stability of the micropropagated plants and plants retrieved from cryopreservation, (b) establishment of genetic uniformity / deviation of the *in-vitro* raised plants using biochemical (isoenzyme, protein, alkaloids, etc.) and molecular tools and techniques (RFLP, RAPD, PCR); (c) evaluation of the overall performance of field-established plants derived from *in-vitro* cultures and cryopreservation.

METHODS OF *IN-VITRO* CONSERVATION

There are two basic methods that are followed to maintain germplasm collections *in-vitro:* (i) slow growth, and (ii) cryopreservation.

The *in-vitro* cultures are grown under modified conditions enabling them to be stored for longer periods before transfer to the fresh medium constitute **slow growth systems** (Rajdan, 2002). The advantage of this approach is that cultures can be readily brought back to normal culture conditions to produce plants on demand. However, the need for frequent subculturing may pose a great disadvantage including contamination of cultures as well as imposition of selection pressure with subsequent change in the genetic make-up due to somaclonal variation.

In **cryopreservation** the plant material is frozen and maintained at the temperature of liquid nitrogen (LN), which is around -196°C. At this temperature the cells stay in almost completely inactive state i.e. in non-dividing or zero metabolism state (Bhojwani and Rajdan, 1996). This method offers the possibility for long-term storage with maximal phenotypic and genotypic stability (Steponkus, 1985). This method being relatively convenient and economical, large number of genotypes and variants could be conserved and thus maximize the potential for storage of genetically desirable material.

In a way similar to seed genebanks, where orthodox seed storage is carried out for the medium-term or long-term as 'active' and 'base' collections, respectively, the Advisory Committee

Table 1: Examples of successfully preserved *in-vitro* cultures under slow growth

Species	Normal culture conditions	Storage conditions	Storage period
Dioscorea alata *D. rotundata* *Ipomea batatas*	26°C; Standard medium	16-20°C; 3% mannitol	8-24 months
Solanum stenotomum *S. goniocalyx* *S. chaucha* *S. juzepejukii* *S. tuberosum* *S. curtilobum*	22°C; 16 hr photoperiod; 3% sucrose	12°C /16 hr day, 6°C/8 hr night; 8% sucrose	12 months
Solanum species (as above)	22°C; standard medium; 20 ml medium	22°C; 20mg/l ABA; 60 ml medium	12 months
Solanum species (as above)	22°C; standard medium	22°C; 6% mannitol	12 months
S. tuberosum	20-22°C; 4000 lx; standard medium	10°C; 2000 lx; 50mg/l B-nine (N-dimethylsuccinamic acid	24 months
Trifolium repens	25°C; 16 hr photoperiod;	5°C; dark	10 months
Vitis amurensis *V. berlandieri* *V. caribea* *V. champini* *V. labrusca* *V. rupestris* *V. vinifera* *V. hybrids*	26°C/15 hr day, 20°C/9 hr night	9.5°C; 15 hr photoperiod	6 months

Source: Withers (1985); Rajdan (2002).

Table 2: Examples of successfully cryopreserved shoot tips/meristems/embryos

Method	Shoot tip/meristem	Embryo (zygotic/somatic)
Slow freezing	*Arachis hypogea, Beta vulgaris, Cicer arietinum, Dianthus caryophyllus, Fragariaxananassa, Lycopersicon esculentum, Mentha arvensis, Phoenix dactylifera, Pisum sativum, Prunus* sp., *Pyrus communis, Solanum tuberosum, Trifolium repens.*	*Daucus carota, Nicotiana tabacum, Zea mays*
Rapid freezing after desiccation	*Actinidia deliciosa, Asparagus officinalis,*	*Camelia sinensis, Cocos nucifera, Coffea arabica, Cucumis melo, Daucus carota, Juglans regia, Poncirus trifolium.*
Vitrification	*Ipomea batatas, Malus domestica, Mentha* sp., *Morus bombycis, Pyrus communis, Trifolium alexandricum.*	*Asparagus officinalis*

Source: Bhojwani and Rajdan (1996).

on *in-vitro* storage of IBPGR (1986) recommended following terminologies:

(i) *In-vitro* active genebank (IVAG) – for *in-vitro* active collections as slow growth cultures;

(ii) *In-vitro* base genebank (IVBG) – for *in-vitro* base collections as storages under cryopreservation;

(iii) *In-vitro* genebank – comprising *in-vitro* active and *in-vitro* base collections, as defined above.

The **materials used for *in-vitro* conservation** may be (i) protoplasts, (ii) cell suspensions, (iii) callus cultures, (iv) meristem tips, (v) propagules at various stages of development, or (vi) organized plantlets. The materials in an undifferentiated phase (protoplasts, cell suspensions and callus cultures) pose two main problems: (a) genetic instability *in-vitro* is very common, (b) totipotency i.e potentiality to produce complete plants is not exhibited in many cases or it is lost after some time. Therefore, they are not the ideal systems for germplasm conservation and organized structures and material such as the shoot apices, embryos and young plantlets are preferred (Bhojwani and Rajdan, 1996).

An **outline of the strategy** for the *in-vitro* conservation of plant genetic resources (Grout, 1990) is given below:

1. Select the plant material to be conserved.
2. Disease indexing, Characterization (morphological and biochemical analysis), and Evaluation (based on yield, biomass production and chemical constituents).
3. Develop protocols for rapid *in-vitro* clonal multiplication.
4. *In-vitro* conservation
    a. For short to medium-term conservation, develop strategies to maintain *in-vitro* cultures under minimal growth conditions
    b. For long-term conservation, develop techniques/protocols for cryopreservation in liquid nitrogen (-196°C)
5. Demonstration of viability and regeneration of plants retrieved from *in-vitro* cultures/ cryopreservation, their field trials and re-characterization.

Slow Growth Cultures

Maintenance of *in-vitro* cultures under normal growing conditions involves frequent subculturing, usually, every four to twelve weeks, which is laborious, costly and poses risks of losses due to error or contamination. For *in-vitro* conservation the growth of the cultures has to be slowed down to extend the period between the successive transfers. Therefore, the basic objective of maintenance of cultures under slow or minimal growth conditions is to reduce the requirements of subculturing with all its attendant risks (Mandal, 1995; Staritsky, 1997).

The various approaches to achieve this are: (i) Reduction of temperature, (ii) Use of growth retardants, (iii) Starvation of cultures, (iv) Reduction of oxygen, (v) Encapsulation and Dehydration.

Reduction of temperature

This method is simple, easy to use which may be applicable to a wide range of genotypes, and therefore is the most commonly used method for slow growth culture. The principle behind this technique is that when plant cultures are maintained at low and non-freezing temperatures, metabolic activities are affected and growth is restricted. Storage temperature is species specific. Usually the temperate plants are stored at 1-10°C, whereas the tropical ones are stored within the range of 15-22°C (Mandal, 1995). Since the first report of Galzy in 1969 on *in-vitro* storage, this approach has been used to store the germplasm of a large number of plants. However, uninterrupted maintenance of reduced temperature may be difficult and expensive.

Use of growth retardants

In this method some natural or artificial growth retardants (regulators) are used in the medium to reduce the growth of the cultures and thereby the subculture interval is enhanced. In potato, storage of plantlets at 22°C, alongwith use of abscisic acid (5-10 mg/l), dimethylamino succinamide (50mg/l), phosphon D (1 or 3 mg/l) and maleic hydrazide (10 mg/l) enhanced subculture duration up to six months. The advantage of the application of growth retardants in the medium is that the cultures can be stored at normal room temperature alongwith other normally growing cultures. This obviates need for the low temperature facility. However, its application may result a number of physiological and genetic aberrations and therefore the application of growth retardants is not advisable (Staritsky, 1997).

Starvation of cultures

Starvation of cultures means maintaining cultures at a reduced nutritional status. This can be achieved particularly by alteration of carbon source by two means: (i) reducing as nutrient factor, and/or (ii) increasing as osmotic factor. Use of sucrose either decreased (0.5–2%) or increased (5–10%) can reduce the growth rate of cultures. In both cases the culture is starved. In the first case, it is because carbon source is present in low quantity that restricts growth. In the second case, it is because the **osmolarity** of the medium is raised and the plant tissue absorbs water and other nutrients with more difficulty that ultimately leads to restriction in growth. Osmolarity can also be raised by inclusion of non-metabolizable carbon source e.g. sugar alcohols particularly mannitol or sorbitol (0.5-6%), which has been quite effective in restricting the growth in several plant species *in-vitro*.

Raised osmolarity in combination with reduction of temperature has been more rewarding. Therefore, this combination has been found to be the most realistic and cost effective method because it shows synergistic effect in prolonging the shelf life of cultures in a number of taxa.

Reduction of oxygen

There are three ways through which reduction of oxygen level is achieved: (i) low pressure storage (LPS), (ii) low oxygen storage (LOS), and (iii) mineral oil overlay.

In low pressure storage (LPS), the atmospheric pressure surrounding the tissue cultures is reduced, resulting in partial decrease of pressure created by contact with the plant material. On the other hand, in low oxygen storage (LOS), whereas the atmospheric pressure is not reduced but the inert gases (particularly nitrogen) are combined with oxygen to create low oxygen pressure (Rajdan, 2002). Using this technique, tobacco and chrysanthemum plantlets could be stored for six weeks under 13% oxygen conditions and oil palm somatic embryos for 4 months at 1% oxygen level (Mandal, 1995). The results obtained from the LPS and LOS have been found comparable regardless of plant species used (Bridgen and Staby, 1981). However, the major limitations are: (i) desiccation of medium limit in the storage period; and (ii) poor growth responses of stored material on regeneration (Rajdan, 2002).

In mineral oil overlay, for reducing oxygen, the cultures are covered with it. The technique was initially used for storages of calluses of carrot and *Catharanthus* but it has also been used for organized structures. However, the main problem is that vitrification is often observed and when returning to standard conditions, partial or complete necrosis of explants is sometimes observed (Mandal, 1995).

Encapsulation and Dehydration

Encapsulated buds and somatic embryos in alginate beads could be stored for 45 days at +4°C. Partial desiccation was also found to be useful in restricting growth and conserving somatic embryos at room temperature for 8 months when water content was reduced to 15% (Mandal, 1995).

Cryopreservation

Cryopreservation or storage in liquid nitrogen at a temperature of -196°C, at which all the cells are in a state of suspended animation, is the most promising method of *in-vitro* germplasm storage (Stushnoff and Fear, 1985). Cryopreservation has been successfully applied to callus, protoplast, pollen, meristems, zygotic and somatic embryos and suspension cultures of several crop species. The first report of successful cryopreservation of plant cell suspension (Quatrano, 1968) and regeneration of somatic embryos from cryopreserved cells (Nag and Street, 1973), led to numerous studies on cryopreservation of plant system (Kartha, 1985; Steponkus, 1985; Rajdan and Cocking, 1997, 2000).

Cyropreservation of living cells demands considerable technical skill. It is essential that cells during freeze-preservation be protected against **cryogenic injuries**. The potential sources of cell damage are: (i) formation of large ice crystals inside the cell, leading to the rupture of organelles and the cell itself, and (ii) intracellular concentration of solutes increasing to toxic level, (iii) loss of vital solutes through leakage during the freezing process (Bhojwani and Rajdan, 1996). There are other factors which affect the viability of the frozen plant material, such as: (i) the nature and physiological condition of the plant material, (ii) pre-freezing treatments, and (iii) application of cryoprotectants.

The morphological and physiological conditions of the **plant material** considerably influence the viability. In general, small, richly cytoplasmic, meristematic cells survive better than the larger, highly vacuolated cells (Nag and Street, 1975; Withers and King, 1980). In comparably larger tissues (such as shoot apices, embryos, and plantlets), highly vacuolated cells are severely damaged, and re-growth occurs from the meristematic cell component. Young, globular embryos survive better than the older ones (Dumet *et al.*, 1997).

Cryopreservation broadly involves five steps: (i) Pre-freezing treatments, (ii) Freezing, (iii) Storage, (iv) Thawing, and (v) Reculture.

Pre-freezing treatments

Freezing of plant cells implies conversion of its water, free water (water involved in solvation) as well as bound water (associated with macromolecule), to ice, whereas thawing is reversal of this transition. Formation of ice in cells is injurious, particularly due to freezing of bound water. Therefore, any mechanism by which less of free water is made available for freezing will provide higher chances of survival of the cells or tissues in post-freezing. Pre-freezing treatments provide capability to withstand freezing injury. Following are some of the strategies, which provide protection from freezing and thawing injury:

Cryoprotectants

Since viability loss is not expected to occur at ultra-low temperatures, protection of the cells during freezing and thawing becomes of paramount importance. There are several compounds such as glycerol, dimethyl sulphoxide (DMSO), ethylene glycol (EG), polyethylene glycol (PEG), sorbitol, mannitol, glucose sucrose, proline, etc. either alone or in combination which protect living cells during freezing and thawing (Ulrich, 1985). However, for plants, dimethyl sulphoxide (DMSO) has been found to be the most effective cryoprotectant.

This is generally amorphous in solid state and dissolves in water, preventing the formation of large ice crystals inside the cell. This may be due to their colligative property, which causes depression in freezing point and can thus alter the size/property of crystals of ice during freezing.

As a result of dehydration of cells before or after freezing, the intra-cellular concentration of solutes in the protoplasm increases to the toxic level. Cryoprotectants, being neutral solutes, prevent the accumulation of other solutes to the injurious level and thus also protect cells from the toxic solution effect.

Pre-culture

It is the conditioning of plant material prior to freezing at ultra-low temperature. This is achieved by incubation of *in-vitro* grown shoot tips or meristems for a period of 2 to 3 days or so in the culture medium supplemented with the suitable cryoprotectant. Pea and strawberry meristems pre-cultured on respective shoot regeneration media supplemented with 5% DMSO for a period of two days and frozen at slow cooling rates exhibited enhanced survival (Kartha, 1985). The freshly isolated cells and tissues also exhibit higher survival, when these are cultured for 2 to 3 days before cryopreservation.

Cold acclimation

Plant species which are capable of withstanding severe freezing stress in nature have evolved mechanism to accomplish this by allowing the water to either remain super-cooled or by permitting the outflow of free water to extra-cellular space by exosmosis where they freeze resulting in the promotion of freeze-induced dehydration. The freeze-induced dehydration further depresses is the freezing point of the remaining cellular water thus facilitating the cells and tissues to withstand exposure to even lower temperatures including that of LN (Kartha, 1985). Thus, the cold acclimated plants are more suitable material in cryopreservation.

In order to exploit the capability of cold acclimation two approaches are made in cryopreservation: (i) utilization of explants from plants which are naturally cold acclimated, and

(ii) induction of cold induction either (a) in the donor plants from which the explants are to be taken or (b) in the *in-vitro* cultures prior to freezing. Shoot apices, from apple, gooseberry, currant, raspberry and pear, preconditioned by low temperature in winter, exhibit better freeze tolerance (Kartha, 1985). Cold treatment of donor carnation plants at 4°C for 3 days or more resulted in a doubling of the survival. Similarly, cold acclimation of *in-vitro* cultures at 2-4°C have shown better results.

Desiccation

This process involves exclusion of freezable water from the cells before freezing. This increases rate survival during freezing and an extensive programmable freezing apparatus is not required. Desiccated somatic embryo of carrot with 3% water and that of melon with 11.8%, after ABA treatment, could survive direct immersion in LN (Bhojwani and Razdan, 1996). Similarly, calcium-alginate encapsulated shoot tips of kiwi-fruit, dehydrated by storage with silica gel, showed very high survival after direct immersion in LN. The dehydration optimum varies with species and tissues, e.g. 3% for somatic embryo of carrot, 11.8% for somatc embryo of melon, 29% for calcium-alginate encapsulated shoot tips of kiwi-fruit and 13% for embryonal axes of tea.

Application of abscisic acid

Abscisic acid, the stress hormone in plant, has been found to provide freezing tolerance in plants and *in-vitro* cultures. It circumvents the requirements for cold acclimation.

Freezing

The material suspended in the culture medium and treated with suitable cryoprotectant(s), is transferred to sterile, polypropylene ampoules with screw cap and frozen. Since the sensitivity of the plant cells to low temperature varies with species several methods of freezing have been developed. These are: (i) Slow freezing method, (ii) Rapid freezing method, (iii) Step-wise freezing method, (iv) Vitrification, (v) Encapsulation / Dehydration, and (vi) Droplet freezing

Slow freezing method

In this method, the material is frozen at slow rates 0.5-4°C /min, starting from 0°C until the temperature reaches –100°C and finally it is transferred to LN. This procedure has been successful especially with cells from suspension cultures.

Rapid freezing method

In this method, the materials contained in vials are placed directly into LN. The temperature is decreased rapidly at the rate of >1000°C/min. This method has been found suitable for dehydrated plant materials i.e. plant materials with cells containing low water content. Bring very simple method and has been applied for a large number of plants.

Stepwise freezing method

This method combines the advantages of the slow and rapid freezing. Initially, the plant material is cooled slowly and stepwise (ca. 1-5°C/min) to an intermediate temperature, maintained at that temperature for 30 minutes, and then rapidly cooled by plunging it into LN.

Vitrification

Vitrification is an effective mechanism against freezing injury wherein a highly concentrated cryprotectant solution supercools to very low temperatures and solidifies into metastable glass without undergoing crystallization (Fahy *et al.*, 1984). In this method, cells or meristems are sufficiently dehydrated with a highly concentrated vitrification solution (generally, a glycerol-based combination of cryprotectants) at 25°C or 0°C without causing injury.

Encapsulation/Dehydration

In this method, the cells/tissues are trapped into calcium alginate beads followed by incubation in 0.85M sucrose for 4-16 hours and rapid freezing in LN. In this process sucrose is the only cryoprotectant which induces dehydration. Prior to immersion of the beads in LN, these are desiccated in a laminar air flow chamber and/or with silica gel. The drying process increases the sugar concentration in the beads, which results in the glass transition during cooling to -196°C (Dereuddre *et al.*, 1991).

Droplet freezing

In this method, as applied to cassava meristems and should other taxa (Kartha *et al.*, 1982), cryoprotected specimens are kept in 2-3μl droplets on a aluminum foil contained in a Petridish and then frozen by slow cooling (0.5°C/min) to a temperature between -20 and -40°C prior to immersion in the liquid nitrogen.

Storage

Maintaining the frozen material at the correct temperature is as important as proper freezing itself (Bhojwani and Razdan, 1996). Storage of material is best carried out in liquid nitrogen immersion phase (-196°C) or in its vapour phase (-150°C). Temperature above -130°C may cause ice crystal formation inside the cells and hence reduce their viability. A LN refrigerator holding about 4000 ampules of 2 ml each is estimated to consume 20-25 l of LN per week. In theory, as long as a regular supply of LN is ensured it should be possible to maintain the frozen material with little further care (Bhojwani and Razdan, 1996).

Thawing

Rapid thawing at 500-750°C/min has been found suitable for most of the cryopreservation protocols. It is accomplished by removing the specimen from liquid nitrogen and immersing it in a warm (37-40°C) water bath for about 1-2 min or until the phase change occurs and ice is transformed into water. Subsequently, the vials are transferred to a water bath at room temperature and maintained there until recultured or its viability is tested. Slow thawing at room temperature has proved fatal. Rapid thawing protects cells from damaging effects of ice crystal formation, which may occur during slow thawing.

Reculture

Before the thawed material is recultured, it is washed several times to remove the cryoprotectants. The viability of the thawed material may then be tested by a staining technique, such as fluorescein diacetate (FDA), Evan's blue, or 2,3,5-triphenyltetrazolium chloride (TTC). However, regrowth is the most accurate criterion for assessing viability of the thawed material, since only a small proportion of the cells, found viable in staining test, enters into division phase (Engelmann, 2004). Various factors, such as, handling of specimens, incorporation of certain additives in regrowth medium and physical environment during early regrowth are important.

IN-VITRO CONSERVATION: FUTURE PERSPECTIVES

In-vitro techniques have been found to be useful in the propagation of a large number of threatened plants (Dhar *et al.*, 2000). Conservation of ornamental germplasm also takes the advantage of innovative techniques of *in-vitro* techniques (Ozudogru *et al.*, 2010). The use of *in-vitro* techniques in germplasm conservation is increasing, but there is great potential for further development (Sarasan *et al.*, 2006). *In-vitro* active genebank (IVAG) has been developed at several prominent centres of the world but the efforts made for the establishment of *in-vitro* basebank (IVBG) are far from satisfactory. However, the potential value of cryopreservation as a conservation tool has been extolled by many authors, and it is currently the only available technique for the safe long-term storage of species with recalcitrant seeds or those that are vegetatively propagated (Engelmann, 2004).

In India, to meet the conservation needs the National Bureau of Plant Genetic Resources (NBPGR) has established a National Facility for Plant Tissue Culture Repository (NFPTCR) to serve as a national *in-vitro* genebank. The associated aspects, such as monitoring of genetic stability through morphological, biochemical and cytological techniques, also receive adequate emphasis in this national programme. For *in-vitro* conservation, major thrust has been assigned to tuberous and bulbous crops, spices, plantation and new industrial crop plants, fruit crops as well as medicinal, aromatic and endangered plant species. The CSIR institutions and ICAR institutions with plant tissue culture facilities have established IVAGs for specific crops.

REFERENCES

Bhat, SR (1995). *In-vitro* conservation – practical considerations. *In:* Rana RS, Chandel KPS, Bhat SR, Mandal BB, Karihaloo JL, Bhat KV & Pandey R (Eds), Plant Germplasm Conservation: Biotechnological Approaches. NBPGR, New Delhi

Bhojwani, SS & Rajdan, MK (1996). Plant Tissue Culture: Theory and Practice. A Revised Edition. Elsevier, Amsterdam

Bridgen, MP & Staby, GL (1981). Low pressure and low oxygen storage *Nicotiana tabacum* and *Chrysanthemum* x *Morifolium* tissue culture. *Plant Sci Lett* **22:** 177-186

Chandel, KPS & Pandey, R. (1991). Plant Genetic Resources Conservation: Recent Approaches. *In:* Paroda RS & Arora RK (eds.), Plant Genetic Resources Conservation and Management: Concepts and Approaches. International Board for Plant Genetic Resources, New Delhi

Dereuddre, J, Hassen, N, Blandin, S & Kaminski, M (1991). Resistance of alginate coated somatic embryos of carrot (*Daucas carota* L.) to desiccation and freezing in liquid nitrogen 2: Thermal analysis. *Cryo-Letter* **12:** 135-148

Dhar, U, Upreti, J & Bhatt, ID (2000). Micropropagation of *Pittosporum napaulensis* (DC.) Rehder & Wilson–a rare, endemic Himalayan medicinal tree. *Plant Cell Tissue Organ. Cult.* **63:** 231-235

Dumet, D, Berjak, P & Engelmann, F (1997) Cryopreservation of zygotic and somatic embryos producing recalcitrant and intermediate seeds. *In:* Rajdan MK & Cocking EC (Eds), Conservation of Plant Genetic Resources *In-vitro*, Vol. **1:** General Aspect *in-vitro*. Science Publishers, Inc., Enfield, New Hampshire, England

Engelmann, F (1997). *In-vitro* conservation research activities at the International Plant Genetic Resources Institute (IPGRI). *Plant Tissue Cult. Biotech.* **3:** 46-52

Engelmann, F (2004). Plant cryopreservation: progress and prospects. *In-vitro Cell Dev. Biol. Plant* **40:** 427-433

Fahy, GM, Macfarlance, DR, Angell, CA & Meryman, HT (1984). Vitrification as an approach to cryopreservation. *Cryobioloy* **21:** 407-426

Grout, BWW (1990). *In-vitro* conservation of germplasm. *In:* Bhojwani, SS (Ed.), Plant Tissue Culture: Application and Limitations. Elsevier Science Publishers, Amsterdam

IBPGR, (1986). Advisory Committee on *in-vitro* storage: Design, Planning, and Operation of *in-vitro* Genebanks. Report of a subcommittee meeting, Rome

Kartha, KK (1985). Cryopreservation of plant cells and organs. CRC Press, Boca Raton, Florida, USA

Mandal, BB (1995). Methods of *in-vitro* conservation: principles, prospects and constraints. *In:* Rana, RS, Chandel, KPS, Bhat, SR, Mandal, BB, Karihaloo, JL, Bhat, KV & Pandey, R (Eds.), Plant Germplasm Conservation: Biotechnological Approaches. NBPGR, New Delhi

Nag, KK & Street, HE (1973). Carrot embryogenesis from frozen cultured cells. *Nature* **245:** 270-272

Ozudogru, EA, Previati, A & Lambardi, M (2010). *In-vitro* conservation and cryopreservation of ornamental plants. *In:* Jain, SM & Ochatt, SJ (Eds), Protocols for *in-vitro* propagation of ornamental plants, Methods in molecular biology, Vol 589. Humana Press, New York

Prescott-Allen, R & Prescott-Allen, C (1981). *In-situ* conservation of crop genetic resources: A report. IBPGR, Rome

Quatrano, RS (1968) Freeze preservation of cultured flax cells utilizing dimethyl sulfoxide. *Plant Physiol.* **43:** 2057-2061

Razdan, MK (2002). Introduction to Plant Tissue Culture 2/e. Oxford & IBH, New Delhi

Razdan, MK & Cocking, EC (1997). Conservation of Plant Genetic Resources *In-vitro*, Vol 1: General Aspect. Science Publishers, Enfield, New Hampshire, England

Razdan, MK & Cocking, EC (2000). Conservation of Plant Genetic Resources *In-vitro*, Vol 2: Application and Limitations. Science Publishers, Enfield, New Hampshire, England

Sarasan, V, Cripps, R, Ramsay, MM, Atherton, C, Mcmichen, M, Prendergast, G, & Rowntree, JK (2006) Conservation *in-vitro* of threatened plants–progress in the past decade. *In-vitro Cell Dev. Biol. Plant* **42:** 206-214

Staritsky, G (1997). Background and Principles of *In-vitro* Conservation of Plant Genetic Resources. *In:* Rajdan, MK & Cocking, EC (Eds), Conservation of Plant Genetic Resources *In-vitro*, Vol 1: General Aspects. Science Publishers, Enfield, New Hampshire, England

Steponkus, PL (1985) Cryobiology of isolated protoplasts: Applications to plant cell cryopreservation. *In:* Kartah, KK (Ed.), Cryopreservation of plant cells and organs. CRC Press, Boca Raton, Florida, USA

Stushnoff, C & Fear, C (1985). The potential use of *in-vitro* storage for temperate fruit germplasm. IBPGR Status Report, International Board for Plant Genetic Resources, Rome, Italy

Ulrich, JM (1985). Storage and shipping of plant cells. *In:* Kurstak, E (Ed.), Techniques in setting up and maintenance of tissues and cell cultures, *Cell Bio.* Vol Cl. Elsevier, Ireland

Withers, LA & Alderson, PG (1986). Plant tissue culture and its agricultural applications. Butterworths, London

Withers, LA & Engels, JMM (1990). The test-tube genebank—a safe alternative to field conservation. *IBPGR Newsletter for Asia and the Pacific* **3:** 1-2

Withers, LA & King, PJ (1980). A simple freezing unit and cryopreservation method for plant cell suspensions. *Cryo-Letters* **1:** 213-220

Withers, LA & Williams, JT (1985). IBPGR research highlights-*In-vitro* conservation. IBPGR, Rome, Italy

29

An Appraisal of the Diversity of Arborescent Plants of Doon Valley, NW Himalaya

PS Negi

Scientist, Wadia Institute of Himalayan Geology, 33, General Mahadev Singh Road, Dehradun–248 001. India.

ABSTRACT

Tree wealth of Doon Valley in North-western Himalaya is represented by 674 taxa, 92 families and 368 genera. Total species include 591 trees, 35 shrubs, 12 climbers, 15 palms and 21 bamboos. Out of these, gymnosperms are represented by 42 species, monocots by 36 species and dicots by 596 species. As regards their member, the dicotyledons hold 88.42 per cent share whereas the monocotyledons and gymnosperms contribute 5.34 and 6.24 per cent share in the entire floristic composition of the valley. Interestingly, Doon Valley perpetuates suitable climatic conditions for luxuriant growth of both tropical species (*Tectona grandis, Shorea* spp., *Phoenix* spp., etc.) and temperate species (*Pinus* spp.). The cosmopolitan structure of flora is created by introduction or invasion of 31.15% exotics. Significant contribution to the local flora is recorded by the addition of 256 species to the existing literature. Present study also contributes the addition of 11 species to the existing list of Forest Research Institute Arboretum plants. Conservation potential and prospects of the flora have been appraised by identification and listing of 18 species which are declared as threatened according to the International Union of Conservation of Nature and Natural Resource (IUCN) red list for the year 2003. Progenitors of 17 cultivated and economic plants have been listed for conservation of genetic resource. During field investigations 13 species have been identified threatened due to habitat loss, whereas 7 species have already been reported threatened due to massive exploitation for various socio-economic reasons.

Keywords: Doon valley, tree wealth, threatened plants, exotics, North-west Himalaya.

INTRODUCTION

The Himalayan mountain encompasses unique and rich tree wealth and its diversity treasure forms an important centre of plant origin among the 8 biogeographic realms of World (Udvardy 1975) and 8 biogeographic zones of India (Rodgers 1985). Doon Valley is one of the picturesque localities in Asia at the foothills of the beautiful North-western Himalayan range of India. It represents a transitional zone between the upper Gangetic plains and the outer Himalaya. It is a unique micro-geomorphic unit with parallelogram shaped structure and stretches in the NW-SE direction following the main Himalayan trend. Geographically the valley lies between latitudes 30°00' N and 30°30' N longitudes 77°36' E and 78°18' E, covering an area of about 2244.78 sq km. The present study covers the area of valley proper only which ranges from an elevation of 315m to 1000m and is about 28% of the total geographical area of the valley. The interaction of physical and natural factors has created a unique type of geomorphic structure in the valley, which in turn has produced a special type of climate entirely different from the higher hills as well as from the plains. The average annual temperature of the area during last 42 years (1961-2002) has been 19.62°C and the average annual rainfall is 1950.24mm. Botanically the Doon Valley is of great interest since it perpetuates suitable climatic conditions for luxuriant growth of both temperate (*Pinus* spp.) and tropical species (*Tectona grandis*, *Shorea robusta, Phoenix rupicola,* etc.) and harbours an interesting mixture of plant species common in the Indian plains and lower hills. (Negi and Gupta 1987, Negi *et al.*, 1991).

PREVIOUS WORK

The history of plant exploration in the valley goes back to 18th centuary and may be divided in three successive periods.

First Period (up to 1840): T. Hardwick (1751-1835), a European, was the first to collect plants from North-western (NW) Himalaya. With the establishment of residencies at Dehra Dun, Mussoorie, Shimla and other places along the trade routes and after restoration of Botanic Garden at Saharanpur by Hastings in 1820, there has been an extensive exploration of NW Himalaya and surroundings plains. The first two superintendents, Govan and Royle, particularly the latter, made the earliest collections. J.F. Royle (1799-1858), the second Suprintendent of Saharanpur Garden, was a pioneer phytogeographer of NW Himalaya. He collected plants from Dehra Dun which he described in his excellent treatise, *Illustration of the Botany of the Himalayan Mountains* (1833-1840). Hooker and Thomson (1855) mentioned this as "the only book dealing with the rich flora of the Himalaya except Wallich's *Tentamen Flora Napalensis* and contains the first and only authentic attempt to demonstrate prominent features of geographical distribution of North Indian plants in reference to elevation and climate they inhabit and to the botany of surrounding countries. Other explorers such as N. Wallich (1824-1826) and V. Jacquemont (1935) etc., also collected plants from this region during this period.

Second Period (1840-1905): As the collections were getting accumulated by various explorers from different parts of India and other adjacent regions and also with the collections made by J.D. Hooker and T. Thomson over a period of 3-5 years (1848-1851), particularly from Eastern Himalaya, the publication of flora of India was initiated. The first outcome was *Flora Indica*

(1855) by J.D. Hooker and T. Thomson. Later, J.D. Hooker in collaboration with other contemporary taxonomists published *Flora of British India* in 7 Volumes (1872-1897). Since 1876, when J.F. Duthie was appointed as successor of W. Jameson at Saharanpur Garden and after the establishment of a Forest School at Dehra Dun by W. Schlich and D. Brandis in 1878-1881, there has been a new impetus to the development of phytogeography of the Sub-Himalayan tracts in particular and of North India in general. Brandis explored Dehra Dun and adjacent areas extensively and along with Stewart (1874), he published the *Forest Flora of NW and Central India*, wherein frequent references were made to Dehra Dun collections. A detailed account of Dehra Dun forests was also given in his article "The forest of the outer NW Himalaya" which appeared in *Indian Forester* (13:51-70, 1887). In the year 1882, Sir George King published a list of plants of Garhwal, Jaunsar-Bawar and Dehra Dun. J.F. Duthie (1906) was one of the ablest plant explorers and phytogeographers of N and NW India and also made intensive collection in and around Dehra Dun. He published (1903-1929) in 3 Volumes his *Flora of the Upper Gangetic Plain and of the Adjacent Siwalik and Sub-Himalayan Tracts*, in which Dehra Dun formed the northern most boundry.

This was the only comprehensive flora dealing with the plants of entire Gangetic delta, in which he included and described plants from Dehra Dun, based upon his own collection as well as collections made by Royle, Falconer, Gamble, Kanjilal, Gammie, Bell, King, Brandis and Sanders, etc. In spite of its own drawbacks of erroneous identification and illegitimate names etc., this flora was and still is the authentic and readymade reference for Plants of Dehra Dun.

Among other important collectors of this period were W. Govan (1841-1876), E. Sanders (1841), and D. Brandis (1862-1881). R. Fischer. (1869), G. King (1869-1870), J.S. Gamble (1877-1879). G.A. Gammie (1894), and U. Kanjilal. In 1901 Kanjilal published *Forest Flora of School Circle,* which was another hallmark to the phytogeography of Dehra Dun. This was revised and enlarged by B.L. Gupta (1928) as *Forest Flora of Chakrata, Dehra Dun and Saharanpur Forest Divisions of United Provinces.* Thus, this period may be described as Duthie's Period.

Third Period (1905 and onwards): During this period, Forest Research Institute, Dehra Dun became an active centre of botanical researches in India, especially in dealing with exploration and publications of forest floras. Several batanists made valuble collections and published their work on this region. Those who actively worked at FRI are R.N. Parker, C.E. Parkison, N.L. Bor, M.B. Raizada and K.C. Sahni, etc. Some of the important and noteworthy publications appeared during this period were Raizada's (1976) Supplementary list's to the *Flora of the Upper Gangetic Plain*, in which he dealt with the Dehra Dun plants extensively on the basis of his own collection as well as collections gathered by P.C. Kanjilal, R.N. Parker, B.L. Gupta, Sri Ram, Sohan Lal, H.G. Champion. Later in the year 1978 Raizada with Saxena published *Flora of Mussoorie* which turned out to be a significant contribution to the Botany of this region. The other important plant collectors of this area are Krishnaswamy and Mathuda (1954), Raizada and Sahni (1966), Rao (1969), Babu (1977), Hajra (1983), Gaur and Raiwani (1995), Gaur (1999), etc., and various other botanists from BSI, FRI and HNBGU Teachers and Research Scholars, etc. It is evident from a perusal of the literature and comparison with the present investigation that no up to date account of exotic and indigenous woody flora of Doon Valley is available. Therefore, the present work is likely to serve as a supplement to the earlier publications.

METHODOLOGY

Detailed and extensive field surveys were conducted regularly for three years in the study area in all four seasons, i.e., summer, rainy, winter and spring. Two field trips were made for specimen collection in every season to cover flora of all topographical zones and habitats in their best flowering and fruiting stages. The area of investigation remains restricted up to 1000m elevation and covers about 28% area of total geographical area of the valley.

For the convenience of field work and plant collection from all topographic zones, the area of study was divided into; (a) northern tropical deciduous forest; (b) vegetation of ravines and slopes; (c) vegetation of Siwalik hilltops, slopes and plateaus; and (d) planted/introduced trees in various institutions, parks, gardens and other places. Routine herbarium practices (Jain & Rao, 1977) have been followed for preserving the specimens. Indigenous species were identified with the help of field data, existing F.R.I.(DD) and B.S.I.(BSD) herbarium specimens and regional floristic works, e.g., Kanjilal (1901), Duthie (1903-1929), Osmaston (1927), Brandis 1874, 1906), Raizada & Saxena (1978) Gaur (1999). For the identification of exotic species, flora of Baily (1928, 1949), Daniel (1871), De Wit (1966-67), Graf (1981, 1982 and 1992), Grierson *et al.* (1983), Hutchinson (1967), Kunkel (1978), Matthew (1969), Oliver *et al.* (1868-1937), Ouden *et al.* (1954), Ress (1802-1820), Sharma (1968), Thunberg (1784), Tutin *et al.* (1964) and Verdcourt (1968), have been consulted.

For deciding the correct nomenclature, recommendations of the latest ICBN (International Code of Botanical Nomenclature, ed. 1978), have been followed along with the latest literature available. The recent name changes of taxa have been taken from Bennet (1987). The collected specimens were matched with existing specimens in DD/BSD and living specimens in Arboretums/ Gardens for confirmation of the identity. The references of authors and herbarium specimens are given along with list of species. Statistical synopsis of all the collected species has been prepared for families, genera and species and presented in Table 1.

In order to find out the number of species not covered by earlier workers, the list of the collected specimens has been compared with flora of Kanjilal (1901)/Gupta (1928, rev. ed.). This (Gupta, 1928) being the latest and only available flora that covers maximum number of species from the study area. After comparison, additional species have been marked by asterisk (*) in the enumeration as a contribution to existing record. Among these newly contributed species, indigenous taxa have been futher marked by double asterisk (**), whereas the exotics retained the single asterisk. Specimens collected from the campus of Forest Research Institute (FRI) have been compared with existing list (Gupta *et al.*, 1988) of Arboretum and additional species have been marked by @ as an updation to the present literature. The total number of taxa for five dominant families has been compared with the Flora of British India.

Finally, the research output has been discussed and analysed under subheadings of tree wealth diversity, local flora contribution and conservation prospects. Floristic list of the valley has also been compared with the recent IUCN (Internatonal Union for Conservation of Nature and Natural Resources) Red List (Anonymous 2003) of threatened species in order to search prospects and potential of regional flora for global biodiversity conservation. The progenitors of cultivated and economic plant species have also been identified for conservation of genetic resources.

Fig. 1. *Shorea robusta.*

Fig. 2. *Phoenix rupicola.*

Fig. 3. *Butea monosperma.*

Fig. 4. *Pinus roxburghii.*

DISCUSSION AND ANALYSIS OF TREE WEALTH

Since woody plant diversity represents multidimensional aspects of conservation, therefore, present findings have been discussed and analysed in following subheadings. (a) Tree wealth diversity (b) Contribution to local flora and its relevance, (c) Prospects and potential for conservation of genetic resources and global diversity.

(a) **Tree wealth diversity:** The study area perpetuates suitable climatic condition for luxuriant growth of both tropical species like *Tectona grandis*, *Shorea* spp. (Fig. 1), *Phoenix rupicola* (Fig. 2) and *Butea monosperma* (Fig. 3), etc., and temperate species like *Pinus* spp. (Fig. 4). The total floral diversity is represented by 674 taxa, 92 families, and 368 genera. The total taxa include 591 trees, 35 shrubs, 12 climbers, 15 palms and 21 bamboos. Under angiosperms the dicots are represented by 596 species and monocots by 36 species, whereas the gymnosperms contribute 42 species. A detailed statistical synopsis of woody flora and their percentage has been presented in Table 1.

Table 1: Statistical synopsis and percentage of floral diversity

Groups	Families		Genera		Species	
	Numbers	Per cent	Numbers	Per cent	Numbers	Per cent
Angiosperms	–	–	–	–	–	–
(i) Dicotyledons	81	88.05	340	92.40	596	88.42
(ii) Monocotyledons	3	3.26	14	3.80	36	5.34
Gymnosperms	8	8.69	14	3.80	42	6.24
Total	92	100.00	368	100.00	674	100.00

The dominant families in terms of number of species are Caesalpiniaceae (37), Euphorbiaceae (37), Fabaceae (37), Myrtaceae (32) and Mimosaceae (30) in comparison to Leguminosae (Caesalpiniaceae + Fabaceae + Mimosacea), Rubiaceae, Euphorbiaceae and Acanthaceae in the Flora of British India (Hooker 1872-1897).

(b) **Contribution to local flora and its relevance:** In order to prepare an up to date record and systematic account on indigenous and exotic woody flora, the present investigation has contributed 256 new records of species, which have not been covered by earlier workers like Kanjilal (1901) and Gupta (1928), etc.

These new floristic records are mainly due to the introduction/naturalization of 210 exotic species and 48 of the indigenous species having been overlooked during previous explorations. After comparison of species collected from FRI campus (Arboretum, Botanical garden, etc.) and existing list (Gupta *et al.*, 1988) of Arboretum, it is revealed that 11 species turned additional to the existing record and may be added as updation of Arboretum publication.

These new additions to the local flora are of multifarious importance and, therefore are bound to provide useful information to enrich regional knowledge pertaining to biodiversity, applied botany, ecology, forestery and other allied disciplines of sustainable development, economic amelioration and resource management.

(c) **Prospects and potential of tree wealth for conservation of Biodiversity and genetic resources:** The present as well as earlier investigation (Negi, 2005) reveals that the valley harbours 31.15% of exotics and 68.85% of the indigenous species and some of the designated as threatened in their native countries. These threatened species are traded internationally for their specific and multifarious significance and presently facing anthropogenic and ecological threat at global level. Their sustenance and conservation is important for human life but they have acquired the status of threatened species in recent red list of IUCN. Since these species are already growing well in the valley as introduced or indigenous therefore, this area may be identified as a centre for biodiversity conservation. These species are *Acacia ferruginea, Araucaria angustifolia, Cryptomeria japonica, Cupressus cashmeriana, Cycas revoluta, Delonix regia, Dipterocarpus turbinatus, Ginkgo biloba, Hydnocarpus kurzii, H. laurifolia, Juniperus barbadensis, J. procera, Khaya senegalensis, Mangifera indica, Phoenix rupicola, Pinus merkusii, Pittosporum eriocarpum and P. nepaulense*, etc.

Intrestingly the valley also abounds in progenitors of cultivated and economic plants and therefore these species may be conserved as genetic resources for plant wealth. These wild relatives include *Citrus aurantium, C. limon, Cordia dichotoma, Ficus palmata, Grewia asiatica, Mangifera indica, Morus australis, Phoenix acaulis, P. sylvestris, Prunus cerasoides, Punica granatum, Pyrus pashia, Rubus ellipticus, Rubus niveus, Spondias pinnata, Vitis parvifolia*, and *Ziziphus oenoplia*, etc.

Detailed field and systematic study reveals that some of indigenous species of high importance are under pressure due to massive exploitation for different domestic and commercial uses or due to loss of habitat by developmental activities and other anthropogenic pressure. These species are either found in much deteriorated vegetative structure or are representated with very low spatial density in comparison to their basic known habit of abundance. Main indigenous species found threatened due to massive exploitation are *Acer oblongum, Pittosporum eriocarpum, Abrus precatorius, Cornus controversa, Quercus leucotrichophora, Sophora mollis and Stephania glabra.* Species like *Grewia optiva, Pyrus pashia, Leucomeris spectabilis, Diploknema butyracea, Diospyros cordifolia, D. exsculpta, Carallia brachiata, Abrus precatorius, Azadirachta indica, Berberis asiatica, Itea nutans, Rauvolfia serpentina and Ficus palmata,* etc., are identified as threatened due to habitat loss. These species need urgent attention for biodiversity conservation and sustainable exploitation.

Tree Wealth Diversity (Enumeration)

Table 2: Gymnosperms

Names of the Species	Authors specimen numbers	BSD/DD specimen numbers
Agathis robusta (Moore) F.M. Bailey*	584	136415 (DD)
Araucaria angustifolia Bertal.*	708, 750	N.A.
Araucaria bidwilli Hook.*	104	122281, 145145 (DD)
Araucaria columnaris (Forst.*) Hook.	702	119595, 110028 (DD)
Araucaria cunninghamii Sweet.*	102	155117, 149434 (DD)
Cephalotaxus harringtonia Koch.*	255, 509	N.A.
Cryptomeria japonica L. f.	708, 750	N.A.
Cunninghamia lanceolata Lam.*	151	40163 (BSD)
Cupressus arizonica Greene*	690	38959, 140853 (DD)
Cupressus cashmeriana Royle ex Carrie*	907	148364, 131025 (DD)
Cupressus corneyana Knight & Perr	149	113515, 79200 (DD)
Cupressus lusitanica Mill.*	619	57405, 73985 (DD)
Cupressus pisifera Koch.*	1411	4439, 37405 (DD)
Cupressus sempervirens L.	652	148363, 24012 (DD)
Cupressus torulosa D. Don.	91	131026, 128651 (DD)
Cycas circinalis L.	157	6216 (DD)
Cycas revoluta Thunb*.	468	41291 (BSD)
Ginkgo biloba L.*	114	141941 (DD)
Juniperus barbadensis Pilg. & Ekman*	739	25533, 122245 (DD)
Juniperus chinensis L.*	447	138353, 149622 (DD)
Juniperus communis L.*	906, 911	N.A.

Table Contd.

* Additional species to the flora; ** Indigenous additional species to the flora; @ Addition to FRI, Arboretum

Table Contd.

Names of the Species	Authors specimen numbers	BSD/DD specimen numbers
Juniperus phoenicea L.*	221	113507, 12246 (DD)
Juniperus procera Hochst.*	689	163668, 12236 (DD)
Juniperus scopulorum Sarg.*	1343	77143, 79142 (DD)
Pinus canariensis Sm.*	154, 155	N.A.
Pinus caribaea More.*	701	39160, 149580 (DD)
Pinus densiflora Sieb. & Zucc.*	103	73237, 140315 (DD)
Pinus halepensis Mill.*	95	150599, 45513 (DD)
Pinus hartwegii Lindl.*	1067	103416, 103417 (DD)
Pinus khasya Royle*	224	145605, 149336 (DD)
Pinus merkusii Jungh & de Vriese	Junn. & Diur.*	100 131592, 130855 (DD)
Pinus oocarpa Lamb.	Schiede*	693 150639 (DD)
Pinus patula Schide & Deppe*	153	148406 (DD)
Pinus roxburghii Sarg.	152	31211, 1721 (BSD)
Pinus thunbergii Parl.*	99	79421, 24871 (DD)
Podocarpus gracilior Pilger*	105	45368, 94635 (DD)
Podocarpus macrophyllus Pilger*	744	126640, 74523 (DD)
Podocarpus nerifolius D. Don	Don.*	156 94633, 134638 (DD)
Podocarpus wallichianus Presl.*	748	102462, 102463 (DD)
Taxodium distichum Rich.*	699, 160	N.A.
Taxodium mucronatum Tenore*	223	79147, 74600 (DD)
Thuja occidentalis Linn.*	44, 53	N.A.
Thuja orientalis L.*	07	37687, 149830 (DD)

Table 3: Angiosperms

Names of the Species	Author's specimen numbers	BSD/DD specimen numbers
Ablemoschus manihot (L.) Medik.	492	N.A.
Abrus precatorius L.	530	2707 (DD)
Abrus pulchellus Wall.	501	18407 (DD)
Abutilon indicum Sweet.	636	N.A.
Acacia auriculiformis A. Cunn. ex Benth.*	16	70566 (DD)
Acacia catechu (L.f.) Willd.	819	77883, 22727 (BSD)
Acacia farnesiana (L.f.) Willd.	71	34642, 3963 (BSD)

Table Contd.

Table Contd.

Names of the Species	Author's specimen numbers	BSD/DD specimen numbers
Acacia ferruginea DC.*	821	69526 (DD)
Acacia gageana Craib.	539	11672 (DD)
Acacia karroo Heyne*	17	79649 (DD)
Acacia lenticularis Buch.–Ham.*	784	69529 (DD)
Acacia modesta Wall.	1083	69229 (DD)
Acacia nilotica (L.) Willd. @	540	3433 (BSD)
Acacia pennata (L.) Willd.	541	34758, 37293 (BSD)
Acacia polycantha Willd.**	826	64406 (DD)
Acacia pseudo-eburnea J.R.Drumm.	175	13970 (BSD)
Acacia robusta Burch.*	18	59847 (DD)
Acacia sinnuta. (Lour.) Merr.	205	13886, 35245 (BSD)
Acacia torta (Roxb). Craib.	19	27829 (BSD)
Acer oblongum Wall. ex DC.	129	8089 (BSD)
Acrocarpus fraxinifolius Wight & Arn.**	131	35380 (BSD)
Acronychia pedunculata (L.) Miq.	638	5592 (BSD), 30846 (DD)
Adenanthera microsperma Teijem.*	983	109256 (DD)
Adenocalymma alliaceum Miers.*	846	35359, 53750 (BSD)
Adhatoda zeylanica Medik.	705	3928, 32274 (BSD)
Aegle marmelos (L.) Corr.	126	11976, 19797 (BSD)
Aesculus assamica Griff.	351	11976, 19797 (BSD)
Aesculus indica (Colebr. ex Camb) Hiernin	449	79428, 113582 (DD)
Aglaia spectabilis (Miq.) Jain & Bennet**	555	145074, 13534 (DD)
Ailanthus altissima (Mill.) Swingle	185	109259 (DD)
Ailanthus excelsa Roxb.	1430	6749 (DD)
Alangium chinense (Lour.) Hard.	213	133391 (DD)
Albizia amara Boivin	827	11312 (DD)
Albizia chinensis (Osb.) Merr.	206	110665 (DD)
Albizia lebbek Benth.	25	15538, 27614 (BSD)
Albizia lucidior Steud.	785	20122, 15455 (BSD)
Albizia odoratissima (L.f.) Benth.	26	71531 (DD)
Albizia procera (Roxb). Benth.	26 (a)	19785 (BSD)
Aleurites moluccana Willd.	407	42410, 80926 (DD)

Table Contd.

Table Contd.

Names of the Species	Author's specimen numbers	BSD/DD specimen numbers
Allamanda cathartica L.	837	1243 (BSD)
Alnus nitida Endl.	880	108825, 30366 (DD)
Alstonia scholaris R. Br.	687	70811 (DD)
Ampelocissus divaricata (Wall. ex Laws.) Planch.	547	31287 (BSD)
Amphilobium mutissi HB. & K*	547	35357, 2829 (BSD)
Annona reticulata L.*@	65	N.A.
Anogeissus acuminata (Roxb. ex DC.) Guill. & Perr.**	570	58166 (DD)
Anogeissus latifolia (Rox. ex DC.) Wall.	373	155642 (DD)
Anogeissus leiocarpus Guill.**	268	57206 (DD)
Anogeissus pendula Edgew**	29	149943 (DD)
Anogeissus sericea Brandis**	1033	69271 (DD)
Anthocephalus chinensis (Lam.). A. Rich. ex Walp.	1304	77895(BSD), 37725(DD)
Antidesma acidum Retz.	486	3115, 28178 (BSD)
Aphanamixis polystachya (Wall.) Parker**	601	23989, 25766 (BSD)
Ardisia thyrsiflora D.Don	395	27625 (DD)
Ardisia solanacea Roxb.	650	148436 (DD)
Artabotrys hexapetalus (L.f.) Bhandari	520	35356 (DD)
Artocarpus chamba Buch.-Ham.	170	149760, 106130 (DD)
Artocarpus heterophyllus Lam.	625	18576, 57956 (DD)
Artocarpus lakoocha Roxb.	327	38558, 144866 (DD)
Asclepias curassavica L.	909	147355 (DD)
Aspidopterys nutans (Roxb. ex DC.) Juss.	553	38366(BSD), 21077(DD)
Atalantia monophylla DC.**@	598	N.A.
Averrhoa carambola L.*	125	69226, 22052 (DD)
Azadirachta indica A. Juss.	188	37773, 15467 (BSD)
Baliospermum montana (Willd.) Muell.	937	136458, 69232 (DD)
Bambusa arundincea Willd.	305	75109, 153702 (DD)
Bambusa balcooa Roxb.**	304	Number less (DD)
Bambusa burmanica Gamble*	237	94290, 147832 (DD)
Bambusa glaucescens (Willd.) Sieb.*	243, 252	N.A.
Bambusa longispiculata Gamble*	232	N.A.
Bambusa nutans Wall.	230	150150, 43718 (DD)

Table Contd.

Table Contd.

Names of the Species	Author's specimen numbers	BSD/DD specimen numbers
Bambusa pallida Munro**	228	29240, 29252 (DD)
Bambusa tulda Benth**	226	156440, 94286 (DD)
Bambusa vulgaris Schard*	227	Number less (DD)
Barringtonia acutangula L.	295	32230, 12614 (BSD)
Bauhinia galpini N.E. Br.	965	58606 (DD)
Bauhinia malabarica Roxb.	537	101737 (DD)
Bauhinia purpurea L.	645	73972 (DD)
Bauhinia racemosa Lam.	283	80897 (DD)
Bauhinia roxburghiana Voigt.	566	97094 (DD)
Bauhinia vahlii Wight & Arn.	362	11063 (DD)
Bauhinia variegata L.	302	136310 (DD)
Beilschemiedia roxburghina Nees.**	326	119792 (DD)
Berberis asiatica Roxb. ex DC.	491	34537 (DD)
Berchemia floribunda (Wall.) Brongn.	349, 497	N.A.
Bischofia javanica Bl.	427	3902 (BSD)
Bixa orellana L.*	548	4719, 57406 (DD)
Boehmeria macrophylla D. Don.	429	89715, 123962 (DD)
Boehmeria penduliflora Long.	591	661, 887 (DD)
Boehmeria rugulosa Wedd.	624	29284, 23725 (BSD)
Bombax ceiba L.	119	25648(BSD),42393 (DD)
Boswellia serrata Roxb. ex Coleb.@	186	79805 (DD)
Bougainvillea glabra Choisy	855	123895, 137138 (DD)
Bougainvillea spectabilis Willd.	709	60413, 57060 (DD)
Brachychiton acerifolium F. Mull.*	440	110659, 106275 (DD)
Brassaiopsis aculeata (Buch.-Ham. ex Don) Seem	82	140013 (DD)
Breynia vitis-idaea (Burn. f.) Fisch.*	928	15915 (BSD)
Bridelia monoica (Lour) Merr.	588	125402 (D)
Bridelia squamosa (Lam.) Gehrm.	936	1880, 30711 (BSD)
Bridelia verrucosa Hains	954	35436 (BSD)
Broussonetia papyrifera Vent.	329	1979, 5551 (BSD)
Brunfelsia calycina Benth.*	1006	N.A.
Brunfelsia densiflora Krug.*	845	N.A.

Table Contd.

Table Contd.

Names of the Species	Author's specimen numbers	BSD/DD specimen numbers
Brunfelsia latifolia Hort. ex Steud.	416	105576, 39663 (DD)
Buchanania lanzan Spreng.	130	20149 (BSD)
Buddleja madagascariensis Lam.	759	33876, 19762 (BSD)
Buddleja neemda Buch.-Ham.	844	147295 (DD)
Butea monosperma (Lam.) Taub.	277	60970 (DD)
Butea parviflora Roxb.	642, 70	N.A.
Byrsonia crassifolia Lunan, ex Gresib.*	444	75101, 36368 (DD)
Caesalpinia bondoc (L.) Roxb.	567	127141 (DD)
Caesalpinia decapetala (Roth). Alston	505	91130 (DD)
Caesalpinia ferra Mart.	989	81230 (DD)
Caesalpinia pulcherrima (L.) Swartz*	990	123960 (DD)
Calamus tenuis Roxb.	908	126785, 99227 (DD)
Calliandra haematocephala Hassk.	648	19079, 25097 (DD)
Calliandra houstaniana Benth.*	711	46519, 123452 (DD)
Callicarpa arborea Roxb.	1057	148057 (DD)
Callicarpa macrophylla Vahl.	518	18543, 33410 (BSD)
Callistemon brachyandrus Lindl.*	991	7751 (DD)
Callistemon citrinus (Curt.) Skeel*	993	13738 (DD)
Callistemon rigidus R.Br.*	1039	433644 (DD)
Callistemon viminalis (Soland.) Cheel	462	1546 (BSD)
Callistris columellaris F. Muell.	741	118139 (DD)
Camellia japonica Cham.	661	N.A.
Camellia sinensis (L.) O. Kuntze	710	N.A.
Capparis zeylanica L.f.	118	5282 (DD), 3232 (BSD)
Carallia brachiata (Lour.) Merr.	78,36	N.A.
Careya arborea Roxb.	209	84480 (DD)
Carissa opaca Stapf.	396	39193, 36858 (BSD)
Caryopteris bicolour (Hard.) Mobbeley	165	74079, 74145 (DD)
Caryota mytis Lour.**	962	94447, 94448 (DD)
Caryota urens L.**	1119	118135, 118134 (DD)
Cascabella thevetia (L.) Lippod.	219	34643, 38789 (BSD)
Casearia elliptica Willd.	05	37914, 13876 (BSD)

Table Contd.

Table Contd.

Names of the Species	Author's specimen numbers	BSD/DD specimen numbers
Casearia graveolens Dalz.	1305	N.A.
Casimiora edulis La Llave.*	554	24595, 145057 (DD)
Cassia alata L.**	986	99168 (DD)
Cassia excelsa HB.& K.	1047	123427 (DD)
Cassia fistula L.	204	1176 (DD)
Cassia floribunda Cav.	219	42595 (DD)
Cassia javanica L.*	969	25762 (DD)
Cassia multijuga Rich.	980	N.A.
Cassia nodosa Buch.-Ham.*	285	113560 (DD)
Cassia occidentalis L.	538	74560 (DD)
Cassia roxburghii DC.*	363	74021 (DD)
Cassia siamea Lam.	646	151397 (DD)
Cassia sophora L.	1073	2489 (DD)
Cassia suranttensis Burm. f.	568	124017 (DD)
Cassine albens (Retz.) Kostm.	716	317 (BSD)
Castanea sativa Mill.	315	15979, 138349 (DD)
Castnospermum australe A. Cunn.*	627	N.A.
Casuarina cunninghamiana Miq.*	622	149186, 29393 (DD)
Casuarina littoralis L.*	704	148361, 34986 (BSD)
Catalpa longissima Sims*@	1069	60245, 112998 (DD)
Catunaregam spinosa (Thunb.) Triuv**	84	33224, 37905 (BSD)
Catunaregam uliginosa (Retz.) Sivaranjan**	220	222769, 3224 (BSD)
Celastrus paniculata Willd.	448	8035, 31271 (BSD)
Celtis australis L.	1037	109090, 74086 (DD)
Celtis sinensis Pers.*	428	51954 (DD)
Celtis tetranda Roxb.	960	25646 (DD)
Cephalostachyum pergracile Munro*	57	89596, 5790 (DD)
Ceriscoides turgida (Roxb.) Trivengum.	214	28764 (DD)
Cestrum diurnum L.*	1077	113574, 63613 (DD)
Cestrum nocturnum L.	483	5586, 113597 (DD)
Choerospondias axillaries (Roxb.) Burtt.	196	N.A.
Chonemorpha fragrans (Moon.) Alston.	798	92182 (DD)

Table Contd.

Table Contd.

Names of the Species	Author's specimen numbers	BSD/DD specimen numbers
Chorisia speciosa St. Hill.	595	70672, 50410 (DD)
Chrysophyllum oliviforme Lam.*	793	27721 (DD)
Chuckrassia tabularis A. Juss.	743, 808	N.A.
Cinnamomum camphora Nees & Ebrem.	248	11477, 18553 (BSD)
Cinnamomum glaucescens Nees**	519	140717, 78943 (DD)
Cinnamomum tamala (Buch.-Ham) Nees & E. berm.	857	97923, 74251 (DD)
Chionanthus intermedia (W.) Bedd.	42, 48	N.A.
Citrus aurantium L.	265	N.A.
Citrus grandis (L.) Osbeck	183	N.A.
Citrus limon (L.) Burm.	342	8021, 33206 (BSD)
Citrus maxima (Burm) Merr.	183	N.A.
Clausena lansium (Lour.) Skeels	266	143202, 79654 (DD)
Clerodendrum indicum (L.) O. Ktze	915	33357, 36477 (BSD)
Clerodendrum philippinum Scher	477	1536, 3058 (BSD)
Clerodendrum phlomoidis L.f.*	07	112719 (DD)
Clerodendrum splendens G.Don*	771	34895 (BSD)
Clerodendrum ugandense Prain.*	853	37778 (BSD)
Clerodendrum viscosum Vent.*	325	1906, 15312 (BSD)
Clytostoma callistegioides Bur.*	748	56089, 9063 (DD)
Cocculus hirsutus L.	117	N.A.
Cocculus laurifolius DC.	259	N.A.
Coffea bengahalensis Heyne ex Roem.	673	13862 (BSD)
Colebrookia oppositifolia Sm.	246	136330, 147417 (DD)
Combretum roxburghii Spreng.	30	19074, 3857 (BSD)
Cordia abyssinica R.* @	422	109263 (DD)
Cordia dichotoma Forst.**	765	13872, 2473 (BSD)
Cordia vestita Hook. f.	245	2324 (BSD)
Cornus controversa Hem. ex. Prain	297	2659 (BSD)
Cornus oblonga Wall.	680	30792 (BSD)
Crateva adansonii DC.	260	57143 (DD)
Crotalaria spectablis Roth.	643, 674	N.A.
Crotalaria tetragona Andr.	603	1287 (DD)

Table Contd.

Table Contd.

Names of the Species	Author's specimen numbers	BSD/DD specimen numbers
Croton roxburghii Balak.*	255	N.A.
Cratoxylon cochinensis (Lour.) Bl.**	338	56189, 74548 (DD)
Cryptocarya amygadalina Nees.	806	69238, 273990 (DD)
Cryptolepis buchanani Roem. & Sch.	317	3247, 15570 (BSD)
Cryptomeria japonica L.f.*	1120	148365, 113508 (DD)
Cuponiopsis anacardioides (A. Rich.) Radlk.	1447	6754 (DD)
Dalbergia assamica Benth.	278	N.A.
Dalbergia cultrata Garh.	355, 358	N.A.
Dalbergia lanceolaria L.f.	356, 814	N.A.
Dalbergia sericea G.Don	357	1171 (DD)
Dalbergia sissoo Roxb.	359	113568 (DD)
Delonix regia (Bojer ex Hk.) Rafin*	817	15388 (BSD)
Dendrocalamus giganteus Munro*	403	97919, 97994 (DD)
Dendrocalamus hamiltonii Nees	510	10186, 70900 (DD)
Dendrocalamus longispathus Kurz*	1135	45638, 150152 (DD)
Dendrocalamus membranaceus Munro*	585	149332, 89594 (DD)
Dendrocalamus strictus (Roxb.) Nees	749	91874, 26054 (DD)
Derris indica (Lam.) Bennt.*	368	39964 (DD)
Derris robusta Bent.**	280	65833 (DD)
Desmodium caudatum (Thunb.)DC.	532	89723 (DD)
Desmodium concinnum DC.	560	N.A.
Desmodium gangeticum (L.) DC.	198	3614 (DD)
Desmodium oojeinense (Roxb.) Ohasi	199	11262, 20154 (BSD)
Dendrolobium triangulare (Retz.) Schindler	502	3063 (DD)
Dichrostachys cinerea (L.) W. & A.*	1095	25010 (DD)
Dillenia indica L.	893	32083, 37726 (BSD)
Diospyros exsculpta Buch.-Ham.	467	N.A.
Diospyros malabarica (Ders.) Kostl.*	833	2708, 34965 (BSD)
Diospyros montana Roxb.	732	8034 (BSD)
Diploknema butyracea (Roxb.) Lam.	685	149757 (DD)
Dipterocarpus tuberculatus Roxb.**	894	129331 (DD)
Dipterocarpus turbinatus Gaertn. f.**	900	112795, 143215 (DD)

Table Contd.

Table Contd.

Names of the Species	Author's specimen numbers	BSD/DD specimen numbers
Docynia indica Decne.**	984	136400 (DD)
Dodonaea viscosa N. Jack.	271	742123 (DD)
Dolichandrone platycalyx Baker*	656	119138, 123972 (DD)
Dombeya cayuseii Andr.*	120	45376, 57751 (DD)
Dombeya mastersii Hook. f.*	121	57750 (DD)
Drypetes assamica (Hook) f.	706	2664, 89718 (DD)
Drypetes roxburghii (Hook. f.) Roxb.	927	2148 (BSD)
Duabanga grandiflora Roxb. ex DC.*	31	123455 (DD)
Duranta erecta L.	863	53709, 990 (BSD)
Dysoxylum binectariferum (Roxb.) Hook. f.**	528, 810	N.A.
Ehretia acuminata R. Br.	767	3967, 37756 (BSD)
Ehretia laevis Roxb.	768	36846, 14823 (BSD)
Embelia tojeriam-cottam DC.	466	34677, 40160 (BSD)
Emblica officinalis Gaertn.	169	33467, 31718 (BSD)
Engelhardtia spicata Lech.	244	27631, 137781 (DD)
Eranthemum pulchellum Andrs.	159	35050, 13812 (BSD)
Eriobotrya japonica (Thunb.) Lindl	72	123458 (DD)
Eriolaena hookeriana Wt. & Arn.**	182	N.A.
Erythrina blakei Hort.	360	N.A.
Erythrina suberosa Roxb.	200	27161 (DD)
Eucalyptus alba Reinw. ex Blume.*	1301	79722 (DD)
Eucalyptus bicolor A. Cunn.*	06	257770 (DD)
Eucalyptus botryoides Sm.*	1040	24602 (DD)
Eucalyptus camaldulensis Dehn	115	149764 (DD)
Eucalyptus citriodora Hook.	678	149765 (DD)
Eucalyptus deglupta Bl. *	381	59859 (DD)
Eucalyptus drepanophylla F. & M. ex Benth.	679	145196 (DD)
Eucalyptus grandis Maiden*	683	5402 (DD)
Eucalyptus maculata Hook.*	381 (a)	147577 (DD)
Eucalyptus microcorys F. & M.*	1301 (a)	73987 (DD)
Eucalyptus paniculata Sm.*	614	141300 (DD)
Eucalyptus propinqua D. & M.*	137	5435 (DD)

Table Contd.

Table Contd.

Names of the Species	Author's specimen numbers	BSD/DD specimen numbers
Eucalyptus punctata DC.*	138	22445 (DD)
Eucalyptus resinifera Sm.*	373	78217 (DD)
Eucalyptus robusta Sm.*	383	140075 (DD)
Eucalyptus rudis Endl.	698	N.A.
Eucalyptus sideroxylon A. Cunn.*	374	39048 (DD)
Eucalyptus tereticornis Sm.	75	154394 (DD)
Eucalyptus torelliana F.V.M.*	80	663 (DD)
Eugenia uniflora L.	1326	48535 (DD)
Euphorbia pulcherrima Willd. ex Klotz.	935	90614 (DD)
Euphorbia royleana Boiss.	926	140018 (DD)
Euphorbia tirucalli Brandis*	833	N.A.
Euphoria longan (Lour.) Steud.	718	26088 (DD)
Fernado adenophylla (G.Don.) Steenis	1007	57342 (DD)
Ficus auriculata Lour.	938, 993	N.A.
Ficus benghalensis L.	515	15575, 15944 (BSD)
Ficus elastica Roxb.	659	150614, 107877 (DD)
Ficus hispida L.f.	173	3968, 34947 (BSD)
Ficus microcarpa L.f.	257	134044, 150620 (DD)
Ficus oligodon Miq.	939	38870 (BSD)
Ficus palmata Forsk.	940	77908, 3236 (BSD)
Ficus pumila L.	660	50295, 13896 (BSD)
Ficus racemosa L.	941	15963 (BSD)
Ficus religiosa L.	488	3999, 31276 (BSD)
Ficus rigida Jack.	884	150615, 74212 (DD)
Ficus rumphii Blume	113	1526 (BSD)
Ficus sarmontosa Buch.-Ham.	858	19167 (BSD)
Ficus semicordata Buch.-Ham.	963	1739, 31280 (BSD)
Ficus subincisa Buch.-Ham.	859	32213, 27018 (BSD)
Ficus tinctoria (Willd.) Corner	174	8092, 2658 (BSD)
Ficus virens Ait.	434	150616, 3929 (DD)
Firmiana colorata R. Br.**	122	35394 (BSD),154149 (DD)
Firmiana fulgens (Wall. ex Master) Corner.	261	28853 (BSD),140630 (DD)

Table Contd.

Table Contd.

Names of the Species	Author's specimen numbers	BSD/DD specimen numbers
Firmiana simplx (L.) Wight.	775	41877, 103414 (DD)
Flacourtia indica (Burm. f.) Merr.	179	35037 (BSD), 18076 (DD)
Flacourtia jangomas (Lour.) Raeusch.	522	2661 (BSD), 477 (DD)
Flemingia chapper Buch.-Ham.	605	21888 (DD)
Flemingia macrophylla (Willd.) O. Kuntze	606, 610	N.A.
Flemingia strobilifera (L.) R. Br.	607	107580 (DD)
Flindersia australis R. Br.*	899	116194 (DD)
Galphimia gracilis Brate*	597	32298 (BSD), 27454 (DD)
Garcinia livingstoni T. Andr.**	774	33165, 113609 (DD)
Gardenia angusta Merr.*	830	6192 (DD)
Gardenia latifolia Ait.**	730	112717 (DD)
Gardenia resinifera Roth.*	1064	78925 (DD)
Gardenia spatulifolia Stapf.*	465	123432 (DD)
Garuga pinnata Roxb.	346	15672 (DD)
Gigantochloa albociliata (Munro) Kurz*	399	156513, 149337 (DD)
Gigantochloa atter Kurz*	51, 54	N.A.
Gigantochloa hasskarlina (Kurz) Baker*	398	5809, 2482 (DD)
Gigantochloa verticellata Munro*	55, 52	N.A.
Gleditsia macarantha Desf.*	132	35393 (BSD),39194 (BSD)
Gleditsia sinensis Lam.*	981	79723 (DD)
Gleditsia tricanthos L.*	1091	80170 (DD)
Glochidion ellipticum Wight	Hook.f	172 80267, 42805 (DD)
Glochidion velutinum Wt.	168	31286, 31253 (BSD)
Glycosmis arborea (Roxb.) DC.	599	3879, 999 (BSD)
Gmelina arborea Roxb.	166	35218 (BSD)
Gongrenema nepalense Dcne.	842	22922 (DD)
Gossypium arboreum L.	593	81093, 95250 (DD)
Gossypium hirsutum L.	637	34682 (BSD),109977 (DD)
Gouania leptostachya DC.	556	19164, 8054 (BSD)
Grevillea robusta A. Cunn.	410	113612, 139687 (DD)
Grewia asiatica L.	341	986, 984 (DD)
Grewia elastica Royle	526	2688 (BSD), 22484 (DD)

Table Contd.

Table Contd.

Names of the Species	Author's specimen numbers	BSD/DD specimen numbers
Grewia helicterifolia Wall.	496	67492, 72146 (DD)
Grewia optiva J.R. Drumm.	340	3415 (BSD), 113813 (DD)
Grewia sapida Roxb.	776, 714	N.A.
Grewia sclerophylla Roxb.	443	21658, 9569 (DD)
Grewia serrulata DC.	525	3237 (BSD), 113581 (DD)
Guazuma ulmifolia Lam.	262	62560, 14213 (DD)
Gymnocladus canadensis Lam.*	647	69029 (DD)
Gynocardia odorata R. Br.*	335	107688, 5426 (DD)
Haematoxylon campechianum L.*	982	143521 (DD)
Haldina cordifolia (Roxb.) Ridsdale	143	2488, 31225 (BSD)
Hamelia patens Jacq.*	680	123474 (DD)
Hedera nepalensis Koch	902	99856, 152382 (DD)
Helicteres isora L.	441	11958(BSD),113732 (DD)
Helinus lanceolatus Brandis	670	1019, 1929 (BSD)
Heteropanax fragrans (Roxb.) Seem	728	3873 (BSD)
Hibiscus arnottianus Gray*	439	56927 (BSD)
Hibiscus rosa- sinensis L.	523	33499 (BSD)
Hibiscus schizopetalus Master. f.	524	35397 (BSD),104336 (DD)
Himalrandia tetrasperma (Roxb.) Yamazaki	215	3288 (BSD)
Hiptage benghalensis (L.) Kurz	124	2177, 8049 (BSD)
Holarrhena pubescens (Buch.-Ham.) Wall.	30	10400, 263 (BSD)
Holmskioldia sanguinea Retz.	657	12823, 1151 (BSD)
Holoptelia integrifolia (Roxb.) Planch.	931	Number less (DD)
Homalium tomentosum Benth.*	635	N.A.
Hovenia acerba Lindl.	812	23092 (DD)
Hydnocarpus kurzii (King) Warb.**	336	140322, 4222 (DD)
Hydnocarpus laurifolia (Donnst.) Sleum.**	337	74604, 123883 (DD)
Hymenodictyon orixense (Roxb.) Mabb.	578	89788 (DD)
Indigofera cassioides Rottle	675	20131, 1618 (BSD)
Indigofera hamiltoni Grah. ex Duthie	503	38793 (DD)
Ipomoea alba L.	849	145588 (DD)
Ipomoea arborescens Sweet.*	998	145748, 113765 (DD)

Table Contd.

Table Contd.

Names of the Species	Author's specimen numbers	BSD/DD specimen numbers
Ixora coccinea L.	904	N.A.
Itea nutans Royle	28, 787	N.A.
Jacaranda mimosaefolia D.Don.	423	31260, 13898 (BSD)
Jasminum arborescens Roxb.	300	35241, 37767 (BSD)
Jasminum dispermum Wall.	1076, 945	N.A.
Jasminum humile L.	834	1179, 8013 (BSD)
Jasminum multiflorum (Burm.f.) Andr.	651	136238 (DD)
Jasminum roxburghianum Wall. ex DC.**	835	27619, 28157 (BSD)
Jatropha curcas L.	545	124042, 137824 (DD)
Jatropha integerrima Jack.	873	N.A.
Joannesia princeps Vell.*	958	79853, 69002 (DD)
Justicia gendarussa Burm.f.	164	31794, 1854 (BSD)
Khaya senegalensis A. Juss.*	943	N.A.
Kigelia africana (Lam.) Benth.	861	2183 (BSD)
Koelreuteria paniculata Laxn.	558	38857 (BSD)
Kydia calycina Roxb.	550	1677, 6262 (BSD)
Lagerstroemia indica L.	1111	3038, 15465 (BSD)
Lagerstroemia loudonii T. & B.*	39	38137 (DD)
Lagerstroemia microcarpa W.*	796	80902 (DD)
Lagerstroemia parviflora Roxb.	832	3203, 3258 (BSD)
Lagerstroemia reginae Roxb.*	791	15457, 34964 (BSD)
Lagerstroemia siamica Gagnep.*	571	60244 (DD)
Lagerstroemia thorellii Gagnep.*	1117	104432 (DD)
Lannea coromandalica (Houtt.) Merr.	273	3962, 5277 (BSD)
Lantana camara L.	875	6295, 23837 (BSD)
Lawsonia inermis L.	81	15520 (DD)
Leptodermis lanecolata Wall.	681	78743 (DD)
Leucaena latisiliqua (L.) Gill.	370	34780 (BSD)
Leucomeris spectabilis D.Don.	682	74452 (DD)
Ligustrum compactum (Wall. ex DC.) Hook. f.	797	99168 (DD)
Ligustrum lucidum Ait. f.*	804	80924 (DD)
Liquidambar formosana Hance*	734	144834 (DD)

Table Contd.

Table Contd.

Names of the Species	Author's specimen numbers	BSD/DD specimen numbers
Liriodendron tulipiflora L.*	92	N.A.
Litchi chinensis Sonn.	128, 99	N.A.
Litsea glutinosa (Lour.) Robins	865	12244 (DD)
Litsea monopetala (Roxb.) Pers.	418	91302, 14555890 (DD)
Livistona australis Mart.*	868	26467, 24786 (DD)
Livistona chinensis R. Br.*	62	N.A.
Lonicera japonica Thunb.*	628	124040, 53895 (DD)
Macaranga denticulata Muell-Arg.*	249	N.A.
Macaranga pustulata King ex Hook. f.**	254	70824 (DD)
Macfadyena unguis-cati (L.) Gentry*	802	37545, 15472 (BSD)
Maclura cochinensis (Lour.) Corner	942	28161, 12962 (BSD)
Maclura pomifera Schneid	175, 892	N.A.
Madhuca longifolia (Koenig) Mac.	298	26094 (DD)
Maesa indica (Roxb.) A. DC.	684	25799, 1925 (BSD)
Magnolia grandiflora L.*	333	1139 (BSD)
Magnolia liliflora Desr.*	691	74073 (DD)
Magnolia soulangiana Hort.*	116	74073 (DD)
Mallotus philippensis (Lamk.) Muell.-Arg.	631	18103, 31202 (BSD)
Malvaviscus arboreus Cav.*	594	47997, 116390 (DD)
Mangifera indica L.	195	25770 (BSD)
Manihot dichotoma Ule*	1076	70752, 56370 (DD)
Manihot esculenta Crantz.*	623	30693 (BSD)
Manilkara hexandra (Roxb.) Dub.	731	116314 (DD)
Marsdenia tenacissima (Roxb.) Moon	482	22028 (DD)
Maotia puya (Hook. f.) Wedd.	430	970, 144944 (DD)
Maytenus royleanus (Wall ex Laws) Rau	669	41597 (DD)
Melia azedarach L.	628	N.A.
Melia dubia Hiern.	66	74455 (DD)
Melicope lunuankenda (Gaertn.) Hartley*	639	115428 (DD)
Melocalamus compactiflorus Benth.*	401	153666, 153667 (DD)
Mesua ferrea L.**	438	15970 (BSD), 80274 (DD)

Table Contd.

Table Contd.

Names of the Species	Author's specimen numbers	BSD/DD specimen numbers
Michelia doltsopa Buch.-Ham. ex DC.	258	38522, 37971 (BSD)
Michelia fuscata Blume*	330	35080 (DD)
Michelia oblonga Wall.* ex Hook. f. & Thoms.	177	97443 (DD)
Miliusa velutina (Dunal) Hook.	490	3201 (BSD)
Millettia bradisiana Kurz	721	N.A.
Millettia extensa Benth. ex Baker	452	3210, 33219 (BSD)
Millettia peguensis Ali	947	62035 (DD)
Millingtonia hortensis L.f.	665	78134, 70768 (DD)
Mimosa rubicaulis Lam.	542	23857, 23888 (BSD)
Mimusops elengi L.	86	15944, 34964 (BSD)
Mitragyna parvifolia (Roxb.) Korth.	579	26444 (DD)
Moringa oleifera Lam.	720	3882, 31206 (BSD)
Morus alba L.	406	147366 (DD), 9024 (BSD)
Morus australis Poir.	755, 762	N.A.
Morus macroura Miq.	801	35027 (BSD)
Morus serrata Roxb.	163	N.A.
Mucuna nigricans (Lour.) Steud.	947	62035 (DD)
Murraya koenigii (L.) Spreng.	343	28162 (BSD),147216 (DD)
Murraya paniculata (L.) Jack	641	79422, 37637 (DD)
Mussaenda belilla Buch.- Ham.	580	18575, 33479 (BSD)
Mussaenda luteola Delile*	995	1005, 60800 (DD)
Myrsine africana L.	146	30791 (BSD)
Myrsine semiserrata Wall.	144, 85	N.A.
Naringi crenulata (Roxb.) Nicolson	446	2683 (BSD), 113733 (DD)
Neolitsea cuipala (Buch.-Ham.) Kost.*	424	128647, 974 (DD)
Nerium indicum Mill.	400	28170, 34761 (BSD)
Nyctanthes arbor-tristis L.	581	1744, 22766 (BSD)
Ochrosia elliptica Labill.* @	799	75108 (DD)
Olea glandulifera Wall. ex G.Don.	733	25636, 27630 (BSD)
Opuntia vulgaris Mill.	40	35234, 20141 (BSD)
Oroxylum indicum (L.) Vent.	476	N.A.
Osyris wightiana Wall.	409	3345, 1035 (BSD)

Table Contd.

Table Contd.

Names of the Species	Author's specimen numbers	BSD/DD specimen numbers
Oxytenanthera abbysinica Munro*	402, 413	N.A.
Pahudia martabanica Prain	286	102159 (DD)
Pandanus fascicularis Lam.**	1344	45620, 148802 (DD)
Pandanus nepalensis St.**	1318	109094, 10994 (DD)
Parkinsonia aculeata L.*	455	74488 (DD)
Pavetta tomentosa Roxb.	394	5488, 3225 (BSD)
Peltophorum africanum Sond.*	967	94739 (DD)
Peltophorum pterocarpum Backer. ex Heyne*	412	N.A.
Peltophorum vogetianum Walp.*	978	94815 (DD)
Pereskia aculeata Mill.*	1051	6215 (DD)
Pereskia grandifolia Haw.	41	153245 (DD)
Persea duthiei (King. ex Hook. f.) Kostrm.	807	36866, 2667 (BSD)
Persea gamblei (King. ex Hook. f.) Kostrm.	419	N.A.
Persea odoratissimia (Nees) Kostrm.	866	71532, 137723 (DD)
Petrea volubilis Jacq.	854	31612 (BSD)
Phlogacanthus thyrsiformis (Hard.) Mobly	111	1840 (BSD)
Phoebe goalparensis Hutch.**	874	N.A.
Phoebe hainesiana Brandis*	882	N.A.
Phoebe lanceolata Nees	420	1599, 631 (DD)
Phoenix acaulis Roxb.	621	42810 (DD)
Phoenix canariensis Chabaud.*	986	117341 (DD)
Phoenix dectylifera L.*	313, 311	N.A.
Phoenix reclinata Jacq.*	307	N.A.
Phoenix rupicola T. And.**	311	5621 (DD)
Phoenix sylvestris (L.) Roxb.	320, 306	N.A.
Phyllanthus grandifolius L.*@	1036	57042, 11429 (DD)
Pistacia khinjuk Stock	274	1890, 27700 (BSD)
Pithecellobium dulce Roxb.	366	8440, 2471 (BSD)
Pittosporum eriocarpum Royle	180	27118 (BSD), 22793 (DD)
Pittosporum nepaulense (DC.) Rehdr.	437	29695 (BSD), 4407 (DD)
Platanus occidentalis L.*	1432	N.A.
Platanus orientalis L.*	1062	133989, 943739 (DD)

Table Contd.

Table Contd.

Names of the Species	Author's specimen numbers	BSD/DD specimen numbers
Plumeria acuminata Ait.	839	25732 (DD)
Plumeria alba L.*	838	89134 (DD)
Polyalthia longifolia (Sonner.) Thw.*	334	39741 (DD)
Polyalthia suberosa (Roxb.) Thw.**@	178	94054 94054 (DD)
Populus deltoidea Marsh.*	753	N.A.
Porana paniculata Roxb.	587	135898, 78096 (DD)
Premna barbata Wall. ex Sch. DC.	167	322, 3273 (BSD)
Premna mucronata Roxb.	171	8048, 13893 (BSD)
Prosopis juliflora (Sw.) DC.	831	80207 (DD)
Protium serratum Engl.	187	N.A.
Prunus armenica L.	895	127194 (DD)
Prunus ceracifera Ehrhart	76	59086, 39199 (BSD)
Prunus cerasoides D.Don.	786, 367	N.A.
Prunus domestica L.	73, 77	N.A.
Prunus persica (L.) Batsch.	76	12839, 19760 (BSD)
Psidium cattleianum Sabine*	1041	46392 (DD)
Psidium guajava L.	829	55001, 15484 (BSD)
Pterocarpus marsupium Roxb.	562	151181 (DD)
Pterocarya stenoptera DC.*	1420	65350, 58720 (DD)
Pterospermum acerifolium Willd.	263	1116 (BSD), 150169 (DD)
Pterospermum canescens Roxb.*	495	75889, 116555 (DD)
Pterospermum lanceaefolium Roxb.	551	123981, 41142 (DD)
Pterospermum semisagittatum Buch.-Ham.*	442	56373, 80969 (DD)
Pterygota alata (Roxb.) R. Br.**	123	109493, 103269 (DD)
Pueraria tuberosa Roxb. ex Willd.	504	144882 (DD)
Punica granatum L.	385	23721, 27159 (BSD)
Pyracantha crenulata (D.Don) Roem.	369	32256 (BSD)
Pyrostegia venusta (Ker-Gawl.) Miers.*	634	123926, 57756 (DD)
Pyrus pashia Buch.-Ham. ex D. Don	459	18089, 34753 (BSD)
Quercus acutissima Carru.	63	74083, 141312 (DD)
Quercus leucotrichophora A. Cam.	754	3050, 80893 (DD)
Quirivelia frutescens (L.) M.R. & S.M. Almeida	87	113562 (DD)

Table Contd.

Table Contd.

Names of the Species	Author's specimen numbers	BSD/DD specimen numbers
Quisqualis indica L.	37	15468 (BSD)
Rauvolfia densiflora Benth.*	840	146533 (DD)
Rauvolfia serpentina Benth.	90	142100 (DD)
Rhamnus persica Boiss	270	22940 (DD)
Rhamnus triquetra (Wall.) Brandis	529	23742, 33464 (BSD)
Rhamnus virgata Roxb.	671	39782, 3908 (BSD)
Rhus lancea L.f.*	673	57498 (DD)
Ricinus communis L.	920	25660, 35445 (BSD)
Rosa brunonii Lindl.	507	35059, 33380 (BSD)
Rosa cathayensis Bailey*	677	34683 (BSD)
Rosa indica L.*	02	41487 (BSD)
Rosa laevigata Michx*@	207	35390, 35077 (BSD)
Roystonia regia (H.B. & K). Cook.*	752, 760	N.A.
Rubus ellipticus Sm.	74	19168, 25467 (BSD)
Rubus niveus Thunb.	134	31281 (BSD)
Sabia paniculata Edgew. ex Hook. f.	69	8007 (BSD), 1352 (DD)
Sabol blackburniana Glaz.*	60	N.A.
Sageretia parviflora (R. & S.) G.Don.	498	30736, 1173 (BSD)
Salix acmophylla Boiss.	162	22896, 1885 (BSD)
Salix babylonica L.	321	26983 (BSD)
Salix tetrasperma Roxb.	158	142379, 28196 (DD)
Santalum album L.	107	15381, 35400 (BSD)
Sapindus mukorossi Gaertn.	450	3054 (BSD)
Sapium baccatum Roxb.	871, 1886	N.A.
Sapium eugeniifolium Buch.-Ham.**	421	71654, 74511 (DD)
Sapium insigne (Royle) Benth. ex Hook.	658	35230, 12829 (BSD)
Sapium sebiferum Roxb.	885	77832, 15426 (BSD)
Saraca asoca (Roxb). De Wilde	787	21875 (DD)
Saurauia nepaulensis DC.	516	2142, 874 (DD)
Schefflera actinophylla Viguir	912	N.A.
Schefflera venulosa (Wight & Arn.) Harms*	83	41231 (DD)
Schima wallichii (DC.) Korth.**	331	107867 (DD)

Table Contd.

Table Contd.

Names of the Species	Author's specimen numbers	BSD/DD specimen numbers
Schleichera oleosa (Lour.) Oken.	194	71597 (DD)
Securinega virosa (Roxb. ex Willd.) Baill.	919	35270, 28116 (BSD)
Selenicereus grandiflorus Britt. & Rose*	511	36755 (DD)
Semecarpus anacardium L. f.	500	1749, 3246 (BSD)
Sesbania sesban (L.) Merr.	563	33540, 15969 (BSD)
Shorea robusta Gaertn. f.	517	55242, 31256 (BSD)
Shorea roxburghii G.Don	181	150987, 110667 (DD)
Sophora mollis Grah. ex Baker	722, 727	
Spathodea campanulata Beauv.*	851	71923, 73267 (DD)
Spermadictyon sauveolens Roxb.	649	18087, 34984 (BSD)
Spiraea cantoiensis Lour.	27, 35	35083 (BSD)
Spondias pinnata (L.f.) Kurz	276, 197	N.A.
Stephania glabra (Roxb.) Miers.	521	3449, 2107 (DD)
Sterculia rupestris Benth.*	339	142942, 142943 (DD)
Sterculia urens Roxb.**	596	142941, 124048 (DD)
Sterculia villosa Roxb.	264	8023 (BSD), 32032 (DD)
Stereospermum chelonoides (L. f.) DC.	432	70076, 80156 (DD)
Stranvaesia nussia (D.Don) Decne	569, 574	69242 (DD)
Strychnos nux-vomica L.	415	117996 (DD)
Suregada multiflorum Juss.	878	11623, 94293 (DD)
Swietenia macrophylla King*	176	28836 (DD)
Swietenia mahagoni (L.) Jack*	921	8018 (DD)
Syzygium cerasoideum (Roxb.) Raizada*	21	32001 (BSD)
Syzygium cumunii (L) Skeels*	377	27195, 2656 (BSD)
Syzygium jambos (L.) Alston*	384	35263 (BSD)
Tectona grandis L.	876	37536, 34749 (BSD)
Tectona hamiltonia Wall.	1026	79892, 38990 (DD)
Tabernaemontana divaricata (L.) R. Br.	905	28600, 19735 (BSD)
Talauma candollei Bl.*	436	102992, 122242 (DD)
Tamarindus indica L.	287	136807
Tamarix dioica Roxb.	773	91170 (DD)
Tecoma stans (L) HB. & K.	852	147326, 123466 (DD)

Table Contd.

Table Contd.

Names of the Species	Author's specimen numbers	BSD/DD specimen numbers
Tephrosia candida (Roxb.) DC.	564	966, 30768 (BSD)
Terminalia alata Heyne ex Roth.*	546	3392 (BSD)
Terminalia arjuna (Roxb. ex DC.) W. & A.	294, 380	N.A.
Terminalia bellirica (Gaertn.) Roxb.	292	15913, 5454 (BSD)
Terminalia chebula Retg	135	33398, 1610 (BSD)
Terminalia citrina (Gaertn.) Roxb. ex Flem.**	208	N.A.
Terminalia myriocarpa Heruck & Muell.-Arg.**	613	34637, 37538 (BSD)
Terminalia oliveri Brandis*	508	98792 (DD)
Terminalia sericea Burch. ex DC.*	372	43955 (DD)
Tetradium glabrifolium (Champ. ex Benth.) Hartley*	445	81222, 81223 (DD)
Tetrapanax papyrifera (Hook.) Koch*	616	1248 (BSD)
Thespesia lampas (Cav.) Dalz. & Gibs.	549	12526 (BSD),124019 (DD)
Tipuna speciosa Benth.*	948	71441 (DD)
Toddalia asiatica Lam.	680	2654, 3952 (BSD)
Toona ciliata Roem.	348	31252, 2147 (BSD)
Toona sinensis (A. Juss.) M. Roem.	780	95411 (DD)
Trachcycarpus fortuni (Hook.) Wendl.*	405, 470	N.A.
Trachcycarpus martiana H. Wendl.**	469, 475	N.A.
Trachelospermum lucidum (D. Don) K. Sch.	43	3953, 35252 (BSD)
Trema bracteolata Blume*	970	74162, 52421 (DD)
Trema bracteolata Blume	970	74162, 52421 (DD)
Trema orientalis (L.) Blume	932	658, 30154 (DD)
Trema politoria Planch.	487	29285, 15946 (BSD)
Trewia nudiflora L.	930	20145, 2469 (BSD)
Trichilia connaroides (W. & A.) Bentlzn*	943	N.A.
Tristania conferata R. Br.*@	1042	112718 (DD)
Triumfetta pilosa Roth	542, 626	N.A.
Tylophora hirstuta (Wall.) Wight.	1056, 562	N.A.
Vallaris solanacea (Roth.) O. Kuntze	756	26099, 147272 (DD)
Ventilago denticulata Willd.	190	4638, 5496 (BSD)
Vernicea cordata (Thunb.) Airy-Shaw*	870	94300, 102740 (DD)

Table Contd.

Table Contd.

Names of the Species	Author's specimen numbers	BSD/DD specimen numbers
Vernicia fordii Hamel*	589	123994, 25532 (DD)
Vernicia montana Lour.*	952	102590, 102591 (DD)
Viburnum mullaha Buch.-Ham. ex D.Don	903	3051 (BSD)
Vitex negundo L.	485	147394, 150745 (DD)
Vitis heyneana Roem. & Schult.	193	352233 (BSD)
Vitis parvifolia Roxb.	350	8093 (BSD)
Wisteria sinensis (Sims.) DC.	964	79897 (DD)
Woodfordia fruticosa (L) Kurz	142	1872, 35060 (BSD)
Wrightia arborea (Dennst.) Mabb.	805	1284 (BSD)
Wrightia coccinea Sims**	841	97983 (DD)
Wrightia tinctoria R.Br.**	757	75123 (DD)
Xylosma longifolium Clos.	668	37988, 18535 (BSD)
Xylia xylocarpa (Roxb.) Taub.	291	150637, 74447 (DD)
Zanthoxylum armatum DC.	447	149819 (DD)
Ziziphus mauritiana Lam.	602	29696, 33920 (BSD)
Ziziphus oenoplia Mill.	191, 782	N.A.
Ziziphus rugosa Lam.	127	33934, 2770 (BSD)
Ziziphus xylopyra (Retz.) Willd.	192	5437, 3122 (BSD)

ACKNOWLEDGEMENTS

I am thankful to Dr P.K. Hajra, former Director, Botanical Survey of India and Dr B.K. Gupta, ex-Reader, DAV (PG) College, Dehra Dun, for academic support and useful suggestions. My special thanks go to the Library and herbarium incharge of Forest Research Institute and Botanical Survey of India, Northern Circle, Dehra Dun for help in consultation of the literature. Thankfulness is also expressed to Institute colleagues and villagers of Doon Valley for their help and co-operation during the field work.

REFERENCES

Anonymous (2003). *IUCN (International Union for Conservation of Nature) 2003, Red List of Threatned Plant Species.* IUCN, Gland, Switzerland

Babu, C.R. (1977). *Harbaceous Flora of Dehradun.* CSIR, New Delhi

Bailey, LH (1928). *The Standard Cyclopedia of Horticulture.* vols. 1-3. MacMillan, New York

Bailey, LH (1949). *Manual of Cultivated Plants* (Second Ed.) MacMillan, New York

Bentham, G & JD Hooker (1862-1883). *Genera Plantarum, etc.* Vols. 1-3. A Black. London

Bennet, SSR (1987). *Name changes in Flowering plants of India and adjacent regions.* Triseas Publishers, Dehradun

Brandis, D (1874). *The Forest Flora of North-West and Central India,* W.H. Allen Publication, London

Brandis, D (1906). *Indian Tress* (Repr. ed. 1921), Houghton Mifflins Company, London

Daniel, O (1871). *Flora of Tropical Africa.* 2-3. 1. Reev & Co. Henrictta

De Wit, HCD (1966-67). *Plants of the World,* Vols. 1&2, World Scientific Publishing Company, London

Duthie, JF (1906). *Catalogue of the Plants of Kumaon and of the Adjacent Portions of Garhwal and Tibet, based on the Collections of Strachey and Winterbottom during the years* 1846-1849. L. Reeve & Co. London

Duthie, JF (1903-1929). *Flora of the Upper Gangetic Plain and of the Adjacent Siwalik and Sub-Himalayan Tracts.* Vols 1-3. Superintandent of Goverment Printing Press, Calcutta

Gaur, RD (1999). *Flora of the Ditrict Garhwal-North West Himalaya with Ethnobotanical Notes.* Transmedia, *Srinagar*, Garhwal, Uttarakhand, India

Gaur, RD & Raiwani, Sukriti (1995). A preliminary report on the threatened arborescent taxa of Garhwal Himalaya. *J. Indian bot. Soc.* **74:** 283-292

Graf, AB (1981). *Tropica.* Roehrs company, East Rutherford, New Jersey, USA

Graf, AB (1982). *Exotica*, series 4. Pictorial cyclopedia of exotic plants. 2 vols. Roehrs, East Rutherford, New Jersey

Graf, AB (1992). *Hortica.* Roehrs company, East Rutherford, New Jersey, USA

Grierson, AJC & Long, DG (1983). *Flora of Bhutan*, Vols 1-2. Royal Botanic Garden Edinburgh, London

Gupta, BL (1928). *Forest flora of Chakrata, Dehradun and Saharanpur Forest Divisions, United provinces* (3rd ed.) Govt. of India Publications, Calcutta

Gupta, PC, Gaur, RC & Dakshini, (1988). *Arboretum.* Forest Research Institute, Dehradun

Hajra, PK (1983). Is Sahastradhara a threatened type of locality.*POSSCFF, an assessment of threatened plants of India*, pp 35-36. *In:* Jain, SK & Rao, RR (Eds). Botanical Survey of India, Howrah. India

Hooker, JD (1872-1897). *The Flora of British India.* Vols 1- 7. Reev Publication, Oxford

Hooker, JD & Thomson, T (1855). *Flora Indica.* W. Pamplin Publication, London

Hutchinson, J (1959). *Families of Flowering Plants* (3rd ed.), Oxford University Press, London

Hutchinson, J (1967). *Key to the Families of Flowering plants of the World.* Clarendon Press, Oxford

Jacquemont, V (1834)1935. *Letters from India: describing a journey in the British Dominions of India, Tibet, Lahore and Cashmere, during the years 1928-1931.* 2 Vols. Century Publishing, London

Jain, SK & Rao, RR (1977). *A Handbook of Field and Herbarium Methods. Today & Tomorrow Publishers*, New Delhi

Kanjilal, UN (1901). Forest *flora of the Siwalik and Jaunsar Forest Division of United Provinces of Agra and Oudh.* Superintandant of Government Printing, Calcutta, India

King, G (1882). List of Plants of Garhwal, Jaunsar-Bhawar and Dehradun **1:** 299-322, Oxford Publishing Co, London

Krishnaswamy, VS & Mathuda, GS (1954). Phenological behavior of a few forest species at new Forest, Dehradun. *Indian For.* **80(3&4):** 124-153 & 187-206

Kunkel, G (1978). *Flowering Trees in Sub-tropical Gardens.* The Hague

Matthew, KM (1969). *The exotic Flora of Kodaikanal Palni Hills.* Manager Publication, Delhi

Negi, PS (2005). Investigation on Ligneous Flora of Doon Valley. A Unpublished Doctoral Thesis, HNB Garhwal University. *Srinagar*, Uttarakhand

Negi, PS & Gupta, BK (1987). Forest Resource of Surkanda Devi, Garhwal Himalaya, *Ind. J. For.* **10:** 283-289

Negi, PS, Gupta, BK & Hajra, PK (1991). Ligneous Flora of Doon Valley-A preliminary survey. *In Higher plants of Indian subcontinent.* **2:**101-118

Oliver, D, Prain, D & Hill, AW (1868-1937). Flora of Tropical Africa. 1-10. London

Osmoston, AE (1927). *A Forest Flora of Kumaon.* Goverment Press. Allahabad

Ouden, PD & Boom, BK (1954). *Manual of Cultivated Conifers.* Edward Arnold. Ltd., London

Raizada, MB & Sahni, KC (1966). Living Indian Gymnosperms: Part 1 (Cycedales, Ginkgoales, Coniferales). *Indian Forest Record* (N.S.) **5:** 73-150

Raizada, MB (1976). *Supplement to the Duthie's Flora of Upper Gangetic Plain* etc. Bishen Singh-Mahendra Pal Singh, Dehradun

Raizada, MB & Saxena, HO (1978). *Flora of Mussoorie.* Bishen Singh-Mahendra Pal Singh, Dehradun

Rao, MA (1969). Flora *of Upper Gangetic Plain and of the Adjacent Siwaliks and sub.Himalayan tracts*–check list. suppl. No. 2, *Bull. bot. Surv. India* **10:** 25-37

Rees, A (1802-1820). Cyclopedia. 39. Toronto Press, London

Rodgers, WA (1985). Biogeographical and protected area planning in India, pp. 18-75. *In:* Thorsell, J (Ed.) *Conservation Asiaés Natural Heritage*. Today and Tomorrow's Printers and Publishers, New Delhi, India

Royle, JF (1833-1840). *Illustrations of the Botany and other branches of the Natural History of the Himalayan Mountains and of the Flora of Cashmere.* Vols 1, 2. W.N. Allen & Co., London

Sharma, BD (1968). *Exotic Flora of Allahabad.* Botanical Survey of India, Calcuta. India

Steward, JL & Brandis, D (1874). *The Forest Flora of North-west and Central India*, Nabu Press, London

Thunberg, CP (1784). *Flora japonica.* Sophia University Press, Lipsiae

Tutin, TG, & Heywood, VH *et al.* (1964). *Flora Europea.* Cambridge University Press, Cambridge, UK

Udvardy. (1975). *A classification of biogeographaical provinces of the world prepared as contribution to UNISCO?s Man and Biosphere Programme.* Project no. 8. IUCN Occasional paper 18, IUCN, Morges. Switzerland

Verdcourt, B (1968). *Flora of the Tropical East Africa.* Reeve and Co., London

Wallich, N (1824-1826). *Tentamen Florae Nepalensis Illustrate.* Asiatic Lithograph Press, Serampore, India

Index